PLANT AND MICROBIAL BIOTECHNOLOGY RESEARCH SERIES
Series Editor: James Lynch

# Plant Protein Engineering

*Edited by*

**P.R. Shewry**
Long Ashton Research Station, Bristol, UK

*and*

**S. Gutteridge**
Du Pont Company, Wilmington, Delaware, USA

CAMBRIDGE
UNIVERSITY PRESS

Published by the Press Syndicate of the University of Cambridge
The Pitt Building, Trumpington Street, Cambridge CB2 1RP
40 West 20th Street, New York, NY 10011–4211, USA
10 Stamford Road, Oakleigh, Victoria 3166, Australia



First published 1992

Printed and bound in Great Britain by
Butler & Tanner Ltd, Frome and London

*A catalogue record of this book is available from the British Library*

*Library of Congress cataloguing in publication data available*

ISBN 0 521 41761 9 hardback

WS

# Contents

# Contributors List

**M.J. Adang**
Department of Entomology
The University of Georgia
Athens
GA 30602
USA

**N.F. Bascomb**
American Cyanamid
Princeton
NJ 08543
USA

**G.P. Belfield**
The Biological Laboratory
University of Kent
Canterbury
Kent CT2 7NJ

**D. Bosch**
CPRO
Keijenbergseweg 6
NL-6700 AA Wageningen
The Netherlands

**N.J. Bulleid**
Department of Biochemistry and Molecular Biology
University of Manchester
Stopford Building
Oxford Road
Manchester M13 9PT

**A.F Campbell**
Medical Research Council
20 Park Crescent
London W1N 4AL

**D.A. Chisholm**
Experimental Station
E.I. du Pont De Nemours and Co.
Wilmington
Delaware 19880
USA

**D.H. Dean**
Department of Biochemistry
The Ohio State University
484 West 12th Avenue
Columbus
OH 43210-11292
USA

**B.A Diner**
Experimental Station
E.I. du Pont De Nemours and Co
Wilmington
Delaware 19880
USA

**S. Eccles**
Division of Gene Structure and Expression
National Institute for Medical Research
The Ridgeway
Mill Hill
London NW7 1AA

**R. Freedman**
The Biological Laboratory
University of Kent
Canterbury
Kent CT2 7NJ

**J.II. Gould**
Department of Biological Sciences
University of Warwick
Coventry

**S. Gutteridge**
E.I. du Pont De Nemours and Co.
PO Box 80402
Wilmington
Delaware 19880-0402
USA

**F. Hartman**
Biology Division
Oak Ridge National Laboratory and The University of
  Tennessee
Oak Ridge Graduate School of Biomedical Sciences
Oak Ridge
Tennessee 37831
USA

**E. Krebbers**
Plant Genetic Systems N.V.
Jozef Plateaustraat 22
B-9000 Gent
Belgium

**N. Lambert**
AFRC Institute of Food Research
Colney Lane
Norwich NR4 7UA

**B. Larkins**
Department of Plant Sciences
University of Arizona
Tucson
Arizona 85721
USA

**C. Lending**
Department of Biological Sciences
SUNY College at Brockport
Brockport
NY 14420
USA

**R. Lenstra**
Glaxo Institute for Molecular Biology SA
Route des Acacias 46
1211 Geneva 24
Switzerland

**Y. Lindqvist**
Department of Molecular Biology
Swedish University of Agricultural Sciences
Biomedical Center, Box 590
S-751 24 Uppsala
Sweden

**J.M. Lord**
Department of Biological Sciences
University of Warwick
Coventry CV4 7AL

**T. Lubben**
Dept 930
Abbott Laboratories
Abbott Park
Chicago IL 60064
USA

**P. Madgwick**
Biochemistry Department
Rothamsted Experimental Station
Harpenden
Herts

**P.J. Nixon**
Experimental Station
E.I. du Pont De Nemours and Co.
Wilmington
Delaware 19880
USA

**D.P. O'Keefe**
Central Research & Development Department
E.I. du Pont De Nemours and Co.
Wilmington
Delaware 19880-0402
USA

**C.A. Omer**
Merck, Sharp & Dohme
Department of Cancer Biology
West Point
Pennsylvania 19486
USA

**R.W. Pickersgill**
AFRC Institute of Food Research
Reading Laboratory
Shinfield
Reading RG2 9AT

**K.A. Pratt**
AFRC Institute of Food Research
Reading Laboratory
Shinfield
Reading RG2 9AT

**P.T. Richardson**
Department of Biological Sciences
University of Warwick
Coventry CV4 7AL

**L.M. Roberts**
Department of Biological Sciences
University of Warwick
Coventry

**G. Schneider**
Department of Molecular Biology
Swedish University of Agricultural Sciences
Biomedical Center, Box 590
S-751 24 Uppsala
Sweden

**P.R. Shewry**
University of Bristol
Department of Agricultural Sciences
Long Ashton Research Station
Long Ashton
Bristol BS18 9AF

**R.A. Spooner**
Department of Biological Sciences
University of Warwick
Coventry CV4 7AL

**M.F. Tuite**
The Biological Laboratory
University of Kent
Canterbury
Kent

**J. Vandekerckhove**
Fysiologische Scheikunde
Rijkuniversiteit Gent
Ledeganckstraat 35
B-900 Gent
Belgium

**C.T. Verrips**
Unilever Research
Laboratorium Vlaardingen
Olivier van Noortlaan 120
3133 At Blaadingen
The Netherlands

**R. Wales**
Department of Biological Sciences
University of Warwick
Coventry CV4 7AL

**J. Wallace**
Department of Biology
Bucknell University
Lewisburg
Pennsylvania 18737

**J.N. Yarwood**
Department of Biological Sciences
University of Durham
Science Laboratories
South Road
Durham DH1 3LE

# Series Preface
## Plant and Microbial Biotechnology

The primary concept of this Series of books is to produce volumes covering the integration of plant and microbial biology in modern biotechnological science. Illustrations abound, for example the development of plant molecular biology has been heavily dependent on the use of microbial vectors and the growth of plant cells in culture has largely drawn on microbial fermentation technology. In both of these cases the understanding of microbial processes is now benefiting from the enormous investments made in plant biotechnology. It is interesting to note that many educational institutions are also beginning to see things in this way and integrating departments previously separated by artificial boundaries.

Having set the scope of the Series, the next objective was to produce books on subjects which had not been covered in the existing literature and, it was hoped, set some new trends. At an early stage in the planning of the Series, I had the opportunity to discuss with Peter Shewry the potential of plant protein engineering and suggested to him that a book on the subject would be an ideal flagship of the new Series; when he suggested Steve Gutteridge, with his commercial as well as academic interest in the subject, as co-editor we became very excited.

There is no need for me to re-iterate the summary given in the Editors' Preface but for my part I am delighted that Peter and Steve were able to bring together a team of international contributors to make an outstanding foundation volume for the Series. Plant protein engineering is with us, its potential is enormous and I hope that this volume will help students and researchers to realize its potential.

*Jim Lynch*

# Preface
## Why engineer plant proteins?

Steven Gutteridge and Peter R. Shewry

A book that deals with protein engineering relevant to plant systems might be considered premature. Very few plant proteins have been characterized at the molecular level and even less have been subjected to detailed protein engineering studies. In addition, plant cell and molecular biologists are still struggling to develop methods for the routine transformation of agronomically important crop plants. Nevertheless the ability to alter protein structure and function by rational design using *in vitro* techniques is now well established. The natural extension of this is to modify the phenotype of plants through selective alterations to the genetic constitution, and this is the goal of research programmes in many publicly and commercially funded laboratories.

It is possible to identify many desirable traits that would increase the potential of crop plants. Plants have not evolved to satisfy the requirements of the human race and only modern farming practices, with their inherent disadvantages, allow these needs to be satisfied, at least in the developed world. Engineering plants to achieve this potential without high inputs would, therefore, be economically and environmentally advantageous.

A major aim of any plant breeding programme is to increase yield, and achieve yield stability across a range of environmental conditions. This includes resistance to stress such as drought, high or low temperatures, and salinity. Hardly less important is resistance to pests and pathogens, which range from insects and nematodes to fungi, bacteria and viruses. Finally, crop quality is an important consideration, especially when the harvested organ is destined for a complex industrial process such as breadmaking or malting and brewing. In some cases (e.g. resistance to fungal pathogens) it is possible to identify single genes which determine these traits, in other cases (e.g. yield) they are quantitative and presumably determined by the interactions of several genes distributed over the genome.

The classical approach to crop improvement is genetical, via plant breeding. Favourable combinations of genes are selected from the progeny of crosses, and specific genes may be introduced from exotic lines or wild relatives by wide crossing. This procedure is time consuming, and depends greatly on the skill and intuition of the plant breeder and his ability to recognize and select for the phenotype of interest. The latter may be limited by complex interactions between the genes of interest, the environment and the genetic background. The application of DNA restriction fragment length polymorphism (RFLP) analysis offers an opportunity to streamline plant breeding, especially for quantitative traits, but this is essentially an adjunct to the classical methods.

Genetic engineering should enable us to make specific alterations to the crop genotype and phenotype. Introducing mutant or novel genes, or down-regulating endogenous genes by anti-sense (for a review see Krol *et al.*, 1988) or 'gene shear' (Hazelhoff and Gerlach, 1988) approaches, will allow us to produce the same range of genotypes as by classical plant breeding, but more quickly and economically. The ability to introduce genes from other species including animals and micro-organisms will extend the range of variation far beyond that currently available to the plant breeder.

Our success in choosing the most desirable traits will also depend on our understanding of the equilibria that exist among intermediates of the primary and secondary metabolic pathways and how they are controlled. An important application of down-regulation will be to identify the enzymes that determine which pathways operate under different conditions. Determining the functional characteristics of the enzymes catalysing the reactions in the pathways will be an essential pre-requisite for manipulating the rates of flux and concentrations of intermediates. Manipulation of enzyme structure by *in vitro* mutagenesis has provided a powerful method of describing the role of protein structure in determining function, not only at the level of the primary sequence but also, through sub-domain alterations, the contributions of higher level structures. The challenge is to return these proteins with altered characteristics back to the plant, and thus change the phenotype in a predictable fashion.

The ultimate aim is to produce an ideal crop which would be sown at any convenient time, grow with little requirement for agrochemicals (pesticides, fertilisers and growth regulators) and show high resistance to environmental stresses. The harvested organ would be easy to handle, have good storage properties (in the field and after harvest) and have good nutritional and/or technological quality. It should also possess an attractive flavour and aroma, features absent from some commercial fruits. The yield and harvest index of the crop should be high, and the remaining plant parts utilizable on the farm (e.g. as feed) or in industrial processes (e.g. as fibre), or readily degraded *in situ* to enrich the soil for subsequent crops.

In this book we have brought together a number of chapters written by prominent researchers, who are involved in protein engineering relevant to understanding plant protein structure or improving plant properties. The targets include proteins of plant origin, and foreign proteins tailored to enhance particular characteristics. We have not included details of plant transformation methods as these are covered by a number of recent reviews (see, for example *Plant Molecular Biology Reporter*, Volume 13 (1989), issue no. 3).

A number of obstacles have delayed the development of plant protein engineering, including the absence of plant-based expression systems equivalent to those available in bacteria, yeast and cultured animal cells (mammalian and insect). It is, therefore, necessary to express plant proteins in heterologous systems, in which signals for post-translational processing and assembly may not be recognized. The sections on plant proteins are,

therefore, preceded by introductory chapters, by Sarah Eccles and by Graham Belfield and Mick Tuite, which discuss the development and characteristics of expression systems based on *E. coli*, yeast and cultured mammalian cells. In addition Gunter Schneider and Ylva Lindqvist provide a broad introduction to the forces that stabilize protein structure, and how these can be explored by protein engineering.

Although plant genes may be difficult to express in heterologous organisms, equivalent genes may be present in photosynthetic bacteria, allowing studies of protein structure and function which can be extrapolated to the higher plant homologues. In some cases photosynthetic bacteria may themselves serve as substitute hosts for higher plant genes, where a selectable regime can be devised to assess *in vitro* generated mutations. This has been revealing with site specific mutations in proteins that compose the photosystem II complex of the cyanobacterium *Synechocystis* 6803, described by Bruce Diner and colleagues, where the function of the protein is very much dependent on the nature of the complex in which it is associated. No such selectable systems exist for ribulose bisphosphate carboxylase/oxygenase (Rubisco), the primary catalyst of carbon dioxide assimilation. Attempts to engineer cyanobacteria to act as a reporter for mutations in Rubisco structure served mainly to highlight the essential nature of the protein in other metabolic processes of the bacterium (Pierce *et al.*, 1989). Fred Hartman describes how mutagenesis of this enzyme has proceeded, by rational design relying on the identification of critical amino acids with novel active site affinity probes and aided most recently by a high resolution crystal structure for the enzyme. Nevertheless it is frustrating to have identified a region of the protein that is involved in the reaction mechanism yet have no system for rapid selection of more efficient mutants.

Many proteins have evolved in plants that might be exploited for therapeutic use. These include inhibitors of mammalian proteases and α-amylases and Alison Campbell describes the structure of the chymotrypsin inhibitor CI-2 from barley seeds, and how site directed mutations of its reactive (inhibitory) site can be used to alter its activity and specificity for different serine proteases. Another such family of proteins is the protein synthesis inhibitors, the most well-characterised being ricin from castor bean. The lethal function of this protein is being tamed by Mike Lord and colleagues in ways that should prove valuable for the treatment of carcinomas. Enno Krebbers and colleagues have adopted a different approach to producing therapeutic proteins in plants. They have isolated the genes for a group of albumin storage proteins and engineered them to replace a variable region of the encoded protein with peptides of therapeutic value. The engineered storage protein is robust to such alterations, and is synthesized to high concentrations in the seeds of transgenic plants. It is also possible to produce other proteins of animal origin in transgenic plants, including antibodies which are correctly assembled and exhibit biological activity (Hiatt *et al.*, 1989).

Storage proteins have proved to be attractive targets for genetic and molecular analyses, because of their role in determining grain quality. Attempts to improve their quality by protein engineering have been less successful. Craig Lending and colleagues describe elegant experiments of mutagenesis, to improve the lysine content, on the properties and packaging of the zein storage proteins of maize, but the application of these studies to the improvement of seed quality may well be limited by the complexity of the zein genetic system as well as the difficulty in routinely transforming maize. Even more problems are encountered in trying to understand and manipulate the structures of legume and gluten proteins that determine technological quality. The 7S and 11S globulins of legumes undergo extensive post-translational processing (disulphide bond formation, glycosylation and proteolytic nicking), while the functional properties of gluten proteins depend on complex interactions between over 50 individual components. The chapters by Pippa Madgwick and colleagues and by Nigel Lambert and Jenny Yarwood are essentially state-of-the-art accounts of attempts to use protein engineering to get to grips with these daunting systems. Neil Bulleid and colleagues describe an alternative approach to study wheat gluten protein assembly, *in vitro* transcription and translation. Finally, Theo Verrips provides a fascinating account of sweet-tasting proteins, which are yet to find a significant role in the food industry.

An important determinant of yield is the ability of the plant to survive infestations with pests and diseases, and to compete with non-crop species (i.e. weeds). At present the major control of pests, pathogens and weeds is by the application of agrochemicals. Two developments have had an important impact on the use of agrochemicals. The first is the identification of the sites of action of many herbicides, showing that some specifically inhibit one enzyme or protein unique to plants. Dan O'Keefe and colleagues briefly discuss the exploitation of these enzymes that catalyse essential biochemical processes in plants. They then describe in detail the use of cytochrome P450 to detoxify specific herbicides, and the potential for engineering this protein. For this approach to be used for weed control it is necessary to engineer resistance into the crop plant, by the introduction of mutated resistant enzyme or enzymes that detoxify the chemical, or by blocking the enzymes that convert an inactive precursor into the active compound. In this context, the details of enzyme mechanisms elucidated by *in vitro* mutagenesis could lead to *de novo* design of potent and specific inhibitors. Our expanding knowledge of plant gene regulation will allow the expression of the inserted genes to be restricted to specific tissues or stages of development, giving a wide choice of weed control strategies.

A second approach is to provide plants with a form of biological protection against pests. This has been successfully used to provide resistance to viruses, using coat protein (Beachy, 1988; Baulcombe, 1989) or satellite DNA (Harrison *et al.*, 1987; Gerlach *et al.*, 1987), and to insects using a serine protease inhibitor derived from cowpea (Hilder *et al.*, 1987), and other possible strategies involve the use of other protease inhibitors (e.g. CI-2), amylase inhibitors, endochitinases and lysozyme. A particularly successful application, which has included detailed protein engineering studies, is the use of insecticidal toxin derived from *Bacillus thuringiensis*, reviewed by Don Dean.

In many cases, it might be desirable that the proteins and enzymes of interest be transported and localized in specific sub-cellular organelles, e.g. storage proteins, enzymes of photosynthesis or amino acid metabolism. Newell Bascomb and Tom Lubben describe the contribution that protein engineering has made to unravelling the targeting information buried in those proteins destined for specific intraorganellar locations.

We hope that this discussion and the chapters that follow will convince the reader that a book on plant protein engineering is not premature, but timely in that it summarizes the current situation and points the way to a range of exciting applications. There are many examples in the literature of proteins with improved characteristics produced by *in vitro* engineering, but relatively little work of this type on plant proteins. It is therefore possible to cover the whole range of work in the area in one book. We anticipate that the rate of growth will be rapid as many of the obstacles associated with the manipulation of plant systems are surmounted, and a high degree of selectivity will be required should we consider the preparation of future editions.

## References

Baulcombe, D. (1989). Strategies for virus resistance in plants. *Trends in Genetics*. **5**: 56–60.

Beachy, R.N. (1988). Virus cross-protection in transgenic plants. In: *Plant Gene Research: Temporal and spatial regulation of plant genes*, (ed. D.P.S. Verma and R.B. Goldberg), pp. 313–27. Springer Verlag, New York.

Gerlach, W.L., Llewellyn, D. and Haselhoff, J. (1987). Construction of a plant disease resistance gene from the satellite RNA of tobacco ringspot virus. *Nature*. **328**: 802–5.

Harrison, B.D., Mayo, M.A. and Baulcombe, D.G. (1987). Virus resistance in transgenic plants that express cucumber mosaic virus satellite RNA. *Nature*. **328**: 799–802.

Hazelhoff, J. and Gerlach, W.L. (1988). Simple RNA enzymes with new and higher specific endoribonuclease activities. *Nature*. **334**: 585–91.

Hiatt, A., Cafferkey, R. and Bowdish, K. (1989). Production of antibodies in transgenic plants. *Nature*. **342**: 76–8.

Hilder, V.A., Gatehouse, A.M.R., Sheerman, S.E., Barker, R.F. and Bovett, D. (1987). A novel mechanism of insect resistance engineered into tobacco. *Nature*. **300**: 160–3.

Krol, A.R. van der, Mol, J.N.M. and Stuitje, A.R. (1988). Modulation of eukaryotic gene expression by complementary RNA or DNA sequences. *Biotechniques*. **6**: 958–76.

Pierce, J., Carlson, T.C. and Williams, J.G.K. (1989). A cyanobacterial mutant requiring the expression of ribulose bisphosphate carboxylase from a photosynthetic anaerobe. *Proceedings National Academy of Science U.S.A.* **86**: 5753–7.

# **Part I** INTRODUCTION TO PROTEIN EXPRESSION AND STRUCTURE

# 1

# Development of Expression Systems for Eukaryotic Proteins in *E. coli* and Mammalian Cells

Sarah Eccles

## Introduction

During the past two decades technical advances in DNA manipulation *in vitro*, together with the development of efficient transformation systems for a wide variety of prokaryotic and eukaryotic cells, have allowed an extensive study of the expression of genes in both homologous and heterologous systems. Such studies have contributed widely to our understanding of the regulation of gene expression via transcriptional, translational and post-translational mechanisms and have allowed the exploitation of recombinant gene expression for protein production and purification. The increase in protein yield obtained in many recombinant expression systems has facilitated studies of protein structure and biological activity that were formerly limited by the ability to purify sufficient quantities of many biologically active proteins from their native systems. Commercial application of recombinant gene expression technology has grown rapidly and many proteins are now produced commercially using recombinant systems. These range from bacterial proteins such as restriction enzymes, cloned from a variety of bacterial species and expressed in *Escherichia coli*, to human proteins such as insulin and tissue plasminogen activator cloned from human tissue and expressed in *E. coli* or mammalian cells.

Much of the early effort on producing mammalian proteins was invested in the development of expression systems in *E. coli*. *E. coli* has obvious advantages as a host for gene manipulation and protein production, related to our extensive knowledge of its genetics and the ease with which it can be grown to very high cell densities in simple growth media. Indeed, for the production of bacterial proteins it has been very successful. However, attempts to express eukaryotic proteins in *E. coli* have met with a number of problems, the most important of which is the lack of post-translational modification of proteins in bacterial cells. Processes such as glycosylation and phosphorylation are necessary for the biological activity, stability and antigenicity of many mammalian proteins and their production in *E. coli* may not result in a product with all the required properties.

Research efforts have therefore diversified into developing eukaryotic expression systems. Yeast has several advantages over *E. coli* as an expression system. In particular, yeast secretes proteins into the medium which aids protein purification and also allows disulphide bond formation and glycosylation of mammalian proteins to occur. The glycosylation pathways in yeast are not identical to those used in mammalian cells, but human albumin is an example of a protein which has been produced successfully using yeast (Collins 1990). Insect cells have also been used to produce mammalian proteins. High yields of foreign protein can be obtained from *Spodoptera frugiperda* cells when they are infected with recombinant baculoviruses, in which the foreign gene is expressed from the strong viral polyhedrin promoter. Recombinant proteins produced in this system are generally functional (Lucknow and Summers 1988) but their glycosylation patterns are not identical to the authentic proteins. Certain mammalian proteins cannot be produced in heterologous systems because of the requirement for very specific modifications, for example, factor IX, which is subject to γ-carboxylation. Hence, the past decade has witnessed the development of efficient mammalian cell expression systems for mammalian proteins in a variety of cell types. Considering the various problems encountered in expressing mammalian genes in heterologous systems, it seems likely that in many cases these will be the systems of choice for producing fully active therapeutic mammalian proteins in the future. A further extension of the use of mammalian cell expression systems is to express cloned genes in whole animals, and the possibility of using transgenic domestic animals to produce therapeutic human proteins is being explored actively.

Expression of exogenous genes in plants has been successfully achieved using both *Agrobacterium tumefaciens* T-DNA mediated gene transfer and transformation of protoplasts with free DNA. More recent procedures for introducing DNA into intact plant cells include microinjection and the use of particle guns. Together these techniques have allowed the production of transgenic plants from a wide variety of species (Gasser and Fraley 1989). Genes introduced into plants include those conferring herbicide, virus and insect resistance as well as a number of mammalian genes.

The ability to modify genes using a combination of *in vitro* recombination and mutagenesis has led to the growth of the field of protein engineering and the possibility of producing protein products with novel activities. The exploitation of this technology requires that appropriate expression systems are chosen. These should allow efficient transcription and translation of engineered genes and appropriate post-translational modifications (when these are required for the desired characteristics of the engineered protein). In this chapter the development of expression systems for eukaryotic genes in bacteria (*E. coli*) and mammalian cells will be reviewed. The intention is to illustrate the types of problems encountered in achieving efficient expression of cloned genes in both homologous and heterologous systems.

## Expression of eukaryotic genes in *E. coli*

Two fundamental requirements for an efficient expression system are that the chosen gene is efficiently transcribed and translated. Since prokaryotic genes do not generally contain introns, the eukaryotic gene either must be in the form of a cDNA or, alternatively, must be synthesized chemically using the known protein or genomic sequence. Transcription is achieved by cloning the eukaryotic sequence downstream from a strong prokaryotic promoter, most commonly the *E. coli lac* or *trp* promoter or the leftward promoter of phage lambda ($P_L$). These promoters contain conserved regions 35 bp and 10 bp upstream of the transcription initiation site (Rosenberg and Court 1979) that are involved in the binding of *E. coli* RNA polymerase, and may also contain regulatory elements such as the *lac* operator which allows inducible expression through the action of the *lac* repressor.

Efficient translation in *E. coli* requires the presence of a ribosome binding site. This consists of the initiation codon, AUG, together with a sequence (lying 3–12 bases upstream) known as the Shine–Dalgarno (SD) sequence. The SD sequence (3–9 bases) is complementary to the 3′ end of 16S rRNA and it is believed that hybridization of the 5′ end of the mRNA to the 3′ end of 16S rRNA locates the message on the 30S ribosomal subunit (Steitz 1979). Translation also depends on the codon composition of the mRNA. *E. coli* shows a codon preference reflected in the abundance of particular tRNA species (Ikemura 1981). The occurrence in eukaryotic mRNA of codons corresponding to rare *E. coli* tRNA species results in poor translational efficiency. For this reason it is generally advantageous to synthesize genes chemically, incorporating the appropriate *E. coli* codons, rather than to express cDNAs. However, this is not entirely without problems because such chemically synthesized genes may give rise to unpredicted secondary structures in the mRNA, which themselves lead to poor translational efficiency (Bialy 1987). A further potential problem raised by codon bias is the importance of translational pauses in protein folding. It is believed that certain proteins utilize clusters of rare codons to slow down the rate of chain elongation, and hence influence the way in which the protein folds (Purvis *et al.* 1987). This mechanism may contribute to the high frequency of heterologous proteins that are insoluble in *E. coli* (Harris 1983).

Many eukaryotic proteins have been produced in *E. coli* as fusion proteins with *E. coli* polypeptides such as β-galactosidase. In this case the promoter, ribosome binding site and initiation codon all come from the *lacZ* gene, together with varying amounts of the β-galactosidase protein coding sequence. β-galactosidase can be used to protect foreign proteins from degradation in *E. coli*. Somatostatin, which was the first eukaryotic protein to be synthesized in *E. coli*, was produced by fusing a synthetic gene to virtually the entire β-galactosidase coding sequence (Itakura *et al.* 1977) and somatostatin was subsequently isolated by cyanogen bromide cleavage of the fusion protein. An earlier construction, which fused the synthetic somatostatin gene to only the first seven amino acids of β-galactosidase, did not yield any somatostatin activity, suggesting that the galactosidase protein had a role in stabilizing the somatostatin peptide in *E. coli*. The A- and B-chains of insulin were also produced as β-galactosidase fusion polypeptides (Goeddel *et al.* 1979) and in this case, although the hybrid proteins constituted 20% of total cell protein, the protein was insoluble and found in inclusion bodies. This appears to be a general property of many eukaryotic proteins synthesized at high levels in *E. coli*, indeed it has also been seen for normal *E. coli* proteins when they are overexpressed (Prouty and Goldberg 1972). Stringent denaturation and renaturation procedures are required to solubilize proteins from inclusion bodies and to refold them correctly to get an active product (Marston 1986). A recent report suggests that fusing proteins to the conserved 76-amino acid eukaryotic protein, ubiquitin, stabilizes foreign proteins in *E. coli* (Butt *et al.* 1989). Ubiquitin can be exactly removed from the fusion product by the action of the enzyme ubiquitin N α-protein hydrolase, yielding the authentic protein. This offers a major advantage over β-galactosidase gene fusions, where cleavage of the fusion protein may be difficult.

A promising development in *E. coli* expression systems involves secretion of recombinant products into the periplasmic space. In 1980 Talmadge *et al.* demonstrated that preproinsulin expressed in *E. coli* from the β-lactamase promoter was secreted into the periplasm and that the natural signal peptide was cleaved to produce proinsulin. Similarly, when the coding region of the precursor of human growth hormone was placed downstream of an *E. coli trp* promoter, correctly processed mature human growth hormone was located in the periplasmic fraction (Gray *et al.* 1985). Furthermore, the human growth hormone produced in this study was monomeric and correctly disulphide-bonded. However, eukaryotic signal sequences do not always direct protein products to the periplasm.

Sarmientos *et al.* (1989) expressed pretissue plasminogen activator from a *trp* promoter in *E. coli* and found that the tPA accumulated in inclusion bodies in the cytoplasm. Linkage of the tPA coding sequences to a bacterial signal sequence from the bacterial *ompA* gene did not result in efficient periplasmic localization either, and the authors concluded that sequences other than the leader sequence contribute to location of proteins in the *E. coli* periplasm. Nevertheless, antibodies have been successfully secreted into the *E. coli* periplasm using *E. coli* signal sequences. Functionally assembled Fv fragment could be purified from *E. coli* periplasm when synthetic DNA sequences coding for the heavy and light chain variable regions were linked to *E. coli* signal sequences and driven from the inducible *lac* promoter in an artificial dicistronic operon (Skerra and Pluckthun 1988). In this case, the heavy chain sequence was fused to the leader sequence from outer membrane protein A (*ompA*) and the light chain sequence to the alkaline phosphatase (*phoA*) leader sequence. A similar report from Better *et al.* (1988) describes the secretion of an active Fab fragment into the *E. coli* periplasm. In this case both the heavy and light chain DNA sequences were linked to the leader sequence from the pectate lyase (*pelB*) gene of *Erwinia carotovora* and driven from the inducible *Salmonella typhimurium araB* promoter. In both cases the disulphide bonds were correctly formed, the protein fragments formed heterodimers and the resulting truncated antibody molecules had affinity constants equivalent, in the case of Fv, to the whole antibody produced in eukaryotic cells, and to proteolytically produced Fab fragments in the case of Fab. These data contrast with earlier studies using immunoglobulin heavy and light chain cDNAs lacking leader sequences that, when coexpressed in *E. coli*, resulted in the production of insoluble antibody chains in inclusion bodies (Boss *et al.* 1984; Cabilly *et al.* 1984). The *pelB* signal sequence has also been used to secrete single domain antibodies (dAbs) isolated by polymerase chain (PCR) amplification of rearranged heavy chain variable regions from spleen DNA (Ward *et al.* 1989). Since glycosylation is not required for antibody binding, the secretion of antibodies by *E. coli* provides a useful system for expression of antibodies when binding is the only property required.

Although *E. coli* is not generally considered to be an appropriate host for secretion of proteins into the culture medium because of the outer membrane barrier, secretion of foreign proteins has been achieved. Abrahmsen *et al.* (1986) describe the secretion of human insulin-like growth factor I after fusion to part of staphylococcal protein A. *E. coli* does excrete a few proteins into the medium naturally; these include haemolysins and bacteriocins. Bacteriocins are synthesized as mature soluble proteins in the *E. coli* cytoplasm and their release into the medium is dependent on the expression of bacteriocin release protein (BRP), which activates a phospholipase located on the outer membrane, resulting in increased permeability of both the inner and outer membranes. Hsiung *et al.* (1989) described the successful exploitation of this mechanism to secrete human growth hormone (hGH) into the culture medium in a regulated manner. Human growth hormone was linked to the leader sequence of the *ompA* gene and coexpressed in *E. coli* with BRP. Both proteins were controlled by the inducible *lac* promoter and induction with isopropyl-β-D-thiogalactosidase resulted in release of mature human growth hormone into the medium. The yield of hGH was reported to be rather less than for the periplasmic synthesis of hGH using the same *ompA* leader sequence. However, since protein can be purified from the medium very simply without rupturing the cells, the development of continuous production systems using immobilized cells can be envisaged.

The possibility of using Gram positive organisms, such as *Bacillus* species (which lack an outer membrane), for protein secretion has been explored but *Bacillus* is not as readily manipulable as *E. coli*, nor is the knowledge of its genetics so extensive. Furthermore, the plasmids used as cloning vectors in *Bacillus subtilis* are frequently unstable and, because *Bacillus* species secrete proteases into the medium, the heterologous secreted proteins tend to be degraded (Errington and Mountain 1990).

A summary of the strategies used for expressing eukaryotic genes in *E. coli* is given in Figure 1.1. Although some of the early *E. coli* expression studies were hampered by the problem of protein insolubility, the recent development of vectors incorporating sequences that direct proteins to the periplasm, or even the medium, now makes *E. coli* a more attractive system for foreign protein production. However, the problem of the lack of post-translational modification still remains. One possible way of getting around this would be to modify *E. coli*-produced proteins *in vitro*; such modification systems are being studied. An alterna-

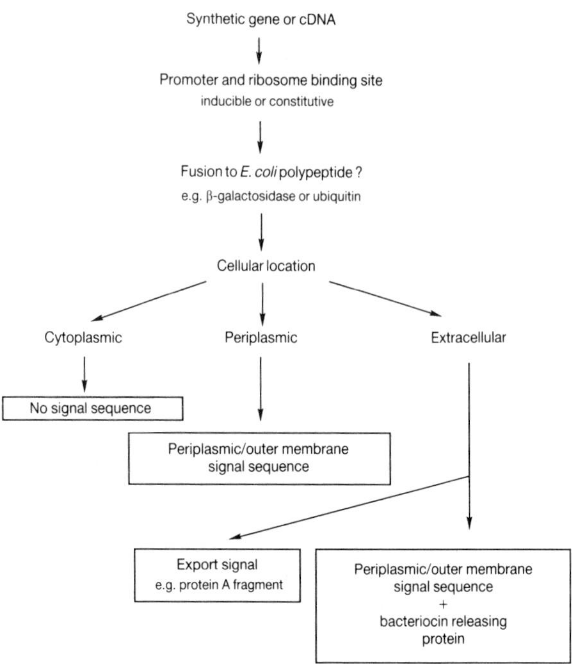



**Fig. 1.1**   Strategies for expressing foreign genes in *E. coli*.

tive, and the only reliable way of making appropriately modified mammalian proteins at the present time, is to express them in mammalian cells.

## Expression of genes in mammalian cells

A variety of factors must be considered in the design of efficient mammalian expression systems. Most expression systems use strong promoter/enhancer combinations to drive transcription. The choice of these will depend partly on the cell type, but viral promoters and enhancers from, for example, SV40 (Mulligan and Berg 1981), Rous sarcoma virus (RSV) (Gorman *et al.* 1982a) and human cytomegalovirus (HCMV) (Boshart *et al.* 1985) have been used widely because they work well in a variety of cell types. When expression in a particular specialized cell type is required, cell type specific promoters and/or enhancers may be preferable. The immunoglobulin heavy chain enhancer (IgH enhancer), for example, directs high level expression in lymphoid cells (Neuberger 1983). Inducible promoters may be used when continuous high level expression of a gene product is toxic to the cell. Examples of inducible promoters that have been used to regulate expression of cloned genes include heat shock protein promoters, which are responsive to temperature shifts (Bendig *et al.* 1987), the metallothionein promoter, which is induced by heavy metal ions and by glucocorticoid hormones (Pavlakis and Hamer 1983) and the glucocorticoid hormone sensitive promoter of the mouse mammary tumour virus LTR (Lee *et al.* 1981). In addition the *E. coli lac* repressor has been introduced into mammalian cell lines allowing inducible expression from genes controlled by the *lac* promoter/operator region (Hu and Davidson 1987). Either genes or cDNAs may be expressed in mammalian cells. Early experiments on the expression of cDNAs in SV40 vectors indicated that the production of stable mRNA was dependent on splicing and that this requirement could be supplied either by a naturally occurring or a heterologous intron (Hamer and Lader 1979, Gruss and Khoury 1980). Vectors that are suitable for expressing cDNAs were therefore designed to include an intron downstream from the cloning site to fulfil the splice requirement. A series of vectors based on SV40 transcription units is described by Berg (1981). One of these vectors, pSV2 (Fig. 1.2), contains the SV40 early region promoter and enhancer, together with the viral replication origin, upstream of the cloning site, and the SV40 small t-antigen intron and viral early region polyadenylation signals downstream of the cloning site. Transcription proceeds from the SV40 promoter, through the cloned DNA, continuing through the intron and polyadenylation signals thus assuring production of mature mRNA. However, it has subsequently been found that certain genes can be efficiently transcribed in the absence of intron sequences, preproinsulin is an example (Gruss *et al.* 1981). What determines the requirement for splicing remains unclear, but the question has been addressed recently by Neuberger and Williams (1988) who showed that, in the case of an immunoglobulin gene, the requirement for an intron is dependent on the promoter used for transcription. When transcription of an immunoglobulin μ gene was driven by an immunoglobulin promoter and enhancer, an intron was required for expression of cytoplasmic immunoglobulin μ mRNA. An intron was also required when a β-globin promoter was used, but expression from a cytomegalovirus or heat shock promoter was not dependent on the presence of an intron. The intron requirement was largely post-transcriptional and therefore presumably has some effect on RNA processing or nuclear

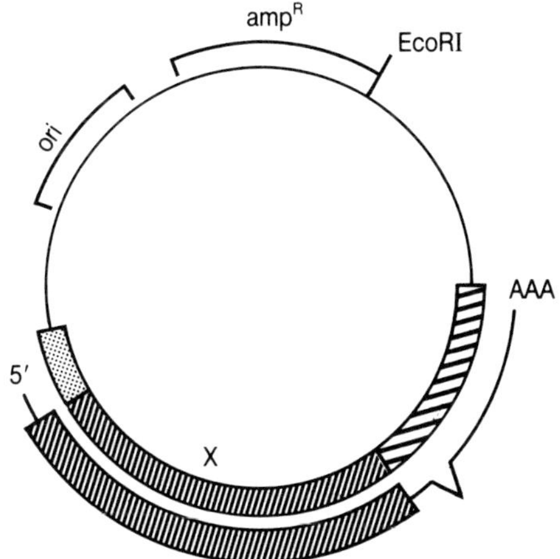

**Fig. 1.2**   The pSV2 vectors. Stippled region, SV40 early region promoter; wide striped region, SV40 polyadenylation site and small t intron; narrow striped region, X = position of test gene (e.g. CAT, β-galactosidase) or selectable marker gene (e.g. *neo*R, *gpt*, *hph*). The transcript derived from the SV40 promoter is indicated below the coding sequences. Additional sequences may be cloned at the single *Eco*RI restriction site. The positions of the pBR322 derived origin of replication (*ori*) and the ampicillin resistance gene (*amp*) are indicated.

export, suggesting a connection between the promoter and one or both of these processes. Since it is, in general, not possible to predict whether an intron will be required for expression, it would seem advisable to include one in the construction of vectors when the expression of a novel cloned gene is to be tested.

## Mammalian selectable markers

Many commonly used eukaryotic vectors are based on pSV2. In addition to the SV40 derived sequences, pSV2 contains an ampicillin resistance gene and the pBR322 plasmid replication origin for selection and replication in *E. coli* (Fig. 1.2). Derivatives of pSV2 containing dominant mammalian selectable markers have been constructed. These include resistance to mycophenolic acid encoded by the *E. coli* xanthine–guanine phosphotransferase gene (*gpt*) (Mulligan and Berg 1981) and resistance to the aminoglycoside antibiotic G418 encoded by the aminoglycoside phosphotransferase gene from Tn5 (*neo*R) (Colbere-Garapin *et al.* 1981). Resistance to the aminocyclitol antibiotic, hygromycin B, encoded by the hygromycin B phosphotransferase gene (*hph*) of *E. coli* has also been used as a dominant selectable marker in mammalian cells (Blochlinger and Diggelmann 1984). Recently Hartman and Mulligan (1988) described two new dominant selectable markers based on expression of an *E. coli* gene involved in the biosynthesis of tryptophan, *trpB*, and a gene from *Salmonella typhimurium* involved in histidine biosynthesis, *hisD*. Expression of these genes in mammalian cells allows growth in medium lacking either tryptophan or histidine. Cells expressing the *hisD* gene could also be selected on the basis of insensitivity to the enzyme precursor, histidinol, which is otherwise toxic to animal cells. These markers have also been incorporated into pSV2 vectors and they may provide inexpensive alternatives to the costly antibiotics G418 and hygromycin B. The thymidine kinase (*tk*) gene of herpes simplex virus can also be used for selection in mammalian cells (Mantei *et al.* 1979) but its use is restricted to cell lines which are deficient in *tk*, for example, derivatives of the mouse L cell line.

## Introduction of DNA into mammalian cells

Introduction of foreign DNA into mammalian cells can be achieved in a number of ways and the choice of procedure is generally dictated by the recipient cell line. Calcium phosphate coprecipitation of DNA (Graham and van der Eb 1973) was one of the first methods to be used and results in highly efficient transfection of a variety of fibroblast cell lines (e.g. mouse L cells, HeLa cells). Stimulation of DNA uptake with DEAE dextran has been used for cells that are poorly transfectable by calcium phosphate. These include a variety of lymphoid cell lines (Sompayrac and Danna 1981). Protoplast fusion, in which bacterial cells containing the cloned gene on a suitable plasmid vector are converted to protoplasts and fused directly with mammalian cells, also results in efficient transfection of cell lines that are refractory to transfection by calcium phosphate (Schaffner 1980). Most recently, electroporation has become the method of choice for transfection of mammalian cells because of its technical simplicity and its application to a wide variety of different cell types (Chu *et al.* 1987).

## Transient and stable gene expression

Expression of genes in mammalian cells can be achieved either transiently or by the generation of stable cell lines that carry the introduced gene. Transient expression occurs during a short period, usually between 12 and 72 h after introduction of DNA into cells, when the plasmid vector is nuclear but not yet integrated into the host cell genome. Examination of mRNA or protein produced during this time can yield information on the contribution of regulatory elements to gene expression. Furthermore, when relatively small amounts of protein are required for analysis, it may be the system of choice for protein production.

The coding sequences of the bacterial enzymes chloramphenicol acetyl transferase (CAT) and β-galactosidase have been used extensively as reporter genes in studies on gene regulation (Gorman *et al.* 1982b; Herbomel *et al.* 1984) and derivates of pSV2 containing the CAT and β-galactosidase genes driven by the SV40 early promoter are widely used as controls in these assays. Recently, the firefly luciferase gene has been found to be a more sensitive reporter gene (DeWet *et al.* 1987) and has the added advantage of giving a linear response over a wide range of concentrations.

Certain transient expression systems have been adapted to give optimal protein yields, for example the SV40/cos cell system. Cos cells are simian CV-1 cells that have been transformed with an origin defective mutant of SV40 that encodes wild type T antigen and supports the replication of defective viruses with deletions in the early region genes (Gluzman 1981). Genes or cDNAs cloned into vectors containing an SV40 origin accumulate to very high copy numbers in these cells, 200–400 000 per cell for a α-globin gene construct (Mellon *et al.* 1981), resulting in a high level of protein expression. The extremely high copy number of these vectors is ultimately toxic to the cells and hence it is not possible to derive stable cell lines using this system (Rigby 1982). However, Whittle *et al.* (1987) describe the detection of 100–500 ng/ml of chimeric antibody in the supernatant of cos cells 72 h after transfection with chimeric heavy and light chain antibody genes. Another transient expression system reported to give sufficient protein yields for preliminary analysis utilizes a vector containing the Rous sarcoma virus LTR in HeLa cells. Browne *et al.* (1988) describe the cloning of a tissue plasminogen activator (tPA) cDNA into the vector pTR312 containing the RSV LTR and SV40 intron and polyadenylation signals. This yielded sufficient protein for estimation of the clearance properties of wild type and mutant tPAs in guinea pigs. Since the RSV LTR is active in a variety of mammalian cells the host range of this vector is not restricted to HeLa cells (Browne *et al.* 1990).

A third transient expression system has been developed using the pox-virus vaccinia. Vaccinia virus has a long (185 kb), linear, double-stranded DNA genome that encodes a complete viral transcription system. The virus replicates in the cytoplasm of a wide variety of mammalian cells. Because of its large size, insertion of foreign DNA into the viral genome is carried out by homologous recombination rather than direct DNA manipulation (Mackett *et al.* 1982). Expression of genes from viral promoters such as the P7.5 promoter, which contains both early and late transcription signals, results in continuous production of foreign proteins for 1–2 days after infection and can result in the production of 1–2 mg of foreign protein per litre of cell culture (Moss and Flexner 1987). Fuerst *et al.* (1986, 1987) report a 15–20-fold increase in expression levels obtainable from this system by incorporating the gene for the highly efficient bacteriophage T7 RNA polymerase into the vaccinia virus genome. This polymerase is used to drive expression of a coding sequence, from a T7 promoter, located either on a plasmid or on a second vaccinia virus in the same cell. These studies used HeLa cells but vaccinia can be propagated on most tissue culture cell lines (Mackett *et al.* 1985) and it can also be used for expression in whole animals. The most extensive use of vaccinia vectors has been for the expression of viral proteins (such as hepatitis B virus surface antigen) in animals, for studies on vaccine development (reviewed by Moss and Flexner 1987). However, vaccinia shows considerable promise as a system for high level transient expression of foreign proteins in cultured cells.

Although transient systems are useful for the preliminary analysis of cloned gene products, production of larger amounts of proteins requires the development of stable cell lines expressing the cloned gene. There are essentially two ways of producing stable transfected cell lines. The first approach utilizes viral vectors to replicate the cloned gene independently from the cellular genome. The viral vectors used initially to introduce cloned DNA into mammalian cells were derived from the genomes of SV40 and polyoma virus (for a review see Rigby 1982). SV40 replace-

ment vectors, in which the cloned DNA replaced part of the viral genome and recombinant viruses were used for the productive infection of cultured cells in the presence of helper virus, were designed. Although this results in the production of large amounts of mRNA and protein in the infected cells, viral infection also causes cell death, so stable cell lines cannot be generated. More recently, viral vectors have been derived from viruses whose genomes are maintained episomally in the nucleus of permissive cells. There are several potential advantages to using episomal vectors. First, the copy number per genome is generally high, potentially increasing the yield of protein per cell. Secondly, since, by definition, these vectors do not normally integrate into the cellular genome they are free from the 'position effects' that act on integrated transfected genes as a consequence of regulatory elements associated with DNA flanking the insert. The two episomal vectors that have been studied most extensively are those derived from bovine papilloma virus and Epstein–Barr virus (EBV), but an episomal vector derived from the human papovirus BK has also been described (Milanesi *et al.* 1984).

Bovine papilloma virus (BPV) is a double-stranded DNA virus with a 7.95 kb circular genome. The virus causes epithelial and mesenchymal tumours *in vivo* and transforms bovine and rodent fibroblasts *in vitro*. The transforming and episomal functions both lie on a 5.4 kb fragment known as the 69% transforming fragment, which has been used in the construction of vectors by linkage to bacterial replicons. These are generally derivatives of pBR222 from which the poison sequences that inhibit replication of SV40 in monkey cells (Lusky and Botchan 1981) have been deleted (Binetruy *et al.* 1982; Sarver *et al.* 1982; Campo and Spandidos 1983). Such vectors are reported to replicate as multiple copies (10–200 per cell) of unrearranged non-integrated monomers in mouse fibroblasts (Sarver *et al.* 1982; Campo and Spandidos 1983) and can be rescued in *E. coli* by transfection with episomal DNA isolated from transformed cells, thus providing a useful method of shuttling DNA between bacteria and mouse cells. Several eukaryotic genes have been successfully expressed using BPV-1 vectors in mouse C127 cells. These include human β-globin (Di Maio *et al.* 1982), human growth hormone (Pavlakis and Hamer 1983), human interferon (Zinn *et al.* 1982, 1983) and human tissue plasminogen activator (Bendig *et al.* 1987). In all cases the yields of these proteins were good, usually between $10^6$ and $10^7$ molecules per $10^6$ cells per day (Stephens and Hentschel 1987), and represented authentic human products as judged by various *in vitro* assays. However, many of these studies describe rearrangements of BPV-1-derived vectors (Zinn *et al.* 1982; Pavlakis and Hamer 1983; Di Maio *et al.* 1984; Green *et al.* 1986). For example, Green *et al.* (1986) describe the stable episomal replication of a human histone gene in a BPV-1 vector in C127 cells but the episomes isolated from these cells contained a 1 kb DNA sequence of unknown origin, not present in the original construct. Other rearrangements include deletions and insertions as well as integrations into the cellular genome. These variations may not be a serious disadvantage, provided that a stable derivative expressing the cloned gene can be isolated (as was the case for the human histone gene). A further disadvantage of BPV-1 vectors is their limited host cell range. Selection of transformants by morphological transformation can only be used in cells capable of focus formation. Dominant eukaryotic selectable markers have been introduced into BPV-1 vectors in an attempt to broaden the host range. Incorporation of the HSV-tk gene allowed transformation of tk$^-$ cell lines of mouse, hamster, rat and human origin to the tk$^+$ phenotype (Lusky *et al.* 1983; Sekiguchi *et al.* 1983; Bostock and Allshire 1986) but although some of these clones contained a few episomes, the majority of the exogenous DNA was incorporated in the cellular genome. BPV-1 vectors incorporating the Tn5 neomycin resistance gene are reported to yield mouse C127 cell transformants containing between 20 and 200 unrearranged episomes per cell (Law *et al.* 1983; Matthias *et al.* 1983; Meneguzzi *et al.* 1984), but the same vector always integrated into the genome in mouse L cells (Bostock and Allshire 1986). Furthermore, Bostock and Allshire (1986) found that the method of introducing DNA into the cell influences the fate of the BPV-1 vector. Microinjection of DNA directly into nuclei appears to cause less damage to the DNA than calcium phosphate mediated transfection and more frequently results in BPV-1 vectors replicating autonomously as unrearranged monomers in C127 cells.

Another series of vectors has been developed based on the genome of EBV (Yates *et al.* 1984, 1985). EBV is a human lymphotropic Herpes virus with a large (172 kb), double-stranded DNA genome that can infect and transform human B lymphocytes. Incorporation of the 1.8 kb origin region, termed *oriP*, into recombinant plasmids is sufficient to direct episomal replication of these plasmids in EBV transformed human lymphoid cells or in cells expressing the EBNA-1 antigen of EBV, which encodes a transacting nuclear factor required for replication (Yates *et al.* 1984). Vectors that include both the *oriP* region and the 2.5 kb EBNA-1 gene from EBV, together with a hygromycin resistance gene, have been developed. These plasmids replicate as multiple episomal copies in lymphoid and non-lymphoid cells from a variety of species including human, monkey, dog and rodent (Yates *et al.* 1985). EBV vectors would therefore appear to represent a more versatile episomal system than the BPV-1 system. However, it was found (in the author's laboratory) that the stability of vectors containing both the EBV origin and the EBNA-1 gene varies dramatically in different human cell lines. Although HeLa cells could be efficiently transfected with the vector p201 (Yates *et al.* 1985) and the resulting transformants contained intact episomes, transfection of the human cell line A431 with the same vector resulted in very few transfectants, none of which contained episomes. Both A431 and HeLa cells could be efficiently transfected with the vector p292, which is similar to p201 except that the EBNA-1 gene is expressed from a stronger promoter (the SV40 early promoter). However, the resulting A431 transfectants contained a large number of rearrangements and chromosomal integrations. The p292 vector was more stable in HeLa cell lines but its presence had adverse effects on the growth of both HeLa and A431 clones after prolonged culture (Vidal *et al.* 1990). Hence, although viral vectors may provide useful expression systems, careful choice of the combination of vector and cell line appears to be essential for the generation of stable episomes in recipient cells.

The second approach to the generation of stable clones takes advantage of the observation that, following transfection, a small proportion of the cells that have taken up DNA stably incorporate some of this DNA into apparently random sites in the genome. By linking the gene of interest to one of the selectable markers described above, such clones can be identified and tested for expression of the cotransfected gene. These clones contain varying numbers of copies of the transfected gene but, in general, the level of expression of the exogenous coding sequence does not correlate directly with its copy number. This variable expression is believed to result from the regulatory influence of DNA flanking the insertion site. However, expression of a transfected gene in an individual clone can be increased by amplifying its copy number *in situ*. This can be achieved by including an amplifiable gene in the vector used for transfection. When cells are grown in the presence of certain toxic drugs, resistant clones arise, in which the gene encoding the drug-sensitive enzyme has been amplified. The first example of this was amplification of the dihydrofolate reductase (DHFR) gene in the presence of its inhibitor, methotrexate (reviewed by Stark and Wahl 1984). The region of DNA amplified in these cells was found to be much larger than the DHFR gene itself, leading to the idea that linkage of a transfected gene to DHFR, followed by selection with increasing concentrations of methotrexate, might result in coamplification of the linked gene. This was indeed the case and copy numbers as high as 2000 per cell have been generated using DHFR vectors (Crouse *et al.* 1983). In general, the increase in copy number results in a roughly proportional increase in gene expression (Bebbington and Hentschel 1987). DHFR coamplification is most conveniently performed in cells that lack endogenous DHFR activity, and a DHFR-deficient Chinese Hamster Ovary (CHO) cell line is used frequently. In this case, the DHFR gene can also be used as a selectable marker for transfection. It is possible to carry out DHFR coamplification in cells that contain active DHFR genes but because both the endogenous and transfected DHFR genes are amplified, very high levels of methotrexate must be used and eventually the solubility of the drug limits the number of gene copies that can be obtained. Examples of proteins that have been produced at high levels in CHO cells after DHFR coamplification include human tissue plasminogen activator (tPA) and human α-interferon (IFN-α) (Kaufman *et al.* 1985; Mory *et al.* 1986). The human tPA from CHO cells had a similar specific activity to native tPA but its glycosylation pattern, although similar, was not identical. The IFN-α consisted of a mixture of glycosylated and unglycosylated species, similar to that obtained from human lymphocytes.

Other genes that have been used both as amplifiable genes and selectable markers include the adenine deaminase (ADA) gene (Kaufman *et al.* 1986) and the glutamine synthetase (GS) gene (Bebbington and Hentschel 1987). These genes offer potential advantages over DHFR because they can be used as dominant selectable markers in a variety of cell types and are not restricted to use in

cell lines that are deficient in the respective enzymes they encode.

The major disadvantage of these gene amplification systems is that it generally takes several months to select appropriate cell lines with sufficiently amplified levels of expression. Furthermore, the amplified sequences are subject to deletion and rearrangement, both during and after amplification, sometimes leading to loss of high level expression. Given sufficient time, however, stable, highly expressing clones can be selected. Figure 1.3 gives an overview of the strategies for designing efficient expression systems in mammalian cells.

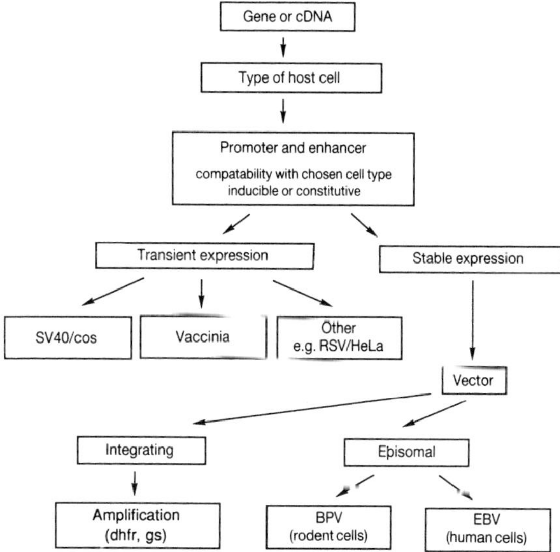


**Fig. 1.3** Strategies for designing efficient expression systems in mammalian cells.

## Future mammalian cell expression vectors

Viral vectors and coamplification of genes with drug resistance markers have both been used successfully to increase gene copy number and protein yield, as can be seen from the examples quoted above, but both systems have disadvantages related to the stability of the multicopy genes and the host cell specificity of the vectors. An alternative way of achieving high level expression is to exploit the regulatory elements associated with genes that are naturally expressed at very high levels in particular cell types. Mouse myeloma cells, for example, are capable of producing very high levels of antibody protein from their endogenous rearranged antibody genes (in the region of 100 μg/ml). However, when rearranged antibody genes are transfected into myeloma cells, the production of recombinant protein is at best generally not more than one-tenth of this level, even when the genes are linked to the strong enhancer from the mouse immunoglobulin heavy chain locus. A possible explanation for this poor level of expression is that the transfected genes integrate randomly into the mouse genome and they may not be located in an appropriate chromosomal environment for high level expression. A possible way of overcoming this 'position effect' is to recombine transfected immunoglobulin genes into the endogenous immunoglobulin loci using homologous recombination. That this may be feasible is suggested by the results of Baker *et al.* (1988), who corrected a mutation in an endogenous immunoglobulin μ constant region gene segment (Cμ) by homologous recombination with a pSV2neo vector containing a wild type Cμ gene segment. The frequency of homologous recombination in this experiment was estimated to be 100- to 1000-fold lower than that of random integration of the vector (non-homologous recombination), hence this procedure is extremely time consuming. A promising finding that may have more general applicability to the development of expression vectors is the identification of sequences of DNA (flanking the human β-globin locus) that confer position-independent expression in erythroid cells. Linkage of these sequences, collectively referred to as a dominant control region (DCR), to a human β-globin gene resulted in the transfected gene being expressed, in both mouse erythroleukaemia (MEL) cells and in transgenic mice, at levels equivalent to that of the endogenous mouse β-globin gene (Grosveld *et al.* 1987; Blom von Assendelft *et al.* 1989). An element with similar properties has been identified downstream of the gene encoding the human T-cell surface protein CD2 (Greaves *et al.* 1989). Both these examples involve genes that are expressed in a tissue specific manner and, in the case of β-globin at least, at a very high level. If dominant control regions are found to be generally associated with highly expressed tissue specific genes (e.g. immunoglobulin genes) their identification and inclusion in expression vectors may allow the expression of these genes at levels that approach or even exceed the expression of the endogenous genes. If this approach is successful it may replace the need for amplification of genes using viral vectors or drug selection. One of the characteristics of transfectants that contain genes linked to a β-globin DCR is that every transfectant that receives the gene and the DCR intact expresses the gene. This property could significantly reduce the screening that is currently carried out to identify clones that express high levels of recombinant protein and may offer particular advantages for poorly transfectable cell lines.

## Expression in transgenic animals

The ability to introduce genes into the germline of animals has opened up a whole new approach to the study of gene regulation and development. Transgenic mice have been used to study the regulation of tissue specific gene expression and to analyse gene function by observing the effect of increasing, decreasing or abolishing expression of particular genes. Transgenic mice have also produced models for human disease. For example, the introduction of oncogenes has provided models for various types of cancer. Although transgenic studies began with mice, it is now possible to introduce genes into a variety of domestic animals including cows, chicken, pigs, rabbits and sheep. A number of studies have been undertaken with the goal of improving the growth characteristics of farm animals. Growth hormone, for instance, has been introduced into pigs (Hammer *et al.* 1985; Purzel *et al.* 1989). Another potential use of transgenic animals is in the production of recombinant proteins, particularly human therapeutics. Both the human α-1-antitrypsin gene and human factor IX gene have been introduced into sheep under the control of a β-lactoglobulin promoter with the intention of isolating the recombinant proteins from milk (Simons *et al.* 1988; Clark *et al.* 1989). This resulted in two transgenic sheep lines that expressed human factor IX, but the level of expression was very low—25 ng/ml—250 times lower than the concentration of factor IX in normal human plasma and about $10^5$ times lower than the concentration of β-lactoglobulin in sheep's milk (Clark *et al.* 1989). The low level of protein obtained did not allow a full analysis of the post-translational modification of factor IX isolated from milk but the protein did show activity in clotting assays, suggesting that at least those post-translational modifications that are essential for function, such as γ-carboxylation, are carried out in the mammary gland. The data presented by Clark *et al.* (1989) suggest that the poor level of expression may be due to inefficient translation and poor transcription, and they note that a sheep expressing α-1-antitrypsin from the same β-lactoglobulin gene fusion is expressed at substantially higher levels (5 μg/ml). These experiments demonstrate the feasibility of producing human therapeutics in domestic livestock and, since milk is produced in large quantities and can be collected without detrimental effect to the animal, if the levels of expression can be improved, transgenic livestock may prove to be a future source of a variety of recombinant proteins.

# References


Abrahmsen, L., Moks, T., Nilsson, B. and Uhlen, M. (1986). Secretion of heterologous gene products to the culture medium of *Escherichia coli. Nucleic Acids Research.* 14: 7487–500.

Baker, M.D., Pennell, N., Bosnoyan, L. and Shulman, M.J. (1988). Homologous recombination can restore normal immunoglobulin production in a mutant hybridoma cell line. *Proceedings of the National Academy of Sciences, USA.* 85: 6432–6.

Bebbington, C.R. and Hentschel, C.C.G. (1987). The use of vectors based on gene amplification for the expression of cloned genes in mammalian cells. In *DNA cloning,* Vol. III, (ed. D.M. Glover), pp. 163–88. IRL Press, Oxford.

Bendig, M.M., Stephens, P.E., Crockett, M.I. and Hentschel, C.C.G. (1987). Mouse cell lines that use heat shock promoters to regulate the expression of tissue plasminogen activator. *DNA.* 6: 343–52.

Berg, P. (1981). Dissections and reconstructions of genes and chromosomes. *Science.* 213: 296–303.

Better, M., Chang, C.P., Robinson, R.R. and Horvitz, A.H. (1988). *Escherichia coli* secretion of an active chimaeric antibody fragment. *Science.* 240: 1041–3.

Bialy, H. (1987). Recombinant proteins: virtual Authenticity. *Biotechology.* 5: 883–90.

Binetruy, B., Meneguzzi, G., Breathnach, R. and Cuzin, F. (1982). Recombinant DNA molecules comprising bovine papillomavirus type 1 DNA linked to plasmid DNA are maintained in a plasmidial state both in rodent fibroblasts and in bacterial cells. *EMBO Journal.* 1: 621–8.

Blochlinger, K. and Diggelmann, H. (1984). Hygromycin B phosphotransferase as a selectable marker for DNA transfer experiments with higher eukaryotic cells. *Molecular and Cellular Biology.* 4: 2929–31.

Blom van Assendelft, G., Hanscombe, O., Grosveld, F. and Greaves, D.R. (1989). The β-globin dominant

control region activates homologous and heterologous promoters in a tissue specific manner. *Cell.* 56: 969–77.

Boshart, M., Weber, F., Jahn, G., Dorsch-Häsler, K., Fleckenstein, B. and Schaffner, W. (1985). A very strong enhancer is located upstream of an immediate early gene of human cytomegalovirus. *Cell.* 41: 521–30.

Boss, M.A., Kenten, J.H., Wood, C.R. and Emtage, J.S. (1984). Assembly of functional antibodies from immunoglobulin heavy and light chains synthesised in *E. coli. Nucleic Acids Research.* 12: 3791–806.

Bostock, C.J. and Allshire, R.C. (1986). Comparison of methods for introducing vectors based on bovine papillomavirus-1 DNA into mammalian cells. *Somatic Cell and Molecular Genetics.* 12: 357–66.

Browne, M.J., Carey, J.E., Chapman, C.G. *et al.* (1988). A tissue-type plasminogen activator mutant with prolonged clearance *in vivo. Journal of Biological Chemistry.* 263: 1599–602.

Browne, M.J., Carey, J.E., Chapman, C.G., Dodd, I., Lawrence, G.M.P. and Robinson, J.E. (1990). The expression of tissue-type plasminogen activator and related enzymes. In *Protein Production: the exploitation of micro-organisms, cells and animals to make useful proteins* (4th Biological Council Biotechnology Symposium), (ed. T.J.R. Harris), Elsevier Science Publishers, London. pp. 117–30.

Butt, T.R., Jonnalagadda, S., Monia, B.P. *et al.* (1989). Ubiquitin fusion augments the yield of cloned gene products in *Escherichia coli. Proceedings of the National Academy of Sciences, USA.* 86: 2540–4.

Cabilly, S., Riggs, A.D., Pande, H. *et al.* (1984). Generation of antibody activity from immunoglobulin polypeptide chains produced in *Escherichia coli. Proceedings of the National Academy of Sciences, USA.* 81: 3273–7.

Campo, M.S. and Spandidos, D.A. (1983). Molecularly cloned bovine papillomavirus DNA transforms mouse fibroblasts *in vitro. Journal of General Virology.* 64: 549–57.

Chu, G., Hayakawa, H. and Berg, P. (1987). Electroporation for the efficient transfection of mammalian cells with DNA. *Nucleic Acids Research.* 15: 1311–26.

Clark, A.J., Bessos, H., Bishop, J.O. *et al.* (1989). Expression of human anti-haemophilic factor IX in the milk of transgenic sheep. *Biotechnology.* 7: 487–92.

Colbere-Garapin, F., Horodniceaunu, F., Kourilsky, P. and Garapin, A.C. (1981). A new dominant hybrid selective marker for higher eukaryotic cells. *Journal of Molecular Biology.* 150: 1–14.

Collins, S. (1990). The production of secreted proteins in yeast. In *Protein Production: the exploitation of micro-organisms, cells and animals to make useful proteins* (4th Biological Council Biotechnology Symposium) (ed. T.J.R. Harris), pp. 61–78. Elsevier Science Publishers, London.

Crouse, G.F., McEwan, R.N. and Pearson, M.L.

(1983). Expression and amplification of engineered mouse dihydrofolate reductase minigenes. *Molecular and Cellular Biology*. 3: 257–66.

DeWet, J.R., Wood, K.V., Deluca, M. *et al.* (1987). Firefly luciferase gene structure and expression in mammalian cells. *Molecular and Cellular Biology*. 7: 725–37.

Di Maio, D., Treisman, R. and Maniatis, T. (1982). Bovine papillomavirus vector that propagates as a plasmid in both mouse and bacterial cells. *Proceedings of the National Academy of Sciences, USA*. 79: 4030–4.

Di Maio, D., Corbin, V., Sibley, E. and Maniatis, T. (1984). High-level expression of a cloned HLA heavy chain gene introduced into mouse cells on a bovine papillomavirus vector. *Molecular and Cellular Biology*. 4: 340–50.

Errington, J. and Mountain, A. (1990). Is Baccilus an alternative expression system? In *Protein Production: the exploitation of micro-organisms, cells and animals to make useful proteins* (4th Biological Council Biotechnology Symposium) (ed. T.J.R. Harris), Elsevier Science Publishers, London. pp. 1–14.

Fuerst, T.R., Niles, E.G., Studier, F.W. and Moss, B.M. (1986). Eukaryotic transient-expression system based on recombinant vaccinia virus that synthesises bacteriophage T7 RNA polymerase. *Proceedings of the National Academy of Sciences, USA*. 83: 8122–6.

Fuerst, T.R., Earl, P.L. and Moss, B. (1987). Use of a hybrid vaccinia virus–T7 RNA polymerase system for expression of target genes. *Molecular and Cellular Biology*. 7: 2538–44.

Gasser, C.S. and Fraley, R.T. (1989). Genetically engineering plants for crop improvement. *Science*. 244: 1293–9.

Gluzman, Y. (1981). SV40-transformed Simian cells support the replication of early SV40 mutants. *Cell*. 23: 175–83.

Goeddel, D.V., Kleid, D.G., Bolivar, F. *et al.* (1979). Expression in *E. coli* of chemically synthesised genes for human insulin. *Proceedings of the National Academy of Sciences, USA*. 76: 106–10.

Gorman, C.M., Merlino, G.T., Willingham, M.C. *et al.* (1982a). The Rous sarcoma virus long terminal repeat is a strong promoter when introduced into a variety of eukaryotic cells by DNA mediated transfection. *Proceedings of the National Academy of Sciences, USA*. 79: 6777–81.

Gorman, C.M., Moffat, L.F. and Howard, B. (1982b). Recombinant genomes which express chloramphenicol acetyltransferase in mammalian cells. *Molecular and Cellular Biology* 2: 1044–51.

Graham, F.L. and van der Eb, A.J. (1973). A new technique for the assay of infectivity of human adenovirus 5 DNA. *Virology*. 52: 456–67.

Gray, G.L., Baldridge, J.S., McKeown, K.S. *et al.* (1985). Periplasmic production of correctly processed human growth hormone in *Escherichia coli*: natural and

bacterial signal sequences are interchangeable. *Gene*. 39: 247–54.

Greaves, D.R., Wilson, F.D., Lang, G. and Kioussis, D. (1989). Human CD2 3′ flanking sequences confer high-level, T cell-specific, position-independent gene expression in transgenic mice. *Cell*. 56: 979–86.

Green, L., Schlaffer, I., Wright, K. *et al.* (1986). Cell cycle-dependent expression of a stable episomal human histone gene in a mouse cell. *Proceedings of the National Academy of Sciences, USA*. 83: 2315–19.

Grosveld, F., Blom van Assendelft, G., Greaves, D.R. and Kollias, G. (1987). Position-independent high-level expression of the human β-globin gene in transgenic mice. *Cell*. 51: 975–85.

Gruss, P. and Khoury, G. (1980). Rescue of a splicing defective mutant by insertion of an heterologous intron. *Nature*. 286: 634–7.

Gruss, P., Efstratiadas, A., Karathanasis, S. *et al.* (1981). Synthesis of a stable unspliced mRNA from an intronless simian virus 40-rat preproinsulin gene recombinant. *Proceedings of the National Academy of Sciences, USA*. 78: 6091–5.

Hamer, D.H. and Leder, P. (1979). Splicing and the formation of stable RNA. *Cell*. 18: 1299–302.

Hammer, R.E., Pursel, V.G., Rexnoad Jr., C.E. *et al.* (1985). Production of transgenic rabbits, sheep and pigs by microinjection. *Nature*. 315: 680–3.

Harris, T.J.R. (1983). Expression of eukaryotic genes in *E. coli*. In *Genetic Engineering* 4 (ed. R. Williamson), pp. 127–85. Academic Press, London.

Hartman, S.C. and Mulligan, R.C. (1988). Two dominant-acting selectable markers for gene transfer studies in mammalian cells. *Proceedings of the National Academy of Sciences, USA*. 85: 8047–51.

Herbomel, P., Bourachot, B. and Yaniv, M. (1984). Two distinct enhancers with different cell specificities co-exist in the regulatory region of Polyoma. *Cell*. 39: 653–62.

Hsiung, H.M., Cantrell, A., Luirink, J. *et al.* (1989). Use of a bacteriocin release protein in *E. coli* for excretion of human growth hormone into the culture medium. *Biotechnology*. 7: 267–71.

Hu, M.C.-T. and Davidson, N. (1987). The inducible *lac* operator-repressor system is functional in mammalian cells. *Cell*. 48: 555–66.

Ikemura, T. (1981). Correlation between the abundance of *E. coli* transfer RNAs and the occurrence of respective codons in its protein genes. *Journal of Molecular Biology*. 146: 1–21.

Itakura, K., Hirose, T., Crea, R. *et al.* (1977). Expression in *E. coli* of chemically synthesised gene for the hormone somatostatin. *Science*. 198: 1056–63.

Kaufman, R.J., Wasley, L.C., Spiliotes, A.J. *et al.* (1985). Coamplification and coexpression of human tissue-type plasminogen activator and murine dihydrofolate reductase sequences in Chinese hamster ovary cells. *Molecular and Cellular Biology*. 5: 1750–9.

Kaufman, R.J., Murtha, P., Ingolia, D.E. *et al.* (1986). Selection and amplification of heterologous genes encoding adenosine deaminase in mammalian cells. *Proceedings of the National Academy of Sciences, USA.* **83**: 3136–40.

Law, M.-F., Byrne, J.C. and Howley, P.M. (1983). A stable bovine papillomavirus hybrid plasmid that expresses a dominant selective trait. *Molecular and Cellular Biology.* **3**: 2110–15.

Lee, F., Mulligan, R., Berg, P. and Ringold, G. (1981). Glucocorticoids regulate expression of dihydrofolate reductase cDNA in mouse mammary tumour virus chimaeric plasmids. *Nature.* **294**: 228–32.

Lucknow, V.A. and Summers, M.D. (1988). Trends in the development of baculovirus expression vectors. *Biotechnology.* **6**: 47–55.

Lusky, M. and Botchan, M. (1981). Inhibition of SV40 replication in simian cells by specific pBR322 DNA sequences. *Nature.* **293**: 79–81.

Lusky, M., Berg, L., Weiher, H. and Botchan, M. (1983). Bovine papilloma virus contains an activator of gene expression at the distal end of the early transcription unit. *Molecular and Cellular Biology.* **3**: 1108–99.

Mackett, M., Smith, G.L. and Moss, B. (1982). Vaccinia virus: a selectable eukaryotic cloning and expression vector. *Proceedings of the National Academy of Sciences, USA.* **79**: 7415–19.

Mackett, M., Smith, G.L. and Moss, B. (1985). The construction and characterisation of Vaccinia virus recombinants expressing foreign genes. In *DNA Cloning*, Vol. II, (ed. D.M. Glover), pp. 191–211. IRL Press, Oxford.

Mantei, N., Boll, W. and Weissmann, C. (1979). Rabbit β-globin mRNA production in mouse L cells transformed with cloned rabbit β-globin chromosomal DNA. *Nature.* **281**: 40–6.

Marston, F.A.O. (1986). The purification of eukaryotic polypeptides synthesised in *Escherichia coli. Biochemical Journal.* **240**: 1–12.

Matthias, P.D., Bernard, H.V., Scott, A. *et al.* (1983). A bovine papilloma virus vector with a dominant resistance marker replicates extrachromosomally in mouse and *E. coli* cells. *EMBO Journal.* **2**: 1487–92.

Mellon, P., Parker, V., Gluzman, Y. and Maniatis, T. (1981). Identification of DNA sequences required for transcription of the human α1-globin gene in a new SV40 host-vector system. *Cell.* **27**: 279–88.

Meneguzzi, G., Binetruy, B., Grisoni, M. and Cuzin, F. (1984). Plasmidial maintainance in rodent fibroblasts of a BPV1–pBR322 shuttle vector without immediately apparent oncogenic transformation of the recipient cells. *EMBO Journal.* **3**: 365–71.

Milanesi, G., Barbanti-Brodano, G., Negrini, M. *et al.* (1984). BK virus-plasmid expression vector that persists episomally in human cells and shuttles into *Escherichia coli. Molecular and Cellular Biology.* **4**: 1551–60.

Mory, Y., Ben-Barak, J., Seger, D. *et al.* (1986). Efficient constitutive production of human IFN-γ in Chinese hamster ovary cells. *DNA.* **5**: 181–93.

Moss, B. and Flexner, C. (1987). Vaccinia virus expression vectors. *Annual Review of Immunology.* **5**: 305–24.

Mulligan, R.C. and Berg, P. (1981). Selection for animal cells that express the *Escherichia coli* gene coding for xanthine-guanine phosphoribosyltransferase. *Proceedings of the National Academy of Sciences, USA.* **78**: 2072–6.

Neuberger, M.S. (1983). Expression and regulation of immunoglobulin heavy chain gene transfected into lymphoid cells. *EMBO Journal.* **2**: 1373–8.

Neuberger, M.S. and Williams, G.T. (1988). The intron requirement for immunoglobulin gene expression is dependent upon the promoter. *Nucleic Acids Research.* **16**: 6713–24.

Pavlakis, G.N. and Hamer, D.H. (1983). Regulation of a metallothionein-growth hormone hybrid gene in bovine papilloma virus. *Proceedings of the National Academy of Sciences, USA.* **80**: 397–401.

Prouty, W.F. and Goldberg, A.L. (1972). Fate of abnormal proteins in *E. coli.* Accumulation in intracellular granules before catabolism. *Nature New Biology.* **240**: 147–50.

Purvis, I.J., Bettany, A.J.E., Santiago, T.C. *et al.* (1987). The efficiency of folding of some proteins is increased by controlled rates of translation *in vivo.* A hypothesis. *Journal of Molecular Biology.* **193**: 413–17.

Purzel, V.G., Pinkert, C.A., Miller, K.F. *et al.* (1989). Genetic engineering of livestock. *Science.* **244**: 1281–8.

Rigby, P.W.J. (1982). Expression of cloned genes in eukaryotic cells using vector systems derived from viral replicons. In *Genetic Engineering 3* (ed. R. Williamson), pp. 83–141. Academic Press, London.

Rosenberg, M. and Court, D. (1979). Regulatory sequences involved in the promotion and termination of RNA transcription. *Annual Review of Genetics.* **13**: 319–53.

Sarmientos, P., Duchesne, M., Denefle, P. *et al.* (1989). Synthesis and purification of active tissue plasminogen activator from *Escherichia coli. Biotechnology.* **7**: 495–501.

Sarver, N., Byrne, J.C. and Howely, P.M. (1982). Transformation and replication in mouse cells of a bovine papillomavirus-pML2 plasmid vector that can be rescued in bacteria. *Proceedings of the National Academy of Sciences, USA.* **79**: 7147–51.

Schaffner, W. (1980). Direct transfer of genes from bacteria to mammalian cells. *Proceedings of the National Academy of Sciences, USA.* **77**: 2163–7.

Sekiguchi, T., Nishimoto, T., Kai, R. and Sekiguchi, M. (1983). Recovery of a hybrid vector, derived from bovine papilloma virus DNA, pBR322 and the HSV *tk* gene, by bacterial transformation with extrachromosomal DNA from transfected cells. *Gene.* **21**: 267–72.

Simons, J.P., Wilmut, I., Clark, A.J. *et al.* (1988). Gene

transfer into sheep. *Biotechnology*. **6**: 179–83.

Skerra, A. and Pluckthun, A. (1988). Assembly of functional immunoglobulin F$_V$ fragment in *Escherichia coli. Science*. **240**: 1038–41.

Sompayrac, Z. and Danna, K. (1981). Efficient infection of monkey cells with DNA of simian virus 40. *Proceedings of the National Academy of Sciences, USA*. **78**: 7575–8.

Stark, G.R. and Wahl, G.M. (1984). Gene amplification. *Annual Review of Biochemistry*. **53**: 257–66.

Steitz, J.A. (1979). Genetic signals and nucleotide sequences in messenger RNA. In *Biological regulation and development 1. Gene Expression* (ed. R.F. Goldberger), pp. 349–99. Plenum Press, New York.

Stephens, P.E. and Hentschel, C.C.G. (1987). The bovine papilloma virus genome and its uses as an eukaryotic vector. *Biochemical Journal*. **248**: 1–11.

Talmadge, K., Kaufman, J. and Gilbert, W. (1980). Bacteria mature preproinsulin to proinsulin. *Proceedings of the National Academy of Sciences, USA*. **77**: 3988–92.

Vidal, M.A., Wrighton, C.J., Eccles, S.J., Burke, J.F. and Grosveld, F.G. (1990). Properties of EBV-based shuttle vectors in human cells. *Biochimica et Biophysica Acta*. **1048**: 171–7.

Ward, E.S., Gussow, D., Griffiths, A.D. *et al.* (1989). Binding activities of a repertoire of single immunoglobulin variable domains secreted from *Escherichia coli. Nature*. **341**: 544–6.

Whittle, N., Adair, J., Lloyd, C. *et al.* (1987). Expression in cos cells of a mouse-human chimaeric B72.3 antibody. *Protein Engineering*. **1**: 499–505.

Yates, J., Warren, N., Reisman, D. and Sudgen, B. (1984). A cis-acting element from the Epstein–Barr viral genome that permits stable replication of recombinant plasmids in latently infected cells. *Proceedings of the National Academy of Sciences, USA*. **81**: 3806–10.

Yates, J.L., Warren, N. and Sudgen, B. (1985). Stable replication of plasmids derived from Epstein–Barr virus in various mammalian cells. *Nature*. **313**: 812–15.

Zinn, K., Mellon, P., Ptashne, M. and Maniatis, T. (1982). Regulated expression of an extrachromosomal human β-interferon gene in mouse cells. *Proceedings of the National Academy of Sciences, USA*. **79**: 4897–4901.

Zinn, K., Di Maio, D. and Maniatis, T. (1983). Identification of two distinct regulatory regions adjacent to the human β-interferon gene. *Cell*. **34**: 865–79.

# 2

# Detection and Analysis of Recombinant Proteins

Graham P. Belfield and Michael F. Tuite

## Introduction

Intense effort, often heartache, is generally required to clone a target gene, even before one starts to think about its structure and the cellular function of its encoded product. Once cloned it is a relatively trivial matter to determine the precise nucleotide sequence of the cloned DNA and to use a microcomputer to predict the primary structure of the encoded product. This is usually followed by a database search in the hope of identifying a nucleotide or amino acid sequence of known function that, at least superficially, shows a similarity to the primary structure of the target gene and/or its putative product. Yet even if one scores a significant 'hit' in such a desktop experiment it does not circumvent the need for experimental confirmation of the predicted structural and biochemical properties of the encoded polypeptide. To conduct such an analysis requires the generation of sufficient quantities of the gene product; exactly how much one needs depends largely on the questions being asked.

The strategies used to engineer cloned genes for overexpression in either prokaryotic or eukaryotic cells are described in detail elsewhere in this volume (see Eccles, Chapter 1). Such strategies require the cutting and splicing of the cloned gene into a suitable expression vector and its introduction into a genetically and biochemically maniputable host cell, which usually bears little resemblance to the cell or cell type from which the gene was originally obtained. The ever increasing knowledge we have about how to orchestrate high efficiency transcription and translation of cloned genes in a heterologous host cell provides an opportunity for generating sufficient quantities of the cloned

gene's product for biochemical analysis. All that remains to be answered before one embarks on a programme of protein purification are two key questions:

(1) Is the gene efficiently expressed at both the RNA and the protein level in its new host?
(2) Is the expressed gene product authentic, i.e. does it have the necessary primary, secondary and tertiary structure for meaningful biochemical studies to be undertaken on it?

The answer to these two questions requires the application of a range of biochemical, molecular biological and immunological techniques and it is these techniques that will be reviewed here.

## Hosts for expressing recombinant proteins

One of the most important decisions to make when seeking to overexpress a cloned gene is the choice of host cell, particularly if one is looking to produce a fully functional and authentic recombinant protein. Table 2.1 lists the important criteria that one would seek to satisfy with the heterologous host of choice. As will be taken for granted throughout this article, it is assumed that the cloned gene is of eukaryotic origin (be it fungal, plant or mammalian) and that its gene product is subject to one or more co- or post-translational modifications, which may or may not be necessary for its full biological activity.

The choice of which host to use is relatively restricted, with only three being well established and widely exploited; the bacterium *Escherichia*

**Table 2.1**   The suitability of *E. coli*, *S. cerevisiae* and cultured mammalian cells as hosts in which to express a cloned eukaryotic gene

| | | Host cell | |
| --- | --- | --- | --- |
| Criteria | *E. coli* | *S. cere-visiae* | Mammalian cells |
| 1. Established host-vector systems | | | |
|    autonomous | + | + | ± |
|    integrated | ± | + | + |
| 2. Growth in the laboratory | | | |
|    rapid growth | + | + | ± |
|    media cost | + | + | ± |
|    high biomass | + | + | ± |
| 3. Available expression vectors | | | |
|    high efficiency | + | + | ± |
|    regulatable | + | + | + |
|    amplifiable | − | ± | + |
| 4. Post-translational modification | | | |
|    glycosylation | − | ± | + |
|    disulphide bonds | − | + | + |
|    tertiary structure | − | ± | + |
| 5. Ability to secrete proteins | | | |
|    periplasmic space | + | + | ? |
|    extracellular | ± | + | + |

+ well established; ± limited; − not available; ? unknown.

**Table 2.2**   The utility of other host cells for the expression of recombinant proteins: second generation hosts

| Host | Examples of expressed heterologous proteins |
| --- | --- |
| *Bacillus subtilis* | Hepatitis B core antigen (Hardy *et al.* 1981), human growth hormone (Honjo *et al.* 1987) |
| *Streptomyces lividans* | Bovine growth hormone (Gray *et al.* 1984), human interferon α1 (Noak *et al.* 1988) |
| *Kluyveromyces lactis* | Bovine prochymosin (van den Berg *et al.* 1990) |
| *Pichia pastoris* | Bovine lysozyme (Digan *et al.* 1989), hepatitis B surface antigen (Cregg *et al.* 1987) |
| *Aspergillus nidulans* | Human tissue plasminogen activator (Upshall *et al.* 1987), bovine prochymosin (Cullen *et al.* 1987) |
| Insect cells | Human β-interferon (Smith *et al.* 1983), tissue plasminogen activator (tPA) (Devlin *et al.* 1989) |

*coli*, the budding yeast *Saccharomyces cerevisiae* and cultured mammalian cells such as Chinese Hamster Ovary (CHO) cells (see Eccles, Chapter 1). Given sufficient resources and manpower, for any single target gene all three of these 'first generation' hosts should be examined side by side because it is often difficult to predict the most suitable host for efficient and authentic expression of the cloned gene. The choice is usually made empirically.

For the reasons detailed in Table 2.1, one would tend, not withstanding our own personal prejudices, to favour *S. cerevisiae* as the host of choice. However, this simple eukaryote has one major drawback as an expression host; the fidelity of its glycosylation of eukaryotic polypeptides (see 'Glycosylation', p. 32), although this is only a significant problem if one is considering the recombinant protein for human pharmacological studies or if the post-translational modification is essential for function.

It must be stressed that these three 'first generation' hosts are not the only ones being used to overexpress eukaryotic genes; Table 2.2 gives an indication of the range of both microbial and multicellular hosts being used by both academic and industrial laboratories to express recombinant proteins. In each case the new 'second generation' expression hosts offer one or more advantages over the more established 'first generation' hosts. For example, filamentous fungi, of which *Aspergillus nidulans* is the most widely exploited in this context, have great potential for efficient secretion of authentic mammalian polypeptides (Saunders *et al.* 1989), while the methylotrophic industrial yeast *Pichia pastoris* can use cheap carbon sources (e.g. methanol) and has the very efficient and tightly controlled alcohol oxidase promoter that can be used to express recombinant proteins (Cregg *et al.* 1987). Few of these alternative hosts have anything like the plethora of fully characterized expression elements (promoters, vectors and secretion signals) necessary for maximizing the success of desired options in expression studies. Neither do they have the benefit of a large research community probing into their mechanisms of gene expression and post-translational modifications, which is so vital if one is to exploit a host cell to its fullest potential.

The following discussion will therefore focus primarily on the three 'first generation' expression hosts, although much of what will be discussed can be applied to other hosts.

# Detecting heterologous gene expression

Without the efficient transcription of a cloned gene in its new host cell there is clearly little chance of detecting the encoded polypeptide. It is therefore important to analyse the expression of a heterologous gene in its new host at the levels of both RNA *and* protein synthesis.

## At the RNA level

Northern blotting techniques allow for both estimation of the size (in kilobase pairs; kb) and the relative abundance of any RNA species for which a DNA or oligonucleotide probe is available (Thomas 1980; Williams and Mason 1985). When used to detect specific transcripts such techniques are sensitive enough to be used with total cellular RNA without prior enrichment for the poly($A^+$) mRNA-containing fraction. Molecular size can be estimated by comparing the electrophoretic mobility of the target heterologous mRNA with RNAs of known size, usually large ribosomal RNAs from either mammalian cells (28S rRNA, 6.3 kb; 18S rRNA, 2.4 kb) or *E. coli* (23S rRNA, 3.6 kb; 16S rRNA, 1.8 kb).

The relative abundance of a heterologous mRNA species can be simply determined by hybridizing with a DNA probe that contains, in addition to the target heterologous sequence, a second DNA sequence that will hybridize specifically to an endogenous mRNA species. Truly quantitative analysis of RNA levels is not essential in studies of heterologous gene expression, but if desired can be achieved by a variety of solution hybridization (Young and Anderson 1985) or filter hybridization (Anderson and Young 1985) methods. The transcription start site(s) can be determined by S1 nuclease mapping (Berk and Sharp 1977) or by primer extension mapping (McKnight *et al.* 1981).

## At the protein level

The presence of a recombinant protein in the host cell can be assayed for at several levels; by its biological activity, by detection with an antibody or by staining after polyacrylamide gel electrophoresis (PAGE). Figure 2.1 provides an overview of these approaches, incorporating techniques in a way that

would provide a quick answer to the question "Is the protein expressed at a detectable level?" While detection of biological activity is perhaps the most direct assay method it must always be supported by physical detection methods because biological activity does not provide a quantitative estimate of the levels of incorrectly folded or otherwise inactive protein. Also, the target recombinant protein may not have an easily assayable activity, be it enzyme

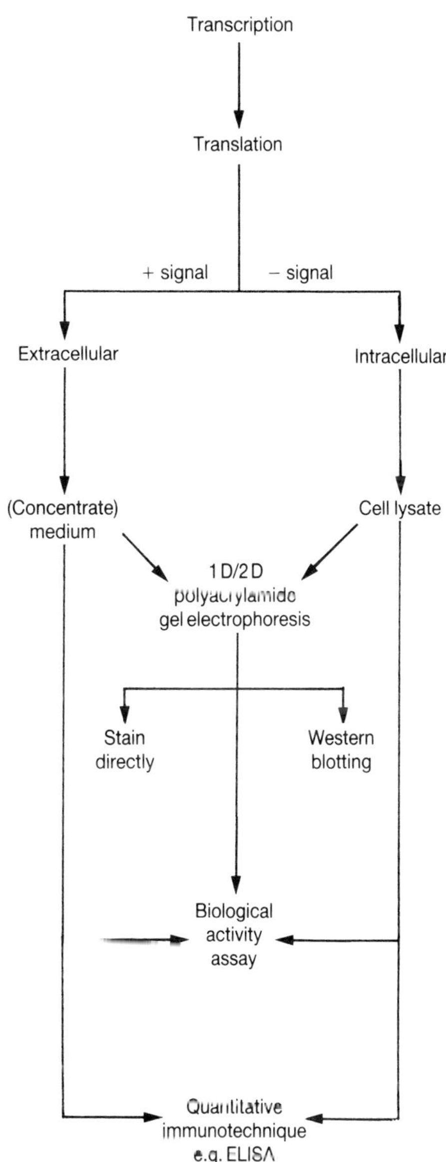


**Fig. 2.1**   Strategies for the detection and analysis of recombinant proteins.

activity or specific ligand binding. Immunological methods, using either monoclonal or polyclonal antibodies raised against the recombinant protein, provide the means of detecting the recombinant protein and any of its post-translationally modified derivatives without the need for the protein to be biologically active. Immunological methods can also be used to quantitate the levels of the recombinant protein using techniques such as radio-immune assay (RIA) and rocket immuno-electrophoresis. Direct staining of polyacrylamide gels can be used when the levels of expression are sufficiently high to allow detection against the background of host proteins, but little information other than the molecular weight of the expressed protein and an approximate estimation of its relative abundance can be gleaned from this strategy.

## One-dimensional polyacrylamide gel electrophoresis (1D-PAGE)

A recombinant protein usually represents a single polypeptide synthesized as part of the normal complex collection of cellular or extracellular proteins of the host. Depending on the efficiency of expression it can represent between 0.01% and 50% of the total cell protein.

1D-PAGE is one of the simplest and most effective means of separating complex protein mixtures on the basis of physical characteristics such as size and charge. 1D-PAGE in the presence of sodium dodecyl sulphate (SDS) also allows for the determination of the molecular weight of a given protein because, under denaturing and reducing conditions, all proteins bind a quantity of SDS proportional to their molecular weight (Pitt-Rivers and Impiombato 1968; See and Jackowski, 1989). This binding facilitates an inversely proportional relationship between the molecular weight of a protein and its mobility on SDS-PAGE. Thus SDS-PAGE is particularly useful in characterizing recombinant proteins because it can be used to demonstrate post-translational modification events, such as glycosylation and signal peptide cleavage, that alter the molecular weight of the protein (see p. 32). Furthermore, it can be used to detect the generally undesirable occurrence of proteolysis.

A number of methods are available for the detection of proteins separated by SDS-PAGE, the most commonly used being staining with Coomassie Brilliant Blue. Proteins separated by SDS-PAGE will bind this stain in amounts directly proportional to their abundance allowing for approximate quantitation of proteins by scanning densitometry of the stained gel. Quantitation with Coomassie Brilliant Blue is most accurate when the stained proteins are present in amounts of between 2–10 μg per band. However, silver staining can be used for detection of proteins present in low levels because this method has an optimal detection range of between 0.5 and 2 ng per band (Merril *et al.* 1981). However, silver staining is not generally used for quantitation by densitometry because it is easy to over-saturate, thus giving a non-linear response.

An alternative to staining proteins separated by SDS-PAGE is to radiolabel the proteins *in vivo*, prior to electrophoresis. After electrophoresis the proteins can be detected by autoradiography. The most commonly used isotope for such analysis is [$^{35}$S]-methionine, although the low abundance, or even absence of methionine from many cellular proteins means that alternative isotopes such as [$^{14}$C]- or [$^{3}$H]-labelled amino acids are more suitable, particularly where one is interested in quantitating the resulting autoradiograph by densitometry.

Analyses of recombinant protein synthesis by silver staining of polyacrylamide gels or by radiolabelling procedures are technically more difficult than staining with Coomassie Brilliant Blue and constitute a greater health risk. Furthermore, the advantage of more accurate quantitation by densitometry makes SDS-PAGE coupled with Coomassie Brilliant Blue the most favourable means of analysing recombinant protein expression.

One dimensional native PAGE (i.e. electrophoresis in the absence of SDS) can also be employed in the analysis of recombinant proteins because it allows the electrophoresed polypeptides to retain their natural conformation (Goldenberg 1989). Native PAGE can therefore be a useful indicator of whether an expressed recombinant polypeptide is authentically folded in its new host (see p. 29).

A further 1D-PAGE technique that can be used to study recombinant proteins is isoelectric focusing (IEF), which fractionates proteins according to their pIs in a continuous pH gradient generated and stabilized in the polyacrylamide gel by carrier ampholytes (Righetti 1983). This is a very sensitive method for differentiating between variously modified forms of the same polypeptide, e.g. phosphorylation, acetylation, etc. (see p. 35), because these will alter the pI but not necessarily the molecular weight of a polypeptide sufficiently to be detected by SDS-PAGE.

## Two-dimensional polyacrylamide gel electrophoresis (2D-PAGE)

One of the major drawbacks of 1D-PAGE methods is that they are unable to resolve two different proteins of effectively identical molecular weight or pI; this problem is often encountered when studying complex protein mixtures. However, two-dimensional methods that exploit both molecular weight and pI differences in the separation of complex protein mixtures are available. These allow for the successive separation of proteins in the first dimension based on pI differences and in a second perpendicular direction based on molecular weight differences (O'Farrell 1975; O'Farrell *et al.* 1977). As with 1D-PAGE, both staining and isotopic methods can be used to detect separated proteins.

To allow for analysis of the wide spectrum of proteins within a cellular lysate, a broad pI range (3–10) can be used in the first dimension, although a variation of the relative concentrations of the carrier ampholytes in the first dimension can narrow the effective separation range to two or three pH units. This is particularly useful if one knows the pI of the expressed recombinant protein, allowing for greater resolution of polypeptides with similar pIs. The original method of 2D-PAGE described by O'Farrell (1975) is not very effective at separating basic proteins and a modification of the original method, called non-equilibrium pH gradient gel electrophoresis (NEPHGE) has been described for this purpose (O'Farrell *et al.* 1977).

## Immunological techniques

Immunological techniques are widely employed in the detection and analysis of recombinant proteins but require that an antibody, capable of specifically recognizing the recombinant protein, is available. Since recombinant proteins are generally expressed from genes of known DNA sequence, the amino acid sequence of the encoded protein can be predicted with a high degree of certainty. Antibodies that can potentially recognize the recombinant protein can therefore be raised either against the protein overexpressed in a bacterial host such as *E. coli* (Eccles, Chapter 1) or against synthetic peptides, which may be only 8–12 amino acids in length, corresponding to a segment of the protein

that has a predicted high antigenic potential, e.g. covering a region exposed on the surface of the protein. For expression in bacteria, the usual strategy is to express the protein as part of a fusion protein containing *E. coli* β-galactosidase (Ruther and Muller-Hill 1983; Stanley and Luzio 1984). Neither the protein fusion nor the synthetic peptide strategies are foolproof; for example, difficulty in expressing the recombinant proteins in *E. coli* may be experienced if the gene product is toxic to the bacterial cell (see Eccles, Chapter 1).

The method used to recover the antigen prior to generating the antibody can also be important. For example, native (i.e. non-denatured) and denatured forms of the same protein may well have different antigenic epitopes—certain epitopes may be buried and therefore inaccessible within the conformation of the native protein—yet be exposed by denaturation. Furthermore, since the most commonly used immunological detection procedure, Western blotting (see below) requires that the proteins are first separated under denaturing conditions by SDS-PAGE, this strongly argues for the use of denatured antigens in generating antibodies for detecting recombinant protein synthesis.

Crude antisera can be employed in most of the immunological techniques described below, yet it is sometimes advantageous to purify the antisera partially, either by anion exchange chromatography (e.g. DEAE-cellulose, DE-52) or by affinity chromatography (e.g. protein A–Sepharose) to isolate the IgG fraction. Further purification of the specific antibodies may be required when dealing with a low titre polyclonal antibody, a step usually achieved by affinity chromatography using the antigen bound to an insoluble matrix such as Sepharose.

Monoclonal antibodies are not usually employed to detect recombinant proteins because they restrict the likelihood of detection by recognizing only a single epitope, whereas polyclonal antibodies recognize multiple epitopes. In addition, the generation of monoclonal antibodies is a long and complex process involving immunization, cell fusion and screening (Kohler and Milstein 1975, 1976). The major advantage of the monoclonal antibody is that it is highly specific and rarely contains the non-specific antibodies that are so often present in crude antisera.

Once armed with a suitable antibody there are a number of immunological techniques that can be used to detect the presence of the corresponding antigens in either crude cellular lysates or specific subcellular fractions. In addition, immunological methods can provide both qualitative and quantitative information on the expressed protein (Scheidtmann 1989).

*Western blotting*
This is a method, first described by Towbin *et al.* (1979), that allows for the rapid detection of a recombinant protein following SDS-PAGE of a cellular lysate and electrophoretic transfer of the separated proteins to a nitrocellulose or nylon membrane. The membrane is first incubated with an antibody raised against the recombinant protein, and then with a second antibody directed against the first antibody and to which an enzyme such as horseradish peroxidase (HRP) or alkaline phosphatase (AP) has been conjugated. In either case the addition of a suitable substrate yields a coloured material on the membrane at the site of the specific antibody–antigen interaction. This method is very sensitive, allowing for the detection of proteins present in nanogram quantities. Even higher sensitivities can be obtained using the biotin–streptavidin signal amplification system (Wilchek and Bayer 1984) or the chemiluminescence system (Whitehead *et al.* 1983).

Western blotting is most commonly used for qualitative analysis, including estimation of molecular weight, identification of modified forms and defining subcellular location of a recombinant protein. It cannot be used quantitatively because both the enzyme-linked assays and the membranes quickly saturate, although the substitution of a second antibody to which [125]I has been attached can give quantitative results provided that the membrane is not saturated.

*Enzyme-linked immunosorbent assay (ELISA)*
This is an immunological method that can be used to quantitate the abundance of a protein to the picogram level (Engvall and Perlmann 1971; Van Weeman and Schuurs 1971). It involves binding the antigen to the wells of a plastic microtitre dish and the subsequent detection of antigen using an enzyme-linked second antibody procedure similar to that used in Western blotting. The colour reaction is measured spectrophotometrically using a series of dilutions of the target antigen to avoid problems with saturation. The method has one major drawback — it only indicates the quantity of immunoreactive material, giving no indication of integrity of the recombinant protein; this must be confirmed by Western blotting.

*Immunoprecipitation*
Under stoichiometric conditions an antibody–antigen complex will form an insoluble aggregate called precipitin. This complex can provide a rapid means for the detection, quantitation and purification of a recombinant protein. For example, precipitin can be visualized directly or by Coomassie Brilliant Blue staining, allowing for detection of the antigen. This is exploited in the technique of rocket immunoelectrophoresis, in which protein extracts are electrophoresed into agarose containing a polyclonal antibody raised against the target protein. Since precipitin will form in the agarose at distances from the point of sample application that are proportional to the abundance of the target protein, this technique can be used to quantitate the levels of a recombinant protein, provided that samples containing known concentrations of the recombinant protein are available for use as standards (Fig. 2.2). This method provides a means for the rapid screening of multiple samples of intracellular or extracellular material.

An extension of the principles of rocket immunoelectrophoresis can be used for the *in situ* detection of recombinant proteins secreted from yeast or *E. coli*. In this case the antibody is added to a rich buffered medium, set with agarose, and cells plated on to the surface of the medium. The secreted recombinant protein forms a halo of precipitin around the growing colony. This *in situ* method is useful for detecting clonal variants or mutants that overexpress the recombinant protein, amongst a population of wild type cells, an application recently used to identify clonal variants of yeast that secrete high levels of human serum albumin (Sleep *et al.* 1990). As with the ELISA method, this method has the drawback that it cannot differentiate between the intact protein and antigenic breakdown products.

Formation of the antigen–antibody complex can also be used to isolate a protein in its native conformation, even from complex mixtures of proteins. This is achieved by adsorbing the complex to an insoluble matrix that can be rapidly sedimented out of the protein mixture and from which the antigen can be subsequently eluted and analysed biochemically. The insoluble matrix used usually contains protein A, to which certain (but not all) antibody classes can bind via their $F_C$ domains (Goding 1978). Protein A is a major cell surface protein of the bacterium *Staphylococcus aureus* and 10% suspensions of *S. aureus* can be used directly to precipitate immune complexes. Alternatively,

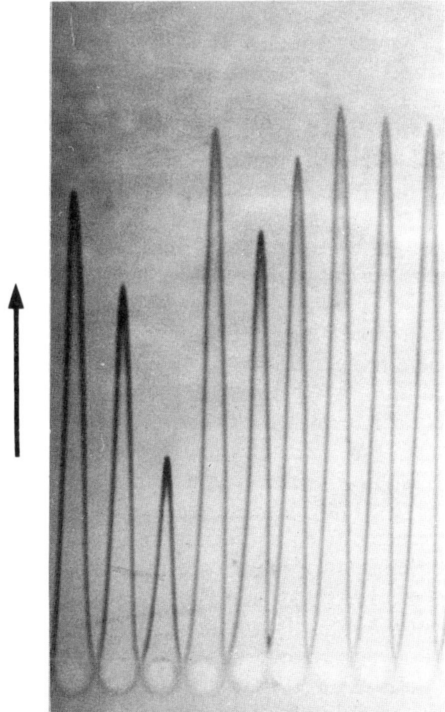


**Fig. 2.2**   Rocket immunoelectrophoretic analysis of secreted recombinant human serum albumin (HSA) from yeast. Lanes 1–3, three known concentrations of HSA; lanes 4–9, culture broths from a number of different yeast strains secreting recombinant HSA. Direction of electrophoresis is indicated by the arrow. (Figure courtesy of Dr Andrew Goodey, Delta Biotechnology.)

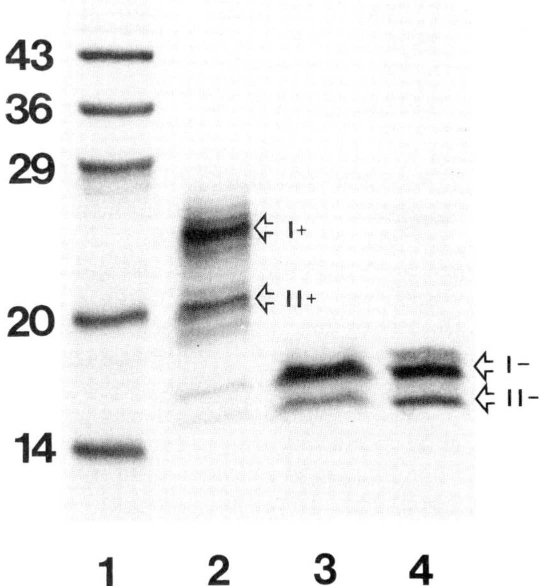


**Fig. 2.3**   Analysis of the glycosylation of recombinant human $\gamma$-interferon ($\gamma$-IFN) secreted from Chinese Hamster Ovary (CHO) cells. Lane 1, molecular weight markers (kDa); lane 2, untreated; lane 3, secreted from cells grown in the presence of tunicamycin; lane 4, secreted from cells grown in the absence of tunicamycin but treated with glycopeptidase F prior to electrophoresis. An autoradiograph of [$^{35}$S]-Met labelled polypeptides, immunoprecipitated with an anti-$\gamma$ IFN antibody is shown. The forms indicated are I$^+$, $\gamma$IFN glycosylated at 2 sites; I$^-$, the same but deglycosylated; II$^+$, $\gamma$IFN glycosylated at one site; II$^-$, a deglycosylated and cleaved form of $\gamma$-IFN. (Figure courtesy of Dr Lis Curling and Dr Nigel Jenkins, University of Kent.)

protein A can be covalently coupled to Sepharose. After binding to the protein A–Sepharose the antigen can be released from the immune complex by a variety of buffers (e.g. buffers of high ionic strength or pH) that retain the native conformation of the antigen.

Immunoprecipitation methods can also be used to study the molecular forms of an expressed recombinant protein to define, for example, the number and range of post-translationally modified forms. In Figure 2.3, the immunoprecipitation of *in vivo* radiolabelled human $\gamma$-interferon, expressed in cultured mammalian cells, is shown following electrophoresis of the immune complex on SDS-PAGE.

### *Immunodetection* in situ

The use of fluorescent dyes, such as fluorescein isothiocyanate (FITC) or rhodamine $\beta$-isothiocyanate, covalently coupled to an antibody that will recognize a primary antibody bound to the target antigen, provides an effective means for the *in situ* detection of recombinant proteins, be they expressed on the surface of a cell or internally, subsequent to permeabilization of the cell to allow the antibodies to enter the cell.

The recombinant protein can also be detected in cut cell sections by employing immunodetection methods similar to those described for Western blotting above, except that the second antibody is usually bound to biotin, which in turn is coupled with colloidal gold. Immunoreactivity in the cell section can be detected by the appearance of electron dense particles that can be clearly visualized by electron microscopy (e.g. Wright and Rine 1989).

## Assaying for biological activity

Many heterologous proteins expressed in microbial hosts have assayable biological activities, including numerous enzymes for which standard biochemical assays exist. Such assays can be carried out on total cell extracts prepared from the expressing host, but they rely upon the host cell not having a similar enzymatic activity that would otherwise mask the recombinant protein's activity. Other assayable recombinant proteins include biological response modifiers, such as lymphokines, hormones, etc., which can be specifically assayed for *in vitro*, using the corresponding target cells and again using total extracts prepared from recombinant strains. However, assaying expressed peptides for their further potential as antigens does require their purification prior to measuring their ability to elicit an antibody response. The purification of such potential vaccines can be greatly facilitated by coupling to an epitope carrier such as hepatitis B viral cores (Valenzuela *et al.* 1985) or virus-like particles, as recently described for yeast (Adams *et al.* 1987).

In order to assay the biological activity of a recombinant protein, it is vital that the cells are gently lysed, to avoid protein denaturation and degradation (see next section) and that they are expressed in a biologically active form, i.e. containing all of the post-translational modifications necessary for biological activity (see pp. 32–5). If one is successful in both of these aims then the biological assay employed clearly depends on the target protein being investigated.

## Stability of recombinant proteins

The successful maintenance of both the structural and functional integrity of an expressed recombinant protein is influenced by a number of factors; the host organism employed, the mode of expression (be it intracellular or extracellular), the physical properties of the recombinant protein itself and the mode of extraction and/or purification employed can all contribute. Not all of these factors can be controlled by the experimenter.

Proteolysis can be detected immunologically as either the appearance in Western blots of one or more discrete polypeptide bands that are *smaller* than the authentic recombinant protein or when proteolysis is severe, through a failure to observe a protein of the predicted molecular weight, but instead observing immunogenic material running with the dye front on the polyacrylamide gel. The use of monoclonal antibodies raised against a specific epitope of a recombinant protein may fail to detect certain proteolytic products of that protein, further stressing the importance of using a polyclonal antibody in immunodetection methods (see p. 21).

Proteolytic degradation by host enzymes is generally considered the major cause of heterologous protein instability, with the degree of proteolysis being largely determined by the host used for expression. For example, filamentous fungi, especially *Aspergillus* species, and some bacterial hosts such as *Bacillus subtilis*, secrete significant levels of proteases into their growth medium; these present problems for other secreted proteins (Ulmanen *et al.* 1985). In addition, a number of other eukaryotic host cells, including *S. cerevisiae*, contain vacuole-associated proteolytic enzymes (Achstetter and Wolf 1985b), which although generally not considered a problem *in vivo* for intracellular expression, can present significant problems when total cell extracts are prepared by procedures that disrupt the vacuolar membrane. In such cases a cocktail of compounds known to inhibit a wide range of proteolytic activities can be employed in the extraction buffer. Table 2.3 lists the protease inhibitors commonly used to counteract a wide range of cellular proteases during protein extraction. Few of these can be used *in vivo* because

**Table 2.3**  Commonly used protease inhibitors and their specificities

| Endopeptidase class | Commonly used inhibitors |
| --- | --- |
| Serine proteases | PMSF (phenylmethylsulphonyl fluoride)<br>TLCK (N-tosyl-L-lysine chloromethylketone)<br>TPCK (N-tosyl-L-phenylalanine chloromethyl ketone) |
| Cysteine proteases | Iodoacetic acid<br>Iodoacetamide<br>N-ethylmaleimide |
| Aspartic proteases | Pepstatin<br>Chymostatin<br>Leupeptin |
| Metalloproteases (Zn) | 1,10-phenanthroline<br>EDTA |

they may inhibit host cell growth, may not be taken up by the cell or may be highly unstable themselves.

Elimination of the problem of recombinant protein proteolysis *in vivo* usually requires the use of host strains that have had one or more endogenous proteases removed by mutation; the *lon* mutation of *E. coli* has been particularly useful in this context (e.g. Boss *et al.* 1984). In spite of a large number of protease-negative mutants having been described for *S. cerevisiae* (Achstetter and Wolf 1985b), there is no clear evidence of improved stability and/or yield of recombinant protein.

The types of proteases to which a recombinant protein is exposed are also affected by the mode of expression. A recombinant protein destined for secretion will encounter signal peptidases and processing enzymes contained within the various compartments of the secretory pathway. In addition, it may encounter naturally secreted proteases *en route*. If intracellular expression results in the recombinant protein forming highly aggregated inclusion bodies (see the next section), this may result in greater resistance to proteolysis (Marston 1986). These points can be illustrated by reference to the expression of human serum albumin (HSA) in *S. cerevisiae*. When the expressed HSA is directed into the secretion pathway by the yeast *MFα1* signal sequence (see pp. 26–7), not only does full length mature HSA accumulate in the growth medium, but an N-terminal fragment of 45 kDa can also be detected (Sleep *et al.* 1990). The relative abundance of this fragment differs with the choice of host strain and signal sequence (Sleep *et al.* 1990). Yet when expressed intracellularly, under otherwise equivalent conditions, no such fragment is observed (Quirk *et al.* 1989). One possible cause of such secretion-dependent cleavage in yeast, and possibly in other eukaryotic cells, is the occurrence of one or more Lys–Arg or Arg–Arg pairs within the mature protein that act as a recognition site for cathepsin B-like proteases such as the *S. cerevisiae KEX2*-encoded protease located in the secretory pathway (Bathurst *et al.* 1987; Fuller *et al.* 1989). Although the susceptible dibasic pairs can be removed by site directed mutagenesis and replaced with different amino acid residues, this immediately renders the recombinant protein 'non-authentic' (see p. 29).

# Cellular location of expressed recombinant proteins

## Intracellular expression

A major problem frequently experienced when a mammalian protein is overexpressed intracellularly in *E. coli* is that it is deposited as insoluble inclusion bodies (Marston 1986). Inclusion bodies (IBs) contain, in addition to the recombinant protein, a number of host proteins and nucleic acids (Hartley and Kane 1988) and can themselves be as large as a bacterial cell. One advantage of inclusion bodies is that, because of their dense nature, they can facilitate rapid purification by low speed centrifugation of whole cell lysates (Marston 1986). However, they present a major technical problem if one wishes to obtain a fully active and soluble recombinant protein. Solubilization can be achieved, although the harsh treatment that is necessary can cause derivatization of amino acids (Marston 1986). Once solubilized, the recombinant protein has to be refolded in the absence of the denaturant.

Inclusion body formation can be reduced, or even eliminated, by changing fermentation conditions, particularly by reducing the culture temperature to 30°C or less (Piatak *et al.* 1988; Schein and Noteborn 1988), although this will present problems if using a thermoinducible $\lambda_{PL}$-based *E. coli* expression vector (see Eccles, Chapter 1). The physical properties of the recombinant protein itself may also determine whether or not it is produced as a soluble protein in the intracellular prokaryotic environment (Schein 1989).

Insolubility of recombinant proteins is much less of a problem in eukaryotic cells, provided that the recombinant protein is expressed in an appropriate host cell compartment that will facilitate correct post-translational modification and folding of the recombinant protein (see pp. 32–5). This is not to say insolubility is never seen; for example Fieschko *et al.* (1987) have reported that human interferon γ expressed in yeast is present partly as insoluble material. Similar observations have also been made with bovine prochymosin (Mellor *et al.* 1983) and human serum albumin (Quirk *et al.* 1989) expressed in yeast.

## Secretory proteins

A more sophisticated alternative to expressing a recombinant protein intracellularly is to direct synthesis into the secretion pathway of the host cell. This strategy will (potentially) produce in the medium a soluble, correctly folded polypeptide that can facilitate sample purification. Furthermore, all the detection methods described on pp. 19–24 can usually be used directly on the supernatant without prior purification.

The transport of proteins into and across membranes in an organized and specific fashion is generally accepted to be an important feature of all cells. However, when attempting to express a naturally secreted recombinant protein, microorganisms have proved to be a popular choice, even when dealing with complex mammalian glycoproteins. This is due largely to ease of handling and potentially higher yields with improved productivity. In addition to directing proteins out of the cell in a correctly folded and soluble form, a number of post-translational modifications can only occur within the secretion pathway. The extent and authenticity of these modifications varies between organisms and may be the basis for deciding on a particular host for secretion (see Table 2.1).

## A brief overview of secretion in eukaryotes

To be secreted, a protein will almost always require an N-terminal signal sequence consisting of a few basic amino acids followed by 15–30 predominantly hydrophobic residues. The primary translation products of many membrane and secreted proteins contain such an N-terminal signal sequence, and this facilitates the insertion of the nascent protein into or across the membrane of the endoplasmic reticulum (e.r.) (Blobel and Dobberstein 1975). In mammalian cells this process is mediated by the signal recognition particle (SRP) (Walter and Blobel 1981), which binds to the signal sequence and arrests translation on the cytoplasmic ribosomes. This complex is then targeted to the SRP receptor in the e.r. membrane and translation resumes, the protein being translocated across the e.r. membrane (Gilmore *et al.* 1982). Several mammalian proteins have been found to cross the e.r. membrane post-translationally (e.g. human glucose transporter, Mueckler and Lodish 1986), although how this is achieved is unclear.

Both co- and post-translational translocation of secretory proteins across the e.r. can occur in yeast (Schatz 1986). The existence of the SRP and its receptor have recently been demonstrated in this lower eukaryote (Hann *et al.* 1989).

In most eukaryotic cells, including yeast, the N-terminal signal sequence is cleaved by a signal peptidase upon insertion into the e.r. membrane (Evans *et al.* 1986) to yield the mature protein. The loss of the signal sequence can usually be detected by SDS-PAGE because it can contribute between 1.5–5 kDa to the molecular mass of a protein and, provided the molecular weight of the protein is less than 60–70 kDa, this will generate a detectable increase in electrophoretic mobility.

Some secretory proteins contain an additional N-terminal sequence that is exposed upon signal cleavage, namely the pro-sequence. The role of the pro-sequence is poorly understood; its length, structure and position can vary from protein to protein (Doherty and Steiner 1982) and not all secretory proteins contain one. It may be necessary to ensure correct folding of the protein or it may act as a further sorting sequence (Bankaitis *et al.* 1986; Johnson *et al.* 1987). It may also form a functional unit with the signal peptide to assure optimal signal function (Wiren *et al.* 1988). Loss of a pro-sequence can usually be detected electrophoretically on SDS-PAGE analysis.

Upon entering the lumen of the e.r. a secretory protein can be subjected to a number of different post-translational modifications, including the addition of sugar molecules to asparagine residues and fatty acylation (see pp. 32–4). The folding of the protein into its native configuration also occurs, beginning simultaneously with the emergence of the nascent polypeptide chain (Bohni *et al.* 1988). Some proteins are retained within the e.r. via a C-terminal retention signal; -KDEL in most eukaryotes, -HDEL in *S. cerevisiae* (Munro and Pelham 1987; Pelham *et al.* 1988). If a protein is not retained in the e.r. it will be exported in vesicles to the Golgi, where it may undergo a number of post-translational modifications. Thereafter, proteins intended for immediate secretion leave the Golgi in secretory vesicles, which fuse with the periplasmic membrane. In yeast this has been shown to occur at the region of the growing bud tip (Schekman and Novick 1982). With appropriate amino acid signals the protein may be retained in the Golgi or routed to other cellular locations, such as the vacuole or peroxisomes (Schekman 1985; Pfiffer and Rothman 1987; see p. 29).

The final barrier to release of a protein into the medium is the cell wall. However, in some instances (for example invertase of *S. cerevisiae* (Schauer *et al.* 1985)), the polypeptide can be retained in the periplasmic space between the periplasmic membrane and the cell wall. This is also the case for heterologous proteins expressed with the invertase signal sequence (Chang *et al.* 1986).

Recombinant proteins secreted from bacteria will undergo none of the above post-translational modifications, except the cleavage of the N-terminal signal sequence by a signal peptidase buried in the periplasmic membrane (Dalbey *et al.* 1987). It is not uncommon for bacterial secretory proteins to utilize a signal located elsewhere in the molecule (Blobel 1980).

## Secretion of recombinant proteins

When attempting to exploit an organism's secretory potential it is essential to prefix the desired mature recombinant protein with an N-terminal signal sequence derived from an endogenous secretory protein. This sequence should enable the protein to be translocated across the host's e.r. or periplasmic membrane, but should also be recognized by the endogenous signal peptidase. It may also be necessary to include a pro-peptide to facilitate the maturation, packaging and compartmentalization of a recombinant protein. The pro-peptide should also be cleaved from the mature polypeptide subsequent to the removal of the signal sequence. Systems for the secretion of recombinant proteins in a variety of micro-organisms have been described, although in many cases the secreted protein remains cell-associated, either trapped in the periplasmic space or the cell wall. In prokaryotic cells it is also often difficult to define what constitutes a signal sequence that will ensure that a protein is secreted because not all such signals reside at the N-terminus (Blobel 1980).

Perhaps one of the most successfully exploited secretory systems in prokaryotes is that of *B. subtilis*. Although early attempts to secrete mammalian proteins from this bacterium met with limited success (e.g. Palva *et al.* 1983), improvements have been made with the use of protease-deficient strains (Honjo *et al.* 1986) and by the introduction of genes from a related species, *B. amyloliquefaciens*, that enhance secretion in such strains. Use of these strains in conjunction with a multicopy plasmid-based expression system facilitated the secretion of human growth hormone to 200 mg/l in a bioreactor (Honjo *et al.* 1987).

The major drawback of using a prokaryotic host for secretion of eukaryotic proteins is the lack of authentic post-translational modifications. The yeast *S. cerevisiae* has therefore become an attractive microbial host for secretion of recombinant proteins due to its eukaryotic-like secretory pathway (Schekman and Novick 1982). A number of signal sequences have been employed for this purpose, one of the most popular being the prepro-sequence encoded by the mating factor alpha gene *MFα1* with levels of secretion in the order of 10–100 mg/l being reported (e.g. Bitter *et al.* 1984; Brake *et al.* 1984; Ernst *et al.* 1987). Heterologous signal sequences can also be efficiently recognized and authentically processed by the yeast secretory pathway (e.g. mouse α-amylase, Thomsen 1983), as can novel fusion signal sequences that combine sequences from homologous and heterologous secretory proteins (Sleep *et al.* 1990).

Filamentous fungi also represent potentially very effective hosts for the secretion of recombinant proteins, being able to secrete endogenous proteins to levels in excess of 20 g/l (e.g. Enari 1983). Unfortunately little is known about the secretory mechanisms in filamentous fungi and in addition this group of organisms have a tendency to secrete high levels of extracellular proteases; this is not conducive to high product recovery.

## Analysis of secreted proteins

Whether or not a recombinant protein has entered the secretory pathway of a host can be determined in several ways relating to how far down the pathway the protein has moved. Establishing whether or not a recombinant protein has crossed into the lumen of the e.r. can be simply achieved by assessing the integrity of the protein in lysates incubated with a protease such as proteinase K. If the protein is membrane-bound or within the lumen of an organelle it should be protease-resistant; if it is protease-sensitive this implies that it is in the soluble cytoplasm (Rothblatt and Schekman 1989). A further test of entry into the e.r. lumen is to determine whether the recombinant protein has undergone any one of a number of e.r.-specific post-translational modifications, e.g. signal sequence cleavage, carbohydrate addition and disulphide bond formation (see pp. 32–5).

For *S. cerevisiae*, the secretion of a polypeptide usually yields a more authentically folded and processed protein than that produced by intracellular expression, however, the concentration of the secreted protein clearly depends upon the volume of the supernatant it resides in. While the actual levels observed will also depend upon a variety of factors, including the expression system employed and the nature of the recombinant protein, the recombinant protein will rarely be present at levels in excess of 100 mg/l and is quite often present at a level of μg/l. The occurrence of the recombinant protein in such low concentrations can make detection difficult, although a number of methods can be used to concentrate the protein, e.g. precipitation with acetone or ethanol or passing the supernatant through a microfiltration membrane with a suitable molecular weight cut-off. Similarly, many of the chromatographic steps necessary for the purification of the recombinant protein will act as a concentration step. A more serious consequence of low product yield is encountered when using complex or ill-defined media such as corn steep liquor or molasses because they will themselves contain high levels of protein that will make analysis and purification of the recombinant protein as difficult as if one were analysing intracellular expression. Such complex media should therefore be avoided in favour of a clearly defined medium.

## Membrane- and organelle-associated proteins

Not all proteins entering the secretory pathway will be deposited extracellularly because many proteins are routed to specific organelles within the secretory pathway or to other cellular compartments. Such peptides are 'sorted' and possess signals and structures often in addition to those required for passage through the secretion pathway.

### Membrane proteins

On completion of synthesis, soluble secretory proteins are usually released into the lumen of the e.r. while membrane proteins retain an amino acid segment spanning the membrane. Such membrane spanning segments usually consist of one or more potential transmembrane regions, identifiable as broad hydrophobic peaks on hydrophilicity plots spanning at least 20 amino acids. While secretory proteins can be exported from the cell without further sorting signals, membrane proteins must contain additional sorting information to route them to the cellular membrane and to ensure their permanent association with the membrane (Coleman 1982).

Membrane proteins fall into two broad classes, being either extrinsic or intrinsic to the lipid bilayer of the membrane. The extrinsic proteins lie outside and do not penetrate the lipid bilayer, although they may be anchored to the membrane by covalently bound lipid, e.g. the transferrin receptor (Omary and Trowbridge 1981). The intrinsic membrane proteins come in a variety of forms, always consisting of regions of the polypeptide spanning the membrane bilayer. Such proteins may have extensive globular domains that lie outside the membrane; they may be predominantly hydrophobic and almost totally buried within the lipid bilayer; or they may be a combination of both. Members of the intrinsic membrane protein class include proteins that translocate other molecules through the membrane, e.g. haemolysin B of *E. coli* (Holland *et al.* 1989) and the higher eukaryotic multiple drug resistance (mdr) proteins (Gerlach *et al.* 1986).

High efficiency expression of heterologous membrane proteins represents a technical challenge because such proteins will often associate themselves into the host's membranes. The resulting structure is usually deleterious to the host, either as a consequence of the binding of important cofactors, such as NAD, or by unbalancing the import/export of molecules or ions (e.g. $H^+$, $Ca^{2+}$), thereby irrevocably disrupting intracellular pH or membrane conductivity. The use of tightly regulated expression vectors provides an obvious advantage when it comes to expressing membrane proteins to overcome the deleterious effects observed with constitutive expression vectors.

Determining the cellular location of an expressed membrane protein can be achieved by isolating the different membrane-containing fractions of the host cell using a combination of high speed centrifugation and fractionation techniques. Immunoelectrophoretic methods, principally Western blotting (see p. 22) can be used to identify the recombinant protein among the separated fractions.

Cellular fractionation techniques usually have to be developed specifically for a given host; for example, much higher centrifugal forces are required to sediment e.r.-derived membranes (i.e. microsomes) from rat liver homogenates than are required to sediment yeast microsomes (Rothblatt and Schekman 1989). A detailed description of cellular fractionation techniques is beyond the scope of this article, but a number of excellent reviews are available on the subject (Tartakoff 1989a,b).

## Organelle-associated proteins

Molecules that enter the early stages of the secretion pathway, only to be retained within organelles or redirected elsewhere must contain within their primary structure further sorting signals. For example, two C-terminal signals have been identified that sort proteins to the peroxisomes (Gould *et al.* 1988), with the primary sequences:

LIKAKKGGKSKL   and   SK/HL

Most mitochondrial proteins are encoded in the nucleus and signals that target proteins to the mitochondria are found in a 20–30 amino acid pre-sequence. For proteins destined for the outer mitochondrial membrane these pre-sequences are not usually cleaved (Hay *et al.* 1984; von Heijne 1986). The use of a mitochondrial-specific pre-sequence can be used to direct heterologous proteins to the correct submitochondrial compartment (Hurt and van Loon 1986), and so contain the signals for targeting to and sorting within the mitochondria. These pre-sequences are generally abundant in hydroxylated and basic amino acids, but show little primary sequence homology (von Heijne 1986; Hurt and van Loon 1986). The pre-sequences of proteins targeted to the chloroplast are less well defined, although at least one such sequence (from the enzyme ribulose bisphosphate carboxylase/oxygenase, Rubisco) can function as a yeast mitochondrial targeting sequence (Hurt *et al.* 1986).

Proteins destined for the vacuole (lysosome) in yeast follow the classic secretion pathway as far as the Golgi, where they are packaged and routed to the vacuole rather than the periplasmic membrane. Such sorting has been attributed to the phosphorylation of mannose residues (man-6-P) in the Golgi (Sahagian *et al.* 1981; Holflack and Kornfeld 1985) although the situation may be much more complex in yeast, because a large number of loci (designated *VPL*) have been identified that, when mutated, can block transport to the vacuole (Rothman and Stevens 1986).

In addition to cell fractionation and Western blotting techniques, the targeting of recombinant proteins to organelles can also be demonstrated using immunogold labelling technology in combination with electron microscopy of cell sections (see pp. 23–4).

The consequences of expressing a recombinant protein targeted to a subcellular compartment such as the vacuole are much less deleterious than those observed with membrane-associated proteins. In addition, depositing the recombinant protein intracellularly, but within a defined and readily isolated cell compartment such as the vacuole, can facilitate its purification. However, a word of caution; the yeast vacuole is a rich store of proteolytic enzymes (Mechler *et al.* 1988).

# Authenticity of expressed recombinant proteins

While the methods described above will tell you whether or not your protein is synthesized and how efficiently it is expressed, they will not tell you how *authentic* (i.e. 'natural') the protein is. Authenticity in this context can be either at the level of the primary amino acid sequence and/or at the level of post-translational modification. A lack of authenticity at either level may not dramatically affect the protein's detectable biological activity, but can have profound effects on its tertiary structure, stability and, most importantly for proteins destined for pharmaceutical use, immunogenicity (Bialy 1987).

## Primary structure

The misincorporation of incorrect amino acids is bound to occur during the process of translation. For example, in wild type *E. coli* a misincorporation rate of $10^{-3}$ to $10^{-4}$ per codon translated has been reported for a number of different endogenous mRNAs (Parker 1989). To put this in context; for an enzyme comprising 1000 amino acids (e.g. β-galactosidase) and assuming that the error rate is constant for all translated codons, this natural level of error would generate between 10% and 100% of the synthesized β-galactosidase molecules with at least one misincorporated amino acid. The problem of mistranslation can only be exacerbated by engineering a cell to translate a heterologous mRNA whose codon usage has evolved in response to its natural host cell's tRNA population (McPherson 1988) rather than to its new host cell's tRNA population. The occurrence of strings of rare codons in an abundant heterologous mRNA will lead to ribosome stalling at the rare codons once the corresponding rare decoding tRNAs become limiting. Such ribosome stalling will have little effect on the overall rate of translation, but will greatly increase the chances of their being decoded by an abundant but non-cognate tRNA. The effect of codon bias on amino acid incorporation has yet to be measured quantitatively for any of the major expression hosts; what data does exist is deduced for mRNAs with an optimized codon bias. However, it is almost certain that most cells have evolved complex proteolytic systems capable of dealing with the mistranslated proteins.

A further aspect of authenticity of the primary amino acid sequence of a recombinant protein comes from the observation that, under certain fermentation conditions, *E. coli* can replace a high proportion of methionine residues in expressed human interleukin 2 (IL-2) with the unusual amino acid norleucine (Lu *et al.* 1988). Such misincorporation arises by the incorporation of *de novo* synthesized norleucine during protein synthesis. The 'misincorporation' of norleucine can, however, be suppressed by adding excess L-methionine and L-leucine to the fermentation broth (Tsai *et al.* 1988).

Amino acid analysis of a purified recombinant protein can be used to assess its overall composition and purity but cannot be used to estimate the authenticity of the amino acid sequence. Furthermore, it can be used to detect the existence of unusual amino acids, such as norleucine.

To determine the amino acid sequence authenticity of an expressed recombinant protein requires the application of a range of protein chemical methods of varying degrees of complexity and reliability. The most direct method is protein sequencing using the chemical degradation method originally described by Edman (Edman 1950). With the recent development of fully automated 'gas-phase' protein sequencers (Hewick *et al.* 1982; Hunkapiller and Hood 1983) the quantity of the target protein necessary for sequencing has been reduced from the picomolar level to the nanomolar level and the technology is now in place for obtaining sequence information from proteins isolated directly from one-dimensional or two-dimensional polyacrylamide gels (Kennedy *et al.* 1988).

However, direct sequencing of isolated proteins by Edman degradation does have a number of drawbacks:

(1) Not all proteins can be sequenced by this method because of the relatively common occurrence of blocked N-termini (for example, by acetylation).
(2) A sequence of only 30–50 amino acid residues can be accurately obtained.
(3) Glycosylated residues can be problematical.

Some of these problems can be overcome by using mass spectrometry (MS), and particularly Fast Atom Bombardment (FAB-MS) techniques (Morris and Greer 1988), which are not only insensitive to blocked N-termini but can be used to identify and assign specific post-translational modifications as well as identifying heterogeneity at the C-terminus, i.e. 'ragged ends'.

The efficacy of both MS and Edman degradation methods in confirming protein sequence authenticity can be improved by using specific peptides derived from the intact polypeptide by either chemical or enzymatic cleavage. Table 2.4 lists the currently popular methods to achieve this; for a more detailed discussion see Carrey (1989). Peptides generated in this manner can also be used to 'fingerprint' a polypeptide, given the specificity of the peptide cleavage site. Initially, peptides generated in this way were separated and compared by 1D SDS-PAGE (Cleveland *et al.* 1977), but while giving a good indication of authenticity of a recombinant protein versus its native counterpart, this remains a relatively low resolution analytical tool; small molecular weight peptides ($< 10\,kDa$) are usually difficult to separate by SDS-PAGE. With the advent of high performance liquid chromatography (HPLC) methods a much higher resolution analytical tool for monitoring the primary sequence of a protein is now available. It allows for the rapid and reproducible separation of very complex mixtures of peptides. Of particular value in this context is reverse phase HPLC (RP-HPLC), which differentiates peptides on the basis of their hydrophobicity. A complete trypsin digest of a polypeptide will generate a complex mixture of peptides between 2 and 20 amino acids in length. These tryptic

**Table 2.4**  Commonly used chemical and enzymatic treatments used for cleaving polypeptides into defined peptides

| Agent | Cleavage site |
| --- | --- |
| Chemical | |
| cyanogen bromide | Met-X |
| Enzymatic | |
| trypsin | Arg-X |
| | Lys-X |
| endoproteinase Arg-C | Arg-X |
| endoproteinase Glu-C | Glu-X |
| (V8 protease) | (Asp-X) |
| endoproteinase Lys-C | Lys-X |

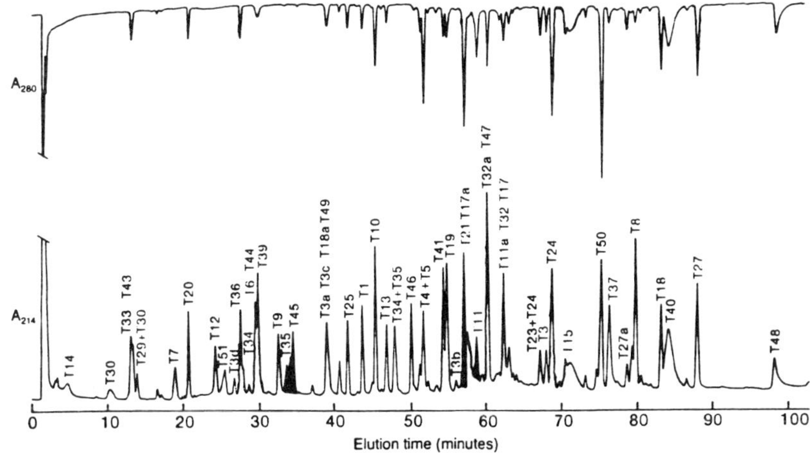

**Fig. 2.4**   The primary sequence 'signature' of recombinant tissue plasminogen activator (tPA) as shown by reverse phase HPLC of tryptic fragments. The tryptic fragments are numbered from the N-terminus and glycopeptides are highlighted in black (from Anicetti *et al.* (1989), printed with permission from Elsevier Trends Journals, publishers).

peptides will each have a distinct hydrophobicity and will thus have different retention times on RP-HPLC. Figure 2.4 shows the primary sequence 'signature' for recombinant tissue plasminogen activator (tPA) generated by RP-HPLC of tryptic peptides (Anicetti *et al.* 1989). RP-HPLC is also relatively insensitive to the presence of carbohydrate side chains, although these can be removed enzymatically prior to analysis (see p. 33). In theory, RP-HPLC can be used to detect single amino acid substitutions, and these would be expected to change the retention time of the peptide on the column. RP-HPLC thus represents possibly the best analytical alternative to direct protein sequencing for monitoring primary amino acid sequence authenticity. The sensitivity of either method in detecting low level misincorporation of amino acids in recombinant proteins has yet to be rigorously tested.

A further problem of primary amino acid sequence authenticity can also arise in microbial hosts, through the retention of the N-terminal methionine residue. In *E. coli* the initiator, Met, is N-formylated (to give fMet); subsequent to the first peptide bond being formed the fMet is deformylated and in some, but not all, cases is cleaved from the growing polypeptide chain. Thus the expression of a eukaryotic protein in *E. coli* can result in the synthesis of a non-authentic variant carrying an additional N-terminal amino acid residue (Goeddel *et al.* 1979). In *S. cerevisiae*, the ability of the enzyme methionine amino peptidase to cleave the

N-terminal Met is inhibited by having a charged residue as the penultimate amino acid (Huang *et al.* 1987). Unwanted retention of the N-terminal Met residue could have adverse effects on the activity and/or immunogenicity of the recombinant protein.

One novel way of ensuring removal of an unwanted N-terminal Met from recombinant proteins expressed in *E. coli* or *S. cerevisiae* has come from the exploitation of the ubiquitin–ubiquitin hydrolase system. The hydrolase cleaves precisely at the C-terminal side of the residues Arg–Gly–Gly, naturally located at the C-terminus of ubiquitin. By creating a precise ubiquitin–interferon γ fusion, Sabin *et al.* (1989) were able to express at high levels the interferon γ in yeast with an authentic N-terminus. The system can also be transplanted into *E. coli* (Miller *et al.* 1989).

The N-terminus of a purified recombinant protein, or derived peptide, can be determined directly by RP-HPLC analysis of the hydrolysed dansylated protein or peptide (Levina and Nazimov 1984). The dansylated amino acid, i.e. the N terminal amino acid, can readily be detected by absorbance at 254 nm. This technique requires up to 100 pmol of the protein or peptide but is susceptible to the presence of a blocked N-terminus.

There are also a number of reports of retained N-terminal Mets being acetylated on recombinant proteins (e.g. Quirk *et al.* 1989). Such modifications result in a blocked N-terminus inhibiting the Edman degradation technique for protein sequencing (see p. 30). The modification can also be identified by RP-HPLC of chemical or proteolytic digests of the recombinant protein.

## Cleavage of pre- and pro-sequences

Proteins destined for secretion or for compartments other than the cytoplasm are directed to these locations by 'signals' encoded within their amino acid sequence (see pp. 26 and 29). Most signals that direct post-translational modifications or compartmentalization are retained as an integral part of the polypeptide. However, the pre- and pro- segments of secreted proteins are usually removed, by proteolytic cleavage. The pre- 'signal peptide' is cleaved by highly specific endopeptidases on translocation of the protein into the lumen of the e.r., the matrix of the mitochondria or, in the case of bacteria, the periplasm, resulting in a detectable reduction in molecular weight of the protein on SDS-PAGE.

In addition to the signal peptide 'pre-' leader, a number of secretory proteins carry an additional 'pro-' region not found in the mature sequence. This region is removed by a peptidase in the Golgi body (Griffiths and Simmons 1986) and/or in the developing secretory granule (Orci *et al*. 1986). This cleavage often occurs after dibasic residues (Achstetter and Wolf 1985a). It would appear that the immediate sequence environment preceding and following the cleavage site is very important in ensuring its presentation to the protease and the subsequent successful removal of the pro-sequence, which may explain why many secretory proteins containing putative dibasic cleavage sites are not extensively cleaved by pro-peptidases. However, this undesirable additional cleavage has been detected for certain proteins secreted from yeast, e.g. β-endorphin (Bitter *et al*. 1984). The pro-sequences of many naturally secreted yeast proteins, e.g. alpha-factor and killer toxin, are cleaved by the $Ca^{2+}$-dependent serine protease Kex2, and with mating alpha-factor (Bussey 1988), the dibasic cleavage site Lys-Arg is followed by the 'spacer' peptide Glu-Ala-Glu-Ala, which is later removed by another peptidase encoded by the *STE13* gene. Sequences containing dibasic amino acid regions in the correct cleavage environment are therefore potentially susceptible to cleavage by pro-peptidases such as Kex2, resulting in the synthesis of truncated forms of the polypeptide detectable by SDS-PAGE. However, instances are known where such secondary cleavage is fortuitous; for example, those proteins requiring cleavage at internal dibasic residues for correct processing, e.g. the expression of mouse neuroendocrine prohormone in a processing-defective mammalian cell line in the presence of coexpressed yeast. Kex2 results in the synthesis of the correctly processed form resulting from Kex2 cleavage (Thomas *et al*. 1988).

The production of authentically processed proteins, with the secretion leaders successfully cleaved from the whole population of molecules can be confirmed by protein sequence determination of the first few amino terminal residues.

## Glycosylation

The addition of carbohydrate groups to proteins is a common property of many biologically important proteins, from yeast to man. Such glycoproteins fall into two basic classes: those that have the carbohydrate linked to the amido nitrogen of asparagine (i.e. an N-glycoside linkage), and those that have the carbohydrate linked to the hydroxyl oxygen of serine or threonine (i.e. an O-glycoside linkage). N-glycoside linkages prevail among mammalian glycoproteins, although a single glycoprotein can possess both O-glycosides and N-glycosides.

Proteins possessing the amino acid sequence motifs Asn–X–Ser (NXS) or Asn–X–Thr (NXT) (where X is any amino acid) are generally subject to N-linked glycosylation on the asparagine residue, although not all potential glycosylation sites are necessarily used within a given protein molecule.

The heterologous expression of plant or mammalian proteins that contain carbohydrate side chains presents a number of problems in terms of analysis, not the least being the authenticity of the modification. In the extreme case, namely in a bacterium such as *E. coli*, the host cell is unable to carry out either O-linked or N-linked carbohydrate addition. In yeast, O-linked and N-linked glycosylations do occur, but there is a tendency to add only mannose residues to the oligosaccharide core (Ballou 1982) often leading to hyperglycosylation of an expressed recombinant glycoprotein (e.g. HTLV-type 1 envelope protein, Kuga *et al*. 1986). Such hyperglycosylation has also been reported in filamentous fungi (Upshall *et al*. 1987). Amongst mammalian glycoproteins there can be many different types of sugar-to-sugar linkages, in addition to mannose, and thus their expression in yeast often leads to the synthesis of a non-authentic product.

However, it should not be assumed that the expression of a recombinant mammalian glycoprotein in a mammalian cell will lead to an authentic glycoprotein because there are a number of examples of cell type-specific N-linked glycosylation patterns in mammalian cells. For example, recombinant human interferon $\beta_1$, produced in mouse epithelial cells (C127), and a human lung adenocarcinoma line (PC8) carried oligosaccharide structures not present in the native human interferon $\beta_1$ (Kagawa *et al*. 1988). Similarly, non-authentic glycosylation of recombinant human interferon $\beta$ was observed in a Chinese hamster ovary (CHO) cell line (Kagawa *et al*. 1988), the cell line of choice for expressing many human recombinant glycoproteins of potential therapeutic value (Eccles, Chapter 1). As yet, little is known about the authenticity of glycosylation of recombinant proteins in the 'second generation' expression hosts listed in Table 2.2, although there is evidence to suggest that hyperglycosylation is not a problem with the methylotrophic yeast *P. pastoris* (Ratner 1989).

An immediate indication that an expressed recombinant protein is glycosylated is by its aberrant migration on SDS-PAGE. Generally, a glycoprotein with more than 10% carbohydrate by mass binds much less SDS per unit weight than the non-glycosylated form (Segrest *et al*. 1971; Leach *et al*. 1980) and thus migrate more slowly on SDS-PAGE, precluding an accurate molecular weight determination. The use of Western blotting techniques (see p. 22) can be used to confirm the identity of the abnormally migrating polypeptide species while its glycoprotein nature can be confirmed using a concanavalin A (Con-A) staining procedure subsequent to transfer of the protein to a nitrocellulose filter (Clegg 1982).

It is not uncommon to detect a number of electrophoretically distinct species of a recombinant glycoprotein, particularly when expressed in yeast. This heterogeneity is generally due to incomplete glycosylation of the expressed protein; for example when mouse granulocyte–macrophage colony stimulating factor (GM-CSF) was expressed in yeast, 50% of the product was unglycosylated or only core glycosylated, with the remaining 50% showing heterogeneity in the extent of the outer chain glycosylation (Ernst *et al*. 1987).

In order to further establish that heterodisperse higher molecular weight forms arise as a consequence of glycosylation, one can use an inhibitor of glycosylation to block both core and outer chain glycosylation. Particularly useful in this context is the antibiotic tunicamycin, which inhibits the transfer of N-acetylglucosamine to dolichol phosphate, the first step in the synthesis of the lipid-linked oligosaccharide. Thus, a glycoprotein expressed in cells grown in the presence of tunicamycin will carry no Asp-linked oligosaccharide, although it does not block O-linked glycosylation (Mahoney and Duskin 1979; Onishi *et al*. 1979). An example is shown in Figure 2.3.

The presence of N-linked oligosaccharides on heterologous proteins can also be demonstrated *in vitro* by using enzymes that cleave the oligosaccharide from the protein. As shown in Figure 2.5, there are three enzymes particularly useful in this context: endoglycosidases F and H and glycopeptidase F (often referred to as N-glycanase). While the two endoglycosidases leave the sugar N-acetylgalactosamine bound to the Asp residue, the glycopeptidase will cleave an intact oligosaccharide from the glycoprotein. The two forms of 'deglycosylated' protein can be readily distinguished from each other and from the fully glycosylated forms by SDS-PAGE (see Fig. 2.3). In addition, by exposing a multiply glycosylated recombinant protein to glycopeptidase F for short periods of time one can identify forms in which only some of the oligosaccharide side chains are removed. An example is shown in Figure 2.6 for the tissue inhibitor of metalloproteases (TIMP) which has two oligosaccharide side chains. In this case fully deglycosylated TIMP can only be obtained by first heat denaturing TIMP prior to exposure to glycopeptidase F. For comparison the fully deglycosylated TIMP produced in *E. coli* is also shown.

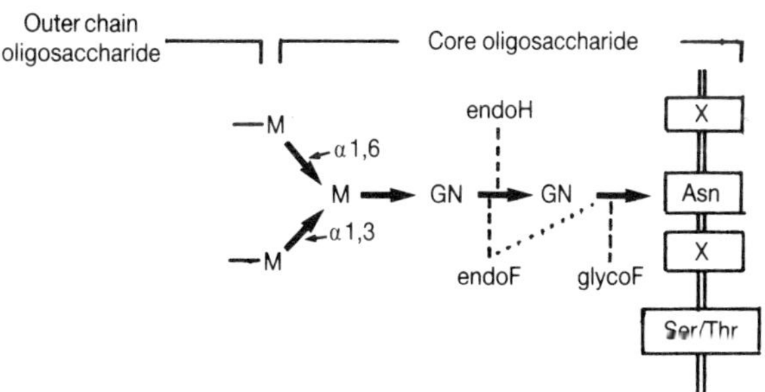


**Fig. 2.5** Substrate specificities of enzymes used to cleave carbohydrates from N-linked glycoproteins. M, D-mannose; GN, N-acetyl-D-glucosamine, X, any amino acid. endoH, endoglycosidase H (endo-β-N-acetyl-glucosaminidase H: EC 3.2.1.96); endoF, endoglycosidase F (endo-β-N-acetylglucosaminidase F: EC 3.2.1 96); glycoF, glycopeptidase F (peptide N-glycohydrolase F: EC 3.2.2.18). NB. Many commercial preparations of endoF are contaminated with traces of glycoF.

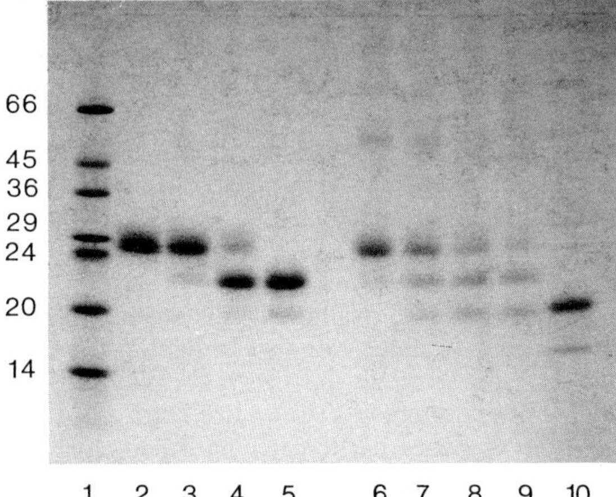

**Fig. 2.6**  SDS-PAGE analysis of *in vitro* deglycosylation of recombinant tissue inhibitor of metalloproteinases (TIMP) with glycopeptidase F. Lane 1, molecular weight markers (kDa); lane 2, non-treated TIMP; lanes 3–5, TIMP incubated with glycopeptidase F for 0, 4 and 24 h, respectively; lanes 6–9, heat denatured TIMP purified incubated with glycopeptidase F for 0, 1, 4 and 24 h, respectively; lane 10, partially purified recombinant TIMP purified from *E. coli*. (Figure courtesy of Dr Richard Williamson and Professor Robert Freedman, University of Kent.)

Once liberated, an oligosaccharide can be examined in the absence of the polypeptide by high pH anion exchange chromatography (Hardy and Townsend 1988; Spellman *et al.* 1989). Alternatively, if sufficient quantities of the purified recombinant protein are available it can be analysed by FAB-MS (Morris and Greer 1988 see p. 30).

## Disulphide bond formation

The folding and assembly of secretory proteins into a native conformation requires, among other steps, the formation of either intra- or intermolecular disulphide bonds between Cys residues. In proteins where more than one disulphide bond forms it is important for the correct functioning of the protein that the correct disulphides form. For example, in a protein where there are eight Cys residues there are potentially 105 different disulphide-linkage isomers. The enzyme responsible for catalysing the formation and isomerization of natural disulphides in secretory proteins is protein disulphide isomerase (PDI), an enzyme associated with the lumenal surface of the endoplasmic reticulum (Freedman 1984; Freedman *et al.* 1989). If one attempts to express a naturally disulphide-bonded protein intracellularly, i.e. in the absence of PDI,

there is no guarantee that the correct disulphides will form; disulphides will form spontaneously *in vitro* in the absence of PDI, although the cytoplasmic environment is thought to be too reducing for disulphide bond formation (Fahey *et al.* 1977).

In *E. coli* it has been suggested that the reason for the failure of a recombinant secretory protein to reach its correct conformation when expressed intracellularly is incorrect disulphide bond formation (Pigiet and Schuster 1986), which in turn may lead to inclusion body formation. In the case of bovine prochymosin, a monomeric secreted protein with three natural disulphides, it is usually found associated with the insoluble fraction when expressed as an intracellular protein in both *E. coli* (Schoemaker *et al.* 1985) and yeast (Mellor *et al.* 1983; Smith *et al.* 1985). Expression of this enzyme as a secreted protein from yeast produces a soluble, fully active protein suggesting authentic disulphide bond formation (Smith *et al.* 1985) and further indicating that the e.r. of yeast must contain a PDI-like activity, an assumption we and others have recently confirmed (Katakura *et al.* 1990; Farquhar *et al.* 1990). In *E. coli*, thioredoxin, an analogous enzyme with a redox active disulphide/dithiol in its active site, may catalyse disulphide bond formation in recombinant proteins (Gleason and Holmgren 1988; Okumura *et al.* 1988).

Full biological activity of a disulphide-bonded recombinant protein is a good indication of correct disulphide bond formation and there are a number of direct and relatively simple physical methods that can be used to address whether or not an expressed protein contains intra- or intermolecular disulphide bonds. The most widely used is comparing the electrophoretic mobility of the recombinant protein on SDS-PAGE in reducing and non-reducing conditions. Disulphide-bonded proteins that are intramolecularly cross-linked unfold to a smaller hydrodynamic volume in SDS than molecules that do not contain formed disulphides (Goldenberg and Creighton 1984). Thus, by carrying out SDS-PAGE analysis of a recombinant protein in non-reducing conditions, but where artefactual disulphide formation and oxidation are blocked, a disulphide-bonded protein will show a greater mobility than it does under reducing conditions. This is illustrated in Figure 2.7, where the differences in mobility in reducing and non-reducing conditions of three disulphide-bonded proteins are shown. Similarly, in non-reducing conditions a disulphide-bonded protein will show greater electrophoretic mobility than a non-disulphide-bonded

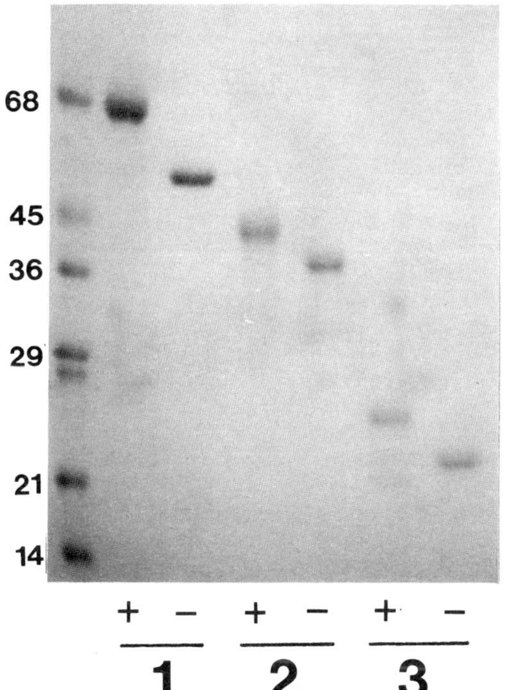



**Fig. 2.7** SDS-PAGE analysis of disulphide-bonded proteins under reducing (+) and non-reducing (−) conditions. The proteins are: 1, bovine serum albumin; 2, ovalbumin; 3, prolactin. The proteins are detected by Coomassie Blue staining. (Figure courtesy of Dr Jan Paver and Professor Robert Freedman, University of Kent.)

protein of the same molecular weight. SDS-PAGE in non-reducing conditions can therefore be used to resolve and distinguish reduced and oxidized forms of a recombinant protein and can thus be used to assay formation of intramolecular disulphide bonds.

A slightly more sensitive method for determining whether authentic disulphide bonds are formed in a recombinant protein is to carry out high resolution peptide mapping of the recombinant protein and an authentic sample of the protein in the presence and absence of reducing agents. Identical peptide maps would be strong evidence for authentic disulphide bond formation, whereas the appearance of new bands in the recombinant protein would strongly suggest disulphide mismatching (Quirk *et al.* 1989).

For a more detailed discussion on the analysis of disulphide bonds in proteins the reader is referred to Creighton (1989).

## Other post-translational modifications

In addition to signal peptide cleavage glycosylation and disulphide bond formation, eukaryotic proteins can be subjected to a wide range of other post-translational modifications that may or may not occur in a heterologous host cell. These include phosphorylation, acetylation, amidation, sulphation and attachment of fatty acids, none of which apparently occur in *E. coli*. Many of these modifications can be detected by *in vivo* radiolabelling methods; for example fatty acid acylation of proteins in yeast can be examined by radiolabelling with [³H]-palmitate (Rothblatt and Schekman 1989) and phosphorylation of a recombinant protein in yeast can be determined by a combination of radiolabelling with $H_3[^{32}PO_4]$ and immunoprecipitation of the recombinant proteins (e.g. Miyamoto *et al.* 1985). Oligomerization of a recombinant protein can be determined using native gel electrophoresis to separate monomer, dimer, tetramer, etc.

## Purifying recombinant proteins

The strategies for purifying recombinant proteins to homogeneity are, to a large extent, dictated by the form of expression used (native or as a fusion protein), the cellular location (intracellular, periplasmic or secreted into the medium), whether the protein is expressed in a soluble or insoluble form and, above all, the physical properties of the recombinant protein itself (mol wt., pI, stability, etc.). Thus, each unique recombinant protein will require the establishment of a purification protocol that will not necessarily parallel strategies devised for the purification of the same protein from its natural source. The reader is referred to three excellent general texts that outline protein purification strategies: Scopes (1982); Janson and Ryden (1989) and Deutscher (1990).

There are a number of general strategies that can be employed to simplify purification of a recombinant protein from either a microbial or a mammalian cell. For example, secretion of the native protein, apart from improving the chances of achieving authentic post-translational modifications (see p. 26), usually results in the protein being obtained in a relatively simple mixture of polypeptides, albeit in a very high dilution. In addition, the use of certain gene fusion strategies can greatly facilitate purification of the non-native portion of the fusion protein if it is able to bind specifically to a chromatographic matrix. The most widely used fusion in this context for *E. coli* expression systems is one that employs protein A, which binds specifically to the Fc region of IgG (Nilsson *et al.* 1985). The single step purification of a protein A fusion can be obtained by using IgG linked to Sepharose as an affinity matrix. The problem with such a fusion strategy is the need to cleave the purified fusion protein to generate the native recombinant protein. Specific cleavage of fusion proteins can be achieved by employing a peptide linker containing a site that is cleavable either chemically or enzymatically. The peptide linker Ile-Glu-Gly-Arg-X is useful in this context, being cleaved at the position indicated, by the blood clotting factor Xa (Nagai *et al.* 1985).

# References


Achstetter, T. and Wolf, D.H. (1985a). Hormone processing and membrane-bound proteinases in yeast. *EMBO Journal.* **4**: 173–7.

Achstetter, T. and Wolf, D.H. (1985b). Proteinases, proteolysis and biological control in the yeast *Saccharomyces cerevisiae*. *Yeast.* **1**: 139–158.

Adams, S.E., Dawson, K.M., Gull, K., Kingsman, S.M. and Kingsman, A.J. (1987). The expression of hybrid HIV: Ty virus-like particles in yeast. *Nature.* **329**: 68–70.

Anderson, M.L.M. and Young, B.D. (1985). Quantitative filter hybridisation. In *Nucleic Acid Hybridisation. A Practical Approach* (ed. B.D. Hames and S.J. Higgins), pp. 73–112. IRL Press Ltd., Oxford.

Anicetti, V.R., Keyt, B.A. and Hancock, W.S. (1989). Purity analysis of protein pharmaceuticals produced by recombinant DNA technology. *Trends in Biotechnology.* **7**: 342–9.

Ballou, C.E. (1982). Yeast cell wall and cell surface. In *Molecular Biology of the Yeast Saccharomyces, Vol. 2: metabolism and gene expression* (ed. J. Strathern, E.W. Jones and J.R. Broach), pp. 335–60. Cold Spring Harbor Laboratory, New York.

Bankaitis, V.A., Johnson, L.M. and Emr, S.D. (1986). Isolation of yeast mutants defective in protein targeting to the vacuole. *Proceedings of the National Academy of Sciences, USA.* **83**: 9075–9.

Bathurst, I.C., Brennan, S.O., Carrell, R.W., Cousens, L.S., Brake, A.J. and Barr, P.J. (1987). Yeast KEX2 protease has the properties of a human proalbumin converting enzyme. *Science.* **235**: 348–51.

Berk, A.J. and Sharp, P.A. (1977). Sizing and mapping of early adenovirus mRNAs by gel electrophoresis of S1 endonuclease digested hybrids. *Cell.* **12**: 721–32.

Bialy, H. (1987). Recombinant proteins: virtual authenticity. *Bio/Technology.* **5**: 883–90.

Bitter, G.A., Chen, K.K., Banks, A.R. and Lai, P.-H. (1984). Secretion of foreign proteins from *Saccharomyces cerevisiae* directed by α-factor gene fusions. *Proceedings of the National Academy of Sciences, USA.* **81**: 5330–4.

Blobel, G. (1980). Intracellular protein topogenesis. *Proceedings of the National Academy of Sciences, USA.* **77**: 1496–500.

Blobel, G. and Dobberstein, B. (1975). Transfer of proteins across membranes. II Reconstitution of functional rough microsomes from heterologous components. *Journal of Cell Biology.* **67**: 852–62.

Bohni, P.C., Deshares, R.J. and Schekman, R.W. (1988). *SEC11* is required for signal peptide processing and yeast cell growth. *Journal of Cell Biology.* **106**: 1035–42.

Boss, M.A., Kenton, J.H., Wood, C.R. and Emtage, J.S. (1984). Assembly of functional antibodies from immunoglobulin heavy and light chains synthesised in *Escherichia coli. Nucleic Acids Research.* **12**: 3791–806.

Brake, A.J., Merryweather, J.P., Lorr, D.G., Heberlein, U.A., Masiarz, F.R., Mullenbach, T., Urdea, M.S., Valenzuela, P. and Barr, P.J. (1984). α-factor directed synthesis and secretion of mature foreign proteins in *Saccharomyces cerevisiae. Proceedings of the National Academy of Sciences, USA.* **81**: 4642–6.

Bussey, H. (1988). Proteases and the processing of precursors to secreted proteins in yeast. *Yeast.* **4**: 17–26.

Carrey, E.A. (1989). Peptide mapping. In *Protein Structure: a practical approach* (ed. T.E. Creighton), pp. 117–44. IRL Press Ltd., Oxford.

Chang, N.C., Matteucci, M., Perry, J., Wulf, J.J., Chen, C.Y. and Hitzeman, R.A. (1986). *Saccharomyces cerevisiae* secretes and correctly processes human interferon hybrid proteins containing yeast invertase signal peptides. *Molecular and Cellular Biology.* **6**: 1812–19.

Clegg, J.C. (1982). Glycoprotein detection in nitrocellulose transfers of electrophoretically separated protein mixtures using concanavalin A and peroxidase: application to arenavirus and flavivirus proteins. *Analytical Biochemistry.* **127**: 389–94.

Cleveland, D.W., Fischer, S.G., Kirschner, M.W. and Laemmli, U.K. (1977). Peptide mapping by limited proteolysis in sodium dodecyl sulphate and analysis by gel electrophoresis. *Journal of Biological Chemistry.* **252**: 1102–6.

Coleman, A. (1982). Cells that secrete foreign proteins. *Trends in Biochemical Sciences.* **7**: 435–7.

Cregg, J.M., Tschopp, J.F., Stillman, C., Siegel, R., Akong, M., Craig, W.S., Buckholz, R.G., Madden, K.R., Killaris, S.A., Davis, G.R., Smiley, B.L., Cruze, J., Torregrossa, R., Velicelebi, G. and Thill, G.P. (1987). High level expression and efficient assembly of hepatitis B surface antigen in the methylotrophic yeast. *Pichia pastoris. Bio/Technology.* **5**: 479–86.

Creighton, T.E. (1989). Disulphide bonds between cysteine residues. In *Protein Structure: a practical approach*, (ed. T.E. Creighton), pp. 155–67. IRL Press Ltd., Oxford.

Cullen, D., Gray, G.L., Wilson, L.J., Hayenga, K.J., Lamsa, M.H., Rey, M.W., Norton, S. and Berka, R.M. (1987). Controlled expression and secretion of bovine chymosin in *Aspergillus nidulans. Bio/Technology.* **5**: 369–76.

Dalbey, R.E., Kuhn, A. and Wickner, W. (1987). The internal signal sequence of *Escherichia coli* leader peptidase is necessary, but not sufficient for its rapid membrane assembly. *Journal of Biological Chemistry.* **262**: 13241–5.

Deutscher, M.P. (ed.) (1990). *Guide to Protein Purification.* Academic Press, New York.

Devlin, J.J., Devlin, P.E., Clark, R., O'Rourke, E.C., Levenson, C. and Mark, D.F. (1989). Novel expression of chimaeric plasminogen activators in insect cells. *Bio/Technology.* **7**: 286–92.

Digan, M.E., Lair, S.V., Brierley, R.A., Siegel, R.S., Williams, M.E., Ellis, S.B., Kellaris, P.A., Provow, S.A., Craig, W.S., Velicelebi, G., Harpold, M.M. and Thill, G.P. (1989). Continuous production of a novel lysozyme via secretion from the yeast *Pichia pastoris. Bio/Technology.* **7**: 160–4.

Docherty, K. and Steiner, D.F. (1982) Post-translational proteolysis in polypeptide hormone biosynthesis. *Annual Review of Physiology.* **44**: 625–38.

Edman, P. (1950). A method for the determination of the amino acid sequence in peptides. *Acta Chemica Scandanavica.* **4**: 283–93.

Enari, T.-M. (1983). Microbial cellulases. In *Microbial Enzymes and Biotechnology*, (ed. W.M. Fogarty), pp. 183–223. Applied Science Publishers, New York.

Engvall, E. and Perlmann, P. (1971). Enzyme linked immuno-sorbent assay (ELISA): quantitative assay of immunoglobulin G. *Immunochemistry.* **8**: 871–4.

Ernst, J.F., Mermod, J.-J., DeLamarter, J.F., Mattaliano, R.J. and Moonen, P. (1987). O-glycosylation and novel processing events during secretion of α-factor/GM-CSF fusions by *Saccharomyces cerevisiae. Bio/Technology.* **5**: 831–4.

Evans, E.A., Gilmore, R. and Blobel, G. (1986) Purification of microsome signal peptidase as a complex. *Proceedings of the National Academy of Sciences, USA.* **83**: 581–5.

Fahey, R.C., Hunt, J.S. and Windham, G.C. (1977). On the cysteine and cystine content of proteins. Differences between intracellular and extracellular proteins. *Journal of Molecular Evolution.* **10**: 155–60.

Farquhar, R., Honey, N., Bossier, P., Murant, S.J., Schultz, L., Montgomery, D., Ellis, R., Freedman, R.B. and Tuite, M.F. (1990). Isolation and characterisation of a yeast gene, *PD11*, encoding protein disulphide isomerase. *Yeast.* **6**: s392.

Fieschko, J.C., Egan, K.M., Ritch, T., Koski, R.A., Jones, M. and Bitter, G.A. (1987). Controlled expression and purification of human immune interferon from high-cell-density fermentations of *Saccharomyces cerevisiae. Biotechnology and Bioengineering.* **29**: 1113–21.

Freedman, R.B. (1984). Native disulphide bond formation in protein biosynthesis: evidence for the role of protein disulphide isomerase. *Trends in Biochemical Sciences.* **9**: 438–41.

Freedman, R.B., Bulleid, N.J., Hawkins, H.C. and Paver, J.L. (1989). Role of protein disulphide-isomerase in the expression of native proteins. *Biochemical Society Symposium.* **55**: 167–92.

Fuller, R.S., Brake, A. and Thorner, J. (1989). Yeast prohormone processing enzyme (Kex2 gene product) is a $Ca^{2+}$-dependent serine protease. *Proceedings of the National Academy of Sciences, USA.* **86**: 1434–8.

Gerlach, J.H., Endicott, J.A., Juranka, P.F., Henderson, G., Sarangi, F., Deuchers, K.L. and Ling, V. (1986). Homology between P-glycoprotein and a bacterial transport protein suggests a model for multidrug resistance. *Nature.* **324**: 485–9.

Gilmore, R., Blobel, G. and Walter, P. (1982). Protein translocation across the endoplasmic reticulum. 1. Detection in the microsomal membrane of a receptor for the signal recognition particle. *Journal of Cell Biology.* **95**: 463–9.

Gleason, F.K. and Holmgren, A. (1988). Thioredoxin and related proteins in prokaryotes. *FEMS Microbiology Reviews.* **54**: 271–98.

Goding, J.W. (1978). Use of staphylococcal protein A as an immunological reagent. *Journal of Immunological Methods.* **20**: 241–53.

Goeddel, D.V., Heyneker, H.L., Hozumi, T., Arentzen, R., Itakura, K., Yansura, D.G., Ross, M.J., Miozzari, G., Crea, R. and Seeburg, P.H. (1979). Direct expression in *Escherichia coli* of a DNA sequence coding for human growth hormone. *Nature.* **281**: 544–8.

Goldenberg, D.P. (1989). Analysis of protein conformation by gel electrophoresis. In *Protein Structure. A Practical Approach* (ed. T.E. Creighton), pp. 225–50. IRL Press Ltd., Oxford.

Goldenberg, D.P. and Creighton, T.E. (1984). Gel electrophoresis studies of protein conformation and folding. *Analytical Biochemistry.* **138**: 1–18.

Gould, S.J., Keller, G.-A. and Subramani, S. (1988). Identification of peroxisomal targetting signals located at the carboxy terminus of four peroxisomal proteins. *Journal of Cell Biology.* **107**: 897–905.

Griffiths, G. and Simons, K. (1986). The *trans*-golgi network: sorting at the exit site of the Golgi complex. *Science.* **234**: 438–43.

Hann, B.C., Poritz, M.A. and Walter, P. (1989). *Saccharomyces cerevisiae* and *Schizosaccharomyces pombe* contain a homologue to the 54-kD subunit of the signal recognition particle that in *Saccharomyces cerevisiae* is essential for growth. *Journal of Cell Biology.* **109:** 3223–30.

Hardy, M.R. and Townsend, R.R. (1988). Separation of positional isomers of oligosaccharides and glycopeptides by high performance anion-exchange chromatography with pulsed amperometric detection. *Proceedings of the National Academy of Sciences, USA.* **85:** 3289–93.

Hardy, K., Stahl, S. and Kupper, H. (1981). Production in *Bacillus subtilis* of hepatitis B core antigen and of major antigen of foot and mouth disease virus. *Nature.* **293:** 481–3.

Hartley, D.L. and Kane, J.F. (1988). Properties of inclusion bodies from recombinant *Escherichia coli*. *Biochemical Society Transactions.* **16:** 101–2.

Hay, R., Boehni, P. and Gasser, S. (1984). How mitochondria import proteins. *Biochimica et Biophysica Acta.* **779:** 65–87.

Hewick, R.M., Hunkapiller, M.W., Hood, L.E. and Dreyer, W.J. (1982). A gas–liquid solid phase peptide and protein sequenator. *Journal of Biological Chemistry.* **256:** 7990–7.

Holflack, B. and Kornfeld, S. (1985). Lysosomal enzyme binding to mouse $P388D_1$ macrophage membranes lacking the 215 kDa mannose-6-phosphate receptor: evidence for the existence of second mannose 6-phosphate receptor. *Proceedings of the National Academy of Sciences, USA.* **82:** 4428–32.

Holland, I.B., Wang, R. and Seror, S.J. (1989). Haemolysin secretion and other protein translocation mechanisms in Gram-negative bacteria. In *Microbial Products: new approaches* (ed. S. Baumberg, I. Hunter and M. Rhodes), pp. 219–54. Cambridge University Press, Cambridge.

Honjo, M., Akaoka, A., Nakayama, A., Shimada, H., Mita, I., Kawamura, K. and Furutani, Y. (1986). Construction of secretion vector and secretion of hIFN-β. In *Bacillus: molecular genetics and biotechnology applications* (ed. A.T. Ganesan and J.A. Hoch), pp. 89–101. Academic Press, New York.

Honjo, M., Nakayama, A., Iio, A., Mita, I., Kawamura, K., Sawakura, A. and Furutani, Y. (1987). Construction of a highly efficient host–vector system for secretion of heterologous protein in *Bacillus subtilis*. *Journal of Biotechnology.* **6:** 191–204.

Huang, S., Elliott, R.C., Liu, P.-S., Korui, R.K., Weickmann, J.L., Lee, J.H., Blair, L.C., Ghosh-Dastidar, P., Bradhsaw, R.A., Bryan, K.M., Einarson, B., Kendall, R.L., Kolacz, K.H. and Saito, K. (1987). Specificity of cotranslational amino-terminal processing of proteins in yeast. *Biochemistry.* **26:** 8242–6.

Hunkapillar, M.W. and Hood, L.E. (1983). Protein sequence analysis: automated microsequencing. *Science.* **219:** 650–9.

Hurt, E.C. and van Loon, A.P.G.M. (1986). How proteins find mitochondria and intramitochondrial compartments. *Trends in Biochemical Sciences.* **11:** 204–7.

Hurt, E.C., Soltanifar, N., Goldschmidt-Clermont, M., Rochaix, J.-D. and Schatz, G. (1986). The cleavable pre-sequence of an imported chloroplast protein directs attached polypeptides into yeast mitochondria. *EMBO Journal.* **5:** 1343–50.

Janson, J.-C. and Ryden, L.G. (1989). *Protein Purification: principles, high resolution methods and applications.* VCH.

Johnson, L.M., Bankaitis, V.A. and Emr, S.D. (1987). Distinct sequence determinants direct intracellular sorting and modification of a yeast vacuolar protease. *Cell.* **48:** 875–85.

Kagawa, Y., Takasaki, S., Utsumi, J., Hosoi, K., Shimuzu, H., Kochibe, N. and Kobata, A. (1988). Comparative study of the asparagine-linked sugar chains of natural human interferon-$\beta_1$ and recombinant human interferon $\beta_1$ produced by three different mammalian cells. *Journal of Biological Chemistry.* **263:** 17508–15.

Katakura, Y., Mizunaga, T., Miura, T. and Maruyama, Y. (1990). The presence of protein disulfide isomerase in the yeast *Saccharomyces cerevisiae*. *Agricultural and Biological Chemistry.* **54:** 1043–4.

Kennedy, T.E., Wager-Smith, K., Barzilai, A., Kandel, E.R. and Sweatt, J.D. (1988). Sequencing proteins from acrylamide gels. *Nature.* **336:** 499–500.

Kohler, G. and Milstein, C. (1975). Continuous cultures of fused cells secreting antibody of predefined specificity. *Nature.* **256:** 495–9.

Kohler, G. and Milstein, C. (1976). Derivation of specific antibody-producing tissue culture and tumor lines by cell fusion. *European Journal of Immunology.* **6:** 511–15.

Kuga, T., Hattori, S., Yoshida, M. and Taniguchi, T. (1986). Expression of human T-cell leukemia virus type I envelope protein in *Saccharomyces cerevisiae*. *Gene.* **44:** 337–40.

Leach, B.S., Collawn, J.F. and Fish, W.W. (1980). Behaviour of glycopolypeptides with empirical molecular weight estimation methods. 1. In sodium dodecyl sulphate. *Biochemistry.* **19:** 5734–41.

Levina, N.B. and Nazimov, I.V. (1984). High performance liquid chromatography of DNS-amino acids in the purity control of peptides. *Journal of Chromatography.* **286:** 207–16.

Lu, H.S., Tsai, L.B., Kenney, W.C., Curless, C.C., Klein, M.L., Lai, P.-H., Fenton, D.M., Altrock, B.W. and Mann, M.B. (1988). Control of misincorporation of *de novo* synthesised norleucine into recombinant interleukin-2 in *E. coli*. *Biochemical and Biophysical Research Communications.* **156:** 733–9.

Mahoney, W.C. and Duskin, D. (1979). Biological activities of the two major components of tunicamycin. *Journal of Biological Chemistry.* **254:** 6572–6.

Marston, F.A.O. (1986). The purification of eukaryotic polypeptides synthesised in *Escherichia coli*. *Biochemical Journal.* **240:** 1–12.

McKnight, S.L., Gravis, E.R., Kingsbury, R. and Axel,

R. (1981). Analysis of transcriptional regulatory signals of the HSV thymidine kinase gene: identification of an upstream control region. *Cell.* **25**: 385–98.

McPherson, D.T. (1988). Codon preference reflects mistranslational constraints: a proposal. *Nucleic Acids Research.* **16**: 4111–20.

Mechler, B., Hirsch, H.H., Muller, H. and Wolf, D.H. (1988). Biogenesis of the yeast lysosome (vacuole): biosynthesis and maturation of proteins yscB. *EMBO Journal.* **7**: 1705–10.

Mellor, J.E., Dobson, M.J., Roberts, N.A., Tuite, M.F., Emtage, J.S., White, S., Lowe, P.A., Patel, T., Kingsman, A.J. and Kingsman, S.M. (1983). Efficient synthesis of enzymatically active calf chymosin in *Saccharomyces cerevisiae. Gene.* **24**: 1–14.

Merril, C.R., Goldman, D., Sedman, S.A. and Ebert, M.H. (1981). Ultrasensitive stain for proteins in polyacrylamide gels shows regional variation in cerebrospinal fluid proteins. *Science.* **211**: 1437–8.

Miller, H.I., Henzel, W.J., Ridgeway, J.B., Kuang, W.J., Chisholm, V. and Liu, C.C. (1989). Cloning and expression of a yeast ubiquitin–protein cleaving activity in *Escherichia coli. Bio/Technology.* **7**: 698–704.

Miyamoto, C., Chizzonite, R., Crowl, R., Rupprecht, K., Kramer, R., Schaber, M., Kumar, G., Poonian, M. and Ju, G. (1985). Molecular cloning and regulated expression of the human *c-myc* gene in *Escherichia coli* and *Saccharomyces cerevisiae*: comparison of the protein products. *Proceedings of the National Academy of Sciences, USA.* **82**: 7232–6.

Morris, H.R. and Greer, F.M. (1988). Mass spectrometry of natural and recombinant proteins and glycoproteins. *Trends in Biotechnology.* **6**: 140–7.

Mueckler, M. and Lodish, H.F. (1986) The human glucose transporter can insert posttranslationally into microsomes. *Cell.* **44**: 629–37.

Munro, S. and Pelham, H.R.B. (1987). A C-terminal signal prevents secretion of lumenal ER proteins. *Cell.* **48**: 899–907.

Nagai, K., Perutz, K.E. and Poyart, C. (1985). Oxygen binding properties of human mutant hemoglobins synthesised in *Escherichia coli. Proceedings of the National Academy of Sciences, USA.* **82**: 7252–5.

Nilsson, B., Abrahmsen, L. and Uulen, M. (1985). Immobilisation and purification of enzymes with staphylococcal protein A gene fusion vectors. *EMBO Journal.* **4**: 1075–80

Noak, D., Geuther, R., Tonew, M., Breitling, R. and Behnke, D. (1988). Expression and secretion of interferon α1 by *Streptomyces lividans*: use of staphylokinase signals and amplification of a *neo* gene. *Gene.* **68**: 53–62.

O'Farrell, P.H. (1975). High resolution two-dimensional electrophoresis of proteins. *Journal of Biological Chemistry.* **250**: 4007–40.

O'Farrell, P.Z., Goodman, H.M. and O'Farrell, P.H. (1977). High resolution two-dimensional electrophoresis of basic as well as acidic proteins. *Cell.* **12**: 1133–9.

Okumura, K., Wakayama, H., Miyake, Y., Murayama, K., Miyake, T., Seto, K., Taguchi, H. and Shimabayashi, Y. (1988). Thioredoxin-catalysed refolding of recombinant prourokinase. *Agricultural and Biological Chemistry.* **52**: 2969–72.

Omary, M.B. and Trowbridge, I.S. (1981). Covalent binding of fatty acid to the transferrin receptor in cultured human cells. *Journal of Biological Chemistry.* **256**: 4715–18.

Onishi, H.R., Tkacz, J.S. and Lampen, J.O. (1979). Glycoprotein nature of yeast alkaline phosphatase. Formation of active enzyme in the presence of tunicamycin. *Journal of Biological Chemistry.* **254**: 11943–52.

Orci, L., Ruvazzolu, M., Amherdt, M., Madseng, O., Perrelet, A., Vassalli, J.-D. and Anderson, G.W. (1986). Conversion of proinsulin occurs co-ordinately with acidifying of mature secretory vesicles. *Journal of Cell Biology.* **103**: 2273–81.

Palva, I., Lehtovaara, P., Kaariainen, L., Sibakov, M., Cantell, K., Schein, C.H., Kashiwagi, K. and Weissmann, C. (1983). Secretion of interferon by *Bacillus subtilis. Gene.* **15**: 43–51.

Parker, J. (1989). Errors and alternatives in reading the universal genetic code. *Microbiology Reviews.* **53**: 273–98.

Pelham, H.R.B., Hardwick, K.G. and Lewis, M.J. (1988). Sorting of soluble ER proteins in yeast. *EMBO Journal.* **7**: 1757–62.

Pfiffer, S.R. and Rothman, J.E. (1987). Biosynthetic protein transport and sorting by the endoplasmic reticulum and Golgi. *Annual Review of Biochemistry.* **56**: 829–52.

Piatak, M., Lane, J.A., Laird, W., Bjorn, M.J., Wang, A. and Williams, M. (1988). Expression of soluble and fully functional ricin A chain in *Escherichia coli* is temperature-sensitive. *Journal of Biological Chemistry.* **263**: 4837–43.

Pigiet, V.P. and Schuster, B.J. (1986). Thioredoxin-catalysed refolding of disulphide-containing proteins. *Proceedings of the National Academy of Sciences, USA.* **83**: 7643–7.

Pitt-Rivers, R. and Impiombato, F.S.A. (1968). The binding of sodium dodecyl sulphate to various proteins. *Biochemical Journal.* **109**: 825–30.

Quirk, A.V., Gelsow, M.J., Woodrow, J.R., Burton, S.J., Wood, P.C., Sutton, A.D., Johnson, R.A. and Dodoworth, N. (1989). Production of recombinant human serum albumin from *Saccharomyces cerevisiae. Biotechnology and Applied Biochemistry.* **11**: 273–87.

Ratner, M. (1989). Protein expression in yeast. *Bio/Technology.* **7**: 1129–33.

Righetti, P.G. (1983). *Isoelectric Focussing: theory methodology and applications.* Elsevier, Amsterdam.

Rothblatt, J. and Schekman, R. (1989). A hitchhiker's guide to analysis of the secretory pathway in yeast. *Methods in Cell Biology.* **32**: 3–36.

Rothman, J.H. and Stevens, T.H. (1986). Protein sorting in yeast: mutants defective in vacuole biogenesis mislocalise vacuolar proteins into the late secretory pathway. *Cell.* **47**: 1041–51.

Ruther, U. and Muller-Hill, B. (1983). Easy identification of cDNA clones. *EMBO Journal.* **2**: 1791–4.

Sabin, E.A., Lee-Ng, C.T., Shuster, J.R. and Barr, P.J. (1989). High level expression and *in vivo* processing of chimaeric ubiquitin fusion proteins in *Saccharomyces cerevisiae. Bio/Technology.* **7**: 705–9.

Sahagian, G.G., Distler, J. and Jourdian, G.W. (1981). Characterisation of a membrane-associated receptor from bovine liver that binds phosphomannosyl residues of bovine testicular β-galactosidase. *Proceedings of the National Academy of Sciences, USA.* **78**: 4289–93.

Saunders, G., Picknett, T.M., Tuite, M.F. and Ward, M. (1989). Heterologous gene expression in filamentous fungi. *Trends in Biotechnology.* **7**: 283–7.

Schauer, I., Emr, S., Gross, C. and Schekman, R.W. (1985). Invertase signal and mature sequence substitutions that delay intercompartmental transport of active enzyme. *Journal of Cell Biology.* **100**: 1664–75.

Schatz, G. (1986). Protein translocation: a common mechanism for different membrane systems? *Nature.* **321**: 108–9.

Scheidtmann, K.H. (1989). Immunological detection of proteins of known sequence. In *Protein Structure. A Practical Approach* (ed. T.E. Creighton), pp. 93–116. IRL Press Ltd., Oxford.

Schein, C.H. (1989). Production of soluble recombinant proteins in bacteria. *Bio/Technology.* **7**: 1141–9.

Schein, C.H. and Noteborn, M.H.M. (1988). Formation of soluble recombinant proteins in *Escherichia coli* is favoured by lower growth temperature. *Bio/Technology.* **6**: 291–4.

Schekman, R.W. (1985). Protein localisation and membrane traffic in yeast. *Annual Review of Cell Biology.* **1**: 115–43.

Schekman, R.W. and Novick, P. (1982). The secretory process and yeast cell surface assembly. In *The Molecular Biology of the Yeast Saccharomyces*, Vol. 2 (ed. J. Strathern, E.W. Jones and J.R. Broach), pp. 361–98. Cold Spring Harbor Laboratory, New York.

Schoemaker, J.M., Brasnett, A.H. and Marston, F.A.O. (1985). Examination of calf prochymosin accumulation in *E. coli*: disulphide linkages are a structural component of prochymosin-containing inclusion bodies. *EMBO Journal.* **4**: 775–80.

Scopes, R.K. (1982). *Protein Purification: Principles and Practice.* Springer-Verlag, New York.

See, Y. and Jackowski, G. (1989). Estimating molecular weights of polypeptides by SDS gel electrophoresis. In *Protein Structure. A Practical Approach* (ed. T.E. Creighton), pp. 1–21. IRL Press Ltd., Oxford.

Segrest, J.P., Jackson, R.L., Andrews, E.P. and Marchesi, V.T. (1971). Human erythrocyte membrane glycoprotein. A re-evaluation of the molecular weight as determined by SDS-polyacrylamide gel electrophoresis. *Biochemical Biophysical Research Communications.* **44**: 390–5.

Sleep, D., Belfield, G.P. and Goodey, A.R. (1990). The secretion of human serum albumin from the yeast *Saccharomyces cerevisiae* using five different leader sequences. *Bio/Technology.* **8**: 42–6.

Smith, G.E., Summers, M.D. and Fraser, M.J. (1983). Production of human β-interferon in insect cells infected with a baculovirus expression vector. *Molecular and Cellular Biology.* **3**: 2156–65.

Smith, R.A., Duncan, M.J. and Moir, D.T. (1985). Heterologous protein secretion from yeast. *Science.* **229**: 1219–24.

Spellman, M.W., Basa, L.J., Leonard, C.K., Chakel, J.A., O'Connor, J.V., Wilson, S. and van Halbeek, H. (1989). Carbohydrate structures of human tissue plasminogen activator expressed in Chinese Hamster Ovary cells. *Journal of Biological Chemistry.* **264**: 14100–11.

Stanley, K.K. and Luzio, J.P. (1984). Construction of a new family of high efficiency bacterial expression vectors: identification of cDNA clones coding for human liver proteins. *EMBO Journal.* **3**: 1429–34.

Tartakoff, A.M. (ed.) (1989a). *Vesicular Transport. Part A (Methods in Cell Biology*, Vol. 31). Academic Press, New York.

Tartakoff, A.M. (ed.) (1989b). *Vesicular Transport. Part B (Methods in Cell Biology*, Vol. 32). Academic Press, New York.

Thomas, P.S. (1980). Hybridisation of denatured RNA and small DNA fragments transferred to nitrocellulose. *Proceedings of the National Academy of Sciences, USA.* **77**: 5201–5.

Thomas, G., Thorne, B.A., Thomas, L., Allen, R.G., Hruby, D.E., Fuller, R. and Thorner, J. (1988). Yeast Kex2 endopeptidase correctly cleaves a neuroendocrine prohormone in mammalian cells. *Science.* **241**: 226–30.

Towbin, H., Staehlin, T. and Gordon, J. (1979). Electrophoretic transfer of proteins from polyacrylamide gels to nitrocellulose sheets: procedure and some applications. *Proceedings of the National Academy of Sciences, USA.* **76**: 4350–4.

Thomsen, K.K. (1983). Mouse α-amylase synthesised by *Saccharomyces cerevisiae* is released into the culture medium. *Carlsberg Research Communications.* **48**: 545–55.

Tsai, L.B., Lu, H.S., Kenney, W.C., Curless, C.C., Klein, M.L., Lai, P.-H., Fenton, D.M., Altrock, B.W. and Mann, M.B. (1988). Control of misincorporation of *de novo* synthesised norleucine into recombinant interleukin-2 in *E. coli. Biochemical and Biophysical Research Communications.* **156**: 733–9.

Ulmanen, I., Lundstrom, K., Lehtovaara, P., Sarvas, M., Ruohonen, M. and Palva, I. (1985). Transcription and translation of foreign genes in *Bacillus subtilis* by the aid of a secretion vector. *Journal of Bacteriology.* **162**: 176–82.

Upshall, A., Kumar, A.A., Bailey, M.C., Parker, M.D., Favreau, M.A., Lewison, K.P., Joseph, M.L., Maraganore, J.M. and McKnight, G.L. (1987). Secretion of active human tissue plasminogen activator from the filamentous fungus *Aspergillus nidulans*. *Bio/Technology*. 5: 1301–4.

Valenzuela, P., Coit, D., Medina-Selby, M.A., Kuo, C.H., Van Nest, G., Burke, R.L., Bull, P., Urdea, M. and Graves, P.V. (1985). Antigen engineering in yeast: synthesis and assembly of hybrid hepatitis B surface antigen – herpes simplex IgD particles. *Bio/Technology*. 3: 323–5.

van den Berg, J.A., van der Laken, K.J., van Ooyen, A.J.J., Renniers, T.C.H.M., Rietveld, K., Schaap, A., Brake, A.J., Bishop, R.J., Schult, K., Moyer, D., Richman, M. and Shuster, J.R. (1990). Kluyveromyces as a host for heterologous gene expression: expression and secretion of prochymosin. *Bio/Technology*. 8: 135–9.

Van Weeman, B.K. and Schuurs, A.H.W.M. (1971). Immunoassay using antigen-enzyme conjugates. *FEBS Letters*. 15: 232–5.

von Heijne, G. (1986). Mitochondrial targetting sequences may form amphiphilic helices. *EMBO Journal*. 5: 1335–42.

Walter, P. and Blobel, G. (1981). Translocation of proteins across the endoplasmic reticulum. III. Signal recognition protein causes signal sequence-dependent and site-specific arrest of chain elongation that is released by microsomal membranes. *Journal of Cell Biology*. 91: 557–61.

Whitehead, T.P., Thorpe, G.H.G., Carter, T.J.N., Groncut, C. and Kricka, L.J. (1983). Enhanced luminescence procedure for sensitive determination of peroxidase-labelled conjugates in immunoassay. *Nature*. 305: 158–9.

Wilchek, M. and Bayer, E.A. (1984). The avidin–biotin complex in immunology. *Immunology Today*. 5: 39–43.

Williams, J.G. and Mason, P.J. (1985). Hybridisation in the analysis of RNA. In *Nucleic Acid Hybridisation. A Practical Approach* (ed. B.D. Hames and S.J. Higgins), pp. 139–60, IRL Press Ltd., Oxford.

Wiren, K.M., Potts Jr., J.T. and Kronenberg, H.M. (1988). Importance of the propeptide sequence of human pre pro parathyroid hormone for signal sequence function. *Journal of Biological Chemistry*. 263: 19771–7.

Wright, R. and Rine, J. (1989). Transmission election microscopy and immunocytochemical studies of yeast: analysis of HMG-CoA reductase overproduction by electron microscopy. *Methods in Cell Biology*. 31: 473–512.

Young, B.D. and Anderson, M.L.M. (1985). Quantitative analysis of solution hybridisation. In *Nucleic Acid Hybridisation. A Practical Approach* (ed. B.D. Hames and S.J. Higgins), pp. 47–72. IRL Press Ltd., Oxford.

Zhu, X., Ohta, Y., Jordan, F. and Inouye, M. (1989). Pro-sequence of subtilisin can guide the refolding of denatured subtilisin in an intermolecular process. *Nature*. 339: 483–4.

# 3

# Structural Constraints on Protein Engineering

Gunter Schneider and Ylva Lindqvist

## Introduction

Proteins are delicate creatures: their biologically active, native conformation differs by only 21–63 kJ/mol from the unfolded state. With such a small energy difference to other, non-native conformations, it is obvious that even small changes at the amino acid level can exert dramatic effects on the three-dimensional structure of the protein. It is these effects the protein engineer has to be aware of and must take into account when redesigning a protein. In nature these experiments have been done during evolution and have resulted in very different amino acid sequences with essentially the same structure in homologous proteins. Unfortunately we cannot yet fully predict what changes in the three-dimensional structure of a protein will be obtained when exchanging one amino acid for another and we are even more at a loss when trying to foresee the effects of multiple substitutions. There will be no remedy to this problem until the major challenge in structural molecular biology—the folding problem—is understood and we will be able to predict three-dimensional structure from an amino acid sequence.

Despite this reservation, there is a wealth of knowledge on the principles of protein structure and the interactions responsible for the folding of the polypeptide chain. One of the major points we will try to make in this chapter is that in order to become good protein engineers, we have to understand, like engineers in real life, what we are doing to our 'machines', e.g. proteins. If the three-dimensional structure of the protein is known, i.e. we not only know the parts of our machine—the protein—but also how they are put together, the chances are that we are able to foresee which parts can be changed without collapsing the machine.

Unfortunately, the number of proteins whose three-dimensional structures have become available through protein crystallography or lately, for small proteins, with NMR techniques, is depressingly small compared to the number of protein sequences available in the databases. However, from the analysis of the approximately 250 independent structures that have been solved today it can be deduced that there is a limited number of different ways in which the protein chains can fold to form domains. The globular domains are, on average, built from approximately 150 amino acids and, if the chain is longer, a protein subunit usually consists of several such domains.

The domains are made up from combinations of secondary elements such as $\alpha$-helices, $\beta$-sheets and turns. Certain combinations of secondary structures are more common than others and form structural motifs or super secondary structures. The simplest example of such a motif is the $\beta$-hairpin loop, and one of the most elaborate structural motifs is the eight-fold $\beta/\alpha$-barrel, which has been found in 18 functionally and evolutionary unrelated enzymes. Based on the topology patterns (that is the combination of secondary structures and structural motifs), a taxonomy of protein domains is slowly emerging.

The limitation of different unique ways to form a domain is a consequence of structural rules that have been formulated from analysis of the available structures, e.g. only certain combinations of the conformational angles $\phi$ and $\psi$ are allowed for steric reasons (Ramakrishnan and Ramachandran 1965). Furthermore, there are certain energetically favourable ways to pack helices against each other (Chothia *et al.* 1977). This is also true for $\beta$-sheet packing (McLachlan 1979; Chothia and Janin 1982) and helix sheet packing (Janin and Chothia 1980). Another factor limiting the ways an α/β domain can fold is the empirical rule that the helix connection in β–α–β motifs is always right-handed. Based on the structural knowledge and rules that we have it is now possible to build a reasonably accurate model of the structure of a protein, if the three-dimensional structure of a homologous protein is known (Chothia and Lesk 1986).

The rational design of proteins with changed or new properties can be described by an interactive procedure, which may be called the 'protein engineering cycle' (Fig. 3.1). We assume here that the gene for the protein in question has been cloned, sequenced and can be expressed, resulting in large quantities of protein, of the order of 100s of mg. The availability of these amounts facilitates the biochemical analysis of native and mutant protein and is often required for successful crystallization. Structural information is a prerequisite for the rational design of proteins with changed properties and the determination of the three-dimensional structure of wild type protein is thus one step in this cycle. Based upon biochemical and structural information, a model of the mutant protein is built with computer graphics. The mutant is then prepared by site-directed mutagenesis and expressed in large quantities. The biochemical analysis of the mutant protein, including the determination of its three-dimensional structure, is the next step in this cyclic procedure. If the properties of the new protein correspond to our expectations, we are satisfied and, in commercial applications, we would

have a candidate for a new product. In most cases, the mutant will not have the desired properties and we will have to repeat the cycle with a new mutant.

Although the successful application of site-directed mutagenesis in the design of new or 'better' proteins often asks for a deeper understanding of protein structure and function than we have today, these methods, in combination with protein crystallographic techniques, are invaluable tools to help to extend this knowledge (Knowles 1987; Matthews 1987). In this chapter we will summarize much of the work where recombinant hybrid DNA techniques, in combination with structural molecular biology, have helped to deepen our understanding of protein structure. We will concentrate on problems associated with proper folding of the polypeptide chain, stability and maintenance of biological function. Furthermore, we will address the assembly of oligomeric proteins. The principles for this are the same for all proteins and we will not restrict ourselves to proteins derived from the plant kingdom but rather try to stress the general applicability of the principles of protein structure as they begin to emerge.

# Forces determining protein structure

## Hydrophobic effect and packing

Among the forces that determine protein structure and stability, hydrophobic interactions are the most important and are the driving force for folding of the nascent polypeptide chain. In fact, it is not favourable interactions between apolar atoms, but the absence of hydrogen bonds between nonpolar atoms and water that is the major contribution to protein stability. The tendency of hydrophobic amino acid side chains to be expelled from the solvent and to pack towards each other to form hydrophobic cores in the interior of proteins led Kauzmann (1959) to compare the folded protein with oil drops in water. The packing of hydrophobic amino acids in the interior of a protein stabilizes the protein in two ways. First, the shielding of non-polar side chains from the aqueous solvent contributes considerably to protein stability through hydrophobic interactions. Although this effect has been known for a long time, it has been difficult to quantify. Nozaki and Tanford (1971) introduced a hydrophobicity scale as a measure for

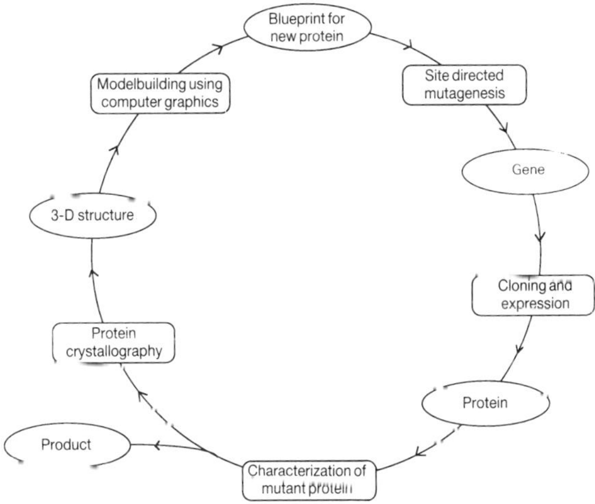



**Fig. 3.1**   The protein engineering cycle.

the strength of hydrophobic interactions. The hydrophobicities of amino acid side chains were estimated by measuring the free energy of transfer of the amino acids from apolar solvent to water. The measured values were normalized with respect to the value for glycine. This concept of hydrophobicity has been modified by a number of workers (for more details see Eisenberg and McLachlan 1986). Furthermore, it could be shown that the hydrophobicity of a given apolar side chain correlates with its accessible surface (Chothia 1974), which is defined as that surface that is in contact with a sphere of radius of a water molecule as it is rolled over the surface (Lee and Richards 1971).

The second effect on protein stability is the tight packing of atoms in the interior of a protein. Hydrophobic cores in a protein have high packing densities. The average packing density of a protein amounts to about 0.75 (Schulz and Schirmer 1979). This has to be compared to a packing density of 0.74 for equal sized hard spheres in closed packing or glasses and oils that have packing values below 0.7, and even below 0.6. This tight packing of proteins maximizes the amount of favourable van der Waals contacts and thus increases protein stability.

Despite the high packing densities of protein cores, the packing of amino acid side chains is not perfect and one can occasionally observe small cavities in the protein interior. The question is whether the stability of proteins can be increased by filling such cavities with bulkier hydrophobic side chains, thus increasing the core density and the number of van der Waals interactions. The two largest cavities in T4 lysozyme (39.0 and 23.4 Å$^3$) have been removed by replacing Leu133 with Phe, and Ala129 with Val (Karpusas *et al.* 1989). Both

mutants are as active, but less stable, than the wild type enzyme. The crystal structure analysis revealed that both side chains are accommodated with very little change in the three-dimensional structure. However, in both cases the introduced side chain adopts a non-optimal dihedral angle $\chi^1$, thus introducing structural strain at the site of mutation. This strain, amplified by a few unfavourable van der Waals contacts, counterbalances the stabilization through higher packing densities. Naturally evolved protein cores seem to be a compromise between the hydrophobic effect, maximizing the packing density and strain occurring due to unfavourable interactions at high core densities.

In an attempt to quantify the importance of hydrophobic effects at individual positions, Matsumura *et al.* (1988) replaced Ile3 in T4 lysozyme with 13 other amino acid side chains by site-directed mutagenesis. In the native T4 lysozyme structure, Ile3 is part of the major hydrophobic core of the C-terminal lobe. The side chain of this residue is in contact with other hydrophobic amino acids; Met6, Leu7, Ile100 and Cys97 (Fig. 3.2). The stability of the mutant proteins was assessed as resistance towards thermal denaturation. For most of the mutants, the changes in stability are directly related to the hydrophobicity of the substituted residue. The analysis confirms that the hydrophobic stabilization is proportional to the reduction of the surface area accessible to solvent on folding. The Ile3Tyr mutant does not follow the observed trends and the crystallographic analysis of this mutant revealed substantial changes in the three-dimensional structure as the cause for this different behaviour. The side chain of Tyr3 does not interact with the hydrophobic core, but rotates $\approx 120°$ so that it is largely exposed to

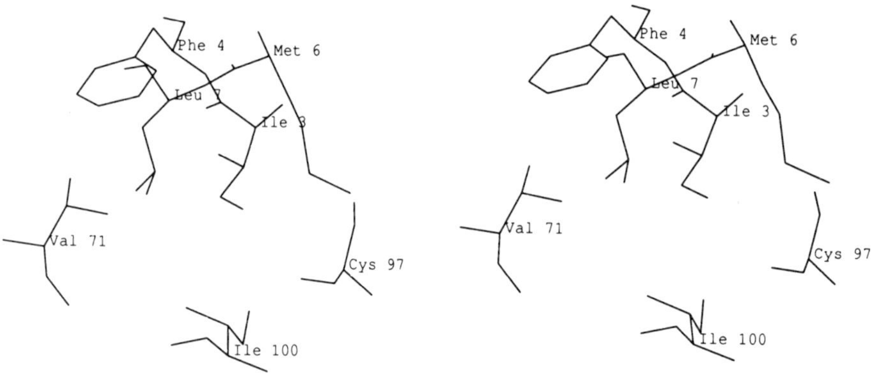

**Fig. 3.2**   Stereopicture showing the hydrophobic core at the C-terminal lobe in T4 lysozyme. The coordinates are from Weaver and Matthews (1987), obtained from the Protein Data Bank.

solvent. The result is that the N-terminal polypeptide (residues 1–9) is shifted 0.6–1.1 Å, compared to the wild type structure.

To further address the magnitude of the hydrophobic effect and to separate the contributions of hydrophobic and packing effects to protein stability, Sandberg and Terwilliger (1989) introduced subtle changes in the hydrophobic core of gene V protein from bacteriophage fl. A comparison of the resistance of the wild type protein towards denaturation with that of the mutants Val35Ile, Ile37Val and Val35Ile/Ile37Val showed that all three mutants were less stable than the wild type. The Val35Ile mutant introduced one methylene group and differed from the wild type protein in both hydrophobicity and packing. Substitution of Ile37 by Val removed one methylene group and resulted in changes both in hydrophobicity and packing. The only change in the double mutant Val35Ile/Ile37Val was in packing, not in hydrophobicity (Fig. 3.3). This mutant was least stable, demonstrating that differences in packing can significantly destabilize a protein. The effect of apolar-to-apolar substitutions at interior residues can only be predicted correctly if packing effects and changes in hydrophobicity are taken into account. Sandberg and Terwilliger (1989) conclude that it will be possible to engineer more stable proteins in

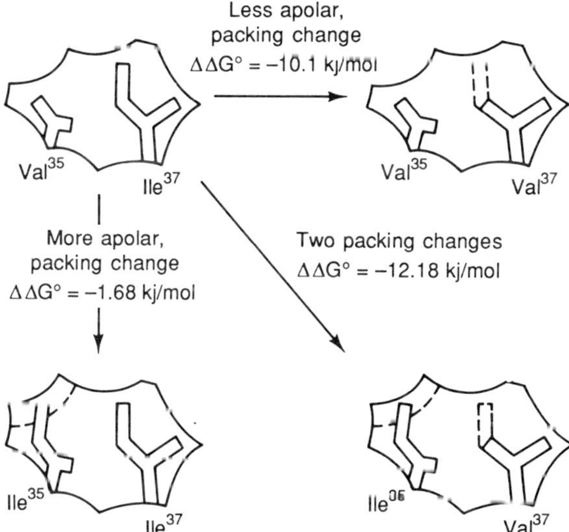


**Fig. 3.3**  Effect of hydrophobicity and packing of hydrophobic amino acids side chains in the hydrophobic core of gene V protein from bacteriophage fl. ΔΔG° is the difference in the free energy of unfolding between the mutants and the wild type protein. (Reproduced with permission from Sandberg and Terwilliger, 1989, Copyright by the AAAS.)

this way but that we will have to learn to repack the interiors of proteins without structural distortion.

Experiments with multiple substitutions in the hydrophobic core of the N-terminal domain of the λ repressor (Lim and Sauer 1989) showed that a mutation at one site can sometimes increase the range of residues allowed at another position and that the volume variation that could be accommodated in the mutated residues increased with the number of mutations. Comparison of the hydrophobic cores of homologous proteins shows that substitutions are rarely simply complementary, but are accommodated by small changes in structure (Chothia and Lesk 1987).

## The hydrogen bond

If hydrophobic interactions provide most of the free energy for stabilizing a protein structure, hydrogen bonds are mainly responsible for the specific conformation of that structure. When the hydrophobic core is formed, the polar atoms of the main chain must make hydrogen bonds, which compensate for those that could be made with the solvent in the unfolded state. This leads to secondary structure formation and other internal hydrogen bonds. Due to their directionality and distance restraints, hydrogen bonds are not only an important determinant of protein structure, but are also responsible for recognition and specificity and are intimately involved in enzyme catalysis. The determination of the energies involved upon hydrogen bond formation has been complicated by the fact that, in aqueous solutions, water competes for available hydrogen bonds.

Using tyrosyl-tRNA synthetase as a model system, site-directed mutagenesis has been employed to explore the contribution of hydrogen bonds to binding of ligands and specificity (Fersht *et al.* 1985). The crystallographic analysis (Brick *et al.* 1989) revealed 11 possible hydrogen bonds between bound substrate, tyrosyl adenylate and the enzyme, eight of which involve amino acid side chains. Systematic mutational analysis of the possible involvement of these hydrogen bonds in substrate binding, and of their energetics has been performed (Fersht *et al.* 1985). From these studies, it was concluded that a hydrogen bond between two uncharged partners is worth 2.1–6.3 kJ/mol. If one of the partners—the hydrogen donor or acceptor—is charged, the hydrogen bond is much stronger, contributing 14.7–18.9 kJ/mol in binding energy. These studies are the first direct assessments of the binding strength of hydrogen bonds in macromolecules and will certainly be followed by others. A proper estimate of the strength of hydrogen bonds is of great importance in model building, energy minimizations and drug design.

## Electrostatic interactions

The most prominent example of an electrostatic interaction is the salt bridge, formed between two differently charged amino acid side chains, e.g. Lys and Asp, or between a charged amino acid side chain and a charged substrate, as observed in many enzyme–substrate complexes. The electrostatic attraction energy is easily calculated using Coulomb's law: it amounts to 21 kJ/mol for an internal salt bridge (dielectric constant $\epsilon = 4$). However, in order to obtain a better estimate of the contribution of salt bridges to protein stability, we must take into account attraction energies of both charges to solvent molecules, shielding effects, entropy effects and so on. Furthermore, the micro-heterogeneity of a protein with respect to a macroscopic entity such as the dielectric constant is difficult to model. The resultant net free energy for salt bridge formation has been estimated to be of the order of 4.2 kJ/mol (Perutz 1970). One would expect that this contribution is significantly higher for buried salt bridges, due to the lower dielectric constant in the interior of the protein and the absence of possible hydrogen bonds to water molecules.

The effects of electrostatic interactions have recently been addressed with two model systems. Subtilisin has been used to redesign substrate specificity (Wells *et al.* 1987) and pH profiles (Russell and Fersht 1987) by modifying charges at the protein surface. These studies allowed the conclusion that the apparent dielectric constant between point charges is 40–50 at low ionic strength, independent of whether the electrostatic interactions are mediated through the protein or through water near the protein surface. For enzyme-substrate complexes, the free energy contribution of an interaction between a charged protein side chain and charged substrate is about 6.3–10.5 kJ/mol.

Another set of experiments has focused on the importance of electrostatic interactions for thermostability, using T4 lysozyme as the model system. It is a common observation that one finds negative charges of aspartic or glutamic acid side chains at the N-terminal end of α-helices in proteins (Fig. 3.4). The negative charge at the N-terminus of the α-helix is thought to stabilize the dipole, which is formed because of the alignment of the hydrogen bond dipoles in the α-helix (Hol *et al.* 1978). Additionally, the side chains of these residues sometimes form hydrogen bonds to the main chain nitrogens of the first helical turn. This hypothesis has been tested by introducing negative charges at the N-terminal ends of two α-helices in T4 lysozyme (Nicholson *et al.* 1988). Three mutants were made: Ser38Asp, Asn144Asp and the double mutant Ser38Asp/Asn144Asp. All three mutants were more stable towards thermal denaturation than wild type lysozyme, with the observed stabilization effects being additive. The increase in the free energy of stabilization for the double mutant relative to the wild type was estimated at 6.7 kJ/mol. The observed stabilization is due to charge interactions, not hydrogen bonds, because the crystallographic analysis for both mutants revealed that no new hydrogen bonds were formed upon substitution.

A similar stabilization of protein structure due to

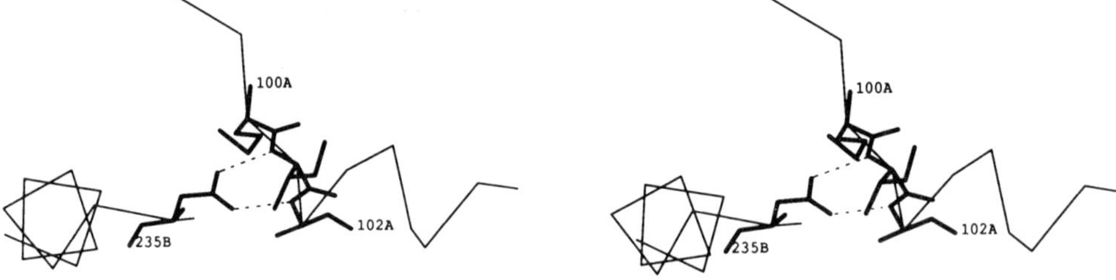



**Fig. 3.4**   Stabilization of protein structure through an interaction of a charged amino acid side chain with the N-terminus of an α-helix. The example shown here illustrates one of the subunit–subunit interactions at the dimer interface of Rubisco from *Rhodospirillum rubrum*. Asp235 forms hydrogen bonds with main chain nitrogens of residues 101 and 102 at the N-terminus of helix αC (Schneider *et al.* 1990).

charge interaction with the dipole of an $\alpha$-helix has been observed in barnase, a small ribonuclease from *Bacillus amyloliquefaciens*. One of the His residues present in the enzyme is located at the C-terminal end of an $\alpha$-helix and has an unusual $pK_a$ value, 1.6 units above the value found in model peptides. By replacing this His residue, it can be shown that this high $pK_a$ value is due to the favourable interaction of the protonated His side chain and the $\alpha$-helix, which stabilizes the native structure by about 8.9 kJ/mol (Sali *et al.* 1988).

## Conformational flexibility versus rigidity

Glycine residues lack the C$\beta$ atom, and therefore have a higher conformational flexibility than other amino acids; the backbone conformation of glycine can often assume $\phi$, $\psi$ angles, which are energetically unfavourable to other residues (Ramachandran and Sasisekharan 1986). The increasing number of conformations that can be adopted by glycines might favour the unfolded state. It has therefore been suggested that the replacement of glycine by alanine might increase the stability of a protein by decreasing the configurational entropy of unfolding (Hecht *et al.* 1986; Matthews *et al.* 1987). In a similar manner, substitution for prolines can be argued because this amino acid has, due to the limited range of allowed conformations, a lower configurational entropy and should stabilize a protein. This argument is, of course, only valid if the amino acid substitution does not introduce unfavourable contacts. Thus, by systematically replacing Gly for Ala and inserting Pro at suitable positions, one should be able to increase protein stability.

Such attempts have been made with the N-terminal domain of $\lambda$-repressor (Hecht *et al.* 1986) and T4 lysozyme (Matthews *et al.* 1987). Two glycines, both located in the third $\alpha$-helix of the N-terminal domain of $\lambda$-repressor, at positions 46 and 48, were replaced by Ala. The single mutants Gly46Ala, Gly48Ala and the double mutant Gly46Ala/Gly48Ala were made (Hecht *et al.* 1986). All mutants are more stable than the wild type protein, with the double mutant having a change in the standard free energy of denaturation of 4.6 kJ/mol. It is noteworthy, that this change in free energy is less than the sum of the two single mutant $\Delta\Delta$G values.

In the case of T4 lysozyme, a similar change,

Gly77Ala, increased the stability of the enzyme, with an increase of the free energy of folding of 2.1 kJ/mol. This glycine residue is part of an $\alpha$-helix and the subsequent crystallographic structure analysis showed that the mutant structure is very similar to that of wild type lysozyme (Matthews *et al.* 1987). In all these experiments, the substituted glycines are located in $\alpha$-helices. Glycine is considered not to prefer $\alpha$-helical conformation and the observed stabilizing effects might be due to the replacement of a poor helix-forming residue by a residue that prefers $\alpha$-helical conformation, thus stabilizing the conformation of the secondary structural element.

Insertions of proline residues in a polypeptide chain are not as straightforward. The proline residue restricts the $\phi$, $\psi$ values at the proline itself and limits those of the preceding residue. A substitution of a residue with proline must be compatible with these restraints. Ala82 in T4 lysozyme can be replaced by Pro, and the crystallographic analysis at 1.7 Å resolution showed no structural changes except at the site of mutagenesis. The mutant is more stable towards thermal denaturation, with an observed free energy of stabilization of 3.4 kJ/mol (Matthews *et al.* 1987).

## Secondary structural elements

The contribution of secondary structural elements themselves to protein stability is two-fold: on one hand, they can sometimes form stable units of folding in themselves and, on the other hand, they build up the core of the protein by packing these stable units towards each other. Although the types of interactions that form these secondary structural elements and determine their packing have been known for a long time, the contributions of these interactions to protein stability have proved difficult to quantify.

### Stabilization of secondary structure

The first and last residues of $\alpha$-helices cannot form intrahelical hydrogen bonds between main chain carbonyl oxygens and amino nitrogens. The analysis of available protein structures (Richardson and Richardson 1988) showed that there is a preference of certain residues at the N- and C terminal ends (the N- and C-caps) of $\alpha$-helices (Table 3.1). These are residues that can make hydrogen bonds to main

**Table 3.1** Stabilization of α-helices through hydrogen bonding at the N-terminus

| Residue type | Stabilization energy relative to Thr* (kJ/mol) | Frequency at the N-terminal end of α-helices** |
|---|---|---|
| Asp | −0.42 | 27 |
| Thr | 0.0 | 21 |
| Glu | 0.8 | 5 |
| Ser | 1.7 | 34 |
| Asn | 5.5 | 34 |
| Gly | 6.3 | 33 |
| Gln | 7.6 | 3 |
| Ala | 9.7 | 10 |
| Val | 9.7 | 1 |

* data from Serrano and Fersht (1989); ** data from Richardson and Richardson (1988).

chain atoms at the N- or C-caps through their side chains. The contributions of such hydrogen bonds to α-helix stability has been analysed by Serrano and Fersht (1989). In a series of site-directed mutagenesis experiments, Thr6 and Thr26 at the N-cap of the two α-helices of the small ribonuclease, barnase, were replaced by a number of different amino acids to obtain more quantitative data on the energetics of such interactions. The contributions of the two Thr residues are two-fold: on one hand, the OH-group forms a hydrogen bond to main chain nitrogens of residue 9 and 29, respectively. Additionally, Cβ and Cγ of the Thr side chains interact with other parts of the protein. Thr6 and 26 were replaced by Gly, Ala, Val, Ser, Asn, Gln, Asp and Glu (Serrano and Fersht 1989). The mutation of either Thr6 or Thr26 at the N-cap of an α-helix gave similar results. Table 3.1 summarizes the effect of the different mutations on the stability of barnase. There is a rough correspondence between the energetics of N-cap stabilization and the observed statistical frequency of a given residue at such a position. Asp gives the highest contribution to stability, probably as a result of the charge interaction with the dipole of the α-helix. Serrano and Fersht (1989) give similar values ( ≈6.3 kJ/mol) for these interactions to those reported by Nicholson *et al.* (1988).

## Change in secondary structure

An attempt to investigate the contribution of secondary structural elements to protein stability has been made by Matthews and co-workers. In T4 lysozyme, Pro86 is at the C-terminal end of an α-helix formed by residues 81–85, with residues 87–91 having $3_{10}$ helix conformation. Pro86 has been replaced by ten other amino acids (Ser, Thr, Cys, Ala, Gly, Leu, Ile, His, Arg, Asp) and seven of these mutants were analysed crystallographically. The conformational changes in the mutant lysozyme result in the formation of a continuous α-helix, which includes residues 81–91. Interestingly, this helix is formed despite the wide range of helical propensities of the residues, introduced at the site of mutation. Even more striking is the fact that none of the mutants stabilizes T4 lysozyme. All mutants are marginally less stable (< 2.1 kJ/mol). Despite differences in hydrogen bonding, charge, etc. between the different mutants, the effects of the mutation on the thermal stability are identical. There is no obvious relation between the helical propensity, length or charge of the side chain, involvement in hydrogen bonding or not and the thermal stability (Alber *et al.* 1988). On the other hand, the replacement of Pro86 by other amino acids led to the formation of a new secondary structural element, thus introducing large (the largest conformational changes in T4 lysozyme mutants studied crystallographically so far) structural changes, extending more than 20 Å away from the site of substitution. A proper separation of all the different effects as a consequence of these relatively large structural changes and their contribution to stability requires further analysis.

## Secondary structures as folding elements

If it is true that the secondary structural elements that form the core of the protein, as opposed to the surface loops, mainly determine the stability of the protein (Goldberg 1985), then the question arises whether the natural given order of secondary structural elements is critical to the folding and stability of proteins. Luger *et al.* (1989) addressed this problem in a series of elegant experiments. The structure of phosphoribosyl anthranilate isomerase from *Escherichia coli*, which is part of a bifunctional enzyme, is a β/α barrel protein (Priestle *et al.* 1987). This folding pattern has been observed in approximately 18 functionally and evolutionarily unrelated enzymes and is considered to provide a very stable structural framework. Luger *et al.* (1989) have circularly permutated the gene of the single domain enzyme from *Saccharomyces cerevisiae* by linking the N- and C-termini and cleavage of a surface loop (Fig. 3.5). Since the active site of the β/α barrel enzyme is located at the carboxyl end

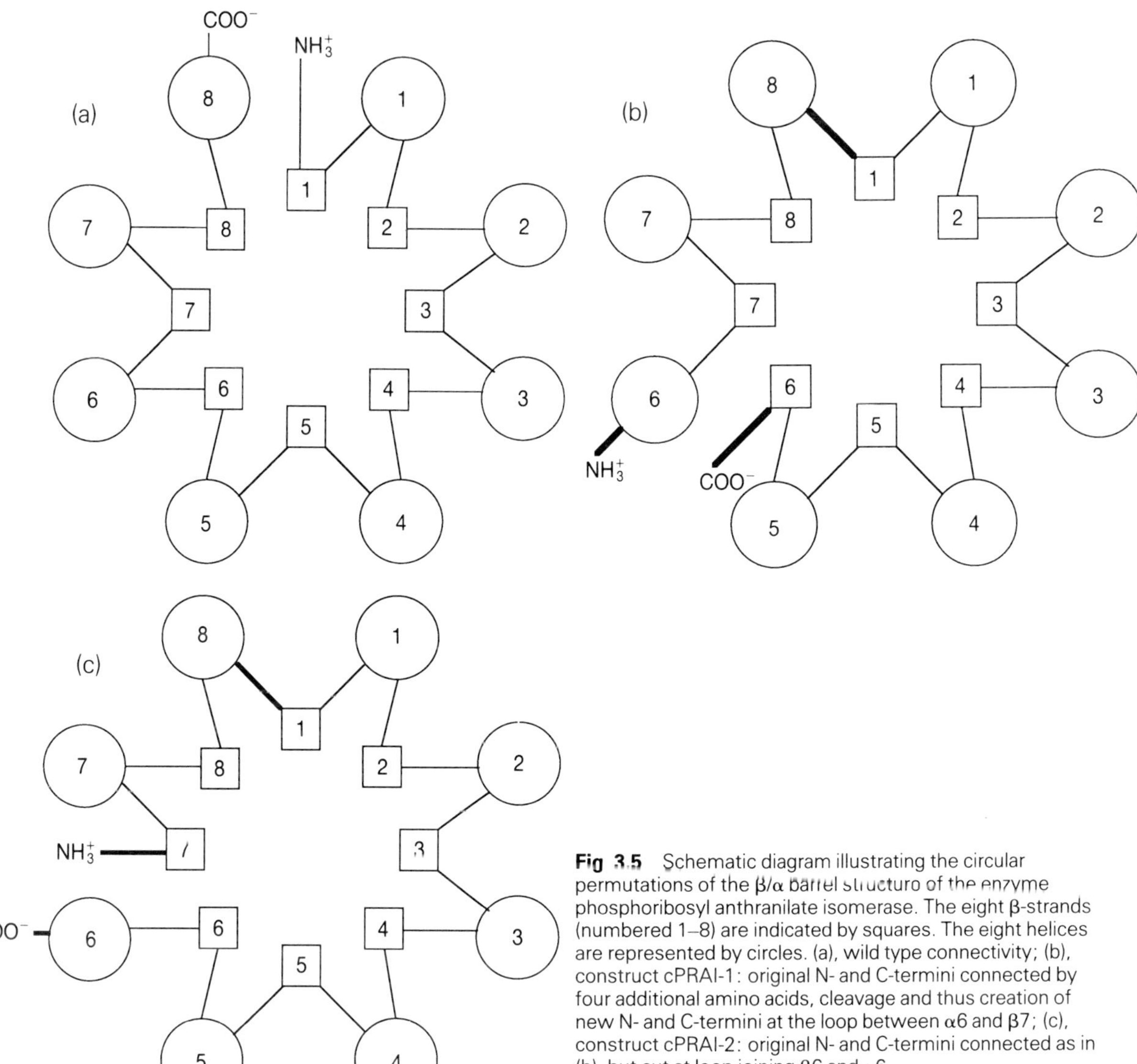



**Fig 3.5**  Schematic diagram illustrating the circular permutations of the β/α barrel structure of the enzyme phosphoribosyl anthranilate isomerase. The eight β-strands (numbered 1–8) are indicated by squares. The eight helices are represented by circles. (a), wild type connectivity; (b), construct cPRAI-1: original N- and C-termini connected by four additional amino acids, cleavage and thus creation of new N- and C-termini at the loop between α6 and β7; (c), construct cPRAI-2: original N- and C-termini connected as in (b), but cut at loop joining β6 and α6.

of the eight parallel β-strands with active site residues positioned at the end of the β-strands, catalytic activity is a sensitive measure of the proper fold of the enzyme. Two such circularly permutated variants of the isomerase were made and expressed in *E. coli*. The expressed mutants were enzymatically active and indistinguishable from the wild type enzyme as probed with a series of biophysical methods. This surprising result demonstrates that neither the original N- or C-termini nor the two surface loops that were cut are essential for folding. These findings might be explained by the existence of multiple folding pathways in which the same structure is generated by the interaction of folding units, consisting of neighbouring secondary structural elements. As Luger *et al.* (1989) point out, one should note that the results of their experiments support the 'thermodynamic hypothesis', which describes the folded, native protein structure as a minimal free energy state (Anfinsen and Scheraga 1975).

# The disulphide bond

The disulphide bond is considered to be a major constituent of protein stability. Disulphide bridges—intermolecular, formed between two different chains, or intramolecular, in one chain—increase the stability of a protein towards thermal denaturation to as much as 31.5 kJ/mol (Scheraga 1963). An analysis of available highly refined structures of proteins, which contain disulphide bridges revealed two families (Richardson 1981; Thornton 1981), characterized by the chirality of the dihedral angle about the sulphur–sulphur bond (Fig. 3.6). A dihedral angle of $\chi^3 \approx -90°$ defines the family of left-handed disulphides, which have a preference for the so called left-handed spiral conformation ($\chi^1 = 60°$, $\chi^2 = -60°$, $\chi^3 = -85°$, $\chi^{2'} = -60°$, $\chi^{1'} = -60°$). The right-handed disulphides, having a dihedral angle of $\chi^3 \approx 90°$, are more variable in the other dihedral angles. The analysis of available coordinates for disulphide bonds gives an average C$\alpha$–C$\alpha$ distance of $5.9\pm0.5$ Å, whereas the C$\alpha$–C$\alpha$ distance in right-handed disulphides is somewhat shorter, $5.07\pm0.7$ Å (Richardson 1981; Katz and Kossiakoff 1986).

The disulphide bond as a device for stabilizing proteins has been engineered into proteins that already contain cysteines, as well as into proteins that contain no cysteines at all. So far, the results obtained are not straightforward and the contribution of the disulphide bridge to protein stability is, in some cases, still a matter of controversy (Wells and Powers 1986; Pantoliano *et al.* 1987). One problem when inserting a disulphide link into a protein is whether the disulphide bond forms in the reducing environment of the cell, e.g. does the

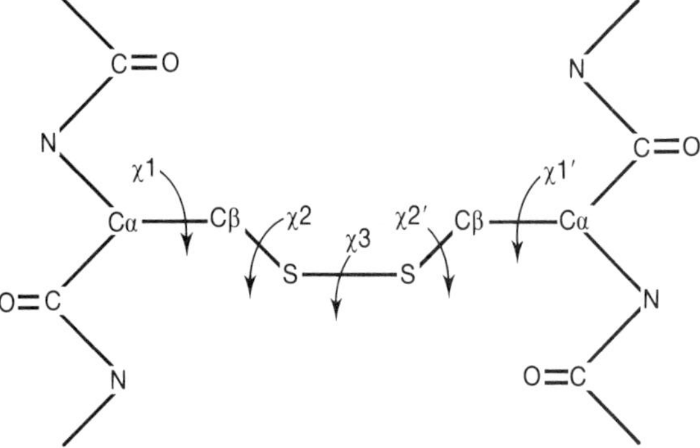

**Fig. 3.6** Schematic diagram of a disulphide bridge. The various dihedral angles are indicated.

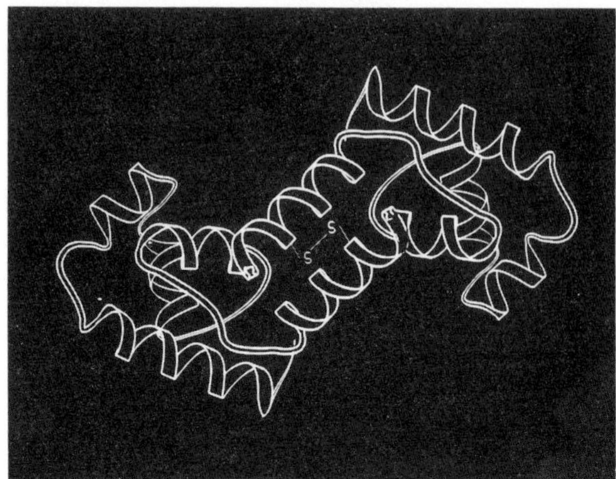

**Fig. 3.7** The interchain disulphide bridge, engineered into the N-terminal domain of λ-repressor. The picture was generated with a program by Priestle (1988), using the Cα coordinates from Pabo and Lewis (1982), Protein Data Bank, and displayed with the graphics program O (Jones *et al.* 1990).

isolated protein contain the disulphide bond or must it be oxidized chemically after the cells have been broken up. In the examples discussed below, only the dihydrofolate reductase (DHFR) gene is not expressed in a foreign host. The disulphide bridge forms *in vivo*, if the engineered enzymes are secreted from the cell. Cytosolic proteins are isolated as the reduced species, with the disulphide bridge sometimes forming during the preparation procedure.

A number of proteins have been engineered into disulphide-containing proteins: subtilisin (Wells and Powers 1986; Pantoliano *et al.* 1987), DHFR (Villafranca *et al.* 1987), the N-terminal domain of λ-repressor (Sauer *et al.* 1986) and T4 lysozyme (Perry and Wetzel 1986; Matsumura and Matthews 1989; Matsumura *et al.* 1989). Among these, the N-terminal domain of λ-repressor is an example of an interchain link: replacements of Tyr 88 at the interface between the two subunits of the repressor by Cys lead to a spontaneously formed disulphide bond (Fig. 3.7). The covalently linked dimer binds DNA more strongly and is more stable towards thermal denaturation or denaturation by urea (Sauer *et al.* 1986).

In the case of lysozyme, all the engineered disulphide bridges seem to stabilize the protein, as observed in a higher resistance towards thermal denaturation. Perry and Wetzel (1986) introduced a cysteine (Ile3Cys) close (in space) to an already existing one (Cys97). The oxidation *in vitro* resulted in a disulphide-containing protein, having 77% of the wild type activity. Wild type T4 lysozyme already contains two cysteines, Cys54 and Cys97. Replacing Cys54 by Thr or Val in the Ile3Cys mutant increased the thermostability of the enzyme further. The heat inactivation of lysozyme results partly from the formation of disulphide-linked oligomers, involving Cys54. Replacement of this residue removes this cause of inactivation and the double mutant Ile3Cys/Cys54Thr is thus more stable.

Matsumura and Matthews (1989) introduced a disulphide bond at the active site of T4 lysozyme by replacing Ser21 and Ser142 with Cys. This mutant was designed to couple the enzyme activity to the redox potential of the solution as a means to control the enzyme activity. A disulphide bond formed during preparation of the enzyme and the obtained oxidized form of the mutant was inactive. However, activity could be restored upon exposure to reducing agent. The inserted disulphide bond also increased the thermostability of the protein.

To investigate the effect of multiple disulphide bonds, Matsumura *et al.* (1989) introduced two or three disulphides in T4 lysozyme. Figure 3.8 shows the location of the engineered disulphide bonds and Table 3.2 summarizes the results. As can be seen, the double mutant is more heat stable than the single mutants. The triple mutant has a melting temperature of 23.1°C higher than wild type lysozyme. The effects of the stabilizing disulphides are approximately additive. It should be noted that

**Table 3.2**   Stability of engineered disulphide containing T4 lysozyme[1]

| Mutant | $T_m$ (°C)[2] | $\Delta T_m$ (mut.-wt*) (°C)[2] |
| --- | --- | --- |
| wt | 41.9 | |
| wt*[3] | 41.9 | |
| Ile3Cys/Cys54Thr | 46.7 | 4.8 |
| Ile9Cys/Leu164Cys/wt* | 48.3 | 6.4 |
| Thr21Cys/Thr142Cys/wt* | 52.9 | 11.0 |
| double 3-97/9-164 | 57.6 | 15.7 |
| double 9-164/21-142 | 58.9 | 17.0 |
| triple 3-97/9-164/21-142 | 65.5 | 23.4 |

[1] data from Matsumura *et al.* 1989; [2] $T_m$ is the temperature at the midpoint of the thermal denaturation transition determined at pH 0–2.0. $\Delta T_m$ is the difference between mutant and pseudo wild type lysozyme; [3] wt* pseudo wild type T4 lysozyme, made cysteine-free through mutation of Cys54Thr/Cys97Ala.

the double disulphide mutant Ile3Cys-Cys97/Ile9Cys-Leu164Cys is fully active and retains its activity at substantially higher temperatures than wild type lysozyme. This result is especially encouraging because it demonstrates that the stability of an enzyme can be substantially increased without loss of enzymatic activity.

In dihydrofolate reductase (DHFR), Pro 39 was replaced by a cysteine residue. In the native structure, this residue is close to Cys85 and a disulphide bond might form. The crystallographic analysis of the expressed protein showed the two sulphurs in van der Waals contact, but no disulphide link had been formed. *In vitro* oxidation resulted in the formation of the disulphide bridge, with a conformation close to the left-handed spiral (Villafranca *et al.* 1987). The engineered protein is more stable with respect to unfolding by guanidine hydrochloride, but shows no increase in stability towards thermal denaturation. The crystallographic analysis revealed further that the structural differences between the two forms are very small. Surprisingly, however, the insertion of the disulphide bond between positions 39 and 85 seemed to increase the thermal motion of atoms close to this site, as judged from the analysis of the crystallographic temperature factors. The comparison of the crystal structures of native and oxidized DHFR shows that in the oxidized mutant structure, the formation of the disulphide bond results in the loss of some favourable contacts, especially between Cys39 and Val88. The lesson to be learned from this observation is that disulphide bridges can only add to the protein's stability if they compensate for possible losses in other interactions, such as

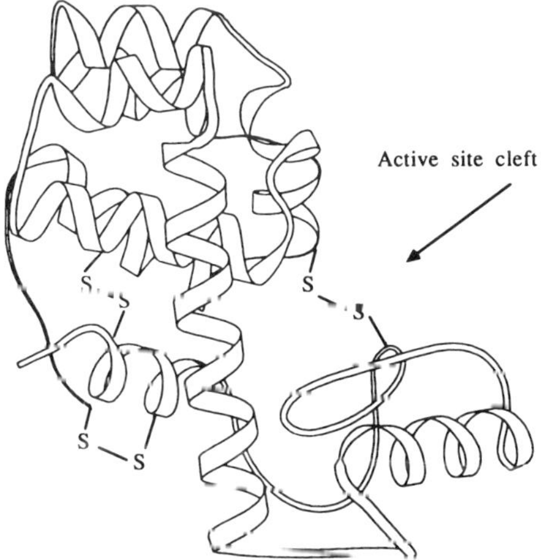


**Fig. 3.8**   Schematic drawing of T4 lysozyme made with a program by Priestle (1988) based on coordinates from Weaver and Matthews (1987), Protein Data Bank. The locations of the engineered disulphide bridges are indicated.

van der Waals interactions or hydrogen bonds (Villafranca *et al.* 1987).

Subtilisin has been the target of several attempts to increase the stability by inserting disulphide bonds. Both the groups at Genex and Genentech substituted Thr22 and Ser87 by cysteines. The Genentech group has also changed Ser24 to Cys. All the expected disulphide bonds formed *in vivo*. The crystallographic analysis of the two mutants Ser22Cys/Ser87Cys and Ser24Cys/Ser87Cys revealed atypical bond parameters in comparison with other disulphide bonds in proteins. The disulphide bonds in these two mutants represent strained, high energy conformations. Nevertheless, the accommodation of the disulphide bridge results in only minor local changes in main chain conformation (Katz and Kossiakoff 1986).

There is some controversy as to the stabilization of subtilisin as a result of the disulphide linkages introduced. Attempts to quantify the thermal stabilities of the mutants have been complicated by aggregation and autolysis. Wells and Powers (1986) report no stabilization of subtilisin towards thermal inactivation by any of the engineered disulphide bridges. The mutant with the disulphide bond Cys22/Cys87 is described to be less stable than the wild type subtilisin. However, Pantoliano *et al.* (1987) find an increase in stability towards thermal inactivation for this mutant. The reason for this controversy is not quite clear. A possible cause of concern might be the fact that the subtilisin used in Wells and Powers study has a third mutation, introduced as part of an unrelated study. The Tyr21 to Ala change, very close to the disulphide bridge might destabilize the mutant to such an extent that this effect cannot be completely reversed by the insertion of the covalent cross link.

Despite some unsettled questions, and keeping in mind that our data base is rather limited, one can draw some conclusions on the disulphide bridge:

(1) Disulphide bonds can be engineered into proteins even with atypical dihedral geometries or Cα–Cα distances. The a priori constraints on a disulphide linkage that is going to be inserted thus are rather weak.
(2) The formation of the disulphide bond is accompanied by only small, subtle changes of the protein structure close to the site of substitution. Structural changes upon insertion of a disulphide bridge are thus local.
(3) In most, if not all, cases studied so far, disulphide bonds in a protein do increase its

stability towards thermal denaturation and/or denaturation by guanidine hydrochloride.

## The origin of thermostability

Proteins are usually inactivated or denatured at temperatures above 50°C. For many industrial applications of enzymes, one would like to extend the stability and activity to higher temperatures. Proteins isolated from thermophilic organisms have therefore been studied extensively with the aim of establishing the origins of thermostability. Unfortunately, a mere comparison of the amino acid sequences of mesophilic and thermophilic proteins will not reveal those amino acid changes responsible for the change in physical properties because they will be masked by changes in the amino acid sequence due to genetic drift. One possible route to increase the thermostability—the insertion of disulphide bonds—has been discussed above. However, for a number of reasons, one would like to extend the arsenal of tools to engineer protein stability. Most of the thermophilic proteins do not contain disulphide bonds, thus demonstrating that there are other means for proteins to maintain their native conformation at high temperature. There are different approaches to this question, and clearly all of them will have to contribute to establish the causes of thermostability and to extract rules that will be applicable in protein engineering.

One way is to transfer essential genes from mesophilic to thermophilic organisms and select for increasing stability of the protein. From this, one should be able to identify the changes accompanying thermostability. Secondly, the increasing number of mutant proteins of known three-dimensional structure will certainly help to understand the molecular causes of increased stability. The crystallographic analysis and comparison of native and mutant ferredoxins and haemoglobins (Perutz and Raidt 1975), thermolysins (Matthews *et al.* 1974) and T4 lysozymes (Grutter *et al.* 1979; Hawkes *et al.* 1984) reveal that differences in the thermostability of these proteins are not due to large changes in the three-dimensional structure, but rather to small, subtle changes such as hydrophobic interactions, hydrogen bonds or salt bridges.

An example of increased thermostability as a result of improved hydrogen bonding pattern is the Asn218Ser mutation in subtilisin (Bryan *et al.*

1986). This mutant, created by random mutagenesis and selected for increased thermostability, is significantly more stable than the wild type enzyme, with an increased half life for thermal inactivation from 59 to 223 min and a 2.4–3.9°C higher midpoint in the thermally induced transition from the folded to the unfolded state. The crystallographic analysis of both wild type and mutant at 1.8 Å resolution shows that the hydrogen bonding parameters for residues in the vicinity of the site of mutation are more favourable due to the fact that two antiparallel β-strands can move closer to each other. This allows shorter and thus stronger hydrogen bonds.

Decreased thermostability can be caused by weakened subunit interactions due to changes at interface regions. One example is the molecular cause of a human disease, triose-phosphate isomerase (TIM) deficiency. This disease manifests itself clinically by chronic haemolytic anaemia and neuromuscular disorders (Valentine *et al.* 1983). The cause of this disease is a decreased level of triose-phosphate isomerase activity, which is thermolabile. The sequence comparison of the normal gene for human TIM and the gene from an individual having TIM deficiency, revealed a single amino acid change of Glu104 to Asp as the reason for thermolability (Daar *et al.* 1986). An inspection of the known crystal structure (Banner *et al.* 1975)

reveals that the amino acid substitution seems to disrupt an intrinsic network of salt bridges and hydrogen bonds located in a hydrophobic pocket close to the subunit–subunit interface of the dimer (Fig. 3.9).

Thermal denaturation not only involves the unfolding of the polypeptide chain, but is accompanied by chemical damage, such as the oxidation of methionine to the sulphoxide β-elimination of cysteine and deamidation of asparagine and glutamine. In attempts to engineer more stable proteins, several groups have tried to make proteins more resistant to these chemical processes.

In the case of subtilisin, Met222 has been identified as the target of oxidation with subsequent inactivation (Stauffer and Etson 1969). This amino acid has been replaced by all other 19 amino acid residues by site-directed mutagenesis to render the enzyme resistant to chemical damage through oxidation (Estell *et al.* 1985). Not surprisingly, all the mutations decrease the catalytic activity—with Met222, the site of mutagenesis is close to Ser221, which is part of the catalytic triad typical of serine proteases. Mutants, which contained non-oxidizable amino acids such as Ala, Ser and Leu, were resistant to oxidation, with the Met222Ala mutant having 80% of the $k_{cat}$ of the wild type enzyme.

Another experiment focused on the effect of

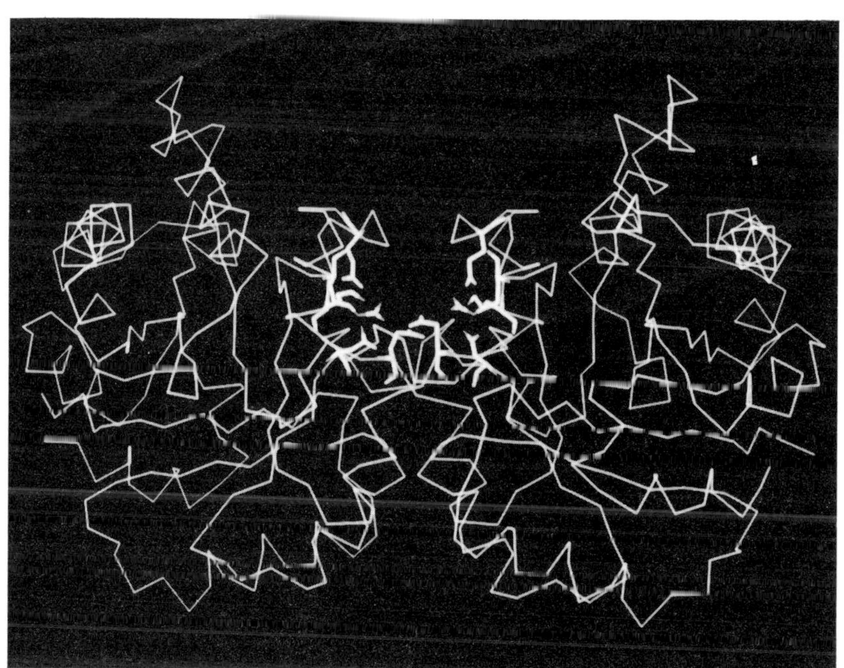

**Fig. 3.9** Structural basis of human triose-phosphate isomerase deficiency. Cα tracing of the dimer of triose phosphate isomerase (Banner *et al.* 1975). The surroundings of the mutated Glu104, giving rise to the decreased thermostability, at the dimer interface are shown. Coordinates obtained from the Protein Data Bank.

changing asparagine residues, located at the sub-unit interface of triose-phosphate isomerase. The replacement of these 'weak' links in the protein should increase the thermostability in a rather straightforward manner (Ahern *et al.* 1987). Three asparagine residues, Asn14, 65 and 78 are located at the dimer interface, which upon deamidation would be converted to aspartic acid residues, and promote dissociation by the repulsion of negative charges. Mutant forms with single substitutions at Asn78 (to Ile, Thr and Asp) and a double mutant (Asn14Thr and Asn78Ile) were made. The replacement of Asn78 by Ile or Thr increased the half life of the enzyme at 100°C by 25%. The double mutant has a half life twice that of the wild type enzyme. As would be expected, the control mutant Asn78Asp is less stable than the wild type TIM.

Another cause of instability or unwanted side reactions might be reactive thiol groups at the surface of the protein. Iso-1-cytochrome *c* from *Saccharomyces cerevisiae* contains such a reactive thiol group at Cys107. This sulphydryl group complicates functional studies by dimer formation through an intermolecular disulphide bond in the absence of reducing agents. Iso-1-cytochrome *c* autoreduces much more rapidly than cytochromes that lack the cysteine residue. The replacement of this cysteine residue by a threonine side chain via site-directed mutagenesis created a mutant protein with biophysical properties undistinguishable from the wild type but with an improved stability due to the lack of unwanted side reactions (Cutler *et al.* 1987). The mutant thus provides a more suitable system for the study of mechanistic and physical aspects of this protein.

## Quaternary structure

The amino acid sequence of a polypeptide chain not only contains the information needed for the folding into the native, biologically active conformation, but also the information on whether the protein is going to be a monomer or to assemble into an oligomer. It has been, and still is, an open question as to why proteins assemble into oligomers when there is no obvious reason such as cooperativity and regulatory control. Mutations at the surface of a protein, site-directed or random, are not necessarily, as often believed, less disastrous than mutations at or near the active site. The following examples are intended to illustrate that mutations at the surface of proteins can have dramatic effects on quaternary structure and biological function.

## Sickle-cell haemoglobin

The classic example of a point mutation at the surface of a protein, causing changes to its quaternary structure and at the same time with major impacts on human health, is of sickle-cell haemoglobin. The amino acid sequence and the conformation of the α-chain is unchanged, but in the β-chain, Glu6 is mutated to a Val, introducing a hydrophobic, 'sticky' side chain at the surface (Ingram 1959). In the deoxygenated form of haemoglobin, this residue interacts with a hydrophobic pocket of a second haemoglobin molecule, giving rise to the formation of polymers (Padlan and Love 1985). These polymers form long fibres, causing the characteristic deformation of the erythrocytes, which thus name the disease.

## The quest for a faster insulin

One of the most convincing successes of modern biochemistry and medicine is the introduction of insulin as a therapeutic agent in the treatment of diabetes. However, there are still problems associated with the use of insulin. One such major issue is the slow absorption from subcutaneous tissue at meal times, when the rapid increase of insulin in blood has to be mimicked. Thus, one goal in the development of more suitable insulins is to provide a more physiological insulin profile at the time of meal consumption. Insulin, as provided in the neutral solutions used in therapy, is found as hexamers (Blundell *et al.* 1972). Conversion into monomers and transport are believed to limit the rate of absorption. Attempts have therefore been made to make monomeric and biologically active insulins by single amino acid substitutions (Brange *et al.* 1988). The strategy chosen was to introduce charge repulsion at the monomer–monomer interface. Negative charges were introduced at positions close to already existing carboxyl groups (Fig. 3.10). A number of different single or double mutants were made and almost all of them turned out to be essentially monomeric. With one exception, the monomeric mutants retained their biological activity *in vivo*. Most importantly, the initial rate of absorption from the subcutaneous deposit after injection is approximately three times faster

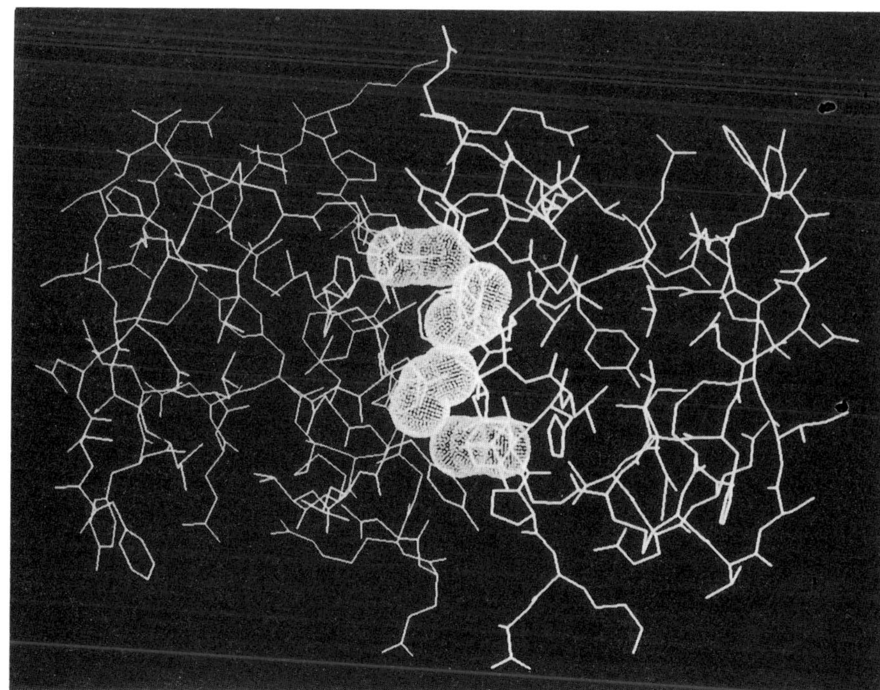

**Fig. 3.10**  Engineering monomeric insulin: the insulin dimer with B13Glu and mutated B9Asp superimposed in dots indicating the van der Waals surface of the carboxyl group atoms. The distances between the four individual carboxyl groups are approximately 4 Å. (Reproduced with permission from Brange *et al.* 1988.)

for one of the mutants than for human insulin (Brange *et al.* 1988). These monomeric insulins are candidates for a quicker delivery of insulin and are of potential use in the therapy of diabetes.

## The assembly of Rubisco

Rubisco, the catalyst of the initial steps in both photosynthetic carbon dioxide fixation and photorespiration, is, with a few well established exceptions, a complex of eight large (L) and eight small (S) subunits (Andrews and Lorimer 1987). Point mutations at the surface of both the large and the small subunit can therefore easily affect interaction areas at the L–L, or the L–S interface and thus disturb assembly. Mutations, affecting transport or binding to the Rubisco binding protein, a protein involved in Rubisco folding and assembly (Hemmingsen *et al.* 1988), will not be considered here. A naturally occurring mutant, with a mutation in the L chain (changing Ser112 to Phe) has been found in tobacco. In the mutant the polypeptides are synthesized, but the assembled holoenzyme is lacking (Avni *et al.* 1989). In all $L_8S_8$ type L sequences known, the serine residue is conserved. The crystal structure of the spinach enzyme (Andersson *et al.*

1989; Knight *et al.* 1989) shows that this residue is part of the L–S interface region. The insertion of the bulkier Phe side chain prevents association of the L and the S subunit and thus the holoenzyme cannot assemble.

When probing for catalytic function of the conserved Lys166 residue at the active site of the dimeric Rubisco from *Rhodospirillum ribrum*, Lee *et al.* (1987) changed this residue to an aspartic acid. This mutant no longer formed dimers, but resulted in inactive monomers. The side chain of Lys166 is part of an intricate network of charges and hydrogen bonds at the subunit–subunit interface in the dimer (Schneider *et al.* 1990). Replacement of the positive charge of Lys166 with the negative charge of the Asp side chain disrupts this network and introduces a repulsion, which results in the formation of monomers.

What gives rise to the formation of oligomers, built up from active building blocks? In a study, intended to analyse the function of the C-terminal helices in Rubisco from *Rhodospirillum rubrum*, Ranty *et al.* (1990) truncated the polypeptide chain at different positions, giving rise to mutants lacking more and more amino acids at the C-terminal end of the chain. Surprisingly, the mutant truncated at position 411 formed an oligomer consisting of eight L-chains (Ranty *et al.* 1990). Unlike the native enzyme, the truncation had exposed three hydrophobic residues (Leu374, Leu424 and Phe437), which introduced a hydrophobic patch at the poles of the dimer. By head-to-tail association, an $L_8$ structure consisting of four dimers could be formed by interactions between these hydrophobic patches. To test this hypothesis, a mutation of Leu424 to Asn was introduced to disrupt this hydrophobic patch. This truncated mutant had the dimeric wild type quaternary structure and did not assemble to higher oligomers. The insertion of a few hydrophobic nucleation points, such as in sickle-cell haemoglobin or the truncated Rubisco mutants, can thus completely change the quaternary structure of these proteins. This raises the question as to whether at least some oligomeric proteins are formed during evolution by single point mutations and are biological curiosities rather than necessities.

## Conclusions

Amino acid replacements via site-directed mutagenesis result in changes in hydrophobicity, hydrophilicity, size and charge. The consequences of such changes are dependent on the structural context in which they occur and can result in the destabilization of the protein to the extent that unfolding occurs. In other cases, structural changes might be restricted to small, local structural changes in the proximity of the site of mutation. What the result will be can be reasonably well rationalized given knowledge of the three-dimensional structure of the protein under study, or at least knowledge of the structure of a homologous protein from which a model of the structure can be built. Unfortunately, this information is not usually available and the protein engineer is restricted to make very conservative amino acid changes if he/she wants to be able to predict or even interpret the effect of the changes made. One example from the plant proteins are the seed storage proteins, where great effort is invested to engineer nutritionally more valuable seeds by introducing more essential amino acids by protein engineering. Clearly, it will be very difficult to perform such mutations without interfering with either the internal packing of the protein chains, the tight packing of the chains in protein bodies or the transport of the proteins in the cell without knowledge of the structure and a deeper understanding of the intracellular routing mechanisms. Another example is Rubisco, where the goal is to obtain a more efficient enzyme and/or an enzyme with an improved specificity for $CO_2$ in comparison with $O_2$. In this case the structure is determined to high resolution but a higher level of understanding has to be reached concerning the enzyme mechanism (with the help of site-directed mutagenesis) before a blueprint for a better Rubisco can be made.

## References


Ahern, T.J., Casal, J.I., Petsko, G.A. and Klibanov, A.M. (1987). Control of oligomeric thermostability by protein engineering. *Proceedings of the National Academy of Science, USA.* **84**: 675–9.

Alber, T., Bell, J.A., Sun, D.-P., Nicholson, H., Wozniak, J.A., Cook, S. and Matthews, B.W. (1988). Replacements of Pro86 in Phage T4 lysozyme extend an α-helix but do not alter protein stability. *Science.* **239**: 631–5.

Andersson, I., Knight, S., Schneider, G., Lindqvist, Y., Lundqvist, T., Brändén, C.-I. and Lorimer, G. (1989). Crystal structure of the active site of ribulose-1,5-bisphosphate carboxylase. *Nature.* **337**: 229–34.

Andrews, J. and Lorimer, G. (1987). Rubisco: structure, function and prospects of improvement. In *Biochemistry of Plants*, Vol. 10 (ed. M.D. Hatch), pp. 131–218. Academic Press, Orlando, Florida.

Anfinsen, C.B. and Scheraga, H.C. (1975). Experimental and theoretical aspects of protein folding. *Advances of Protein Chemistry.* **29**: 205–300.

Avni, A., Edelman, M., Rachailovich, I., Aviv, D. and Fluhr, R. (1989). A point mutation in the gene of the large subunit of ribulose-1,5-bisphosphate carboxylase/oxygenase affects holoenzyme assembly in *Nicotiana tabacum. EMBO Journal.* **8**: 1915–18.

Banner, D.W., Bloomer, A.C., Petsko, G.A., Phillips, D.C., Pogson, C.I., Wilson, I.A., Corran, P.H., Furth, A.J., Milman, J.D., Offord, R.E., Priddle, J.D. and Waley, S.G. (1975). Structure of chicken muscle triose phosphate isomerase determined crystallographically at 2.5 Å resolution using amino acid sequence data. *Nature.* **255**: 609–14.

Blundell, T., Dodson, G., Hodgkins, D. and Mercola, D. (1972). Insulin: the structure in the crystal and its reflection in chemistry and biology. *Advances of Protein Chemistry.* **26**: 279–402.

Brange, J., Ribel, U., Hansen, J.F., Dodson, G., Hansen, M.T., Havelund, S., Melberg, S.G., Norris, F., Norris, K., Snel, L., Sörenson, A.R. and Voigt, H.O. (1988). Monomeric insulins obtained by protein engineering and their medical implications. *Nature.* **333**: 679–82.

Brick, P., Bhat, T.N. and Blow, D.M. (1989). Structure of tyrosyl-tRNA synthetase refined at 2.3 Å resolution. *Journal of Molecular Biology.* **208**: 83–98.

Bryan, P.N., Rollence, M.L., Pantoliano, M.W., Wood, J., Finzel, B.C., Gilliland, G.L., Howard, A.J. and

Poulos, T.L. (1986). Proteases of enhanced stability: characteristation of a thermostable variant of subtilisin. *Proteins: Structure Function and Genetics*. **1**: 326–34.

Chothia, C. (1974). Hydrophobic bonding and accessible surface area in proteins. *Nature*. **248**: 338–9.

Chothia, C. and Janin, J. (1982). Orthogonal packing of β pleated sheets in proteins. *Biochemistry*. **21**: 3955–65.

Chothia, C. and Lesk, A.M. (1986). The relation between the divergence and structure in proteins. *EMBO Journal*. **5**: 823–6.

Chothia, C. and Lesk, A.M. (1987). The evolution of protein structures. *Cold Spring Harbor Symposium in Quantitative Biology, L11*. 399–405.

Chothia, C., Levitt, M. and Richardson, D. (1977). Structure of proteins: packing of α-helices and pleated sheets. *Proceedings of the National Academy of Sciences, USA*. **74**: 4130–4.

Cutler, R.L., Pielak, G.J., Mauk, A.G. and Smith, M. (1987). Replacement of cysteine-107 of *Saccharomyces cerevisiae* iso-1-cytochrome c with threonine: improved stability of the mutant protein. *Protein Engineering*. **1**: 95–9.

Daar, I.O., Artymiuk, P., Phillips, D.C. and Maquat, L.E. (1986). Human triose-phosphate isomerase deficiency: a single amino acid substitution results in a thermolabile enzyme. *Proceedings of the National Academy of Sciences, USA*. **83**: 7903–7.

Eisenberg, D. and McLachlan, A. (1986). Solvation energy in protein folding and binding. *Nature*. **319**: 199–203.

Estell, D.A., Graycar, T.P. and Wells, J.A. (1985). Engineering an enzyme by site-directed mutagenesis to be resistant to chemical oxidation. *Journal of Biological Chemistry*. **260**: 6518–21.

Fersht, A.R., Shi, J.-P., Knill-Jones, J., Lowe, D.M., Wilkinson, A.J., Blow, D.M., Brick, P., Carter, P., Waye, M.M. and Winter, G. (1985). Hydrogen bonding and biological specificity analyzed by protein engineering. *Nature*. **314**: 235–8.

Goldberg, M.E. (1985). The second translation of the genetic message: protein folding and assembly. *Trends in Biochemical Sciences*. **10**: 388–91.

Grutter, M.G., Hawkes, R.B. and Matthews, B.W. (1979). Molecular basis of thermostability in the lysozyme from bacteriophage T4. *Nature*. **277**: 667–9.

Hawkes, R.B., Grutter, M.G. and Schellman, J. (1984). Thermodynamic stability and point mutations of bacteriophage T4 lysozyme *Journal of Molecular Biology*. **175**: 195–212.

Hecht, M.H., Sturtevant, J.M. and Sauer, R.T. (1986). Stabilisation of λ-repressor against thermal denaturation by site-directed gly–ala changes in α-helix 3. *Proteins. Structure, Function and Genetics*. **1**: 43–6.

Hemmingsen, S.M., Woolford, C., van der Vies, S., Tilly, K., Dennis, D.T., Georgopoulos, C.P., Hendrix, R.W. and Ellis, R.J. (1988). Homologous plant and bacterial proteins chaperone oligomeric protein assembly. *Nature*. **333**: 330–4.

Hol, W.G.J., van Duijnen, P.T. and Berendeson, H.J.C. (1978). The α-helix dipole and the properties of proteins. *Nature*. **273**: 443–6.

Ingram, V.M. (1959). Abnormal human haemoglobins. III. The chemical difference between normal and sickle cell haemoglobins. *Biochimica et Biophysica Acta*. **36**: 402–11.

Janin, J. and Chothia, C. (1980). Packing of α-helices onto β-pleated sheets and the anatomy of α/β proteins. *Journal of Molecular Biology*. **143**: 95–128.

Jones, T.A., Bergdoll, M. and Kjeldgaard, M. (1990). O: a macromolecular modelling environment, 1. An overview. In *Molecular Modelling*, Springer Verlag.

Karpusas, M., Baase, W.A., Matsumura, M. and Matthews, B.W. (1989). Hydrophobic packing in T4 lysozyme probed by cavity-filling mutants. *Proceedings of the National Academy of Sciences, USA*. **86**: 8237–41.

Katz, B.A. and Kossiakoff, A. (1986). The crystallographic determined structure of atypical disulfides engineered into subtilisin. *Journal of Biological Chemistry*. **261**: 15480–5.

Kauzmann, W. (1959). Some factors in the interpretation of protein denaturation. *Advances of Protein Chemistry*. **14**: 1–63.

Knight, S., Andersson, I. and Bränden, C.-I. (1989). Reexamination of the three-dimensional structure of the small subunit of Rubisco from higher plants. *Science*. **244**: 702–5.

Knowles, J.R. (1987). Tinkering with enzymes: what are we learning? *Science*. **236**: 1252–8.

Lee, B. and Richards, F.M. (1971). The interpretation of protein structures: estimation of static accessibility. *Journal of Molecular Biology*. **55**: 379–400.

Lee, H.H., Soper, T.S., Mural, R.J., Stringer, C.D. and Hartman, F.C. (1987). An intersubunit interaction at the active site of D-ribulose-1,5-bisphosphate carboxylase/oxygenase as revealed by cross-linking and site-directed mutagenesis. *Biochemistry*. **26**: 4599–604.

Lim, W.A. and Sauer, R.T. (1988). Alternative packing arrangements in the hydrophobic core of λ-repressor. *Nature*. **339**: 31–6.

Luger, K., Hommel, U., Herold, M., Hofsteenge, J. and Kirschner, K. (1989). Correct folding of circularly permuted variants of a β/α barrel enzyme *in vivo*. *Science*. **243**: 206–210.

Matsumura, M. and Matthews, B.W. (1989). Control of enzyme activity by an engineered disulfide bond. *Science*. **243**: 792–4.

Matsumura, M., Becktel, W.J. and Matthews, B.W. (1988). Hydrophobic stabilisation in T4 lysozyme determined directly by multiple substitutions of Ile 3. *Nature*. **334**: 406–10.

Matsumura, M., Signor, G. and Matthews, B.W. (1989). Substantial increase of protein stability by multiple disulphide bonds. *Nature*. **342**: 291–3.

Matthews, B.W. (1987). Genetic and structural analysis of the protein stability problem. *Biochemistry*. **26**: 6885–8.

Matthews, B.W., Weaver, L.H. and Kester, W.J.

(1974). The conformation of thermolysin. *Journal of Biological Chemistry*. **249**: 8030–44.

Matthews, B.W., Nicholson, H. and Becktel, W.J. (1987). Enhanced protein thermostability from site-directed mutations that decrease the entropy of unfolding. *Proceedings of the National Academy of Sciences, USA*. **84**: 6663–7.

McLachlan, A.D. (1979). Gene duplications in the structural evolution of chymotrypsin. *Journal of Molecular Biology*. **128**: 49–79.

Nicholson, H., Becktel, W.J. and Matthews, B.W. (1988). Enhanced protein thermostability from designed mutations that interact with α-helix dipoles. *Nature*. **336**: 651–6.

Nozaki, Y. and Tanford, C. (1971). The solubility of amino acids and two glycine peptides in aqueous ethanol and dioxane solutions. *Journal of Biological Chemistry*. **246**: 2211–17.

Pabo, C.O. and Lewis, M. (1982). The operator binding domain of lambda repressor. Structure and DNA recognition. *Nature*. **298**: 443–7.

Padlan, E.A. and Love, W.E. (1985). Refined crystal structure of deoxyhemoglobin S. *Journal of Biological Chemistry*. **260**: 8280–91.

Pantoliano, M.W., Ladner, R.C., Bryan, P.N., Rollence, M.L., Wood, J.F. and Poulos, T.L. (1987). Protein engineering of subtilisin BPN': enhanced stabilisation through the introduction of two cysteines to form a disulfide bond. *Biochemistry*. **26**: 2077–82.

Perry, L.J. and Wetzel, R. (1986). Unpaired cysteine-54 interferes with the ability of an engineered disulfide to stabilize T4 lysozyme. *Biochemistry*. **25**: 733–9.

Perutz, M.F. (1970). Stereochemistry of cooperative effects in hemoglobin. *Nature*. **228**: 726–39.

Perutz, M.F. and Raidt, H. (1975). Stereochemical basis of heat stability of bacterial ferredoxins and in hemoglobin A2. *Nature*. **255**: 256–9.

Priestle, J.P. (1988). A stereo cartoon drawing program for proteins. *Journal of Applied Crystallography*. **21**: 572–6.

Priestle, J.P., Grutter, M.G., White, J.L., Vincent, M.G., Kania, M., Wilson, E., Jardetzky, T.S., Kirschner, K. and Jansonius, J.N. (1987). Three-dimensional structure of the bifunctional enzyme N-(5'-phosphoribosyl) anthranilate isomerase-indole-3-glycerol-phosphate synthase from *Escherischia coli*. *Proceedings of the National Academy of Sciences, USA*. **84**: 5690–4.

Ramakrishnan, C. and Ramachandran, G.N. (1965). Stereochemical criteria for polypeptide and protein-chain conformation. *Biophysical Journal*. **5**: 909–33.

Ramachandran, G.N. and Sasisekharan, V. (1968). Conformation of peptides and proteins. *Advances of Protein Chemistry*. **23**: 283–437.

Ranty, B., Lundqvist, T., Schneider, G., Madden, M., Howard, R. and Lorimer, G. (1990). Truncation of ribulose-1,5-bisphosphate carboxylase from *Rhodospirillum rubrum* affects the assembly and the activity of the protein. *EMBO Journal*. **9**: 1365–73.

Richardson, J. (1981). The anatomy and taxonomy of protein structure. *Advances of Protein Chemistry*. **34**: 167–339.

Richardson, J. and Richardson, D.C. (1988). Amino acid preferences for specific locations at the ends of α-helices. *Science*. **240**: 1648–52.

Russell, A.J. and Fersht, A.R. (1987). Rational modification of enzyme catalysis by engineering surface charge. *Nature*. **328**: 496–500.

Sali, D., Bycroft, M. and Fersht, A.R. (1988). Stabilization of protein structure by interaction of α-helix dipole with a charged side chain. *Nature*. **335**: 740–3.

Sandberg, W.S. and Terwilliger, T.C. (1989). Influence of interior packing and hydrophobicity of a protein. *Science*. **245**: 54–7.

Sauer, R.T., Hehir, K., Stearman, R.S., Weiss, M.A., Jeitler-Nilsson, A., Suchanek, E.G. and Pabo, C.O. (1986). An engineered intersubunit disulfide enhances the stability and DNA binding of the N-terminal domain of λ-repressor. *Biochemistry*. **25**: 5992–8.

Scheraga, H.A. (1963). Intramolecular bonds in proteins II. Noncovalent bonds. *Proteins*. **1**: 477–594.

Schneider, G., Lindqvist, Y. and Lundqvist, T. (1990). Crystallographic refinement and structure of ribulose-1,5-bisphosphate carboxylase at 1.7 Å resolution. *Journal of Molecular Biology*. **211**: 989–1008.

Schulz, G.E. and Schirmer, R.H. (1979). *Principles of Protein Structure*. Springer Verlag, New York.

Serrano, L. and Fersht, A.R. (1989). Capping and α-helix stability. *Nature*. **342**: 296–9.

Stauffer, C.E. and Etson, D. (1969). The effect on subtilisin activity of oxidizing a methionine residue. *Journal of Biological Chemistry*. **244**: 5333–8.

Thornton, J.M. (1981). Disulphide bridges in globular proteins. *Journal of Molecular Biology*. **151**: 261–87.

Valentine, W.N., Tanaka, K.R. and Paglia, D.E. (1983). In *The Metabolic Basis of Inherited Disease* (ed. J.D. Stanbury, J.B. Wyngarden, D.S. Frederickson, J.L. Goldstein and M.S. Brown), pp. 1606–82. McGraw-Hill, New York.

Villafranca, J.E., Howell, E.E., Oatley, S.J., Xuong, N.H. and Kraut, J. (1987). An engineered disulfide bond in dihydrofolate reductase. *Biochemistry*. **26**: 2182–9.

Weaver, L.H. and Matthews, B.W. (1987). Structure of bacteriophage T4 lysozyme refined at 1.7 Å resolution. *Journal of Molecular Biology*. **193**: 189–99.

Wells, J.A. and Powers, D.B. (1986). *In vivo* formation and stability of engineered disulfide bonds in subtilisin. *Journal of Biological Chemistry*. **261**: 6564–70.

Wells, J.A., Powers, D.B., Bott, R.R., Graycar, T.P. and Estell, D.A. (1987). Designing substrate specificity by protein engineering of electrostatic interactions. *Proceedings of the National Academy of Sciences, USA*. **84**: 1219–23.

# Part II  ANALYSIS OF PLANT METABOLISM

# 4

# Structure–Function Relationships of Ribulosebisphosphate Carboxylase/Oxygenase as Suggested by Site-directed Mutagenesis

Fred C. Hartman

## Introduction

During the past 20 years, ribulosebisphosphate carboxylase/oxygenase (Rubisco) has perhaps received as much attention from the scientific community as any other protein. Some international conferences have been devoted to it entirely (e.g. a Royal Society Discussion Meeting, see Ellis and Gray, 1986), and others have spotlighted the enzyme (e.g. NATO Workshops) (see von Wettstein and Chua 1987; Aresta and Schloss 1990). One monumental observation, above all others, intensified interest in Rubisco—namely that the enzyme, denoted as a carboxylase for the two decades after its discovery (Quayle *et al*. 1954; Weissbach *et al*. 1954), is also an oxygenase. The independent, pioneering studies by Ogren and by Tolbert and their respective colleagues (Bowes *et al*. 1971; Andrews *et al*. 1973; Lorimer *et al*. 1973) that led to this conclusion provided immediate enlightenment. Ribulosebisphosphate (ribulose-$P_2$) is a substrate for both carboxylation and oxygenation. Carboxylation gives rise to two molar equivalents of D-3-phosphoglycerate (net synthesis of carbohydrate from atmospheric carbon dioxide); oxygenation gives rise to one molar equivalent of D-3-phosphoglycerate and one molar equivalent of 2-phosphoglycolate (no net synthesis) with consequential lowering of the steady-state level of ribulose-$P_2$ and the efficiency of carbohydrate synthesis (Fig. 4.1). The former reaction initiates the photosynthetic reductive pathway, the latter initiates the energy-wasteful, presumably functionless, photo-respiratory pathway. Thus, a bifunctional enzyme catalyses the first step in competing metabolic pathways! After a 100-year dormancy, the empirical observation of Otto Warburg, that growth of C3 plants is greatly stimulated under atmospheres of low $O_2$ or high $CO_2$ tension, was clarified and the enormous potential for elevating biomass yields through modulation of the carboxylase/oxygenase ratio was appreciated (for discussion, see Hardy *et al*. 1978).

Apart from the problematical issue of whether a superior Rubisco, and ultimately a more efficient plant, can be designed, many facets of the genetics, physiology, molecular biology and biochemistry of the enzyme are intellectually intriguing and provide insights into life processes that transcend any single macromolecule. Indeed, even a cursory perusal of the voluminous carboxylase literature reveals a fascinating odyssey through many crevices of modern biological sciences, for example:

(1) Regulation of gene expression—because of the abundance of Rubisco, and hence its mRNA, an attractive system is provided for observing regulation at the transcriptional level and the response to photon flux (Rodermel and Bogorad 1985; Nagy *et al*. 1986; Klein and Mullet, 1990).

(2) Coordination of translation—the most prevalent form of Rubisco consists of two gene products, one nuclear encoded and one chloroplastic encoded. Communication between organelles and coordination of levels of

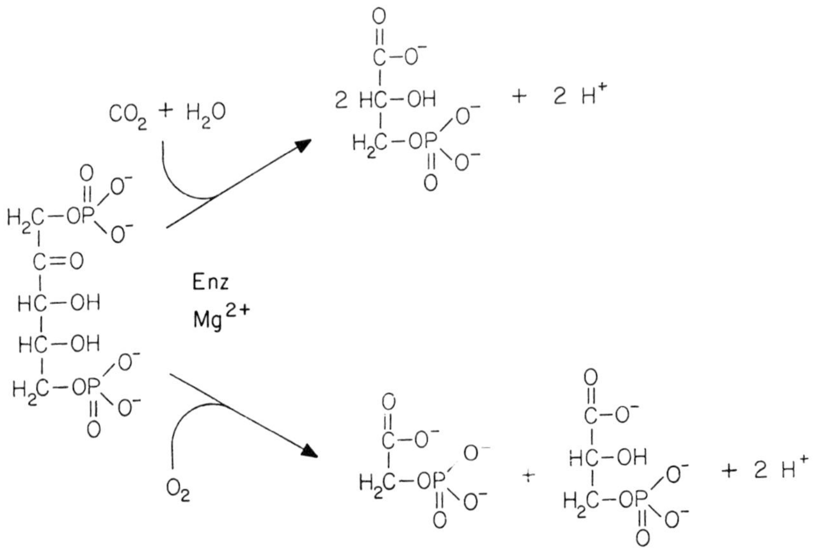

**Fig. 4.1** Carboxylation and oxygenation of D-ribulose-1,5-bisphosphate as catalysed by Rubisco.

biosynthesis of the two subunits become pertinent issues (Rodermel *et al.* 1988).

(3) Processing, translocation and assembly—the carboxylase subunits are derived from larger polypeptide precursors that must undergo processing by specific proteases (Highfield and Ellis 1978; Chua and Schmidt 1979; Langridge 1981; Robinson and Ellis 1984; Gatenby *et al.* 1985). Furthermore, the small subunit precursors must be targeted to the chloroplasts. Folding and assembly of the native hexadecameric enzyme do not occur spontaneously but require a helper or 'chaperonin' protein (Hemmingsen *et al.* 1988; Roy *et al.* 1988). Even in the case of the simpler homodimeric carboxylase, as found in *Rhodospirillum rubrum*, proper subunit–subunit association is facilitated by a chaperonin (Goloubinoff *et al.* 1989).

(4) Regulation of enzyme activity—a complicated hierarchy for *in vivo* regulation of the carboxylase has been described. At the molecular level, an obligatory activation step entails the reaction of $CO_2$ with a particular lysyl $\epsilon$-amino group to form a carbamate, which is stabilized by $Mg^{2+}$ (Miziorko 1979; Lorimer 1981). Unless it is carbamylated, the protein is catalytically incompetent as a carboxylase or as an oxygenase. Although activation is non-catalytic, the equilibrium favours the non-activated form of the enzyme under physiological conditions of pH, $[CO_2]$, and $[Mg^{2+}]$. Carbamylation *in vivo*

is facilitated by another protein—activase. Although the mechanism of action remains obscure, the process requires ATP, thereby partially explaining light activation of the carboxylase (Portis 1990). An independent mode of light regulation is mediated by the thylakoid membrane (Campbell and Ogren 1990). In some species of plants, the activation state of the carboxylase is modulated by a low molecular weight metabolite, 2-carboxyarabinitol-1-phosphate, whose concentration is light-dependent (Gutteridge *et al.* 1986; Berry *et al.* 1987). A specific phosphatase for removing this metabolite has recently been identified (Gutteridge and Julien 1989; Seemann *et al.* 1990).

(5) Post-translational modification—phosphorylation, acetylation, methylation and transglutamination of the carboxylase have all been observed, but the physiological significance of these modifications has not been elucidated (Soll and Buchanan 1983; Mulligan *et al.* 1988; Houtz *et al.* 1989; Margosiak *et al.* 1990).

(6) Protein–protein interactions—the carboxylase provides a model for studying protein–protein interactions within two contexts. First, dimerization of two large subunits (chloroplastic encoded) is required to form a catalytically functional active site, as this comprises interacting domains on adjacent subunits (Larimer *et al.* 1987; Chapman *et al.* 1988; Andersson *et al.* 1989). Secondly, association of small

subunits with large is required for proper conformation of the active site and optimization of catalytic activity (Andrews 1988).

(7) Enzyme mechanisms—studies of the carboxylase mechanism may ultimately enhance general understanding of enzyme catalysed reactions. The overall carboxylase reaction entails diverse chemistry: proton abstraction, enolization, addition of electrophile to enediol, hydration of ketone, carbon–carbon scission, inversion of configuration of carbanion and protonation of carbanion. In comparison to most enzyme catalysed reactions, the $k_{cat}$ for carboxylation is very slow—only $2$–$5\,s^{-1}$; factors constraining this rate are unclear. As an oxygenase, the enzyme is unusual in its lack of dependence on cofactors typically associated with oxygenases—flavin, iron or cobalt.

(8) Protein structure—elucidation of the high resolution structure of the carboxylase by X-ray diffraction shows that most of the active site is contained within an $\alpha/\beta$-barrel region (Andersson *et al.* 1989). The diversity of catalysts that can be elaborated from this prevalent tertiary motif is further illuminated. As more structural information depicting conformational differences between the homodimeric enzyme (large subunits only) from *R. rubrum* and the heterohexadecameric enzyme (large and small subunits) from plants becomes available, a better appreciation of long distance communication between subunits should be gained.

Three excellent, current reviews of the carboxylase literature, which include most of those topics just mentioned, are available, and no useful purpose would be served to mimic them here (Andrews and Lorimer 1987; Incharoensakdi and Takabe 1987; Gutteridge 1990). Rather, the present article will concentrate on recent structural and functional insights about Rubisco that have been provided by site-directed mutagenesis. Since most of the mutagenesis work was conceptualized and executed prior to the elucidation of the three-dimensional structure of the carboxylase, an assessment of the compatibility of conclusions drawn from the two approaches serves to define the kinds of issues that can be addressed by mutagenesis in the absence of crystallographic data.

## Random mutagenesis as an approach to a superior Rubisco

As all investigators in the carboxylase field are tantalized, stimulated and prodded by the prospects of enhancing biomass yields through manipulation of Rubisco, a brief look at relevant examples of random mutagenesis will precede coverage of site-directed mutagenesis.

The primary goals of most site-directed mutagenesis heretofore directed towards the carboxylase have been to address questions about the reaction mechanism or assembly (subunit–subunit interactions); enhancement of the carboxylase/oxygenase activity ratio has not been an overtly stated goal because few clues are available to guide the choice of amino acid substitutions to be incorporated in order to attain such a goal. In contrast, the general genetic approach of searching for mutants by phenotypic selection of desired traits has been unabashedly applied to the challenge of eliciting a superior carboxylase. An oxygen-sensitive mutant of the C3 plant *Arabidopsis thaliana* provided a basis for selection of altered carboxylase (Somerville and Ogren 1979). The mutant is deficient in phosphoglycolate phosphatase; thus, the oxygenase activity of Rubisco leads to a build-up of phosphoglycolate, which inhibits carbon metabolism (probably at the level of triosephosphate isomerase). Any carboxylase mutant with lowered oxygenase activity should exhibit less sensitivity to $O_2$; regrettably, mutants with the desired properties were not found. The *Arabidopsis* system has nevertheless been invaluable in that a nuclear gene mutant (*rca*), deficient in carboxylase activity, was uncovered during the search for photorespiratory mutants. The *rca* mutant expressed wild type Rubisco that was unable to undergo activation in the light at ambient concentrations of $CO_2$ (Somerville *et al.* 1982). It was this mutant that led to the discovery of activase (Salvucci *et al.* 1985), which may serve as a potential focal point for enhancing biomass yields via more efficient activation of Rubisco.

Random mutagenesis *in vivo* as an avenue for generating Rubisco with altered specificity has been skilfully pursued with the green alga *Chlamydomonas reinhardtii* (Spreitzer and Ogren 1985). These investigations are especially significant because they provide the first examples of alteration of the carboxylase/oxygenase activity ratio through induced changes in covalent structure and thereby offer encouragement that improvement in the enzyme can be eventually effected. The kinetic characteristics of the enzyme, which determine relative rates of carboxylation and oxygenation ($v_c/v_o$), can best be appreciated as a ratio of the catalytic efficiencies ($V_{max}/K_m$) for the two activities ($V_c/K_c)/(V_o/K_o)$. This latter term defines the substrate specificity factor, denoted as the constant $\tau$ (Laing *et al.* 1975) and readily allows $v_c/v_o$ to be expressed as a function of the relative concentrations of the two gaseous substrates: $v_c/v_o = \tau \cdot ([CO_2]/[O_2])$.

*C. reinhardtii* is an attractive organism for mutagenesis for several reasons. As a unicellular eukaryote, it elaborates an $L_8S_8$ carboxylase that can be viewed as a prototype for the higher plant enzyme. Furthermore, it contains a single chloroplast whose genome can be rendered essentially haploid by treatment with 5-fluorodeoxyuridine; recessive chloroplast mutants can then be generated by standard techniques of mutagenesis. In the presence of acetate as a carbon source, the photosynthetic apparatus is retained even in the dark, so that photosynthetic-deficient mutants (including those with a deficient Rubisco) can be maintained as acetate-requiring strains.

A number of temperature-sensitive, photosynthetic-deficient mutants of *C. reinhardtii* 2137 $mt^+$, produced by mutagenesis with methyl methanesulphonate, have been identified and partially characterized. Two of these, denoted 68-4PP and 45-3B, have carboxylases of altered specificities (Chen *et al.* 1988; Chen and Spreitzer, 1989). The $\tau$-value ($V_c K_o/V_o K_c$) of the carboxylase purified from 68-4PP is about 15% lower than that for the wild type enzyme (54 versus 62). DNA sequencing identifies substitution of Leu290 by Phe as the basis for the reduced $\tau$-value. Interestingly, $K_o/K_c$ of the mutant was increased, which would argue for a beneficial increase in $\tau$, but this was more than offset by an unfavourable decrease in $V_c/V_o$. The temperature-sensitive phenotype of 68-4PP was attributed to low levels of the carboxylase protein due to holoenzyme instability. Structural and functional consequences of amino acid substitutions at 290 is not surprising because neighbouring residues have been assigned to the active site; these include Arg292, which interacts with the C5 phosphate of bound ribulose-P$_2$ and His298 of unspecified function.

The $\tau$-value of the carboxylase isolated from mutant 45-3B is reduced by 37% relative to that of wild type protein. Here again, a potentially favourable increase in $K_o/K_c$ is negated by an even greater relative decrease in $V_c/V_o$. Also, in analogy with the 68-4PP mutant, the altered kinetic parameters of the carboxylase from 45-3B reflect an amino acid substitution in the active-site region—Val331 is replaced by Ala. The key role of Lys334 at the active site will be thoroughly discussed. A second site revertant (R40-9D) was identified, in which the consequences of Val331→Ala were partially compensated by Thr342→Ile. The kinetic parameters of the wild type, 68-4PP, 45-3B and R40-9D carboxylase are shown in Table 4.1.

Quite unexpectedly, a revertant phenotype of the 68-4PP mutant has been found to reflect a mutation of a nuclear gene (Chen *et al.* 1990). Even more astonishingly, the suppression does not involve mutation of either small subunit gene (*rbc* S1 and *rbc* S2) present in *C. reinhardtii*. Both genes were cloned and completely sequenced, yet no differences were observed between them and their wild type counterparts. The carboxylase isolated from the suppressor strain exhibits enhanced $V_c$ and $\tau$ values relative to those of the 68-4PP carboxylase. One then faces the dilemma of two purified carboxylases with distinctly different kinetic parameters but without differences in primary structure. The authors speculate that differential expression of the two small subunit genes, differences in unspecified post-translational modification, or alterations in subunit assembly could explain this particular suppressor mutation.

As well as illustrating the feasibility of altering the specificity of Rubisco by random mutagenesis, the studies with *C. reinhardtii* direct attention towards amino acid residues, for future systematic replacement by site-directed mutagenesis, that would not have been likely to be examined based solely on considerations of the three-dimensional structure.

**Table 4.1**   Kinetic parameters of wild type and mutant carboxylases of *Chlamydomonas*

| Kinetic parameter | Wild type | 68-4PP | 45-3B | R40-9D |
|---|---|---|---|---|
| $V_c K_o/V_o K_c$ | $62 \pm 3$ | $54 \pm 2$ | $39 \pm 1$ | $52 \pm 3$ |
| $K_c$ ($\mu$M $CO_2$) | $31 \pm 2$ | $37 \pm 2$ | $166 \pm 19$ | $47 \pm 8$ |
| $K_o$ ($\mu$M $O_2$) | $405 \pm 9$ | $664 \pm 26$ | $3198 \pm 280$ | $903 \pm 80$ |
| $V_c$ ($\mu$mol/h/mg protein) | $115 \pm 4$ | — | $8 \pm 1$ | $38 \pm 2$ |
| $K_o/K_c$ | 13 | 18 | 19 | 19 |
| $V_c/V_o$ | 4.8 | 3 | 2.1 | 2.7 |

To increase the sheer number of mutants of the carboxylase that can be generated and screened *in vivo*, Pierce (1988) has argued the merits of transformation systems. Accordingly, *Synechocystis* 6803, a readily transformable cyanobacterium that can be grown both photoautotrophically and heterotrophically with glucose as a carbon source, has been tailored to serve as a selection system for altered carboxylases (Pierce *et al.* 1989). Through the use of plasmids, the approach was to replace the *Synechocystis* gene for carboxylase with a foreign one by insertional inactivation. The gene chosen for introduction was that of *Rhodospirillum rubrum*. Reasons for this choice included the fact that the *R. rubrum* carboxylase has the lowest specificity for $CO_2$ yet observed (a $\tau$ value of 10 in comparison to 80 for the higher plant enzyme, Jordan and Ogren 1981), so relative ease of improvement might be predicted. Also, the *R. rubrum* enzyme has not likely been subjected to environmental selection for improved $CO_2$ specificity, because the organism is an anaerobic photoautotroph.

Biochemical characterization confirmed that the transformant, denoted 'cyanorubrum', produced *R. rubrum* but not *Synechocystis* carboxylase. As predicted from the relative $V_{\mathrm{max}}$ values of the two pure enzymes, the growth rate of the transformant was only half that of the wild type cyanobacterium, even in an atmosphere of 5% $CO_2$/0% $O_2$. Far more dramatic, the growth rate of the latter remained the same at 0.1% $CO_2$/21% $O_2$, conditions under which cyanorubrum was unable to grow at all. Photoautotrophic growth in air was restored by transforming the cyanorubrum mutant with a plasmid carrying the wild type *Synechocystis* gene for carboxylase large subunit.

In the future, the tailored organism might be used directly for *in vivo* mutagenesis and subsequent selection for *R. rubrum* carboxylase mutants with decreased sensitivity to oxygen. Alternatively, carboxylase genes from any source could be subjected to saturation mutagenesis *in vitro* and then used to transform the cyanorubrum mutant; transformants with enhanced growth characteristics would be selected. The cyanorubrum mutant could also prove useful for *in vivo* evaluation of any site-directed mutant carboxylase gene that is concluded to be superior on the basis of kinetic properties of the encoded protein first characterized *in vitro*. A similar experimental system based on *Alcaligenes eutrophus* is under development (Andersen and Wilke-Douglas 1984; personal communication cited in Pierce 1988).

Exploitation of these systems is anxiously awaited.

## General considerations

As with any enzyme, studies of the structure, chemistry, kinetics and reaction pathway of Rubisco are intended ultimately to elucidate a detailed mechanism of action. In turn, the carboxylase/oxygenase mechanism may reveal the feasibility of structural intervention as a route toward optimization of the enzyme's specificity for $CO_2$. An important aspect of establishing structure–function correlations and elucidation of reaction mechanisms of enzymes is active-site characterization. Unquestionably, the most powerful tool for active-site characterization is X-ray crystallography, because it allows direct visualization of features at the atomic level provided that the enzyme–substrate or appropriate enzyme–inhibitor complexes can be crystallized. In those cases in which only non-liganded protein can be crystallized, localization of the active site is frequently dependent upon guidance provided by chemical studies.

Historically, chemical modification has proved exceptionally versatile in labelling and thereby identifying active-site residues. A specialized version of chemical modification, affinity labelling, greatly enhances selectivity of modification of a given site and thereby simplifies interpretation of observed consequences. This method entails incorporating into a protein reagent, via chemical synthesis, structural features of a substrate that endow the reagent with affinity for the substrate binding site. Affinity labelling of Rubisco was instrumental in identifying active-site regions, which have become guideposts for site-directed mutagenesis studies (Hartman *et al.* 1986).

Site-directed mutagenesis has revolutionized or supplanted chemical modification (depending on one's perspective or bias) as a means of testing and establishing structure–function relationships (Leatherbarrow and Fersht 1986; Gerlt 1987; Knowles 1987; Shaw 1987). In the long term site-directed mutagenesis offers prospects for customizing the specificities (Higaki *et al.* 1989; Scrutton *et al.* 1990), efficiencies (Blacklow and Knowles 1990) and stabilities (Mitchinson and Wells 1989) of enzymes for particular applications. It is entirely appropriate to consider site-directed mutagenesis as an extension of chemical modification because both procedures result in known changes in covalent structure of proteins. However, mutagenesis is especially attractive because it enables systematic alteration of protein structure (amino acid sequence) with absolute specificity for any preselected site, thereby overcoming two major weaknesses of traditional chemical modification. Additional advantages include the completeness of the structural change (disregarding rare errors in translation, every protein molecule expressed from the mutant gene will contain the intended change) and the variety of structural changes that can be made routinely (any amino acid residue can be replaced by any of the other 19 so that effects of side chains that are isosteric, larger, smaller, oppositely charged, more or less hydrophobic, more or less hydrophilic, or more or less nucleophilic can be examined).

It is even possible to incorporate unnatural amino acids and thereby extend the diversity of functional groups in mutant proteins. For example, by resorting to an *in vitro*, cell-free transcription/translation system, a suppressor tRNA, acylated with the unnatural amino acid, can deliver that unnatural amino acid to any predetermined position within a protein provided that the normal codon is replaced by a suitable amber codon. In this way, *p*-fluorophenylalanine, *p*-nitrophenylalanine and homophenylalanine were substituted for Phe66 of β-lactamase (Anthony-Cahill *et al.* 1989). Chemical modification, in conjunction with site-directed mutagenesis, can also expand the availability of novel proteins (Profy and Schimmel 1988); application to Rubisco is discussed later in this chapter.

A distinction of direct from indirect effects of structural alterations, whether introduced by chemical modification or mutagenesis, is required to interpret their functional consequences. Specifically, is the impaired function (e.g. diminution of catalytic activity or weakened ligand binding) reflective of alteration or substitution of a residue intimately required for that function, or is it reflective of an induced conformational change? Many would consider conformational changes to be the 'Achilles' heel' of chemical modification. Although the problem is less worrisome with mutagenesis because of specificity and subtlety of the 'modification', extreme caution should be exercised in drawing conclusions until the three-dimensional structures of both wild type and mutant protein have been determined. Although

the three-dimensional structure of Rubisco has been elucidated, none of the many purified mutants have been characterized by crystallography. In the cases of staphylococcal nuclease (Hibler *et al.* 1987) and tyrosyl-tRNA synthetase (Fersht *et al.* 1987), some detrimental effects of single amino acid substitutions are due to conformational perturbations. Particularly disconcerting are those that may be so subtle as to elude detection by X-ray crystallography and be seen only by comparative analyses of free-energy relationships (Lowe *et al.* 1987).

Despite the perceived quantitative nature of modification of protein structure by site-directed mutagenesis as noted above, the lack of absolute fidelity in translation can lead to erroneous conclusions. Schimmel (1989) has cautioned that very low levels of apparent activity ($\leq 0.1\%$) associated with 'mutant enzymes' might reflect contamination by wild type enzyme, because error rates in translation of $10^{-3}$ are not uncommon. He offers a straightforward way to address this issue that does not require tedious isolation of a putative minor species or meticulous determination of kinetic parameters; it takes advantage of the degeneracy of the genetic code and replaces the pertinent codon with another one that codes for the same amino acid. It is highly unlikely that two distinct codons would be mistranslated to the same extent. That this consideration is not merely mental gymnastics is vividly illustrated by mutagenesis studies of β-lactamase. Replacement of active-site Ser68 (encoded by AGC) by glycine (encoded by GGC) leads to β-lactamase with 0.1% the activity of wild type (Toth *et al.* 1988). However, when the same glycyl residue is incorporated via a GGA codon, the mutant protein is totally devoid of lactamase activity (Schimmel 1989). Hence, the initially observed low level of activity can only be explained by expression of wild type enzyme.

## Construction of site-directed mutants

A detailed description of the history and techniques for the construction of site-directed mutant proteins is beyond the scope of this brief article and the reader is referred to superb accounts elsewhere (Kramer and Fritz 1987; Zoller and Smith 1987). Generally speaking, there are two main approaches: primer extension utilizing a single-stranded template into which the target gene has been subcloned, and cassette mutagenesis utilizing a double-stranded plasmid containing the gene of interest. The former method, pioneered by Zoller and Smith (1983), is universally applicable and has armed standard enzymology laboratories, including those that lack any particular skills in molecular biology, with the weaponry of recombinant DNA technology to alter protein structure by design at will. Typically, the gene of known sequence to be mutated is subcloned into the replicative form of the bacteriophage M13 genome. The single-stranded DNA isolated from the bacteriophage then serves as the template for mutagenesis. A synthetic oligonucleotide, which is complementary to a region of the template but contains the desired codon change (as one or two mismatched bases), is used to prime replication of the template. The closed circular heteroduplex is then used to transfect a suitable strain of *Escherichia coli*, and the resulting plaques are screened for mutant DNA. The mutant yield can be greatly enhanced by using a template that contains dU rather than dT (synthesized in a host that is deficient in dUTPase and uracil glycosylase); the parental strand (template) of the heteroduplex will be preferentially degraded in a normal *E. coli* host and hence not propagated to produce wild type plaques (Kunkel *et al.* 1987).

Single primer extension has been used primarily to construct mutants with single amino acid substitutions. Cassette mutagenesis is more versatile and rapid for the replacement of a segment of polypeptide or for multiple substitutions of one or more residues (Lo *et al.* 1984; Kramer and Fritz 1987). The process entails removal of a double-stranded segment of plasmid DNA, which encompasses the region of the cloned gene to be altered, and ligation of a synthetic duplex that encodes the desired information. A drawback is that convenient restriction sites for effecting the exchange of segments may not pre-exist in the cloned gene; they then must be introduced via silent mutations generated by primer extension in a single-stranded vector.

A schematic representation of the construction of site-directed mutants of Rubisco by the primer extension method is illustrated in Figure 4.2.

# Features of Rubisco pertinent to site-directed mutagenesis

Nature elaborates two architecturally distinct forms of Rubisco. The most prevalent is a hexa-

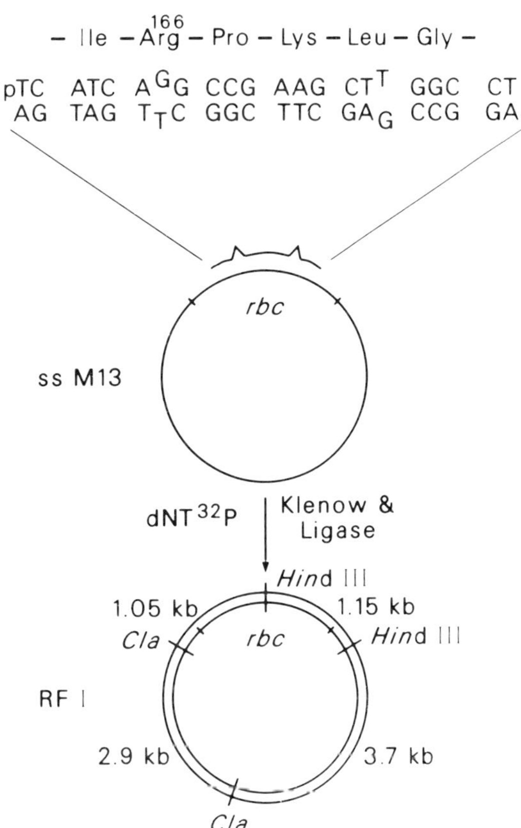



**Fig. 4.2**   Oligonucleotide-directed mutagenesis of the carboxylase gene. The synthetic oligonucleotide shown introduces two mutations. One that alters the codon for Lys166 to a codon for arginine and the other (a silent mutation) that creates a new restriction site and thereby facilitates subsequent plaque screening. The newly-introduced restriction site is the *Hind*III site within the carboxylase gene; the other *Hind*III and *Cla* sites are present in the original M13 vector.

decamer ($L_8S_8$) comprising eight subunits of each of two gene products. Large subunit (approximately 55 kDa) is encoded by the chloroplast genome and provides the catalytic site. Small subunit (approximately 15 kDa) is nuclear encoded. Although the small subunit does not contribute directly to the catalytic site, octamers of large subunit are severely deficient catalytically (Andrews 1988). The carboxylase from *Chromatium vinosum* appears to be an exception, and an active $L_8$ form has been reported (Torres-Ruiz and McFadden 1985). The other form of the carboxylase is a homodimer of large subunits (approximately 50 kDa), apparently restricted to purple, non-sulphur, photosynthetic bacteria like *Rhodospirillum rubrum* (Tabita and McFadden 1974; Schloss *et al.* 1982). Despite these striking differences in quaternary structure, and only 30% overall sequence identities between the *R. rubrum* subunit and the large subunit of the $L_8S_8$ enzyme (Hartman *et al.* 1984; Nargang *et al.* 1984), both types require carbamylation to become catalytically competent, possess inherent oxygenase activity, utilize the same reaction pathway and exhibit qualitatively similar ligand binding patterns. Furthermore, those residues for which sound chemical evidence has invoked functional importance are species invariant. Thus, the structurally simpler carboxylase from *R. rubrum* is a perfectly valid prototype for studies of mechanism of action.

The first cloned carboxylase gene to be expressed as a functional protein was that from *R. rubrum* (Somerville and Somerville 1984). Although the gene product was a fusion protein (containing a 25-residue $NH_2$-terminal appendage derived from β-galactosidase), its kinetic parameters were indistinguishable from those of wild type enzyme. Furthermore, it is the fusion protein whose three-dimensional structure has been determined at high resolution. The appendage, not seen in the electron density map, is apparently unstructured and without consequence to the folding and function of the carboxylase. Prior to the crystallographic analysis, concerns about possible synergism between the 'foreign' appendage and structural perturbations introduced by site-directed mutagenesis prompted reconstruction of the original clone to one that expressed wild type enzyme (Larimer *et al.* 1986). Mutants of *R. rubrum* carboxylase have been generated from the fusion gene and two independent reconstructions that eliminate the appendage (Somerville and Somerville 1984; Larimer *et al.* 1986; Gutteridge *et al.* 1988).

The *R. rubrum* enzyme is particularly convenient when issues of active-site features and mechanisms by site-directed mutagenesis are being addressed because it only requires a single gene product. Obviously, if the goals are to unravel the role(s) of the small subunit and to understand folding and assembly pathways, one of the $L_8S_8$ enzymes must be chosen. Development of efficient expression systems for the $L_8S_8$ enzyme lagged because of the more stringent requirements for cosynthesis and assembly of two gene products. However, high yield expression of active $L_8S_8$ enzyme of several origins has been achieved so that mutagenesis can be carried out with either the structurally simpler or more complex carboxylase (reviewed by McFadden and Small 1988).

A consideration of the intricate details of the carboxylation reaction pathway (Schloss 1990) (Fig. 4.3) that have emerged illustrates some of the unresolved mechanistic questions potentially approachable by site-directed mutagenesis. Three acid–base groups provided by amino acid side chains and a divalent cation are shown as the minimal requirements for catalysing the formation of two equivalents of D-3-phosphoglycerate from ribulose-$P_2$ and $CO_2$. In addition to polarizing the carbonyl group of ribulose-$P_2$ and coordinating with the carboxylate group of the six-carbon reaction intermediate (VIa), the $Mg^{2+}$ cation stabilizes the protein carbamate by direct coordination with its oxyanion (not shown in the figure). NMR and EPR studies (Miziorko and Sealy 1984; Styring and Brändén 1985; Pierce and Reddy 1986) correctly deduced some of the metal ion ligands, which were later seen directly by crystallography (Andersson *et al.* 1989). The identities of the three acid–base groups at the active site have been explored, in part, by site-directed mutagenesis.

The extent and diversity of chemistry required for substrate transformation as catalysed by the enzyme is rather staggering: deprotonation and enolization (intermediates II, III and IV), carbanion development and stabilization (intermediate V), carboxylation (or oxygenation, which is not shown) of the C2 carbanion to give the six-carbon intermediate VIa $\equiv$ VIb (a hydroperoxide is the presumed counterpart during oxygenation), hydration, carbon–carbon scission and stereospecific protonation of a carbanion with inversion of configuration. Perhaps these extreme demands placed on the enzyme for stabilizing multiple transition states account for its exceedingly low $k_{cat}$ of only 2–5 s$^{-1}$.

Experimental data support each of the discrete steps outlined in Fig. 4.3. The initial proton abstraction at C3 with concomitant formation of a carbanion was demonstrated by the exchange of the C3 proton (labelled with tritium or deuterium) with solvent (Saver and Knowles 1982) and the occurrence of deuterium and tritium isotope effects (Fiedler *et al.* 1967; Saver and Knowles 1982; Sue and Knowles 1982; Van Dyk and Schloss 1986). The enediols were detected as acid-labile phosphate (Jaworowski and Rose 1985) and as a monophosphate ester (Schloss 1990) in rapid quench experiments. Addition of $CO_2$ to C2 of ribulose-$P_2$ implied a C2 carbanion (Müllhofer and Rose 1965), as did a carbon isotope effect (Roeske and O'Leary 1985). Rapid quench experiments also indicated tautomerization of the enediolates to the C2 carb-

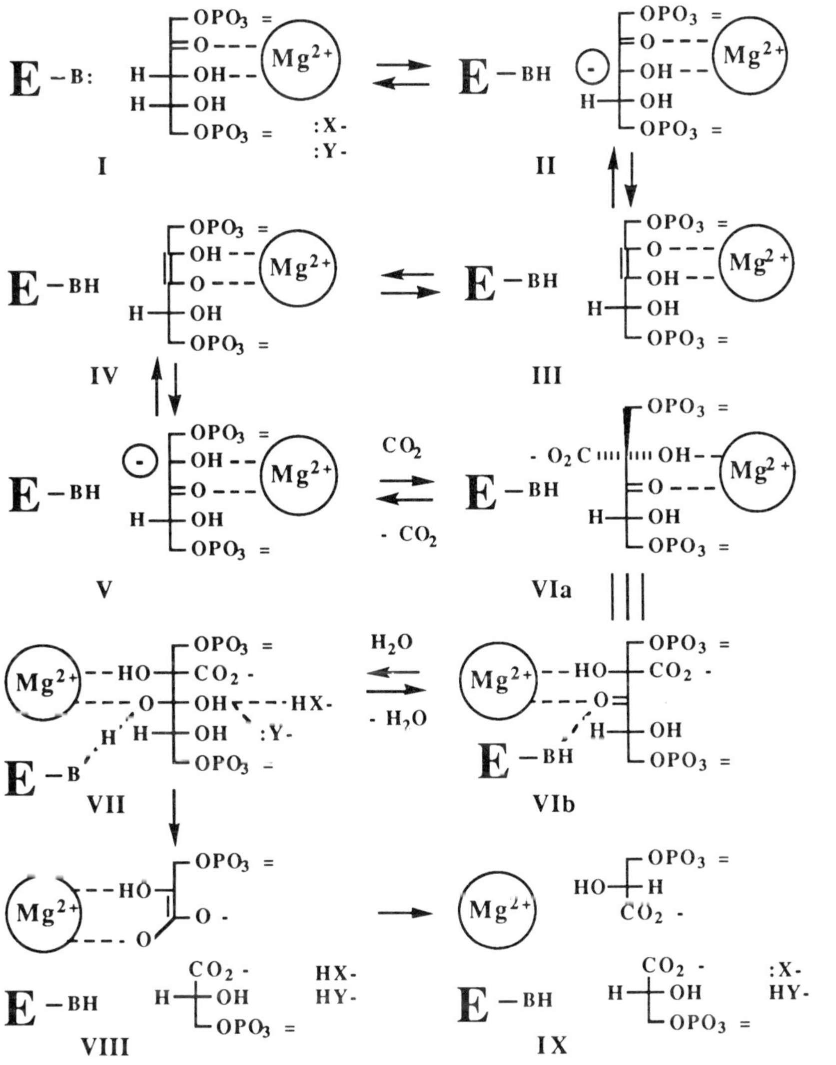

**Fig. 4.3** Reaction pathway for the carboxylase reaction. Reproduced from Schloss (1990) with permission of author and publisher.

anion to be a discrete, partially rate-limiting step, as the rate of carboxylation of enediol exhibited saturation with respect to concentration of $CO_2$ (Schloss 1990). The six-carbon reaction intermediate (VIa, VIb) was first identified by rapid, chemical quench with sodium borohydride to yield carboxyarabinitol-$P_2$ (Schloss and Lorimer 1982). Based on distribution of isotopic label, provided as [3-$^{18}$O]-ribulose-$P_2$, between borohydride-reduced six-carbon intermediate and solvent after rapid quench, it was concluded that the enzyme-bound intermediate must exist predominantly in the hydrated form (VII). In fact, Cleland (1990) argues that the enediol (III) is attacked in concerted fashion by water at C3 and $CO_2$ at C2 so that the free ketone is not an enzyme-bound intermediate. Carbon–carbon scission and concomitant release of phosphoglycerate derived from C3, C4 and C5 of ribulose-$P_2$ leaves a C2 carbanion of phosphoglycerate at the active site, as evidenced by the stereospecific incorporation of solvent hydrogen at C2 of phosphoglycerate derived from $CO_2$ and C1 and C2 of ribulose-$P_2$ (Fiedler *et al.* 1967; Sue and Knowles 1982; Pierce *et al.* 1986b; Gutteridge *et al.* 1984a).

In conjunction with X-ray crystallography, site-directed mutagenesis may reveal at which step in the overall process each active-site residue participates and to what extent that residue facilitates the relevant step(s).

## Guidance in selection of mutants for construction

Ideally, the three-dimensional structure and atomic coordinates of a given protein will be available prior to initiation of mutagenesis studies. In this circumstance, very intricate and quantitative questions can be addressed, such as the precise contribution (in kJ/mol) of a particular hydrogen bond within the enzyme–substrate complex, to stabilization of the transition state. The pioneering studies of Fersht *et al.* (1988) with tyrosyl-tRNA synthetase exemplify the wealth of mechanistic information that can be gleaned by site-directed mutagenesis when guided by high resolution structures.

In stark contrast to the ideal situation, most of the site-directed mutagenesis of Rubisco undertaken thus far proceeded in the absence of meaningful high resolution data; chemical modification and comparative amino acid sequences have been the primary guides.

Chemical modification has revealed five disparate segments of primary structure that constitute the active site. The amino acid sequence of each segment is highly conserved among evolutionarily diverse carboxylases, suggesting a functional relevance of the residues that are subject to labelling. In this connection, the homodimeric carboxylase from *R. rubrum* provides a stringent test for sequence conservation as only 30% of its residues are identical to those of the catalytic subunit of the heterohexadecameric carboxylase.

One of the active-site segments includes the lysyl residue, the $\epsilon$-amino group of which reacts with $CO_2$ to form a carbamate, that is obligatory for enzyme activation (Lorimer and Miziorko 1980). Although the carbamate is stabilized by the essential divalent cation ($Mg^{2+}$), it remains too labile to permit direct characterization. However, the carbamylated enzyme forms an exchange–inert complex with the reaction-intermediate analogue carboxyarabinitol-$P_2$ (Miziorko and Sealy 1980). Treatment of the complex with diazomethane results in esterification of the carbamate to form a stable derivative (Lorimer 1981). Although the esterification is not specific, only one site becomes radioactive, provided that the complex has been prepared with $^{14}CO_2$. Subsequent characterization of labelled peptides in proteolytic digests reveals that the carbamate-forming lysyl residue occupies position 201 in the spinach enzyme (Lorimer 1981) and position 191 in the *R. rubrum* enzyme

(Donnelly *et al.* 1983):

|  | 201 |
| --- | --- |
| (spinach) | D–F–T–K–D–D–E |

|  | 191 |
| --- | --- |
| (*R. rubrum*) | D–F–I–K–N–D–E |

Clearly, one function of the carbamate is to coordinate the $Mg^{2+}$ ion required for catalysis; however, additional functions are likely, as will be discussed later.

Collective data provided by several different affinity labels (Fig. 4.4) have revealed the presence of two additional lysyl residues at the active site (for a review, see Hartman *et al.* 1986). These lysines occupy positions 175 and 334 of the large subunit in the heterohexadecameric enzyme from spinach; and in the homodimeric enzyme from *R. rubrum*, they occupy positions 166 and 329:

|  | 175 |
| --- | --- |
| (spinach) | T–I–K–P–K–L–G–L–S |

|  | 166 |
| --- | --- |
| (*R. rubrum*) | I–I–K–P–K–L–G–L–R |

|  | 334 |
| --- | --- |
| (spinach) | G–K–L–E–G–E |

|  | 329 |
| --- | --- |
| (*R. rubrum*) | G–K–M–E–G–E |

One of the affinity labels (Fig. 4.4) shows that the active-site domain also includes a region of the polypeptide near the $NH_2$-terminus. This reagent labels both His44 and Cys58 of the *R. rubrum* enzyme in a mutually exclusive fashion (Herndon and Hartman 1984); neither residue can be essential because they are not species invariant:

(spinach)

56                                    70
G–A–A–V–A–A–E–S–S–T–G–T–W–T–T–V–W–T

(*R. rubrum*)

44                                    58
A–A–H–F–A–A–E–S–S–T–G–T–N–V–E–V–C–T

However, the high degree of homology within this region indicates a prominent role for one or more of the conserved residues.

Selective modification of His298 of the spinach carboxylase by diethylpyrocarbonate provides the basis for assigning a fifth segment of polypeptide to the active site (Igarashi *et al.* 1985):

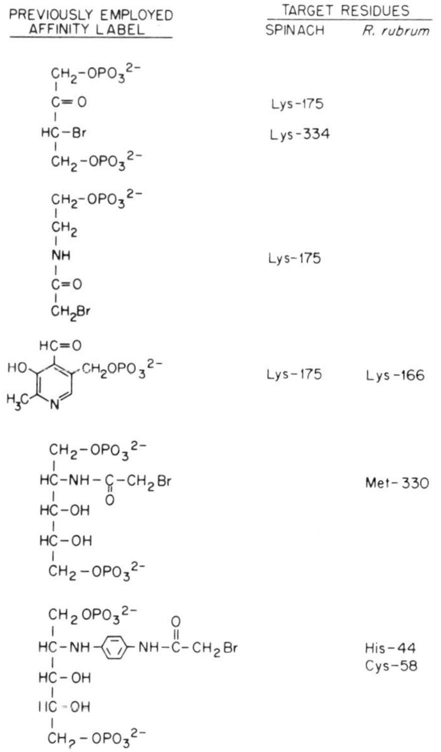

PREVIOUSLY EMPLOYED
AFFINITY LABEL
TARGET RESIDUES
SPINACH    R. rubrum
CH₂-OPO₃²⁻
C=O
HC-Br
CH₂-OPO₃²⁻
Lys-175
Lys-334
CH₂-OPO₃²⁻
CH₂
NH
C=O
CH₂Br
Lys-175
HC=O
HO    CH₂OPO₃²⁻
H₃C    N
Lys-175    Lys-166
CH₂-OPO₃²⁻
HC-NH-C-CH₂Br
HC-OH    O
HC-OH
CH₂-OPO₃²⁻
Met-330
CH₂OPO₃²⁻
HC-NH    NH-C-CH₂Br
HC-OH    O
HC-OH
CH₂-OPO₃²⁻
His-44
Cys-58

**Fig. 4.4** Affinity labels for Rubisco and the sites of modification.

| | 298 |
|---|---|
| (spinach) | L–H–I–H–R–A–M–II–A–V I |
| | 291 |
| (*R. rubrum*) | L–H–Y–H–R–A–G–H–G–A–V–T |

The majority of single amino acid substitutions effected by mutagenesis involve these five segments.

Close scrutiny of Lys166, Lys329, Glu48 and His291 of the *R. rubrum* carboxylase was justified by additional observations. These included extreme acidity and enhanced nucleophilicities of the ε-amino groups of Lys166 and Lys329, intrasubunit proximity of these two ε-amino groups, intersubunit proximity of the Lys166 ε-amino group and the NH₂ terminal segment of the polypeptide that encompasses Glu48 and kinetic data that were consistent with either Lys166 or His291 serving as the base that enolizes ribulose-P₂.

Trinitrobenzenesulphonate arylates Lys166 of the *R. rubrum* enzyme and Lys334 of the spinach enzyme with a very high degree of selectivity (Hartman *et al.* 1985). Based on the pH-dependence of inactivation, a single reagent,

applied to the carboxylase from two different organisms, can provide the $pK_a$ values and the intrinsic reactivities ($k_o$) for two different active-site lysines (Table 4.2). Especially noteworthy are the enhanced nucleophilicities of the two protein ε-amino groups compared to acetyllysine (note deviation from the Bronsted relationship), despite their stronger acidities. These unusual properties are highly suggestive of catalytic functionality of the active-site lysines. His298 of the spinach carboxylase (which corresponds to His291 in the *R. rubrum* enzyme) exhibits a $pK_a$ of 6.8 on the basis of the pH-dependence of inactivation by diethylpyrocarbonate (Paech 1985).

Both His291 and Lys166 have been proposed as the base that initiates catalysis by promoting the enolization of ribulose-P₂. Irrespective of residue assignment, this essential base exhibits a $pK_a$ of 7.5 based on the pH-dependence of the deuterium isotope effect with [3-²H]-ribulose-P₂ as substrate (Van Dyk and Schloss 1986). Judged by the insensitivity of its $pK_a$ to the dielectric constant of the solvent, the base could be either a lysyl or histidyl residue. As these data do not allow a distinction to be made between His291 or Lys166 as the essential base, a resolution has been sought by site-directed mutagenesis.

In addressing the question of a role in substrate binding versus a role in catalysis for the active-site Lys166 and Lys329, several observations argue against the former. The enhanced nucleophilicities of both residues are more consistent with a catalytic involvement than with a function in binding. The incomplete protection of both enzymes against inactivation by saturating levels of carboxyribitol bisphosphate (a simple competitive inhibitor) is inconsistent with those lysine residues, which are targets for arylation, forming salt linkages with phosphate groups of ribulose-P₂. The inactivated spinach carboxylase, in which Lys334 has been arylated, is still able to form the quaternary com-

**Table 4.2** pH dependencies of modifications of carboxylases by trinitrobenzenesulphonate

| | | $k_o$ (M⁻¹min⁻¹) | |
|---|---|---|---|
| Sample | $pK_a$ | Observed | Predicted (by Bronsted relationship) |
| *N*-α acetyllysine | 10.8 | 1250 | 1200 |
| Lys166 (*R. rubrum* carboxylase) | 7.0 | 670 | 120 |
| Lys334 (spinach carboxylase) | 9.0 | 1500 | 280 |

plex with $CO_2$, $Mg^{2+}$ and carboxyarabinitol-$P_2$. With respect to Lys166 in the *R. rubrum* carboxylase, its extreme acidity ($pK_a = 7.9$) appears incompatible with effective utilization as a phosphate-binding site.

Although the unusual acidity and nucleophilicity can account for the selectivity of the arylating reagent for Lys166 and Lys334, their similar reactivities at pH 8.0 raise the question as to why only one of them is modified in each species of carboxylase examined. The explanation offered before the three-dimensional structure appeared was that the two lysines are juxtaposed within the catalytic site so that derivatization of both is precluded on steric grounds. This postulate was confirmed (Lee *et al.* 1986) by formation of an intrasubunit cross-link between the respective ε-amino groups of Lys166 and Lys329 with 4,4′-diisothiocyano-2,2′-disulphonate stilbene, which spans 12 Å. Significantly, the cross-link is not formed in the absence of $CO_2$ and $Mg^{2+}$, conditions under which the deactivated enzyme prevails. Presumably the two lysines are further apart in the deactivated enzyme and a conformational change induced by carbamate formation is crucial to proper alignment of catalytic groups.

Another cross-linking reagent that proved exceedingly revealing is 4,4′-difluoro-3,3′-dinitro-diphenylsulphone, which spans 9 Å. In contrast to the results just described, this reagent forms a cross-link between subunits in which Lys166 and Cys58 (also a target of an affinity label) participate (Lee *et al.* 1987). This finding of an intersubunit contact in the vicinity of a well documented essential residue (Lys166) indicates that the active site is positioned at an interface between subunits. Confirmation of this supposition was provided by site-directed mutagenesis and by crystallography.

The reasons for inspecting Glu48 by site-directed mutagenesis were less compelling. Glu48 represents the only acid–base group, and hence is a candidate for participation in catalysis, within the conserved region flanked by His44 and Cys58, which are targets of an affinity label as noted earlier (see Fig. 4.4). Other residues that have been singled out for substitution without guidance from crystallography included Met330, a non-conserved residue that is selectively alkylated by an affinity label (Fraij and Hartman 1983), Asp188 due to its possible role in binding of $Mg^{2+}$ and Lys191 because of its role in activation.

Residues that have been replaced because of information provided by crystallography include Asp193, Glu194, His287, Lys168 and Arg295. Alterations of both the $NH_2$- and COOH-terminal regions have also been reported; site-directed mutagenesis studies of the small subunit of the $L_8S_8$ enzyme are also progressing rapidly.

From the foregoing, it should be evident that a number of issues can be approached, at least qualitatively, by site-directed mutagenesis even in the absence of three-dimensional structural information. These include, but are not necessarily limited to validation (or invalidation) of prior chemical modification deductions, location of the active site (inter- or intrasubunit) in multimeric proteins, identification of subunit–subunit interactions, function of active-site residues and assignment of active-site residues to particular partial reactions in an overall catalytic pathway. Furthermore, one may be able to ascertain if substrate specificity is sensitive to changes in the microenvironment of the active site.

## General strategies for distinguishing among potential causes of inactivation

Drastic effects are anticipated in those cases in which putative active-site residues are singled out for substitution; however, a protein may be rendered inactive in a variety of ways. Thus, a general challenge is to determine whether catalytic deficiencies of mutant proteins are due to improper folding of polypeptide, failure of subunits to associate, inability to undergo carbamylation as required for activation, failure to bind substrates, or loss of a group that participates in catalysis. The mutant proteins that fall into this latter class offer the greatest potential for extracting mechanistic information. This type of systematic characterization, which includes the prerequisite of protein purification, is tedious and laborious. Given these realities and inherent haste to publish, mechanistic inferences are sometimes drawn on the basis of superficial examination of an activity (or lack thereof) in a crude extract; such inferences may prove risky and not withstand the test of time.

If mutation of a single codon leads to substantially lower expression levels, in comparison to the wild type, improper folding of the polypeptide and consequential proteolysis is likely. In the case of mutants of the *R. rubrum* carboxylase, distinction of monomers (reflective of subunit–subunit association problems) from dimers merely requires semi-quantitative estimation of molecular weight by gel permeation chromatography or polyacrylamide gel electrophoresis under non-denaturing conditions. However, problems in assembly are far more difficult to discern with the $L_8S_8$ enzyme. The reaction-intermediate analogue, 2-carboxyarabinitol-$P_2$, provides an extremely convenient and powerful tool for ascertaining whether a catalytically inactive mutant can nevertheless form a carbamate and bind phosphorylated ligands. Only the activated form of the wild type enzyme binds the analogue with sufficient tenacity to form a stable quaternary complex composed of equimolar amounts of enzyme subunit, $CO_2$, $Mg^{2+}$ and carboxyarabinitol-$P_2$. This complex is readily isolated by gel filtration (Miziorko and Sealy 1980). Its stability is emphasized by a consideration of the exchange rates between bound ligand and excess free ligand added to the complex: $t_{1/2}$ (*R. rubrum* enzyme) = 20 h (Smith *et al.* 1988b); $t_{1/2}$ (plant enzyme) = 530 days (Schloss 1988). Because tight binding of each of the three ligands is dependent on the presence of the other two, mere demonstration of complex formation proves competence in activation chemistry and binding of phosphorylated ligands (ribulose-$P_2$ by inference). However, the kinetics of exchange must be determined in order to evaluate the relative affinities of mutant and wild type proteins for carboxyarabinitol-$P_2$. Failure to form a stable complex does not equate with failure of the mutant protein to recognize phosphoesters, because the tight binding of carboxyarabinitol-$P_2$ to form the exchange–inert complex is a two-step process— rapid, reversible binding followed by a slow conformational change that 'locks' the ligands in place. A mutant protein might bind carboxyarabinitol-$P_2$ but not undergo the subsequent conformational change.

Any catalytically deficient mutant (carrying a single amino acid substitution) that assembles properly, forms a carbamate and binds substrate is deduced to be deficient because of the absence of a key residue. However, the possibility that the detrimental consequences are due to a subtle, localized reorientation of catalytic groups cannot be rigorously excluded without high resolution X-ray crystallographic analysis. This shortcoming can be ameliorated by dissection of the carboxylase pathway into its component parts. Lorimer *et al.* (1987) predicted that mutant proteins devoid of overall carboxylation activity might be competent in catalysing some of the discrete partial reactions illus-

trated in Figure 4.3. Such mutants have been encountered, thereby minimizing concerns that deficiencies reflect substantial conformational perturbations. More importantly, an avenue is provided for pinpointing the specific step in the overall pathway at which an active-site residue participates.

The enolization of ribulose-$P_2$ is readily assayed on the basis of exchange of solvent protons with the C3 proton of ribulose-$P_2$ (Saver and Knowles, 1982; Sue and Knowles, 1982):

$$\text{Enz}-\text{B:} \begin{array}{c} R \\ | \\ C=O \\ | \\ H-C-OH \\ | \\ R' \end{array} \rightleftharpoons \underset{\text{E}-\text{BT}}{\overset{\text{E}-\text{BH}}{T_2O \; \big\| \; H_2O}} \quad \underset{R'}{\overset{R}{C}} \underset{\text{OH}}{\overset{O^-}{C}} \rightleftharpoons \text{Enz}-\text{B:} \begin{array}{c} R \\ | \\ C=O \\ | \\ T-C-OH \\ | \\ R' \end{array}$$

The six-carbon reaction intermediate (2-carboxy-3-ketoarabinitol-$P_2$ is sufficiently stable to be isolated and partially purified; synthesis entails rapid acid quench after mixing essentially equimolar amounts of ribulose-$P_2$ and the carboxylase in the presence of $^{14}CO_2$. Availability of the intermediate (labelled at the carboxyl group) allows its conversion to product (3 phosphoglycerate) and its decomposition, presumably to the enediol, to be monitored independently of overall carboxylation activity. The former is observed as an increase in acid-stable radioactivity, and the latter is equated with a decrease in radioactivity that can be stabilized by borohydride.

$$
\begin{array}{ccccc}
H_2C-OPO_3^{\ominus} & & H_2C-OPO_3^{\ominus} & & \\
| & & | & & \\
C-OH & & HO-C-{}^{14}CO_2^{\ominus} & & {}^{14}CO_2^{\ominus} \\
\| & \xrightarrow{\;{}^{14}CO_2\;} & | & \xrightarrow[\;H_2O\;\;\;2H^+\;]{} & | \\
C-OH & & C=O & & 2\;H-C-OH \\
| & & | & & | \\
H-C-OH & & H-C-OH & & H_2C-OPO_3^{\ominus} \\
| & & | & & \\
H_2C-OPO_3^{\ominus} & & H_2C-OPO_3^{\ominus} & &
\end{array}
$$

A large expenditure of enzyme is necessary to prepare the six-carbon reaction intermediate (only 1.2 micromoles are obtained by exhaustion of 0.5 g of the *R. rubrum* enzyme), and only Lorimer and colleagues (Pierce *et al.* 1986a) have been willing to make the sacrifice necessary to study its enzymatic turnover. Surprisingly, the non-activated form of the wild type enzyme (but not the activated form) catalyses the decarboxylation of the intermediate. Although the physiological significance of this reaction is obscure, it demonstrates an important feature of the normal reaction, i.e. once carboxylation of enediol has occurred, the process is committed in the forward direction (3-phosphoglycerate production) with only insignificant reversal (Pierce *et al.* 1986a).

Finally, if a mutant protein were deficient in the terminal protonation step, dissociation of the C2 carbanion of phosphoglycerate from the active-site might lead to pyruvate via β-elimination of the phosphate group (Andrews and Lorimer 1987). Indeed, this event occurs in about one in every one hundred turnovers with wild type enzyme (T.J. Andrews, personal communication), and a mutant protein with elevated capacity to form pyruvate from ribulose-$P_2$ has been observed in my laboratory.

Prior to reviewing what specifically has been learned about the carboxylase from site-directed mutagenesis, the usage of the term 'essential residue' deserves comment. Traditionally, a residue was defined as essential if its chemical derivatization resulted in total loss of enzymatic activity. As chemical reactions rarely proceed to 100% completion, activity losses of >90% were typically equated with 100%, and the target residue was concluded to be essential. Site-directed mutagenesis has shown that few, if any, residues can be considered truly 'essential' under the strict criterion of replacement leading to total abolishment of enzyme activity. Consider a case in which an enzyme enhances the rate of a reaction by a factor of $10^{12}$ compared to the uncatalysed rate. Replacement of an active-site residue might decrease the catalytic rate by $10^6$-fold, whereby the mutant protein would be only 0.0001% as active as wild type (zero activity as judged by historical standards of chemical modification). However, the mutant protein would still enhance the non-catalysed reaction rate by one million-fold! An example is provided by mutagenesis studies on serine proteases; certainly no one would debate the importance of the active-site serine that serves as the acyl acceptor during hydrolysis of peptide or ester substrates. However, the replacement of the active-site serine of subtilisin (Ser221) by alanine gives a mutant protein with residual activity equal to $5 \cdot 10^{-7}$ of wild type (Carter and Wells 1988). The issue then becomes the extent to which an active-site residue facilitates the reaction catalysed, rather than whether or not it is essential.

In the case of Rubisco, its low $k_{cat}$ ($2$–$5\ s^{-1}$) and lack of an analogous non-enzymatic counterpart constrains quantification of the facilitative contribution of an active-site residue. Given the low $k_{cat}$ and relative insensitivity of the carboxylase

assay, it is difficult to measure, with confidence, levels of activity that fall below 0.01% of wild type. A carboxylase mutant might actually enhance the non-catalysed reaction rate by many orders of magnitude but yet appear devoid of carboxylase activity.

## Small subunit mutagenesis

Unlike the large subunit of the carboxylase in which ~80% of the amino acid residues are species invariant, the sequence homology between the small subunits from higher plant and cyanobacterial carboxylases is only ~40% (see Andrews and Lorimer 1987). Among the small subunits (104–124 residues long) stretches of high homology occur near the $NH_2$-termini, middle and COOH-termini of the polypeptide chains. However, a 12-residue segment (positions 55–66 in the spinach numbering system) within the middle region is altogether lacking in the cyanobacterial enzymes, prompting the postulate that in the higher plant enzymes this segment is crucial for chloroplast import or assembly of small and large subunits. Wasmann *et al.* (1989) approached this question by constructing chimeras in which the pertinent coding region of the pea small subunit gene was inserted into the cyanobacterium *Anacystis nidulans* small subunit gene to fill the deletion. By use of *in vitro* synthesized small subunits and isolated chloroplasts, they showed that imported wild type *A. nidulans* small subunits would not assemble with pea large subunits, whereas the chimeric *A. nidulans* small subunits (that contained the insert from pea small subunit) did form holoenzyme. Thus, in higher plant carboxylases, one conserved region that plays a role in proper small–large subunit interaction has been identified.

Single amino acid substitutions have also been introduced into conserved regions of the small subunit of cyanobacterial Rubiscos. Replacement of two tryptophanyl residues of the *A. nidulans* enzyme, which correspond to Trp67 and Trp70 in spinach, by Phe does not interfere with holoenzyme formation, but reduces $k_{cat}$ 2.5-fold (Voordouw *et al.* 1987). In contrast, non-conservative replacement of Trp67 in *Anabaena* carboxylase by Arg prevents formation of stable holoenzyme (Fitchen *et al.* 1990). Substitutions of Glu13 by Val, Pro73 by His, and Tyr98 by Asn also disrupted small–large subunit interactions. In each of these cases, disruptions could be reconciled with interface interactions in wild type enzyme as revealed by X-ray crystallography.

Detrimental consequences also occur from double substitutions in the *A. nidulans* enzyme within the conserved region near the $NH_2$-terminus (residues 10–20), but the preliminary data do not distinguish effects on assembly and catalysis (McFadden and Small 1988).

## Truncation of the large subunit

Through manipulation of its gene, the carboxylase large subunit of *A. nidulans* was effectively shortened from the $NH_2$-terminus by 12 residues (the first six actually deleted and the next six replaced by an unrelated sequence encoded by a *lacZ* gene polylinker (Kettleborough *et al.* 1987). The truncated large subunit assembled properly and the resultant holoenzyme had a $k_{cat}$ about one-half that of wild type enzyme and a $K_m$ for ribulose-$P_2$ that increased by 10-fold. Although crystallography identifies a close subunit–subunit interaction between residues 20–22 and the COOH-terminal domain of the adjacent subunit, it does not include the truncated region, which is quite remote from the catalytic site. The altered kinetic parameters would appear then to reflect minor conformational perturbations. Comparable truncations of large subunits of wheat and spinach carboxylases have been effected by limited tryptic digestions with similar consequences (Gutteridge *et al.* 1986; Mulligan *et al.* 1988). The conclusions in common from these collective studies is that the $NH_2$-terminal, 15-residue segment of the large subunit is not needed for normal assembly, activation, or catalysis.

The homodimeric *R. rubrum* carboxylase has been systematically truncated at the COOH-terminus by site-directed mutagenesis (Ranty *et al.* 1990). Proteins lacking the unstructured tail (residues 458–466), the tail and the end α-helix (residues 450–466), or the tail and the two end α-helices (residues 441–466) were constructed and compared. One reason for truncating at position 458 was the observed sequence similarity of residues 456–466 and a conserved region (residues 10–20) in small subunits of $L_8S_8$ enzymes. Is this similarity fortuitous or reflective of small subunit function contained within the large subunit of the *R. rubrum* enzyme? The former seems to apply, because this particular truncated enzyme did not differ significantly from wild type. Interestingly, removal of the last or last two helices lowered $k_{cat}$ by a factor of $10^3$ and resulted in an octameric quaternary struc-

ture. Based on electron microscopy, the arrangement of subunits differed from that which occurs in the $L_8S_8$ enzymes. The aggregation state was explained by exposure of a hydrophobic region to solvent in the truncated proteins; proof was provided by 'reversion' to an $L_2$ structure upon replacing Leu424 with Asn. Both truncated proteins underwent carbamylation based on demonstration of binding, albeit weak, of carboxyarabinitol-$P_2$. The three-dimensional structure of the wild type enzyme suggested a model whereby the executed truncations transmit conformational changes to the active site.

## Mutagenesis of the activation domain of the active site

Speculation that the conserved Asp188, in addition to the carbamate oxyanion at position 191 of the *R. rubrum* carboxylase, might be a metal-ion ligand provided the logic for replacing it with Glu in the first site-directed mutant carboxylase constructed (Gutteridge *et al.* 1984b). The substitution proved benign, both spectroscopically and kinetically. This study illustrates the risk of embarking on site-directed mutagenesis on the basis of a presumed function, because subsequent elucidation of the three-dimensional structure showed that the Asp188 side chain was pointing away from the active site.

After X-ray crystallography identified the β-carboxylate of Asp193 as a metal-ion ligand, it was replaced by Asn (Gutteridge *et al.* 1988). As a consequence of weakened interaction with $Mg^{2+}$, the mutant was deficient in carbamylation and binding of carboxyarabinitol-$P_2$ and unable to catalyse either the overall carboxylation reaction or the enolization of ribulose-$P_2$. The mutant retained catalytic competence in processing the six-carbon reaction intermediate, 2-carboxy-3-ketoarabinitol-$P_2$. Formation of 3-phosphoglycerate was favoured in the presence of high $Mg^{2+}$ concentrations; in the absence of $Mg^{2+}$, only decarboxylation was catalysed, as observed for non-activated, wild type enzyme. The authors conclude that Asp193 is the base that enolizes ribulose-$P_2$ by abstraction of the C3 proton. However, taking into account the diverse evidence implicating Lys166 in this step, a more plausible explanation is that failure to undergo normal activation and proper coordination of $Mg^{2+}$ is more devastating to enolization than to turnover of the six-carbon intermediate.

The requirement for nearby Glu194, which is

not coordinated directly to metal ion, is also stringent; replacement by Gln abolishes all activities other than decarboxylation of the six-carbon intermediate. No precise function of Glu194 was invoked (Gutteridge *et al.* 1988).

Another early mutagenesis study (Estell *et al.* 1985) logically asked if a protein carboxylate could functionally replace the carbamate. Accordingly, Lys191 (the carbamate-forming lysyl residue) was replaced with Glu. The only activity exhibited by this mutant is that of decarboxylation of the six-carbon intermediate (Gutteridge *et al.* 1988). Deficiencies probably reflect an inability of the protein to bind $Mg^{2+}$, because the $\gamma$-carboxylate would be unable to extend to the position occupied by the carbamate oxyanion as visualized by crystallography.

More recently, Lys191 was replaced by Cys (Lorimer *et al.* 1987; Smith *et al.* 1988b) with the idea that the newly-introduced sulphydryl could be chemically modified to resemble the carbamate. For example, carboxypropylation should place a carboxylate at the location normally occupied by the carbamate oxyanion:

$$\diagdown \!\!\!\diagup \!\! HC_\alpha CH_2CH_2CH_2CH_2NHCOO^-$$

carbamate

$$\diagdown \!\!\!\diagup \!\! HC_\alpha CH_2SCH_2CH_2CH_2COO^-$$

carboxypropylcysteinyl residue

If the only role of the carbamate is to anchor $Mg^{2+}$, the derivatized cysteinyl mutant might exhibit carboxylase activity. Not surprisingly, the Cys191 mutant is devoid of both carboxylase and enolization activity (Hartman and Lee 1989). Unexpectedly, however, this mutant forms a more stable quaternary complex (protein–$CO_2$–$Mg^{2+}$–carboxyarabinitol-$P_2$) than does the wild type enzyme ($t_{1/2} = 100$ h in comparison to $t_{1/2} = 20$ h). The absence of carboxylase activity, despite competence in 'activation' is indirectly suggested of some specific function for the nitrogen atom of the carbamate. This suggestion must be viewed as tentative, because the bound species within the quaternary complex of Cys191 is bicarbonate rather than $CO_2$ as concluded from NMR (G.L. Lorimer, personal communication). One can envision interaction of bicarbonate with the sulphydryl at position 191 through H-bonding,

whereby the oxyanion of bicarbonate could approximate the normal location of the oxyanion of the carbamate.

Regrettably, approaching the function of the carbamate nitrogen through carboxypropylation of Cys191 has not been possible due to its sluggish reactivity.

## Mutagenesis of other active-site regions

### Lys166 and His291

As both residues had been implicated as the base that initiates catalysis by abstracting the C3 proton from ribulose-$P_2$, mutagenesis provided an avenue for addressing a very specific mechanistic question. Replacement of His291 by Ala reduced $k_{cat}$ for carboxylation by only 60% and increased $K_m$ (ribulose-$P_2$) by 10-fold (Niyogi *et al.* 1986). Although these results were compatible with an active-site location of His291, its small contribution to catalysis clearly excluded a role in proton transfer.

In striking contrast, replacement of Lys166 with a number of other amino acids, including Arg (retention of a cationic side chain), Cys or His (retention of an acid–base group), or Gly (removal of side chain) is highly detrimental (Hartman *et al.* 1987). Each of the four mutant proteins, whose intact dimeric structures were verified, retain <0.2% of the wild type activity. The Gly166 mutant, totally devoid of detectable carboxylase activity, is able to form the quaternary complex, but its stability is reduced relative to the wild type complex ($t_{1/2} = 30$ min compared to $t_{1/2} = 20$ h) (Fig. 4.5). Despite the weakened interaction, complex formation *per se* demonstrates that Lys166 is required neither for the $CO_2/Mg^{2+}$-dependent activation nor for binding of phosphorylated ligands. These observations reinforced the earlier postulate that Lys166 is intimately involved in catalysis, perhaps as the general base that enolizes ribulose-$P_2$. The Arg166 mutant does not form a sufficiently stable quaternary complex to permit detection by gel filtration (Fig. 4.5). The advisability of examining the consequences of multiple, single amino acid substitutions prior to drawing conclusions pertaining to function is thereby illustrated. Further credence to the notion that Lys166 initiates the overall reaction pathway is provided by the catalytic competence of the Gly166 mutant in

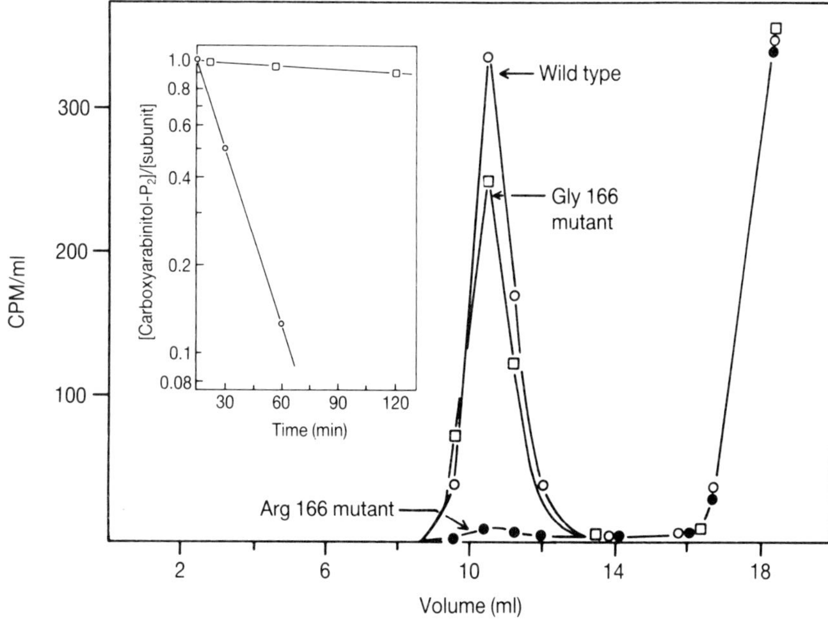



**Fig. 4.5**   Binding of [$^{14}$C]-carboxyarabinitol-$P_2$ to the wild type, Gly166 or Arg166 proteins, in the presence of $HCO_3^-$ and $Mg^{2+}$, as measured by gel filtration. Protein emerges from 10 to 12 ml, coincident with the first peak of radioactivity. Only the leading edge of the large excess of unbound ligand, which elutes in the salt region, is shown. The inset illustrates the exchange of [$^{14}$C]-carboxyarabinitol-$P_2$ from the quaternary complexes of the Gly166 (O) mutant and wild type (□) proteins when challenged with a ten-fold excess of non-radioactive carboxyarabinitol-$P_2$.

the turnover of the six-carbon reaction intermediate (Fig. 4.6) despite its inability to catalyse the enolization reaction (Fig. 4.7) (Lorimer and Hartman 1988).

A complication arises in assigning the role of an active-site residue based on efficacy of mutant proteins to catalyse partial reactions, because some of them are interdependent, or even concerted, rather than cleanly compartmentalized as they are drawn on paper. Although a number of mutant proteins lacking significant carboxylase activity are competent in one or more partial reactions, invariably, the observed rates are depressed relative to wild type. For example, the Gly166 mutant does catalyse turnover of the six-carbon reaction intermediate, but at a rate that is only 10% of wild type. The assignment of Lys166 to enolization is then based on greater disruption of this step than of successive ones. Parenthetically, the $k_{cat}$ for turnover of six-carbon reaction intermediate by wild-type enzyme is only 3% of the $k_{cat}$ for overall carboxylation of ribulose-$P_2$; a plausible explanation for this disparity is a requirement for a slow conformational change in processing of the intermediate (Lorimer and Hartman 1988). An alternative view is that the isolated intermediate (a

free ketone) is not kinetically competent, and only its hydrated form exists in the normal reaction pathway (Cleland 1990).

## Lys329

Substitutions that have been introduced for Lys329 include Gly, Ala, Ser, Cys, Arg and Glu (Soper *et al.* 1988). In each case, the purified mutant protein was shown to be a dimer, demonstrating that these single amino acid replacements are compatible with proper folding and association of subunits. The mutant proteins did not exhibit detectable enzyme activity nor did they form a stable quaternary complex with $CO_2$, $Mg^{2+}$ and carboxyarabinitol-$P_2$. However, based on ligand-selective elution of the mutant proteins from an affinity matrix (green A agarose), they retained the ability to bind substrate analogues. In contrast to the position 166 mutant proteins, replacement of Lys329 did not eliminate enolization activity (see Fig. 4.7) (Hartman and Lee 1989). However, the apparent $k_{cat}$ in the proton exchange reaction of even the more efficient mutant proteins (e.g. Cys329) was only ~5% that of wild type. As the enolization rates are routinely measured using a trace label and are

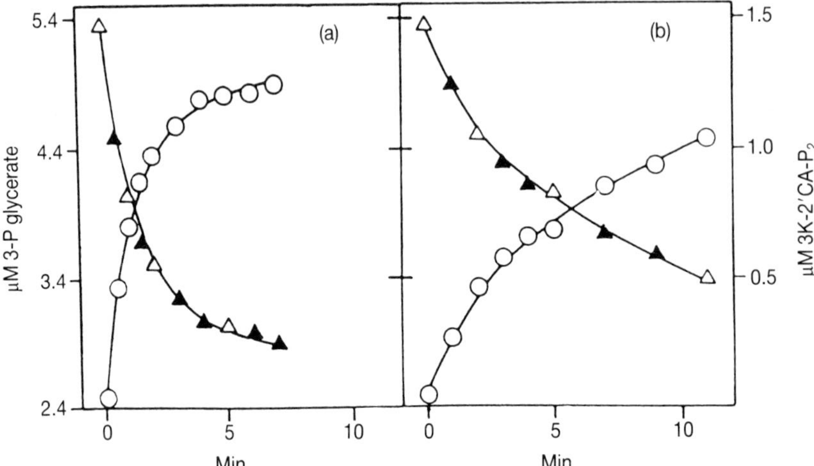

**Fig. 4.6**  Hydrolysis of the six-carbon reaction intermediate, KCA-$P_2$, by (a) the wild type carboxylase and (b) the Gly166 mutant enzyme. Phosphoglycerate (circles) and KCA-$P_2$ (triangles) were determined by the difference method of Pierce *et al.* (1986). Open symbols represent experimentally determined values; closed symbols represent calculated values for the concentration of KCA-$P_2$ assuming a stoichiometry of 2 moles of phosphoglycerate formed per mole of KCA-$P_2$ consumed. Reproduced from Lorimer and Hartman (1988) with permission of publisher.

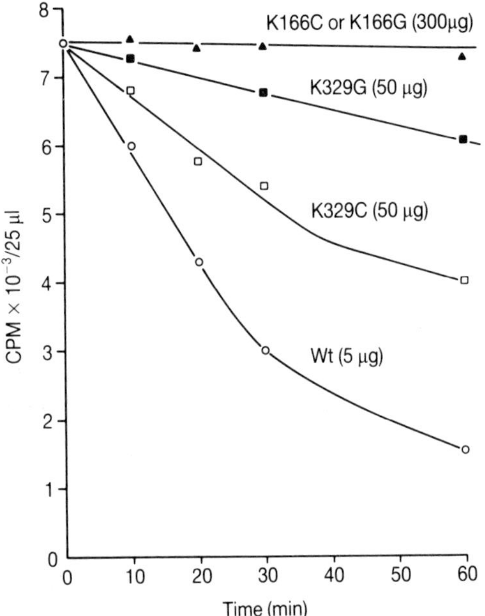

**Fig. 4.7**  Enolization of [3-$^3$H]-ribulose-$P_2$ by wild type and mutant proteins as measured by the decrease in non-volatile radioactivity. The μg-values denoted on the curves represent the amount of protein present in each 200 μl-reaction mixture.

revealed about the same rate as did proton exchange with tritiated ribulose-$P_2$ (E.H. Lee and F.C. Hartman, unpublished observation). It would then appear that the initial C–H bond breaking is not the rate-limiting step in the overall exchange process; exchange of enzyme-bound proton with solvent, transfer of enzyme-bound proton to enediol, or a protein conformational change represent possibilities for the rate-limiting step.

Despite the simplicity of the proton exchange experiments, they are revealing in several ways. Retention of enolization activity by position 329 mutant proteins, lacking in detectable carboxylase activity, shows that this step can proceed independently of others in the overall pathway. The $CO_2$/Mg-dependence of enolization excludes Lys329 as being involved, even indirectly, in the activation process. Reversible enolization of ribulose-$P_2$ is not accompanied by significant decline in its chemical concentration, indicating that misprotonation of the enediol, a minor side reaction demonstrated with wild type enzyme (Edmundson *et al.* 1990), and its dissociation from enzyme rarely occur. Failure of $CO_2$ or $O_2$ to react with the enediol generated by position 329 mutant proteins suggests that neither carboxylation nor oxygenation is spontaneous with the wild type enzyme, but require direct intervention by amino acid side chains.

uncorrected for kinetic isotope effects, the observed rates could be falsely low. However, direct measurement of enolization in $D_2O$ by NMR

The carboxylase-catalysed reaction proceeds by a

Theorell–Chance type of kinetic mechanism: ordered addition and enolization of ribulose-$P_2$ followed by bimolecular reaction of the enediol with gaseous substrate (Pierce *et al.* 1986b; Van Dyk and Schloss 1986). Evidence in favour of this ordered, sequential reaction includes NMR, direct binding, isotope trapping and kinetic analyses indicating that the gaseous substrates do not bind to free enzyme or to enzyme–ribulose-$P_2$. Earlier speculation that the universal oxygenase activity of Rubisco reflects an unavoidable consequence of the inherent reactivity of the enediol is certainly compatible with the elucidated kinetic mechanism (Lorimer and Andrews 1973). Carboxylation or oxygenation of the enediol could then be viewed as non-enzymic. However, the enolization of ribulose-$P_2$, as catalysed by position 329 mutant proteins, without concomitant carboxylation or oxygenation, argues that the enzyme plays an active role in facilitating the addition of gaseous substrate to enediol. Remembering that presteady-state kinetics detect a rate-determining step between enolization and carboxylation of enediol (Schloss 1990), it is reasonable to invoke Lys329 in deprotonation of enediol and development of the nucleophilic centre at C2.

## Met330

Replacement of Met330 of the *R. rubrum* enzyme by Leu was designed to alter its kinetic parameters to resemble more closely those of the $L_8S_8$ enzyme, many species of which carry Leu at position 330 (Terzaghi *et al.* 1986). While the substitution proved relatively benign, e.g. five-fold reduction in $k_{cat}$ and ten-fold increase in $K_m$ values, the specificity factor remained the same.

## Glu48 and Lys168

Prior to the elucidation of the three-dimensional structure of the complex of activated carboxylase with bound carboxyarabinitol-$P_2$ (which shows an intersubunit electrostatic interaction between Lys168 and Glu48), the latter residue had been examined by site-directed mutagenesis as prompted by chemical cross-linking, affinity labelling and comparative sequence considerations as noted earlier in this chapter. More than 99.9% of the carboxylase activity is lost upon replacement of Glu48 with glutamine (Hartman *et al.* 1987), which

removes the acid–base group of the glutamyl side chain without introducing major steric or electronic perturbation. As the mutant protein is a dimer that undergoes carbamylation by $CO_2$ and binds substrate, the drastic reduction in $k_{cat}$ indicates that this residue serves an important function. The corresponding Asp48 mutant exhibits very similar properties, so mere retention of a negative charge at position 48 is insufficient to support normal carboxylase activity (Smith *et al.* 1988a). Like position 329 mutant proteins, the position 48 mutant proteins exhibit significant enolization activity (1–2% of wild type) indicative of preferential interference of overall catalysis at some step beyond the first one. An elevated [pyruvate]/[phosphoglycerate] ratio of products formed from carboxylation of ribulose-$P_2$ might reflect deficiency in the terminal carbanion protonation step (E.H. Lee and F.C. Hartman, unpublished data).

After crystallography identified an electrostatic interaction between Glu48 and Lys168 of adjacent subunits (Andersson *et al.* 1989; Schneider *et al.* 1990), its structural and/or functional significance was further examined by replacement of Lys168 (Mural *et al.* 1990). As in the case of substitutions for Glu48, conservative replacement of Lys168 by Arg did not support carboxylase activity, and removal of a positive charge by replacement with Gln was not disruptive of subunit–subunit association. To continue the analogy, both the Gln48 and Arg168 proteins catalysed enolization of ribulose-$P_2$, discounting the possibility of major conformational perturbations accompanying either of these amino acid replacements. These observations establish that neither the Lys168–Glu48 salt bridge nor the side chains that constitute it are required for subunit association, carbamate formation, substrate binding or enolization activity. The extreme stringency of the dependence of overall carboxylase activity on a lysyl residue at position 168 and a glutamyl residue at position 48 could be explained by catalytic roles at steps subsequent to deprotonation of ribulose-$P_2$ or by subtle alterations in active site topology concomitant with any single amino acid substitution. A distinction between these possibilities must await crystallographic analyses of the pertinent mutant proteins.

## Arg288

Crystallography shows that this arginyl residue (Arg295 in the spinach enzyme and Arg292 in the

*A. nidulans* enzyme) serves as one of the ligands for the C5 phosphate group of ribulose-$P_2$. In the only report in which an active-site residue of an $L_8S_8$ enzyme has been replaced by site-directed mutagenesis, Haining and McFadden (1990) examined some of the properties of the Lys292 and Leu292 mutants of *A. nidulans* Rubisco. Partially purified mutant proteins were devoid of detectable carboxylase activity, but Western blot analyses of crude extracts on non-denaturing gels indicated that the proteins were properly assembled into $L_8S_8$ species. Further characterization subsequent to completed purification will be necessary to pinpoint the deficiencies of these mutant proteins. Interestingly, the lysyl mutant was unable to catalyse carboxylation, even at 50 mM ribulose-$P_2$, so perhaps Arg292 (Arg288 in the *R. rubrum* enzyme) does more than anchor a phosphate group through electrostatic attractions. Thus, conservative changes with retention of those charges found in the wild type enzyme fail to be functionally equivalent for either Lys166, Lys329, Lys168, Glu48 or Arg288.

## Subunit–subunit interactions

The first high resolution structure (but without placement of amino acid side chains) of any Rubisco to be solved was that of the non-activated *R. rubrum* enzyme, which lacked any bound substrate analogue (Schneider *et al.* 1986). Tracing of the polypeptide chain showed two distinct domains: the NH$_2$-terminal domain comprising residues 1–137 and the COOH-terminal domain comprising residues 138–466. Much of the latter domain can be described as an $\alpha/\beta$ barrel, a rather common three-dimensional folding motif observed among globular proteins that can give rise to a variety of catalytic activities. By analogy with other $\alpha/\beta$ barrel proteins and with the foreknowledge of identities of various active-site residues (e.g. Lys166 and Lys191), the active-site of the carboxylase was ascribed to the $\alpha/\beta$ barrel region of the COOH-terminal domain. Without specifying residue numbers, a segment of the NH$_2$-terminal domain was described as close to the active site of the COOH-terminal domain of the adjacent subunit. However, Schneider *et al.* (1986) concluded that the intersubunit distances between domains "are too long for a direct involvement of residues from the NH$_2$-terminal domain in catalysis . . ." Prophetically, the authors also speculated that

"conformational changes such as domain–domain rotations as a consequence of activation of substrate binding might decrease these distances."

On the heels of this crystallographic report, data from chemical cross-linking and mutagenesis prompted the conclusions that "essentiality of Glu48 strongly suggests that the active site of the carboxylase is located at an interface between subunits" (Hartman *et al.* 1987) and a "functional catalytic site requires, directly or indirectly, regions of both subunits at this interface" (Lee *et al.* 1987). Key observations included disruption of subunit–subunit interactions by replacement of active-site Lys166 with a negatively charged aspartyl residue, close proximity of Lys166 and a segment near the NH$_2$-terminus of the adjacent subunit and large facilitation of catalysis by Glu48.

The issue then became to prove or to refute the postulate that a segment of the NH$_2$-terminal domain was a functional, integral part of an active site residing at an interface between subunits. Wente and Schachman (1987) have distinguished between 'independent' and 'shared' active sites in the case of multimeric aspartate transcarbamylase by *in vitro* formation of hybrids of site-directed mutants. Because of difficulties in establishing reversible dissociation conditions for the native *R. rubrum* carboxylase, an *in vivo* procedure has been developed in which genes for the appropriate mutant subunits are coexpressed from separate, but compatible, plasmids (Larimer *et al.* 1987). In retrospect, the failure to achieve efficient, reversible dissociation of the carboxylase *in vitro* was due to the absence of the needed chaperonin, which can be provided by heat shock proteins of *E. coli* (Goloubinoff *et al.* 1989).

If each subunit of the carboxylase contains a complete, functional active site, a heterodimeric hybrid composed of one Gly166 subunit and one Glu48 subunit will be devoid of activity, just like each of the mutant homodimers. However, if a functional active site requires interacting domains of adjacent subunits, the potential exists for restoration of carboxylase activity by hybridization of two different inactive mutant proteins, e.g. Gly166 and Gln48 (Fig. 4.8(a)). Excepting the possibility of a bias in association, the carboxylase population should be 25% of the Gly166 homodimer, 25% of the Gln48 homodimer and 50% of the Gly166/Gln48 heterodimer. Since the latter would contain one functional active site per dimer, the specific activity of the total carboxylase population should be 25% that of wild type.

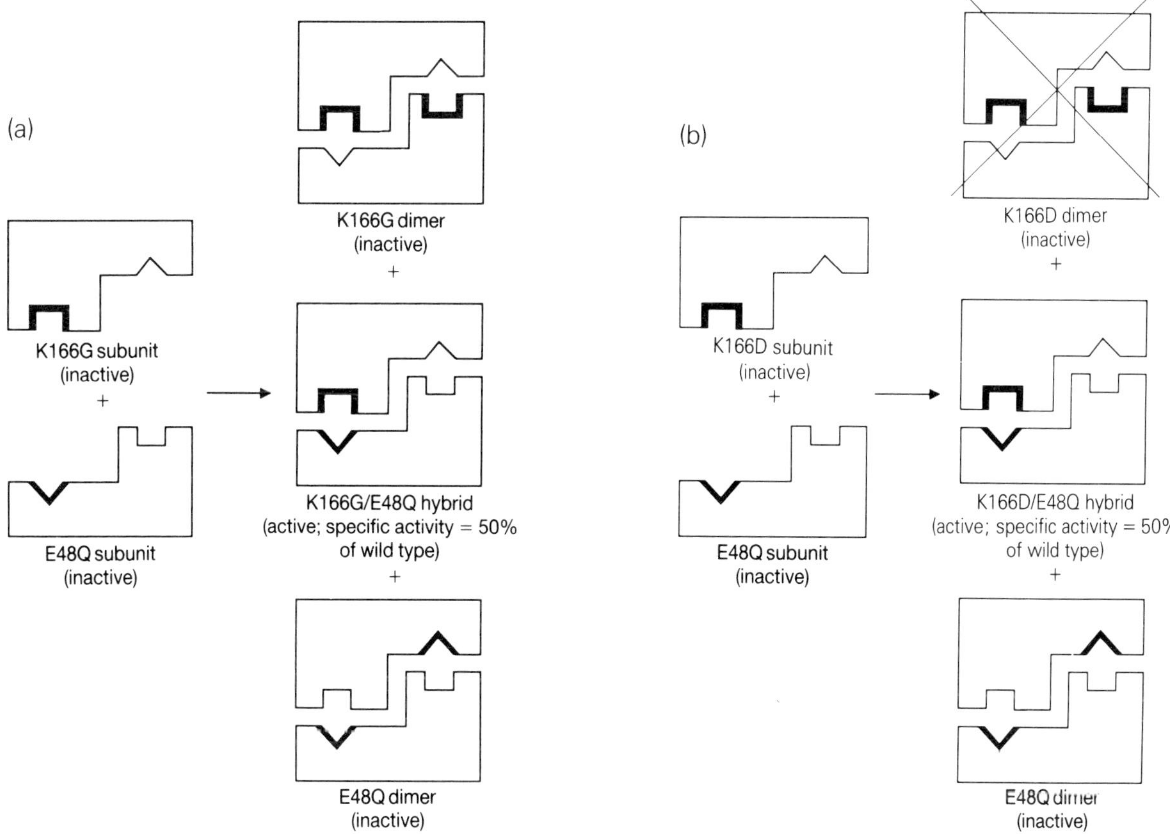



**Fig. 4.8** Schematic diagrams of the different species that can be formed in a cell coexpressing the genes for either Gly166 and Gln48 mutant subunits (a) or the Asp166 and Gln48 mutant subunits (b). The COOH-terminal domain of the active site is illustrated by the rectangular indentations, and the NH$_2$-terminal domain of the active site is illustrated by the triangular indentations. Indentations outlined in black represent non-functional domains due to specific amino acid substitutions. Heterodimer formation from the two mutant subunits generates a species with one wild type active site per dimeric molecule. Although predicted by statistical distribution, experiments (Lee *et al.* 1987) have shown that the K166D mutant protein does not form a stable dimer.

These expectations are fully realized by co-expression in *E. coli* of the genes for the Gly166 and Gln48 mutant proteins (Table 4.3). Analysis of the carboxylase purified from these cells confirms that the activity reflects the presence of a heterodimer (one subunit with the Gly166 substitution and one subunit with the Gln48 substitution) with one active site per molecule (as compared to two in wild type enzyme). This interallelic complementation conclusively demonstrates that domains from each subunit constitute the active site and that side chains from both are crucial, directly or indirectly, for catalysis (Larimer *et al.* 1987). The stringency in requirement for Glu at position 48 and the intersubunit electrostatic interaction between it and Lys168 are consistent with direct involvement of the NH$_2$-terminal domain in catalysis.

Does the conclusion from the hybrid study lessen the credibility of the crystallographic study? Certainly not! As already noted, the authors recognized the possibility of significant structural differences between activated and non-activated forms of the enzyme that were subsequently observed (Schneider *et al.* 1986; Andersson *et al.* 1989; Schneider *et al.* 1990). Furthermore, if electron density maps are properly interpreted, the depicted structure cannot be 'wrong'; it will, however, be static and require extrapolation to function. The strength of the hybridization approach is that it provided a functional demonstration of the requirement for interacting domains of adjacent subunits to create an active site. Thus, the smallest function-

**Table 4.3**   Carboxylase activity of wild type, mutant, and hybrid proteins

| | Crude extracts | | | Purified | |
| | Total protein (mg/ml) | Specific activity ($\mu$mol/min/mg) | Percent activity | Specific activity ($\mu$mol/min/mg) | Percent activity |
|---|---|---|---|---|---|
| Protein | | | | | |
| Wild type | 21 | 0.27 | 100 | 4.4 | 100 |
| Gly166 | 30 | not detectable | — | not detectable | — |
| Gln48 | 26 | not detectable | — | not detectable | — |
| Gly166/Gln48 | 17 | 0.07 | 26 | 0.88 | 20 |
| Asp166 | 24 | not detectable | — | not detectable | — |
| Asp166/Gln48 | 25 | 0.05 | 18 | 0.93 | 21 |
| Glu168 | 23 | not detectable | — | not detectable | — |
| Glu168/Gln48 | 23 | 0.06 | 22 | 0.87 | 20 |

al unit of any Rubisco is a dimer. The intersubunit location of the active site also applies to the $L_8S_8$ enzymes because the $L_8$ core is a tetramer of dimers (Andersson *et al.* 1989).

The facile *in vivo* hybridization of site-directed mutant proteins provides a convenient approach for examining subunit–subunit interactions in general. Replacement of Lys166 with Asp (Lee *et al.* 1987) or Lys168 with Glu (Mural *et al.* 1990) precluded the formation of a stable dimeric protein. The former finding prompted the suggestion that electrostatic repulsion between the newly introduced negative charge at position 166 and nearby negative charges in the $NH_2$-terminal domain of the adjacent subunit were responsible for interference in subunit–subunit association. Identical speculation followed the crystallographic finding of an intersubunit electrostatic interaction between Lys168 and Glu48 (Schneider *et al.* 1990). However, the Lys48 mutant protein, in which two positively charged side chains presumably confront each other, remains as a stable dimer (Mural *et al.* 1990). Intolerance of negative charges at positions 166 or 168 could then reflect conformational changes. Irrespective of electrostatic repulsions or conformational perturbations, a determination of whether a mutant subunit, unable to form a homodimer, can associate with a different mutant subunit to generate a functional enzyme allows a distinction between localized effects and gross misfolding of the polypeptide chain (Fig. 4.8(b)). Mutant subunits with Asp at position 166 or Glu at position 168 form active heterodimers by association with the Gln48 mutant subunit (Table 4.3), demonstrating that overall folding of the individual polypeptide chains is sufficiently 'native-like' to accommodate stable subunit–subunit interactions (Soper *et al.* 1989; Mural *et al.* 1990).

Chemical mutagenesis of tobacco seedlings gave rise to an interesting mutant carboxylase that is unable to assemble into the normal $L_8S_8$ entity (Avni *et al.* 1989). The large subunit appears to interact normally with the chaperonin protein, but this preassembly complex does not proceed to holoenzyme. Cloning and sequencing of the gene for the mutant large subunit indicated a single point mutation with the predicted outcome of replacement of Ser112 by Phe. Ser112 is close to the subunit–subunit interaction that forms the active-site; presumably, introduction of the bulky, hydrophobic side chain interferes with association of large subunits.

## Active-site structure

The three-dimensional structures of several Rubiscos, with and without bound ligands, have been elucidated. These include the non-activated tobacco enzyme (Chapman *et al.* 1988), the non-activated *R. rubrum* enzyme without bound ligand (Schneider *et al.* 1986; Schneider *et al.* 1990) and with bound phosphoglycerate (Lundqvist and Schneider 1988) or bound carboxyarabinitol-$P_2$ (Lundqvist and Schneider 1989), and the activated spinach enzyme with bound carboxyarabinitol-$P_2$ (i.e. the quaternary complex of enzyme–$CO_2$–$Mg^{2+}$–analogue, Andersson *et al.* 1989). A schematic illustration of the active site of the latter is shown (Fig. 4.9); the three-dimensional structure of the quaternary complex is the most revealing mechanistically.

A number of features substantiate earlier conclusions based on spectroscopic, chemical and mutagenesis studies. The divalent metal ion is seen bridging the oxyanion of the carbamate at Lys191 and the carboxylate anion of the bound carboxyarabinitol-$P_2$, as first deduced by EPR and NMR

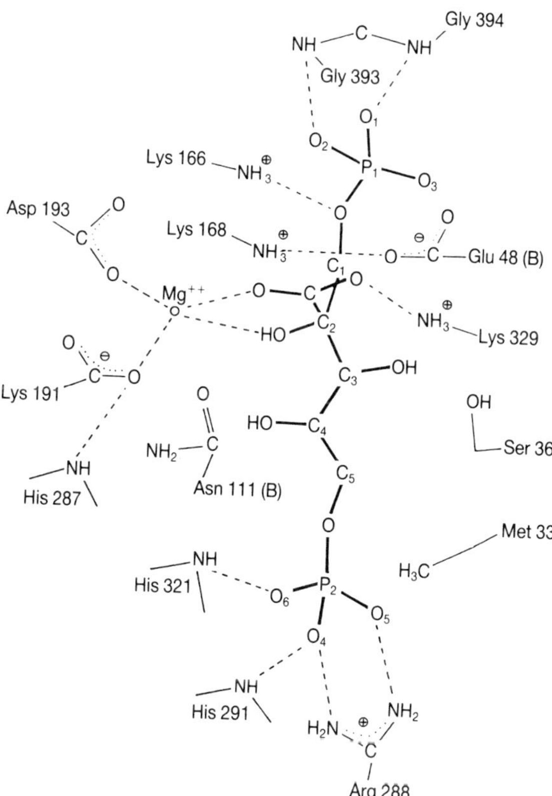



**Fig. 4.9** Schematic diagram of the active-site of ribulose-$P_2$ carboxylase with bound carboxyarabinitol-$P_2$. Reproduced from Andersson *et al.* (1989) with permission of authors and publisher; residue numbers have been altered in accordance with the sequence of the *R. rubrum* enzyme.

(Pierce and Reddy 1986). Lys166, Lys329 and Glu48 are not only found at the active site but are all within contact distances to the bound analogue, entirely consistent with functional roles invoked by chemical modification and site-directed mutagenesis discussed in this article. His291 is also proved to be an active-site residue, even though mutagenesis demonstrated that it contributes little to catalysis. An intersubunit ionic bond is identified between Glu48 and Lys168, verifying the conclusion based on hybridization of mutant proteins that the active site requires interacting domains of adjacent subunits and that residues from both contribute to catalysis (Larimer *et al.* 1987). The ε-amino group of Lys329 is within hydrogen bonding distance of the carboxyl group of carboxyarabinitol-$P_2$, a location consistent with a role in influencing the reactivity of the enediol of ribulose-$P_2$, as suggested by properties of position 329 mutants of the *R. rubrum*

enzyme, which catalyse enolization but not carboxylation of enediol (Hartman and Lee 1989). However, the ε-amino group of Lys166 is closer to the bridge oxygen of the C1 phosphate group than to C3, the location required for it to function as the base that enolizes ribulose-$P_2$. However, "model building outside the electron density could bring the nitrogen atom to within 3 Å from C3" (Andersson *et al.* 1989). Furthermore, these crystallographers point out that the observed structure simulates a stage in the catalytic pathway that is well beyond the initial enolization; in other words, the precise positioning of the ε-amino group of Lys166 in an enzyme–ribulose-$P_2$ complex could differ from that in the quaternary complex. Thus, the three-dimensional structure of the quaternary complex does not exclude Lys166 as the base which abstracts the C3 proton from ribulose-$P_2$, and thereby initiates the overall reaction pathway.

Despite the succinct statements of Andersson *et al.* (1989) and Schneider *et al.* (1990) that crystallographic structures are compatible with Lys166 as the base, some investigators ignore such a possibility (Gutteridge *et al.* 1986; Knaff 1989; Gutteridge 1990). This may be ill-advised on several counts. Although the three-dimensional structure does not prove Lys166 as the proton shuttle group in enolization, neither does the structure reveal a more credible candidate. Possibilities mentioned by Andersson *et al.* (1989), solely on the basis of structural considerations, include His321, Lys329, Ser368 and the nitrogen atom of the carbamate at Lys191. Since position 329 mutant proteins catalyse enolization, Lys329 should be excluded as a possibility. Given the weak nucleophilicity of seryl hydroxyl groups and the observation that the base in question exhibits a p$K_a$ of 7.5 (Van Dyk and Schloss 1986), Ser368 should also be excluded. His321 may be replaced by Asn with retention of considerable activity (F.C. Hartman, unpublished observations), so the contribution of this residue to catalysis is modest. The carbamate remains a possibility, but its low p$K_a$ in model compounds argues against a role in proton transfers. In contrast, and within reach of the C3 proton of substrate, Lys166 possesses the needed nucleophilicity and a p$K_a$ equal to that of the pertinent group seen in kinetic studies. Taking into account the results from site-directed mutagenesis, multiple observations from diverse approaches that lead to the same conclusion must be ruled coincidental if Lys166 were not the base under consideration. Even if further refinement of crystallographic data unequivocally refute

Lys166 as the primary proton acceptor, it is nevertheless required for efficient enolization in some undefined way.

Expectedly, the elucidation of the three-dimensional structure of the quaternary complex of the spinach enzyme provides more, and more detailed, information about side chains in the vicinity of bound ligands than could be gleaned by other approaches collectively. Key groups not identified heretofore included Gly393 and Gly394, the backbone dipoles of which provide the binding site for the C1 phosphate; Asp193, a metal ligand and His321, one anchor for the C5 phosphate. Additional contact residues revealed for the first time were His287, Lys168, Ser368 and Met330 of the COOH-terminal domain and Asn111 of the $NH_2$-terminal domain of the adjacent subunit. Future site-directed mutagenesis studies will no doubt include these residues so that structural information may be extended to more precise mechanistic inferences.

It is perhaps worth reiterating that defining the precise function of an active-site residue is a challenging task, which pushes existing methodologies to their limits. Site-directed mutagenesis, in combination with kinetic studies, chemical modification and comparative sequence analyses, can identify active-site residues in a convincing fashion and can distinguish their roles in substrate binding from participation in catalysis. Indirect influence on catalytic activity through gross distortion of polypeptide folding, disruption of subunit–subunit interactions or interference with activation processes can also be distinguished from direct effects with confidence. However, beyond this, we enter the realm of plausible speculation. The mechanistic conclusions, derived from an array of physical, chemical and kinetic probes, can be considered definitive only when wholly supported by crystallographic observations and interpretations.

Crystallography has its own limitations as regards mechanism, for the structure determined is, of course, of the crystalline form available and extrapolations from a static structure to function is necessitated. Some of these limitations surfaced during initial crystallographic studies of the carboxylase. The non-activated (non-carbamylated) forms of the carboxylase from both *R. rubrum* and tobacco (in addition to the quaternary complex of the spinach enzyme already discussed) have been subjected to three-dimensional structural analyses (Chapman *et al.* 1987; Chapman *et al.* 1988; Lundqvist and Schneider 1988; Lundqvist and Schneider 1989; Schneider *et al.* 1990). Conformational differences between non-carbamylated and carbamylated enzymes, which signal the advisability of exercising caution in extending structural features to mechanistic inferences, are revealed. For example, the structure of the non-activated *R. rubrum* enzyme did not permit the deductions that the active-site is created by subunit–subunit interactions and that side chains from both subunits comprise the active site. Furthermore, this structure, as well as that of non-activated tobacco enzyme, did not show Lys329 as an active-site residue. This lysyl residue is contained within a disordered, flexible loop and its active-site location is only visualized in protein complexed with carboxyarabinitol-$P_2$ (Andersson *et al.* 1989). The activation-state dependence of the inter-residue distance between Lys166 and Lys329 as revealed by crystallography was predicted by earlier cross-linking studies (Lee *et al.* 1986).

Based on the three-dimensional structure of the non-activated *R. rubrum* enzyme, complexed with carboxyarabinitol-$P_2$, both Lys166 and Lys329 are assigned roles as ligands for one of the phosphate groups (Lundqvist and Schneider 1989). Such assignments are not necessarily at variance with those ascribed earlier in this chapter. The carboxylase structure is dynamic and precise positioning of active-site side chains may vary during discrete stages of substrate transformations. Additional questions also arise. What are the structural effects, if any, of the crystals having been grown at pH 5.6, where the enzyme displays little catalytic activity? Is the fact that carboxyarabinitol-$P_2$ binds to this crystalline form of non-activated carboxylase in an upside-down orientation, relative to that observed with carbamylated spinach enzyme, reflective of an altered active-site topology? Some of the questions and speculations presented about functions of active-site residues will no doubt be settled as more structures representing different stages during the catalytic pathway are solved.

## Concerted site-directed mutagenesis and chemical modification

Of the site-directed mutant carboxylases possessing enough activity to allow accurate quantification, none displayed an altered specificity factor ($V_c K_o / V_o K_c$ or $\tau$). To ascertain if structural changes more subtle than amino acid substitutions could alter $\tau$, chemical modification in conjunction with mutagenesis has been used to alter the active-site microenvironment. Examples include conversion of a lysyl residue to an aminoethylcysteinyl residue and conversion of a glutamyl residue to a carboxymethylcysteinyl residue.

$$\text{HC}_\alpha\text{CH}_2\text{CH}_2\text{CH}_2\text{CH}_2\text{NH}_2 \longrightarrow \text{HC}_\alpha\text{CH}_2\text{SH}$$

**Mutagenesis**

$$\xrightarrow{\text{BrCH}_2\text{CH}_2\text{NH}_2} \text{HC}_\alpha\text{CH}_2\text{SCH}_2\text{CH}_2\text{NH}_2$$

**Modification**

$$\text{HC}_\alpha\text{CH}_2\text{CH}_2\text{COO}^- \longrightarrow \text{HC}_\alpha\text{CH}_2\text{SH}$$

**Mutagenesis**

$$\xrightarrow{\text{ICH}_2\text{COOH}} \text{HC}_\alpha\text{CH}_2\text{SCH}_2\text{COO}^-$$

**Modification**

In the former case, the overall structural change is the mere replacement of a lysyl $\gamma$ methylene by a sulphur atom. In the latter case, a sulphur atom, in effect, is inserted between the $\beta$- and $\gamma$-methylene groups of the glutamyl side chain thereby lengthening it by $\sim$1.5 Å.

Lys166, Lys329 and Lys191 of the *R. rubrum* carboxylase have been replaced by aminoethylcysteinyl residues by the approach depicted above (Smith and Hartman 1988; Smith *et al.* 1988b). Beyond the inquiry concerning the feasibility of altering the carboxylase/oxygenase activity ratio, additional considerations justified this strategy with Lys166 and Lys329. When these studies were initiated, the three-dimensional structure of the enzyme had not been determined, and functionality *per se* of the two lysyl residues was open to conjecture; the envisioned chemistry would provide a stringent test of the requirements for these two residues. Their strong acidities and enhanced nucleophilicities presumably reflected a unique microenvironment. If substitutions for Lys166 and Lys329 did not alter the active-site conformation, other reactive side chains at these positions should

also exhibit unusual properties. The cysteinyl mutant carboxylases, devoid of enzyme activity, provided a direct chemical approach to inspecting the conformational integrity of the active site.

Treatment of the Cys166 and Cys329 mutant proteins with 2-bromoethylamine partially restored enzyme activity (Fig. 4.10), presumably as a consequence of selective aminoethylation of the thiol group unique to each protein. Amino acid analyses, isoelectric focusing under denaturing conditions, slow inactivation of the wild type carboxylase by bromoethylamine, and the failure of bromoethylamine to restore activity to the corresponding glycyl mutant proteins supported this interpretation. The observed facile, selective aminoethylations were consistent with active-site microenvironments not dissimilar to that of the native enzyme. Catalytic constants of these novel carboxylases, which contain a sulphur atom in place of a specific lysyl $\gamma$-methylene group, were significantly lower than that of the wild type enzyme. The $k_{\text{cat}}$ for Cys329 was 40% that of wild type, while $k_{\text{cat}}$ for Cys166 was reduced five-fold relative to wild type. Furthermore, the aminoethylated mutant proteins formed isolable complexes with carboxyarabinitol-$P_2$, but with compromised stabilities. These detrimental effects by such a modest structural change underscore the stringent requirement for lysyl side chains at positions 166 and 329. In contrast, the aminoethylated mutant proteins exhibited $K_{\text{m}}$ values that were unperturbed relative to those for the native enzyme. Clearly, major reductions in $k_{\text{cat}}$ with unaltered $K_{\text{m}}$ values argued for direct roles of

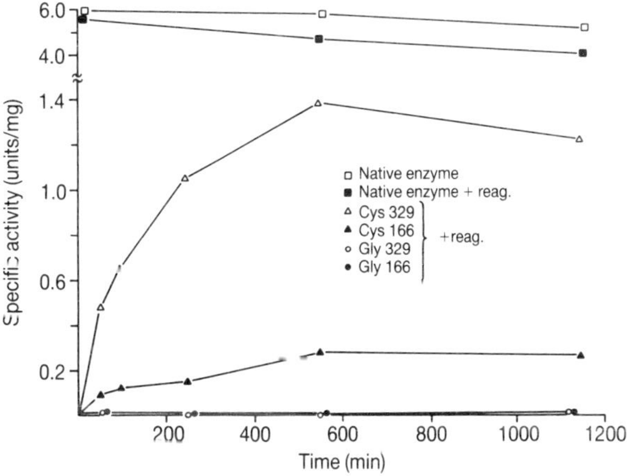


**Fig. 4.10** Carboxylase activity during treatment of wild type and mutant proteins (all at 0.4 mg/ml) with 2-bromoethylamine (100 mM).

Lys166 and Lys329 in catalysis. An analogous strategy has been used to replace Lys191 with aminoethylcysteine, but the more reactive ethylene imine was necessary for the introduction of the aminoethyl side chain, because a sulphydryl group at position 191 is difficult to alkylate. The $k_{cat}$ for the aminoethylated protein was only 4–6% of the wild type value. Despite success in achieving the desired structural alterations with concomitant partial restoration of catalytic activity, the $\tau$-values of the novel proteins were essentially the same as the wild type value.

The consequences of converting Glu48 to a carboxymethylcysteinyl residue were more encouraging as an avenue for altering substrate specificity (Smith *et al.* 1990). The Cys48 mutant protein displayed ~0.05% of wild type carboxylase activity, similar to other position 48 mutants. Treatment of the protein with iodoacetate enhanced the inherent carboxylase activity 100-fold, and several observations supported the notion that the enhancement correlated with carboxymethylation of Cys48. The oxygenase activity was restored to a greater relative extent, with the net outcome being a five-fold decrease in specificity factor. The pronounced relative enhancement of oxygenase activity could be visualized qualitatively by the formation of substantial levels of phosphoglycolate by the carboxymethylated mutant protein at a high concentration of bicarbonate that almost completely suppressed phosphoglycolate formation by wild type protein (Fig. 4.11). This represents the first example in which a change in substrate specificity can be correlated with a programmed structural modification of a known active-site residue. Although the substrate specificity was altered in the 'wrong' direction, the results are encouraging in that they show the sensitivity of $\tau$ to active-site microenvironment and identify a polypeptide segment that influences $\tau$.

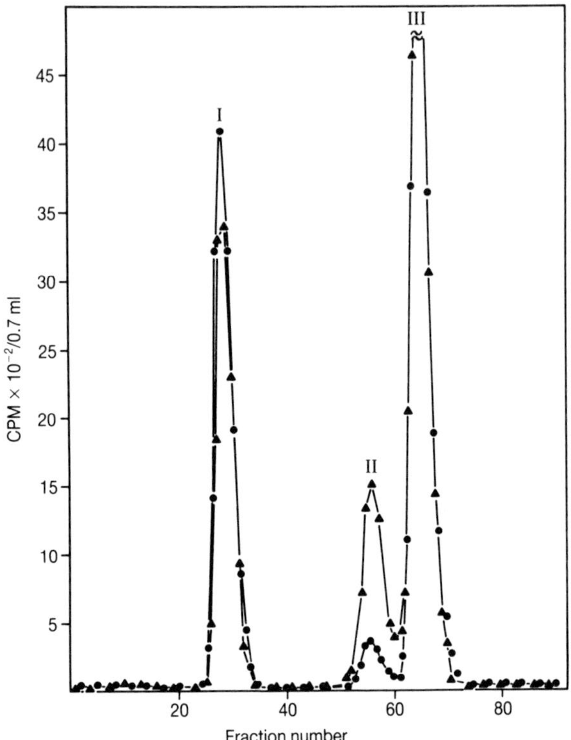

**Fig. 4.11**  DEAE-chromatographic separation of carboxylase and oxygenase products from reactions catalysed by wild type protein (●) and by carboxymethylated mutant protein (▲). [1-$^3$H]-ribulose-P$_2$ was used as substrate so that partitioning of tritium into 3-phosphoglycerate and phosphoglycolate correlated directly to carboxylase and oxygenase activities, respectively; subsequent NADH-coupled reduction of phosphoglycerate to phosphoglycerol then allowed for facile separation and quantitation of the two products. Reaction mixtures were quenched with sodium borohydride to reduce the remaining [$^3$H]-ribulose-P$_2$. Peaks I, II, III correspond respectively to phosphoglycerol (derived from phosphoglycerate), phosphoglycolate, and pentitolbisphosphate (derived from ribulose-P$_2$). Reproduced from Smith *et al.* (1990) with permission of publisher.

# Epilogue

I have attempted to present a balanced view of the utility and limitations of site-directed mutagenesis as applied to proteins with unsolved three-dimensional structures, because investigators will continue to confront structure–function issues in many systems without benefit of crystallographic data. With Rubisco, mutagenesis studies are conveniently subdivided into the pre- and post-three-dimensional structural eras. With the recent emergence of an avalanche of high-resolution structural information, the former era has ended; accrued contributions must be placed in historical perspective. Little more will be learned about the carboxylase from future site-directed mutagenesis studies, unless they are designed to approach new questions elaborated by the three-dimensional structure or to distinguish among multiple, plausible mechanistic details consistent with the three-dimensional structure. Altered properties of mutant proteins will not be fully understood until they too are subjected to crystallographic analyses.

Disappointment may already reign among some quarters because elucidation of the three-dimensional structure of the carboxylase has not yet provided enlightenment for 'tailoring out' the oxygenase activity. Any pessimism should be tempered with the realization that the post-three-dimensional era has just begun. As more structures of the various conformational states of the enzyme become available, and as coordinates are disseminated throughout the community, designed mutagenesis experiments will surely elevate our understanding of the intricate carboxylase/oxygenase reaction mechanisms. Concurrently, collective site-directed mutagenesis in other systems will provide ground rules that relate to determinants of substrate specificities and catalytic efficiencies. Advances in computer graphics, protein modelling, protein dynamics and *de novo* design of catalysts will proceed unabated. With the ever increasing sophistication of tools available to enzymologists, dismissing the possibility of future tailored improvements in the carboxylase, an enzyme whose efficiency is orders of magnitude below the diffusion-limited level, would be premature. The jury is still out.

## Acknowledgements

The research from my laboratory was supported by the Office of Health and Environmental Research, United States Department of Energy, under Contract DE-AC05-84OR21400 with Martin Marietta Energy Systems, Inc. Local colleagues made enormous intellectual and technical contributions to some of the studies described in this article. These researchers include Dr Frank W. Larimer, Dr Richard J. Mural, Dr Thomas S. Soper, Dr Eva H. Lee and Dr Harry B. Smith.

## References

Andersen, K. and Wilke-Douglas, M. (1984) Construction and use of a gene bank of *Alcaligenes eutrophus* in the analysis of ribulose bisphosphate carboxylase genes. *Journal of Bacteriology.* **159**. 973–8.

Andersson, I., Knight, S., Schneider, G., Lindqvist, Y., Lundqvist, T., Brändén, C.-I. and Lorimer, G.H. (1989). Crystal structure of the active site of ribulose-bisphosphate carboxylase. *Nature.* **337**: 229–34.

Andrews, T.J. (1988). Catalysis by cyanobacterial ribulose-bisphosphate carboxylase large subunits in the complete absence of small subunit. *Journal of Biological Chemistry.* **263**: 12213–19.

Andrews, T.J. and Lorimer, G.H. (1987). Rubisco: structure, mechanisms, and prospects for improvement. In *The Biochemistry of Plants, A Comprehensive Treatise, Vol. 10: Photosynthesis,* (ed. M.D. Hatch and N.K. Boardman), pp. 131–218. Academic Press, New York.

Andrews, T.J., Lorimer, G.H. and Tolbert, N.E. (1973). Ribulose diphosphate oxygenase I. Synthesis of phosphoglycolate by fraction-I protein of leaves. *Biochemistry.* **12**: 11–18.

Anthony-Cahill, S.J., Griffith, M.C., Noren, C.J., Suich, D.J. and Schultz, P.G. (1989). Site-specific mutagenesis with unnatural amino acids. *Trends in Biochemical Sciences.* **14**: 400–3.

Aresta, M. and Schloss, J.V. (ed.) (1990). *The Proceedings of NATO ASI on Enzymatic and Model Carboxylation and Reduction Reactions for Carbon Dixodie Utilization.* Kluwer Academic Publishers, Dordrecht.

Avni, A., Edelman, M., Rachailovich, I., Aviv, D. and Fluhr, R. (1989). A point mutation in the gene for the large subunit of ribulose 1,5-bisphosphate carboxylase/oxygenase affects holoenzyme assembly in *Nicotiana tabacum. EMBO Journal.* **8**: 1915–18.

Berry, J.A., Lorimer, G.H., Pierce, J., Seemann, J.R., Meek, J. and Freas, S. (1987). Isolation, identification, and synthesis of 2-carboxyarabinitol 1-phosphate, a diurnal regulator of ribulose-bisphosphate carboxylase activity. *Proceedings of the National Academy of Sciences, USA.* **84**: 734–8.

Blacklow, S.C. and Knowles, J.R. (1990). How can a catalytic lesion be offset? The energetics of two pseudorevertant triosephosphate isomerases. *Biochemistry.* **29**: 4099–108.

Bowes, G., Ogren, W.L. and Hageman, R.H. (1971). Phosphoglycollate production catalyzed by ribulose bisphosphate carboxylase. *Biochemistry and Biophysics Research Communications.* **45**: 716–22.

Campbell, W.J. and Ogren, W.L. (1990). A novel role for light in the activation of ribulosebisphosphate carboxylase/oxygenase. *Plant Physiology.* **92**: 110–15.

Carter, P. and Wells, J.A. (1988). Dissecting the catalytic triad of a serine protease. *Nature.* **332**: 564–8.

Chapman, M.S., Suh, S.W., Cascio, D., Smith, W.W. and Eisenberg, D. (1987). Sliding-layer conformational change limited by the quaternary structure of plant Rubisco. *Nature.* **329**: 354–6.

Chapman, M.S., Suh, S.W., Curmi, P.M.G., Cascio, D., Smith, W.W. and Eisenberg, D.S. (1988). Tertiary structure of plant Rubisco: domains and their contacts. *Science.* **241**: 71–4.

Chen, Z., Chastain, C.J., Al-Abed, S.R., Chollet, R. and Spreitzer, R.J. (1988). Reduced $CO_2/O_2$ specificity of ribulose-bisphosphate carboxylase/oxygenase in a temperature-sensitive chloroplast mutant of *Chlamydomonas. Proceedings of the National Academy of Sciences, USA.* **85**: 4696–9.

Chen, Z. and Spreitzer, R.J. (1989). Chloroplast intragenic suppression enhances the low $CO_2/O_2$ specificity of mutant ribulose-bisphosphate carboxylase/oxygenase. *Journal of Biological Chemistry*. **264**: 3051–3.

Chen, Z., Green, D., Westhoff, C. and Spreitzer, R.J. (1990). Nuclear mutation restores the reduced $CO_2/O_2$ specificity of ribulosebisphosphate carboxylase/oxygenase in a temperature-conditional chloroplast mutant of *Chlamydomonas reinhardtii*. *Archives of Biochemistry and Biophysics*. **283**: 60–7.

Chua, N.-H. and Schmidt, G.W. (1979). Transport of proteins into mitochondria and chloroplasts. *The Journal of Cell Biology*. **81**: 461–83.

Cleland, W.W. (1990). Kinetic competence of enzymic intermediates: Fact or fiction? *Biochemistry*. **29**: 3194–7.

Donnelly, M.I., Stringer, C.D. and Hartman, F.C. (1983). Characterization of the activator site of *Rhodospirillum rubrum* ribulosebisphosphate carboxylase/oxygenase. *Biochemistry*. **22**: 4346–52.

Edmondson, D.L., Kane, H.J. and Andrews, T.J. (1990). Substrate isomerization inhibits ribulosebisphosphate carboxylase-oxygenase during catalysis. *FEBS Letters*. **260**: 62–6.

Ellis, R.J. and Gray, J.C. (ed.) (1986). Ribulose bisphosphate carboxylase-oxygenase. *Philosophical Transactions of the Royal Society of London Series B-Biological Sciences*. **B313**: 303–469.

Ellis, R.J. and Smith, S.M. (1979). Processing of small subunit precursor of ribulose bisphosphate carboxylase and its assembly into whole enzyme are stromal events. *Nature*. **278**: 662–4.

Estelle, M., Hanks, J., McIntosh, L. and Somerville, C.R. (1985). Site-specific mutagenesis of ribulose-1,5-bisphosphate carboxylase/oxygenase. *Journal of Biological Chemistry*. **260**: 9523–6.

Fersht, A.R., Leatherbarrow, R.J. and Wells, T.N.C. (1987). Structure-activity relationships in engineered proteins: analysis of use of binding energy by linear free energy relationships. *Biochemistry*. **26**: 6030–8.

Fersht, A.R., Knill-Jones, J.W., Bedouelle, H. and Winter, G. (1988). Reconstruction by site-directed mutagenesis of the transition state for the activation of tyrosine by the tyrosyl-tRNA synthetase: a mobile loop envelopes the transition state in an induced-fit mechanism. *Biochemistry*. **27**: 1581–7.

Fiedler, F., Müllhofer, G., Trebst, A. and Rose, I.A. (1967). Mechanism of ribulose-diphosphate carboxydismutase reaction. *European Journal of Biochemistry*. **1**: 395–9.

Fitchen, J.H., Knight, S., Andersson, I., Brändén, C.I. and McIntosh, L. (1990). Residues in three conserved regions of the small subunit of ribulose 1,5-bisphosphate carboxylase/oxygenase are required for quaternary structure. *Proceedings of the National Academy of Sciences, USA*. **87**: 5768–72.

Fraij, B. and Hartman, F.C. (1983). Isolation and sequencing of an active-site peptide from *Rhodospiril-*

*lum rubrum* ribulosebisphosphate carboxylase/oxygenase after affinity labeling with 2-[(bromoacetyl)amino]pentitol 1,5-bisphosphate. *Biochemistry*. **22**: 1515–20.

Gatenby, A.A., van der Vies, S.M. and Bradley, D. (1985). Assembly in *E. coli* of a functional multisubunit carboxylase from a blue-green alga. *Nature*. **314**: 617–20.

Gerlt, J.A. (1987). Relationships between enzymatic catalysis and active site structure revealed by applications of site-directed mutagenesis. *Chemical Reviews*. **87**: 1079–1105.

Goloubinoff, P., Gatenby, A.A. and Lorimer, G.H. (1989). GroE heat-shock proteins promote assembly of foreign prokaryotic ribulose bisphosphate carboxylase oligomers in *Escherichia coli*. *Nature*. **337**: 44–7.

Gutteridge, S. (1990). Limitations of the primary events of $CO_2$ fixation in photosynthetic organisms: the structure and mechanism of Rubisco. *Biochimica et Biophysica Acta*. **1015**: 1–14.

Gutteridge, S. and Julien, B. (1989). A phosphatase from chloroplast stroma of *Nicotiana tobacum* hydrolyses 2′-carboxyarabinitol 1-phosphate, the natural inhibitor of Rubisco to 2′-carboxyarabinitol. *FEBS Letters*. **254**: 225–30.

Gutteridge, S., Parry, M.A.J., Schmidt, C.N.G. and Feeney, J. (1984a). An investigation of ribulosebisphosphate carboxylase activity by high resolution $^1$H NMR. *FEBS Letters*. **170**: 355–9.

Gutteridge, S., Sigal, I., Thomas, B., Arentzen, R., Cordova, A. and Lorimer, G.H. (1984b). A site-specific mutation within the active site of ribulose-1,5-bisphosphate carboxylase of *Rhodospirillum rubrum*. *EMBO Journal*. **3**: 2737–43.

Gutteridge, S., Millard, B.N. and Parry, M.A.J. (1986). Inactivation of ribulose-bisphosphate carboxylase by limited proteolysis. *FEBS Letters*. **196**: 263–8.

Gutteridge, S., Parry, M.A.J., Burton, S., Keys, A.J., Mudd, A., Feeney, J., Servaites, J.C. and Pierce, J. (1986). A nocturnal inhibitor of carboxylation in leaves. *Nature*. **324**: 274–6.

Gutteridge, S., Lorimer, G. and Pierce, J. (1988). Details of the reactions catalyzed by mutant forms of Rubisco. *Plant Physiology and Biochemistry*. **26**: 675–82.

Haining, R.L. and McFadden, B.A. (1990). A critical arginine in the large subunit of ribulose bisphosphate carboxylase/oxygenase identified by site-directed mutagenesis. *Journal of Biological Chemistry*. **265**: 5434–9.

Hardy, R.W.F., Havelka, U.D. and Quebedeaux (1978). In *Photosynthetic Carbon Assimilation* (ed. H.W. Siegelman and G. Hind), pp. 165–78. Plenum, New York.

Hartman, F.C. and Lee, E.H. (1989). Examination of the function of active site lysine 329 of ribulosebisphosphate carboxylase/oxygenase as revealed by the proton exchange reaction. *Journal of Biological Chemis-*

*try*. **246**: 11784–9.

Hartman, F.C., Stringer, C.D. and Lee, E.H. (1984). Complete primary structure of ribulosebisphosphate carboxylase/oxygenase from *Rhodospirillum rubrum*. *Archives of Biochemistry and Biophysics*. **232**: 280–95.

Hartman, F.C., Milanez, S. and Lee, E.H. (1985). Ionization constants of two active-site lysyl ε-amino groups of ribulosebisphosphate carboxylase/oxygenase. *Journal of Biological Chemistry*. **260**: 13968–75.

Hartman, F.C., Stringer, C.D., Milanez, S. and Lee, E.H. (1986). The active site of Rubisco. *Philosophical Transactions of the Royal Society of London Series B—Biological Sciences*. **313**: 379–95.

Hartman, F.C., Larimer, F.W., Mural, R.J., Machanoff, R. and Soper, T.S. (1987). Essentiality of Glu-48 of ribulose bisphosphate carboxylase/oxygenase as demonstrated by site-directed mutagenesis. *Biochemical and Biophysical Research Communications*. **145**: 1158–63.

Hartman, F.C., Soper, T.S., Niyogi, S.K., Mural, R.J., Foote, R.S., Mitra, S., Lee, E.H., Machanoff, R. and Larimer, F.W. (1987). Function of Lys-166 of *Rhodospirillum rubrum* ribulosebisphosphate carboxylase/oxygenase as examined by site-directed mutagenesis. *Journal of Biological Chemistry*. **262**: 3496–501.

Hemmingsen, S.M., Woolford, C., van der Vies, S.M., Tilly, K., Dennis, D.T., Georgopoulos, C.P., Hendrix, R.W. and Ellis, R.J. (1988). Homologous plant and bacterial proteins chaperone oligomeric protein assembly. *Nature*. **333**: 330–4.

Herndon, C.S. and Hartman, F.C. (1984). 2-(4-bromoacetamido)anilino-2-deoxypentitol 1,5-bisphosphate, a new affinity label for ribulose bisphosphate carboxylase/oxygenase from *Rhodospirillum rubrum*. *Journal of Biological Chemistry*. **259**: 3102–10.

Hibler, D.W., Stolowich, N.J., Reynolds, M.A., Gerlt, J.A., Wilde, J.A. and Bolton, P.H. (1987). Site-directed mutants of staphylococcal nuclease. Detection and localization by $^1$H NMR spectroscopy of conformational changes accompanying substitutions for glutamic acid-43. *Biochemistry*. **26**: 6278–86.

Higaki, J.N., Evnin, L.B. and Craik, C.S. (1989). Introduction of a cysteine protease active site into trypsin. *Biochemistry*. **28**: 9256–63.

Highfield, P.E. and Ellis, R.J. (1978). Synthesis and transport of the small subunit of chloroplast ribulose bisphosphate carboxylase. *Nature*. **271**: 420–4.

Houtz, R.L., Stults, J.T., Mulligan, R.M. and Tolbert, N.E. (1989). Posttranslational modifications in the large subunit of ribulose bisophosphate carboxylase/oxygenase. *Proceedings of the National Academy of Sciences, USA*. **86**: 1855–9.

Igarashi, Y., McFadden, B.A. and El-Gul, T. (1985). Active site histidine in spinach ribulosebisphosphate carboxylase/oxygenase modified by diethyl pyrocarbonate. *Biochemistry*. **24**: 3957–62.

Incharoensakdi, A. and Takabe, T. (1987). Structure, function, and regulation of ribulose-1,5-bisphosphate carboxylase/oxygenase. *Journal of the Science Society of Thailand*. **13**: 133–57.

Jaworowski, A. and Rose, I.A. (1985). Partition kinetics of ribulose-1,5-bisphosphate carboxylase from *Rhodospirillum rubrum*. *Journal of Biological Chemistry*. **260**: 944–8.

Jordan, D.B. and Ogren, W.L. (1981). Species variation in the specificity of ribulose bisphosphate carboxylase/oxygenase. *Nature*. **291**: 513–15.

Kettleborough, C.A., Parry, M.A.J., Burton, S., Gutteridge, S., Keys, A.J. and Phillips, A.L. (1987). The role of the N-terminus of the large subunit of ribulose-bisphosphate carboxylase/oxygenase investigated by construction and expression of chimaeric genes. *European Journal of Biochemistry*. **170**: 335–42.

Klein, R.R. and Mullet, J.E. (1990). Light-induced transcription of chloroplast genes. *Journal of Biological Chemistry*. **265**: 1895–902.

Knaff, D.B. (1989). Structure and regulation of ribulose-1,5-bisphosphate carboxylase/oxygenase. *Trends in Biochemical Sciences*. **14**: 159–60.

Knowles, J.R. (1987). Tinkering with enzymes: what are we learning? *Science*. **236**: 1252–8.

Kramer, W. and Fritz, H.-J. (1987). Oligonucleotide-directed construction of mutations via gapped duplex DNA. *Methods in Enzymology*. **154**: 350–67.

Kunkel, T.A., Roberts, J.D. and Zakour, R.A. (1987). Rapid and efficient site-specific mutagenesis without phenotypic selection. *Methods in Enzymology*. **154**: 367–82.

Laing, W.A., Ogren, W.L. and Hageman, R.H. (1975). Bicarbonate stabilization of ribulose 1,5-diphosphate carboxylase. *Biochemistry*. **14**: 2269–75.

Langridge, P. (1981). Synthesis of the large subunit of spinach ribulose bisphosphate carboxylase may involve a precursor polypeptide. *FEBS Letters*. **123**: 85–9.

Larimer, F.W., Machanoff, R. and Hartman, F.C. (1986). A reconstruction of the gene for ribulose bisphosphate carboxylase from *Rhodospirillum rubrum* that expresses the authentic enzyme in *Eschericia coli*. *Gene*. **41**: 113–20.

Larimer, F.W., Lee, E.H., Mural, R.J., Soper, T.S. and Hartman, F.C. (1987). Intersubunit location of the active site of ribulosebisphosphate carboxylase/oxygenase as determined by *in vivo* hybridization of site-directed mutants. *Journal of Biological Chemistry*. **262**: 15327–9.

Leatherbarrow, R.J. and Fersht, A.R. (1986). Protein engineering. *Protein Engineering*. **1**: 7–16.

Lee, E.H., Soper, T.S., Mural, R.J., Stringer, C.D. and Hartman, F.C. (1987). An intersubunit interaction at the active site of D-ribulose-1,5-bisphosphate carboxylase/oxygenase as revealed by cross-linking and site-directed mutagenesis. *Biochemistry*. **26**: 4599–604.

Lee, E.H., Stringer, C.D. and Hartman, F.C. (1986). Distance between two active-site lysines of ribulose bisphosphate carboxylase from *Rhodospirillum rubrum*.

*Proceedings of the National Academy of Sciences, USA.* **83**: 9383–7.

Lo, K.-M., Jones, S.S., Hackett, N.R. and Khorana, H.G. (1984). Specific amino acid substitutions in bacterioopsin: replacement of a restriction fragment in the structural gene by synthetic DNA fragments containing altered codons. *Proceedings of the National Academy of Sciences, USA.* **81**: 2285–9.

Lorimer, G.H. (1981). Ribulosebisphosphate carboxylase—amino acid sequence of a peptide bearing the activator carbon dioxide. *Biochemistry.* **20**: 1236–40.

Lorimer, G.H. and Andrews, T.J. (1973). Plant photorespiration—an inevitable consequence of the existence of atmospheric oxygen. *Nature.* **243**: 359–60.

Lorimer, G.H. and Hartman, F.C. (1988). Evidence supporting lysine 166 of *Rhodospirillum rubrum* ribulosebisphosphate carboxylase as the essential base which initiates catalysis. *Journal of Biological Chemistry.* **263**: 6468–71.

Lorimer, G.H. and Miziorko, H.M. (1980). Carbamate formation of the $\epsilon$-amino group of a lysyl residue as the basis for the activation of ribulosebisphosphate carboxylase by $CO_2$ and $Mg^{2+}$. *Biochemistry.* **19**: 5321–8.

Lorimer, G.H., Andrews, T.J. and Tolbert, N.E. (1973). Ribulose diphosphate oxygenase. II. Further proof of reaction products and mechanism of action. *Biochemistry.* **12**: 18–23.

Lorimer, G.H., Gutteridge, S. and Madden, M.W. (1987). Partial reactions of ribulose bisphosphate carboxylase: their utility in the study of mutant enzymes. In *Plant Molecular Biology* (ed. D. von Wettstein and N.-H. Chua), pp. 21–31. NATO ASI Series. Plenum Press, New York.

Lowe, D.M., Winter, G. and Fersht, A.R. (1987). Structure–activity relationships in engineered proteins: characterization of disruptive deletions in the $\alpha$-ammonium group binding site of tyrosyl-tRNA synthetase. *Biochemistry.* **26**: 6038–43.

Lundqvist, T. and Schneider, G. (1988). Crystal structure of the binary complex of ribulose-1,5-bisphosphate carboxylase and its product, 3-phospho-D-glycerate. *Journal of Biological Chemistry.* **263**: 3643–6.

Lundqvist, T. and Schneider, G. (1989). Crystal structure of the complex of ribulose-1,5-bisphosphate carboxylase and a transition state analogue, 2-carboxy-D-arabinitol 1,5-bisphosphate. *Journal of Biological Chemistry.* **264**: 7078–83.

Margosiak, S.A., Dharma, A., Bruce-Carver, M.R., Gonzales, A.P., Louie, D. and Kuehn, G.D. (1990). Identification of the large subunit of ribulose 1,5-bisphosphate carboxylase/oxygenase as a substrate for transglutaminase in *Medicago sativa* L. (Alfalfa). *Plant Physiology.* **92**: 88–96.

McFadden, B.A. and Small, C.L. (1988). Cloning, expression and directed mutagenesis of the genes for ribulose bisphosphate carboxylase/oxygenase. *Photosynthesis Research.* **18**: 245–60.

Mitchinson, C. and Wells, J.A. (1989). Protein engineering of disulfide bonds in subtilisin BPN′. *Biochemistry.* **28**: 4807–15.

Miziorko, H.M. (1979). Ribulose-1,5-bisphosphate carboxylase. Evidence in support of the existence of distinct $CO_2$ activator and $CO_2$ substrate sites. *Journal of Biological Chemistry.* **254**: 270–2.

Miziorko, H.M. and Sealy, R.C. (1980). Characterization of the ribulosebisphosphate carboxylase–carbon dioxide–divalent cation–carboxypentitol bisphosphate complex. *Biochemistry.* **19**: 1167–71.

Miziorko, H.M. and Sealy, R.C. (1984). Electron spin resonance studies of ribulosebisphosphate carboxylase: identification of activator cation ligands. *Biochemistry.* **23**: 479–85.

Müllhofer, G. and Rose, I.A. (1965). The position of carbon–carbon bond cleavage in the ribulose diphosphate carboxydismutase reaction. *Journal of Biological Chemistry.* **240**: 1341–6.

Mulligan, R.M., Houtz, R.L. and Tolbert, N.E. (1988). Reaction-intermediate analogue binding by ribulose bisphosphate carboxylase/oxygenase causes specific changes in proteolytic sensitivity: the amino-terminal residue of the large subunit is acetylated proline. *Proceedings of the National Academy of Sciences, USA.* **85**: 1513–17.

Mural, R.J., Soper, T.S., Larimer, F.W. and Hartman, F.C. (1990). Examination of the intersubunit interaction between glutamate-48 and lysine-168 of ribulosebisphosphate carboxylase/oxygenase. *Journal of Biological Chemistry.* **265**: 6501–5.

Nagy, F., Fluhr, R., Morelli, G., Kuhlemeier, C., Poulsen, C., Keith, B., Boutry, M. and Chua, N.-H. (1986). The rubisco small subunit gene as a paradigm for studies on differential gene expression during plant development. *Philosophical Transactions of the Royal Society of London Series B—Biological Sciences.* **B313**: 409–17.

Nargang, F., McIntosh, L. and Somerville, C. (1984). Nucleotide sequence of the ribulosebisphosphate carboxylase gene from *Rhodospirillum rubrum. Molecular and General Genetics.* **193**: 220–4.

Niyogi, S.K., Foote, R.S., Mural, R.J., Larimer, F.W., Mitra, S., Soper, T.S., Machanoff, R. and Hartman, F.C. (1986). Nonessentiality of histidine 291 of *Rhodospirillum rubrum* ribulose-bisphosphate carboxylase/oxygenase as determined by site-directed mutagenesis. *Journal of Biological Chemistry.* **261**: 10087–92.

Paech, C. (1985). Further characterization of an essential histidine residue of ribulose-1,5-bisphosphate carboxylase/oxygenase. *Biochemistry.* **24**: 3194–9.

Pierce, J. (1988). Prospects for manipulating the substrate specificity of ribulose bisphosphate carboxylase/oxygenase. *Physiologia Plantarum.* **72**: 690–8.

Pierce, J. and Reddy, G.S. (1986). The sites for catalysis and activation of ribulose-bisphosphate carboxylase share a common domain. *Archives of Biochemistry and Biophysics.* **245**: 483–93.

Pierce, J., Andrews, T.J. and Lorimer, G.H. (1986a). Reaction intermediate partitioning by ribulose-bisphosphate carboxylase/oxygenase with differing substrate specificities. *Journal of Biological Chemistry.* **261**: 10248–56.

Pierce, J., Lorimer, G.H. and Reddy, G.S. (1986b). Kinetic mechanism of ribulosebisphosphate carboxylase: evidence for an ordered, sequential reaction. *Biochemistry.* **25**: 1636–44.

Pierce, J., Carlson, T.J. and Williams, J.G.K. (1989). A cyanobacterial mutant requiring the expression of ribulose bisphosphate carboxylase from a photosynthetic anaerobe. *Proceedings of the National Academy of Sciences, USA.* **86**: 5753–7.

Portis, A.R., Jr. (1990). Rubisco activase. *Biochimica et Biophysica Acta.* **1015**: 15–28.

Profy, A.T. and Schimmel, P. (1988). Complementary use of chemical modification and site-directed mutagenesis to probe structure–activity relationships in enzymes. *Progress in Nucleic Acid Research and Molecular Biology.* **35**: 1–26.

Quayle, J.R., Fuller, R.C., Benson, A.A. and Calvin, M. (1954). Enzymatic carboxylation of ribulose diphosphate. *Journal of American Chemical Society.* **76**: 3610–11.

Ranty, B., Lundqvist, T., Schneider, G., Madden, M., Howard, R. and Lorimer, G. (1990). Truncation of ribulose-1,5-bisphosphate carboxylase/oxygenase (Rubisco) from *Rhodospirillum rubrum* affects the holoenzyme assembly and activity. *EMBO Journal.* **9**: 1365–73.

Robinson, C. and Ellis, R.J. (1984). Transport of proteins into chloroplasts. Partial purification of a chloroplast protease involved in the processing of imported precursor polypeptides. *European Journal of Biochemistry.* **142**: 337–42.

Rodermel, S.R. and Bogorad, L. (1985). Maize plastid photogenes: mapping and photoregulation of transcript levels during light-induced development. *The Journal of Cell Biology.* **100**: 463–76.

Rodermel, S.R., Abbott, M.S. and Bogorad, L. (1988). Nuclear-organelle interactions: nuclear antisense gene inhibits ribulose bisphosphate carboxylase enzyme levels in transformed tobacco plants. *Cell.* **55**: 673–81.

Roeske, C.A. and O'Leary, M.H. (1985). Carbon isotope effect on carboxylation of ribulose bisphosphate catalyzed by ribulosebisphosphate carboxylase from *Rhodospirillum rubrum*. *Biochemistry.* **24**: 1603–7.

Roy, H., Cannon, S. and Gilson, M. (1988). Assembly of rIbisco from native subunits. *Biochimica et Biophysica Acta.* **957**: 323–34.

Salvucci, M.E., Portis, A.R., Jr. and Ogren, W.L. (1985). A soluble chloroplast protein catalyzes ribulosebisphosphate carboxylase oxygenase activation *in vivo*. *Photosynthesis Research.* **7**: 193–201.

Saver, B.G. and Knowles, J.R. (1982). Ribulose-1,5-bisphosphate carboxylase: enzyme catalyzed appearance of solvent tritium at carbon 3 of ribulose 1,5-bisphosphate reisolated after partial reaction. *Biochemistry.* **21**: 5398–403.

Schimmel, P. (1989). Hazards of deducing enzyme structure-activity relationships on the basis of chemical applications of molecular biology. *Accounts of Chemical Research.* **22**: 232–3.

Schloss, J.V. (1988). Comparative affinities of the epimeric reaction-intermediate analogs 2- and 4-carboxy-D-arabinitol 1,5-bisphosphate for spinach ribulose 1,5-bisphosphate carboxylase. *Journal of Biological Chemistry.* **263**: 4145–50.

Schloss, J.V. (1990). The kinetic properties of ribulosebisphosphate carboxylase. In *The Proceedings of NATO ASI on Enzymatic and Model Carboxylation and Reduction Reactions for Carbon Dioxide Utilization* (ed. M. Aresta and J.V. Schloss). Kluwer Academic Publishers, Dordrecht.

Schloss, J.V. and Lorimer, G.H. (1982). The stereochemical course of ribulosebisphosphate carboxylase. *Journal of Biology Chemistry.* **257**: 4691–4.

Schloss, J.V., Phares, E.F., Long, M.V., Norton, I.L., Stringer, C.D. and Hartman, F.C. (1982). Ribulosebisphosphate carboxylase/oxygenase from *Rhodospirillum rubrum*. *Methods in Enzymology.* **90**: 522–8.

Schneider, G., Lindqvist, Y., Brändén, C.-I. and Lorimer, G. (1986). Three-dimensional structure of ribulose-1,5-bisphosphate carboxylase/oxygenase from *Rhodospirillum rubrum* at 2.9 Å resolution. *EMBO Journal.* **5**: 3409–15.

Schneider, G., Lindqvist, Y. and Lundqvist, T. (1990). Crystallographic refinement and structure of ribulose-1,5-bisphosphate carboxylase from *Rhodospirillum rubrum* at 1.7 Å resolution. *Journal of Molecular Biology.* **211**: 989–1008.

Scrutton, N.S., Berry, A. and Perham, R.N. (1990). Redesign of the coenzyme specificity of a dehydrogenase by protein engineering. *Nature.* **343**: 38–43.

Seemann, J.R., Kobza, J. and Moore, B. (1990). Metabolism of 2-carboxyarabinitol 1-phosphate and regulation of ribulose-1,5-bisphosphate carboxylase activity. *Photosynthesis Research.* **23**: 119–30.

Shaw, W.V. (1987). Protein engineering. The design, synthesis and characterization of factitious proteins. *Biochemistry Journal.* **246**: 1–17.

Smith, H.B. and Hartman, F.C. (1988). Restoration of activity to catalytically deficient mutants of ribulosebisphosphate carboxylase/oxygenase by aminoethylation. *Journal of Biological Chemistry.* **263**: 4921–5.

Smith, H.B., Larimer, F.W. and Hartman, F.C. (1988a). Substitution of Glu 48 of ribulose bisphosphate carboxylase/oxygenase with S-carboxymethylcysteine and aspartic acid. *Journal of Cell Biology.* **107**: 830a.

Smith, H.B., Larimer, F.W. and Hartman, F.C. (1988b). Subtle alteration of the active site of ribulosebisphosphate carboxylase/oxygenase by concerted site-directed mutagenesis and chemical modification.

*Biochemistry and Biophysics Research Communications.* **152**: 579–84.

Smith, H.B., Larimer, F.W. and Hartman, F.C. (1990). An engineered change in substrate specificity of ribulosebisphosphate carboxylase/oxygenase. *Journal of Biological Chemistry.* **265**: 1243–5.

Soll, J. and Buchanan, B.B. (1983). Phosphorylation of chloroplast ribulose bisphosphate carboxylase/oxygenase small subunit by an envelope-bound protein kinase *in situ. Journal of Biological Chemistry.* **258**: 6686–9.

Somerville, C.R. and Ogren, W.L. (1979). A phosphoglycolate phosphatase deficient mutant of *Arabidopsis. Nature.* **280**: 833–6.

Somerville, C.R. and Somerville, S.C. (1984). Cloning and expression of the *R. rubrum* ribulosebisphosphate carboxylase gene in *E. coli. Molecular and General Genetics.* **193**: 214–19.

Somerville, C.R., Portis, A.R., Jr. and Ogren, W.L. (1982). A mutant of *Arabidopsis thaliana* which lacks activation of RuBP carboxylase *in vivo. Plant Physiology.* **70**: 381–7.

Soper, T.S., Mural, R.J., Larimer, F.W., Lee, E.H., Machanoff, R. and Hartman, F.C. (1988). Essentiality of Lys-329 of ribulose-1,5-bisphosphate carboxylase/oxygenase from *Rhodospirillum rubrum* as demonstrated by site-directed mutagenesis. *Protein Engineering.* **2**: 39–44.

Soper, T.S., Larimer, F.W., Mural, R.J., Lee, E.H. and Hartman, F.C. (1989). Examination of subunit interactions at the active site of ribulose 1,5-bisphosphate carboxylase/oxygenase from *Rhodospirillum rubrum* by hybridization of site-directed mutants. *Journal of Protein Chemistry.* **8**: 239–49.

Spreitzer, R.J. and Ogren, W.L. (1985). *Chlamydomonas* chloroplast mutants with altered ribulose-1,5-bisphosphate carboxylase/oxygenase. In *Molecular Biology of the Photosynthetic Apparatus* (ed. K.E. Steinback, S. Bouitz, C.J. Arntzen and L. Bogorad), pp. 355–60. Cold Spring Harbor Laboratory, New York.

Styring, S. and Brändén, R. (1985). Identification of ligands to the metal ion in copper(II)-activated ribulose-1,5-bisphosphate carboxylase/oxygenase by the use of electron paramagnetic resonance spectroscopy and $^{17}$O-labelled ligands. *Biochemistry.* **24**: 6011–19.

Sue, J.M. and Knowles, J.R. (1982). Ribulose-1,5-bisphosphate carboxylase: fate of the tritium label in [3-$^3$H]ribulose 1,5-bisphosphate during the enzyme-catalyzed reaction. *Biochemistry.* **21**: 5404–10.

Tabita, F.R. and McFadden, B.A. (1974). D-ribulose 1,5-diphosphate carboxylase from *Rhodospirillum rubrum. Journal of Biological Chemistry.* **249**: 3459–64.

Terzaghi, B.E., Laing, W.A., Christeller, J.T., Peterson, G.B. and Hill, D.F. (1986). Ribulose 1,5-bisphosphate, effect on the catalytic properties of changing methionine-330 to leucine in the *Rhodospirillum rubrum* enzyme. *Biochemical Journal.* **235**: 839–46.

Torres-Ruiz, J.A. and McFadden, B.A. (1985). Isolation of $L_8$ and $L_8S_8$ forms of ribulose bisphosphate carboxylase/oxygenase from *Chromatium vinosumn. Archives of Microbiology.* **142**: 55–60.

Toth, M.J., Murgola, E.J. and Schimmel, P. (1988). Evidence for a unique first position codon–anticodon mismatch *in vivo. Journal of Molecular Biology.* **201**: 451–4.

Van Dyk, D.E. and Schloss, J.V. (1986). Deuterium isotope effects in the carboxylase reaction of ribulose-1,5-bisphosphate carboxylase/oxygenase. *Biochemistry.* **25**: 5145–56.

von Wettstein, D. and Chua, N.-H. (1987). *The Proceedings of NATO ASI on Plant Molecular Biology*, Plenum Press, New York.

Voordouw, G., De Vries, P.A., Van Den Berg, W.A.M. and De Clerk, E.P.J. (1987). Site-directed mutagenesis of the small subunit of ribulose-1,5-bisphosphate carboxylase/oxygenase from *Anacystis nidulans. European Journal of Biochemistry.* **163**: 591–8.

Wasmann, C.C., Ramage, R.T., Bohnert, H.J. and Ostrem, J.A. (1989). Identification of an assembly domain in the small subunit of ribulose-1,5-bisphosphate carboxylase. *Proceedings of the National Academy of Sciences, USA.* **86**: 1198–202.

Weissbach, A., Smyrniotis, P.Z. and Horecker, B.L. (1954). Pentose phosphate and $CO_2$ fixation with spinach extracts. *Journal of the American Chemical Society.* **76**: 3611–12.

Wente, S.R. and Schachman, H.K. (1987). Shared active sites in oligomeric enzymes: model studies with defective mutants of aspartate transcarbamoylase produced by site-directed mutagenesis. *Proceedings of the National Academy of Sciences, USA.* **84**: 31–5.

Zoller, M.J. and Smith, M. (1983). Oligonucleotide-directed mutagenesis of DNA fragments cloned into M13 vectors. *Methods in Enzymology.* **100**: 468–500.

Zoller, M.J. and Smith, M. (1987). Oligonucleotide-directed mutagenesis: a simple method using two oligonucleotide primers and a single-stranded DNA template. *Methods in Enzymology.* **154**: 329–50.

# 5

# Isolation and Functional Analysis of Random and Site-directed Mutants of Photosystem II

Peter J. Nixon, Dexter A. Chisholm and Bruce A. Diner

## Introduction

### Redox components and kinetics

The reaction centres of Photosystem I (PSI) and Photosystem II (PSII) are the two sites within the oxygenic photosynthetic electron transport chain in which light energy drives electron transfer (for recent reviews see Golbeck and Bryant 1990; Hansson and Wydrzynski 1990 for PSI and PSII, respectively). In the case of PSII (Fig. 5.1), the subject of this review, excitation energy is transferred from an ensemble of chlorophylls surrounding the reaction centre to the primary electron donor, P680, which probably consists of a pair of chlorophyll a (chl a) molecules. Here P680*, in its lowest excited singlet state (Fig. 5.2), takes 2–3 ps ($t_{1/2}$) (Wasielewski *et al.* 1989) to transfer an electron to the primary electron acceptor, I, a pheophytin a (pheo a) (Klimov and Krasnovskii 1981), generating the charge pair $P680^+I^-$. $I^-$ then reduces the primary plastoquinone electron acceptor, $Q_A$, in 350–500 ps ($t_{1/2}$), generating the semiquinone (Nuijs *et al.* 1986; Schatz *et al.* 1987; Eckert *et al.* 1988; Trissl and Leibl 1987). $P680^+$ is reduced by a secondary donor, Z, a tyrosine (Debus *et al.* 1988b; Metz *et al.* 1989) in 2 30 µs ($t_{1/2}$) (Reinman *et al.* 1981) in the absence, or 20–50 and 260 ns ($t_{1/2}$) (Brettel *et al.* 1984) in the presence, of a cluster of 3-4 manganese (Mn). This cluster is the site of water oxidation, which constitutes the tertiary and terminal electron donor (for review, see Babcock *et al.* 1989). It takes $Q_A^-$ several hundreds of microseconds ($t_{1/2}$) (Robinson and Crofts 1983) to reduce the secondary plastoquinone acceptor, $Q_B$, first to the plasto-

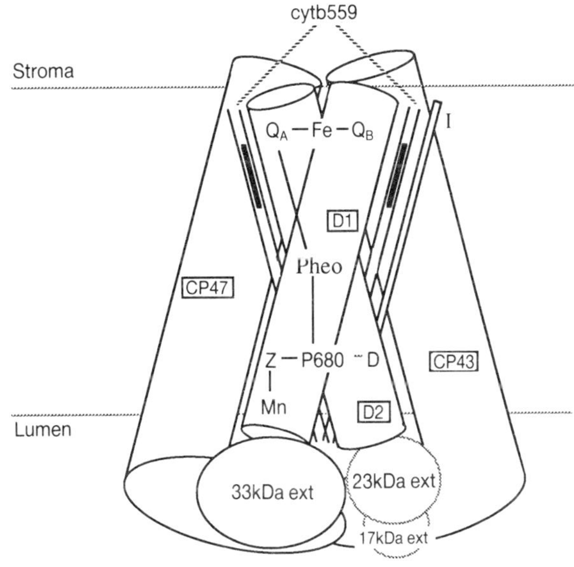



**Fig. 5.1** Model of PSII oxygen-evolving core complex showing polypeptides D1 and D2, which coordinate the primary photoreactants P680 and Pheo, the primary and secondary quinone electron acceptors, $Q_A$ and $Q_B$, and the secondary and tertiary electron donors, Z (D) and the 4 manganese (Mn) of the oxygen-evolving complex (OEC). Also indicated are the intrinsic chlorophyll-protein complexes CP47 and CP43 that constitute the core antenna coordinating about 40 chlorophylls, 2 cytochrome b559 and the *psbI* gene product. The lumenally located extrinsic 33, 23 and 17 kDa (OEE 1, 2 and 3, respectively) polypeptides present in green algae and higher plants are shown. These are peripherally associated with the OEC and promote binding of $Ca^{2+}$ and $Cl^-$, but are not absolutely required for oxygen evolution. Of these only the 33 kDa extrinsic polypeptide has been detected in cyanobacteria. Not indicated are the many subunits of <10 kDa that have recently been found to be associated with the PSII core complex.

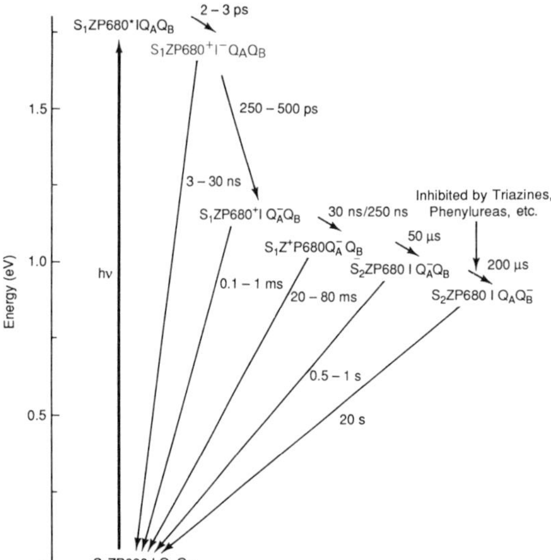


**Fig. 5.2**   Redox states of the PSII reaction centre and the approximate free energy differences between them. The half-times of forward reactions and of charge recombination reactions are also indicated.

semiquinone and then, after a second charge separation, to the plastoquinol. The quinol dissociates from the reaction centre (Diner *et al.* 1984; McPherson *et al.* 1990) and is replaced at the $Q_B$ binding site by a quinone, one of 10 per reaction centre present in the thylakoid membrane. A non-haem iron (for review see Diner and Petrouleas 1987), whose function is not known, is located midway between $Q_A$ and $Q_B$.

These properties and components are believed to be common to PSII reaction centres isolated from all organisms that carry out oxygenic photosynthesis.

## Component polypeptides

All of these electron transfer reactions are thought to occur within a heterodimeric reaction centre complex made up of one copy each of polypeptides D1 and D2, of 38 and 39 kDa, respectively. Cytochrome b559 (cyt b559) is also part of the reaction centre, consisting of two subunits, $\alpha$ and $\beta$, of 9 and 4.5 kDa, respectively, which together bind one haem. There is some controversy concerning the number of cytochrome b559s per centre, with some groups claiming two (de Paula *et al.* 1985; Cramer *et al.* 1986), and others claiming one

(Miyazaki *et al.* 1989). The role of this cytochrome is unknown, although it has been proposed to serve in a protective function, reducing errant oxidizing equivalents that would otherwise lead to photoinhibition (Thompson and Brudvig 1988).

Another subunit, the *psbI* gene product, has been shown by two groups (Ikeuchi and Inoue 1988; Webber *et al.* 1989a) to be part of the reaction centre. Its function is also unknown.

A complex composed of polypeptides D1, D2, cytochrome b559 and the *psbI* gene product, hereafter referred to as the 'PSII reaction centre', has been isolated biochemically (from spinach by Nanba and Satoh 1987; and from pea by Barber *et al.* 1987; Satoh and Nakane 1990). These isolated complexes are still capable of light-driven primary charge separation between P680 and pheophytin I (Danielius *et al.* 1987; Takahashi *et al.* 1987). The rigorous detergent extraction required to isolate this complex strips away both the primary and secondary quinone electron acceptors, $Q_A$ and $Q_B$. Attempts at reassociation of the $Q_A$ quinone have been reported (Chapman *et al.* 1990), although proof of function at the natural binding site has been lacking. As for the other components, there is approximate agreement on the stoichiometry of six chl a: two pheo a: one cyt b559 haem. However, there is disagreement on the number of chlorophylls per reaction centre, with Dekker *et al.* (1989) normalizing to two cyt b559 (thus 10–12 chl a/centre) and Kobayashi *et al.* (1990) normalizing to two pheophytin a (thus six chl a/centre).

There are two chlorophyll–protein complexes, CP47 and CP43, associated with the reaction centre but which are peripheral to the primary electron transfer reactions. Together these bind some 40 chl a, forming the core antenna that transfers light energy to the reaction centre (Bricker 1990). The stoichiometry of the largest subunits, CP47:CP43:D2:D1, is 1:1:1:1 (de Vitry *et al.* 1987; Gounaris *et al.* 1987). CP47 and CP43 stabilize the reaction centre complex against endogenous proteolysis (de Vitry *et al.* 1989; Yu and Vermaas 1990) and are probable binding sites for two extrinsic polypeptides (see below). The reaction centre plus CP47 and CP43 plus an unknown number of recently discovered small subunits have come to be called the PSII core complex (non-oxygen evolving).

In addition to these intrinsic polypeptides there are three, or possibly four, extrinsic polypeptides with apparent molecular weights determined by SDS-polyacrylamide gel electrophoresis of 33, 23

**Table 5.1**   Current PSII genes

| Gene designation[1] | Name of gene product[2] | Higher plants[3] | | *Synechocystis* 6803[4] | | Phenotype of inactivated strain[5] | Location of gene product[6] | |
| | | Location of gene | Size of protein | No. of gene copies | Size | | Wheat | *S. vulcanus* |
|---|---|---|---|---|---|---|---|---|
| *psb A* | D1 polypeptide | CP | 353/343 (38.0) | 3 | 360 | PS$^-$ | RC | RC |
| *psb B* | CP47 apoprotein | CP | 508 (55.6) | 1 | 507 | PS$^-$ | Core | Core |
| *psb C* | CP43 apoprotein | CP | 461/459 (50.1) | 1 | 460 | PS$^-$ | Core | Core |
| *psb D* | D2 polypeptide | CP | 353/352 (39.4) | 2 | 352 | | RC | RC |
| *psb E* | α-subunit cytochrome b-559 | CP | 83/82 (9.3) | 1 | 81 | | RC | RC |
| *psb F* | β-subunit cytochrome b-559 | CP | 39/38 (4.4) | 1 | 44 | PS$^-$ | RC | RC |
| *psb G* | | CP | 284 (32.3) | 2 | 1) 248<br>2) 219 | | nd | nd |
| *psb H* | 10-kDa phosphoprotein or PSII-H polypeptide | CP | 73/72 (7.6) | 1 | 64/63 | | Core | Core |
| *psb I* | PSII-I polypeptide | CP | 36/36 (4.2) | nd | nd | | RC | RC |
| *psb J* | hypothetical polypeptide | CP | 40 | 1 | 39 | PS$^+$ | nd | nd |
| *psb K* | PSII-K polypeptide | CP | 98/37 (4.3) | 1 | 45/37 | | Core | Core |
| *psb L* | PSII-L polypeptide | CP | 38/37 (4.4) | 1 | 39 | | Core | Core |
| *psb M* | PSII-M polypeptide | CP | 34/34 (3.8) | nd | nd | | nd | Core |
| *psb N* | PSII-N polypeptide | CP | 43/43 (4.7) | nd | nd | | nd | Core |
| *psb O* | 33-kDa extrinsic polypeptide | N | 331/247 (26.5) | 1 | 274/246 | PS$^+$ | Core, extrinsic | Core |
| *psb P* | 23-kDa extrinsic polypeptide | N | 267/186 (20.2) | nd | nd | | mem, extrinsic | nd |
| *psb Q* | 16-kDa extrinsic polypeptide | N | 232/149 (16.5) | nd | nd | | mem, extrinsic | nd |
| *psb R* | 10-kDa polypeptide | N | 140/99 (10.8) | nd | nd | | mem | nd |

[1] The nomenclature for the genes of PSII was recently revised following a workshop held in Stockholm at the VIIIth International Congress on Photosynthesis in August 1989. Only this updated nomenclature will be used in this article (see Hallick 1989).   [2] To avoid confusion, only one name for the gene product will be given here. Other names that have been used will be detailed in later sections. An exception will be the *psbH* gene product, which in plants is referred to as the 9-kDa or 10-kDa phosphoprotein, and in cyanobacteria as the PSII-H polypeptide.   [3] The number of amino acid residues present in the products predicted from the chloroplast genes (CP) of tobacco was taken from a recent review by Sugiura (1990). The size of the nuclear genes (N) from spinach was taken from papers by Jansen *et al.* (1987), Lautner *et al.* (1988), and Tyagi *et al.* (1987). Where applicable, the probable number of amino acid residues present in the gene product, after post-translational processing, is given as the second number after the slash. The mass of the mature product in kDa is given in parentheses.
[4] The references for this column will be detailed in the later sections focussing on each gene. The size of the unprocessed gene product and where applicable the mature form, is given as the number of amino acid residues   [5] The phenotype assayed is the ability to grow photoautotrophically (PS$^+$). Only the phenotypes of mutants of *Synechocystis* 6803 with a specifically inactivated gene are given.   [6] RC, present in the minimal PSII reaction centre complex described by Namba and Satoh (1987); Core, present in the $O_2$-evolving PSII core complex described by Ikeuchi and Inoue (1986), but not in the reaction centre complex; mem, found in the PSII-enriched membranes of Berthold *et al.* (1981), but not in the $O_2$-evolving core complex; nd, not detected or not determined. All the polypeptides are considered to be intrinsic membrane components apart from those indicated as extrinsic.

and 17 (1 OEE 1, 2 and 3, respectively) and 10 kDa (for review see Andersson and Åkerlund 1987). Only the first of these has so far been identified in cyanobacteria. They are associated with the oxygen-evolving complex *in vivo* and serve to enhance the binding affinity of $Ca^{2+}$ and $Cl^-$, cofactors required for oxygen evolution (Babcock *et al.* 1987; Coleman 1990). There is evidence to favour binding of the 33 kDa polypeptide to CP47 (reviewed in Bricker 1990) and also to the reaction centre itself (Gounaris *et al.*, 1988) and binding of the 23 kDa subunit to CP43 (de Vitry *et al.* 1989). The reaction centre plus CP47, CP43, the 33 kDa extrinsic subunit, the OEC and a plethora of recently discovered small subunits of less than 10 kDa, described by Ikeuchi *et al.* (1990) and Koike *et al.* (1990) (see Table 5.1), will be referred to as the oxygen-evolving PSII core complex (see also Ghanotakis *et al.* 1987). Altogether close to 20 polypeptides have been attributed to this complex. With a structure this complicated, and the likelihood of partial subunit loss during the course of isolation, it is not surprising that attempts at crystallization of the complex have not been successful. The roles of most of the newly discovered small subunits are unknown.

We will not discuss the light-harvesting chlorophyll a/b proteins in this review. These are peripheral to electron transfer and to assembly of the PSII core complex. They are also not present in the cyanobacteria that constitute a major part of the mutagenesis effort in PSII.

## Comparison with reaction centres from purple non-sulphur photosynthetic bacteria

The purple non-sulphur photosynthetic bacteria (PNSPB) do not evolve oxygen. There is consequently no homology between the secondary and tertiary electron donors of their reaction centres and those of PSII. However, the PNSPB show striking homologies with respect to the remaining redox components of the reaction centre. The primary donor, a pair of bacterio-chl (Bchl) and the electron acceptors bacterio-pheo (Bpheo) and ubiquinone, play homologous roles to the chl a, pheo a and plastoquinones found in PSII. These are bound by polypeptides L and M (Chang *et al.* 1986; Allen *et al.* 1987; Michel and Deisenhofer 1988; Deisenhofer and Michel 1989). Four Bchl, two Bpheo and two ubiquinones are associated with

each centre. Only two or three of the BChl and only one of the Bpheo actually participate in the electron transfer. The pair of Bchl that constitute the primary electron donor and the non-haem iron, located between ubiquinones, $Q_A$ and $Q_B$, lie on a two-fold symmetry axis running through the reaction centre and orientated perpendicular to the plane of the chromatophore membrane.

Like its bacterial counterpart, the PSII reaction centre complex contains more chromophores than actually participate in the primary photochemistry. Also like its bacterial counterpart, it is thought that these components are bound in a bilaterally symmetrical fashion with a like number of chlorophylls and pheophytins associated with each of the two subunits, D1 and D2 (Trebst 1986; Michel and Deisenhofer 1988). The plastoquinones $Q_A$ and $Q_B$ are clearly bound to D2 and D1, respectively, and are, most likely, symmetrically positioned with respect to the non-haem iron (Deisenhofer and Michel 1989). Likewise, secondary donors D and Z are tyrosines (Barry and Babcock 1987) associated with D2 (Debus *et al.* 1988a; Vermaas *et al.* 1988a) and D1 (Debus *et al.* 1988b; Metz *et al.* 1989) respectively, and are thought to be symmetrically positioned with respect to P680.

Polypeptides D1 and D2 have a number of localized homologies with respect to each other and to the bacterial subunits, L and M (Trebst 1986; Michel and Deisenhofer 1988; Komiya *et al.* 1988). They are also thought to have similar folding patterns to the bacterial reaction centre subunits, with five transmembrane α-helices each (Fig. 5.3). In addition to hydropathic considerations (Trebst 1986), the evidence favouring similar folding involves the topology of antigenic determinants (Sayre *et al.* 1986), the identification of phosphorylated threonines at N-termini (Michel *et al.* 1988), the location of mutations bestowing herbicide resistance in a cluster close to $Q_B$ (Erickson *et al.* 1989; Etienne *et al.* 1990; Diner *et al.* 1991), and other point mutations that affect the primary electron donor P680 (Vermaas *et al.* 1987a; W.F.J. Vermaas and B.A. Diner, unpublished results) and the secondary electron donors Z (Debus *et al.* 1988b; Metz *et al.* 1989) and D (Debus *et al.* 1988a; Vermaas *et al.* 1988a). CP47 and CP43 have been proposed to have six to seven transmembrane α-helices (Bricker 1990). Phosphorylation of a threonine at the N-terminus of CP43 would, as for D1 and D2, locate the N-terminus on the stromal side of the thylakoid membrane (Michel *et al.* 1988). The α and β subunits of cyt b559 each

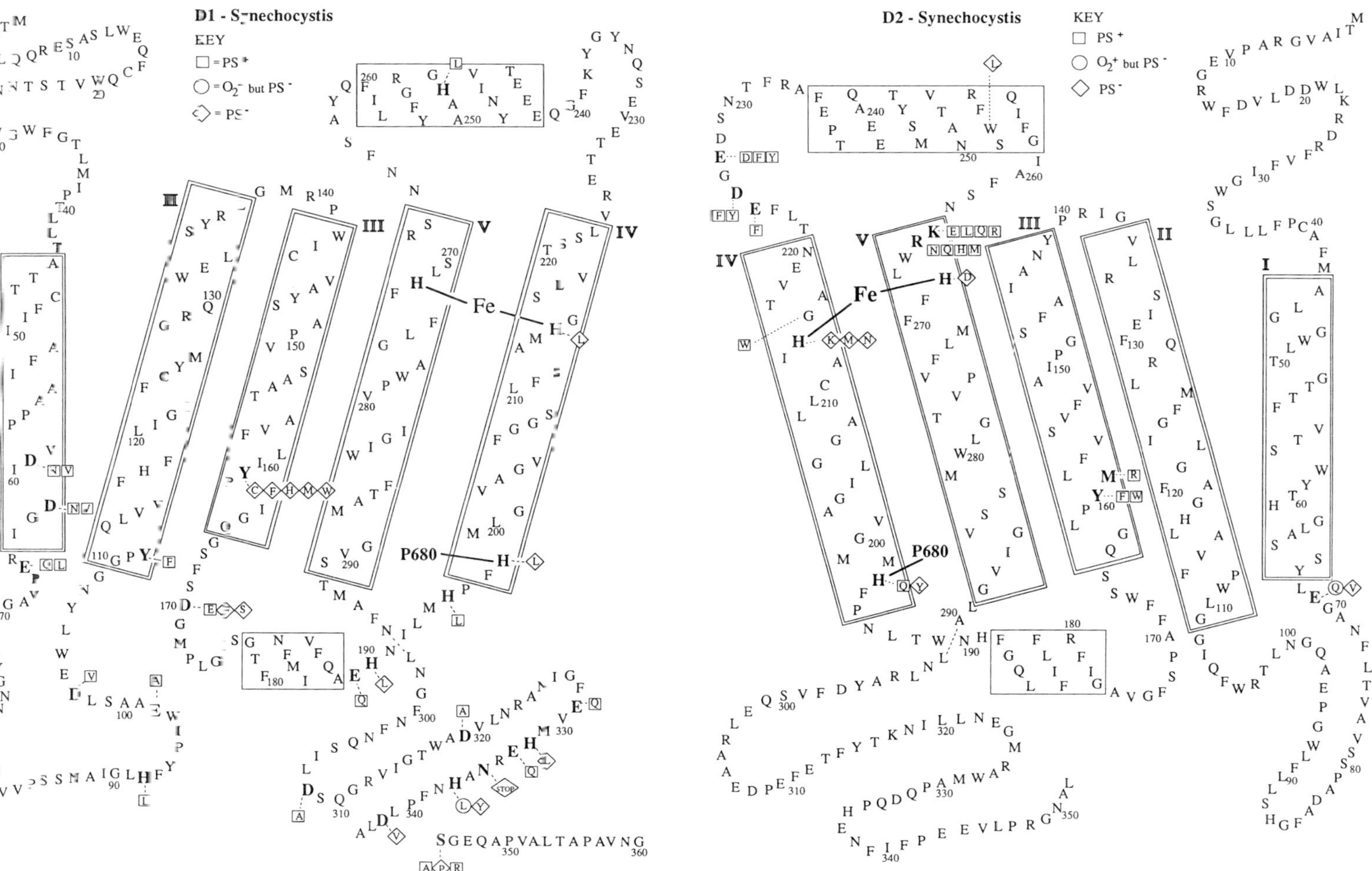


**Fig. 5.3** Probable folding models of (a) polypeptide D1 (*psbA2* and *psbA3* gene products) and (b) polypeptide D2 (*psbD1* and *psbD2* gene products) of *Synechocystis* 6803 showing the locations of the transmembrane α-helices (double-lined rectangles) and membrane parallel α-helices (single-lined rectangles) as suggested by Michel and Deisenhofer (1988) and by Komiya *et al.* (1988) by analogy with the *R. viridis* and *Rb. sphaeroides* reaction centres, respectively. The stromal side of each subunit is on top, the lumenal side on the bottom. Also indicated are the locations of the site-directed mutations of Table 5.3 (in bold) and an indication of whether the mutants are capable of photoautotrophic growth (□), not capable of photoautotrophic growth or oxygen evolution (◇) and not capable of photoautotrophic growth but capable of oxygen evolution (○).

contain one transmembrane α-helix and, like D1–D2, would have their N-termini on the stromal side of the thylakoid membrane (Tae *et al.* 1988). In light of the orientation of the above-mentioned polypeptides, it is likely that the N-terminus of CP47 is also found on the stromal side of the membrane.

No X-ray crystallographic structure exists for the PSII reaction centre. However, the functional and spectroscopic homologies that exist with the reaction centres of *Rhodobacter sphaeroides* and *Rhodopseudomonas viridis* provide some guidance as to the location of redox components and their ligands with respect to the primary structures of D1 and D2. Site-directed mutagenesis in the cyanobacterium, *Synechocystis* sp. PCC 6803 has served to confirm the location and role of ligating residues that have homologous function in the PSII and bacterial reaction centre complexes. While no homology exists with bacterial reaction centres, it has also permitted the location of the secondary electron donors Z (Debus *et al.* 1988b; Metz *et al.* 1989) and D (Debus *et al.* 1988a; Vermaas *et al.* 1988a) and is currently being used by a number of groups in an intensive search for ligands to the Mn cluster responsible for the oxidation of water to molecular oxygen.

Mutagenesis serves as more than a probe of structure. A considerable collection of mutants has been obtained using chemical and radiation-induced mutagenesis, particularly in *Chlamydomonas reinhardtii* (Bennoun and Delepelaire 1982), and more recently with engineered deletions and interruptions in cyanobacterial PSII genes (Vermaas *et al.* 1988b). These have provided a wealth of information on the assembly and stability of the PSII core complex, in the absence or modification of one of the subunits that makes up the core complex (Vermaas *et al.* 1988b; de Vitry *et al.* 1989; Yu and Vermaas 1990). In addition, in PSII (as in the photosynthetic bacteria where crystallographic data exist) site-directed mutants provide a handle on function and mechanism. In the past this kind of information was attainable on a much more limited scale using chemical modification, which is handicapped by incomplete reactions and a limitation in the functional groups affected.

In this chapter we review what is known about the PSII genes and the structure and function of the polypeptides they encode. This information is based primarily on analysis of mutants, although, where relevant, independent biochemical information is introduced. We describe briefly how directed mutants are made, with emphasis on *Synechocystis* 6803, the cyanobacterium that has been most often used for site-directed mutagenesis. We describe the behaviour of a number of these mutants in which particular amino acid residues in polypeptides D1 and D2 have been targeted. Many of these have been isolated in our own laboratory. Finally, we include a description of the more readily available biochemical and biophysical techniques that have been used for the analysis of their function.

# Mutagenesis of PSII

## Random mutagenesis

Two strategies are being used to investigate the connection between structure and function in PSII through mutagenesis. In the first of these, the organism of choice is treated so that a mutation is incorporated randomly into its genome. This can occur spontaneously, but is more efficiently achieved with chemical mutagens (such as nitroso-guanidine and 5-fluorodeoxyuridine) or radiation (such as X-rays and ultra-violet). The resulting mutants can then be screened for a particular phenotype, such as resistance to the herbicide, atrazine, and the nature of the lesion can be determined. Many herbicide resistant strains of plants, algae and cyanobacteria have been generated in this way and have been shown to result from mutations in the *psbA* gene, which encodes the D1 polypeptide (see p. 110).

Although this technique has been useful in certain cases, its limitations lie in the difficulty in diagnosing the cause of certain phenotypes, and the lengthy procedures needed to determine the nature of nuclear lesions. For instance, random mutagenesis of higher plants, such as maize, has yielded a class of mutants that shows elevated levels of fluorescence upon illumination (Miles *et al.* 1985). One subclass of these high fluorescent mutants has been shown to result from the loss of the entire PSII complex from the thylakoid membrane of mesophyll cells (Leto *et al.* 1985). This pleiotropic effect makes a biochemical determination of the precise cause of the defect extremely difficult.

## Site-directed mutagenesis

The second technique now commonly used is that

of site-directed mutagenesis (Rossi and Zoller 1987). In this method, a defined mutation, such as the addition, deletion or substitution of specific nucleotides, is incorporated into the target gene *in vitro*. The effect of this mutation is then usually analysed by returning the modified gene to the organism from which it was obtained (see p. 102). However, if this is difficult, a heterologous expression system could be used (see Chapter 10). The advantages that this methodology offers over the random approach is that a defined mutation, such as changing the codon for a particular amino acid in a protein, can be specified in advance and subsequent subtle or lethal changes in phenotype, which may not have otherwise been detected, can be monitored.

The creation of site-directed mutants of an organism can be thought of as occurring in three stages:

(i) part or all of the target gene is cloned;
(ii) the target gene is mutated;
(iii) the modified gene is returned to the parent organism and stably maintained.

The analysis is made much easier if deletion strains are available to act as recipients for the modified gene, in order first to avoid 'contamination' of the mutant phenotype by wild type copies of the target gene and secondly to prevent potential problems of reversion. Although a deletion strain is preferable, the wild type strain can still be used in situations where the modified gene (or, perhaps, a novel gene), codes for a dominant characteristic, such as herbicide resistance (Brusslan and Haselkorn 1989); or where the modified gene is to replace the wild type copy through homologous recombination.

With the recent, and continuing, development of techniques for the cloning and modification of genes, the first two steps in the generation of mutants have become routine. For instance, there are now several commercially available kits for introducing specific mutations into cloned genes using oligonucleotide-mediated mutagenesis. These systems can work extremely efficiently, achieving high rates of mutation ($>50\%$) by selecting against the wild type strand of DNA, through, for example, biological (Kunkel *et al.* 1987) or biochemical (Vandeyar *et al.* 1988) means. Lately, there are also methods available that exploit the polymerase chain reaction (PCR) (Saiki *et al.* 1988). In these cases, close to 100% efficiency in generat-

ing the mutant product can be achieved (Ho *et al.* 1989).

Such methods are extremely versatile, allowing not only the construction of point mutations, but also specific deletions within a polypeptide, or even the rapid construction of hybrid proteins containing protein domains from several polypeptides (Clackson and Winter 1989). 'Saturation mutagenesis' using oligonucleotide cassettes has proved popular for the *in vitro* mutagenesis of small defined areas of a gene (Wells *et al.* 1985; Hill *et al.* 1986).

The major problem in investigating the phenotypic effects of a site-directed mutation is point (iii) above—the transfer of the modified gene back to the host, and the construction of deletion strains. For studies on PSII there are potentially three types of organism from which to choose: plants, green algae and cyanobacteria. In the next three sections we shall briefly review the capability of each of these organisms to be manipulated through *in vitro* mutagenesis.

**Plants**

Plants represent the least amenable system for generating and analysing site-directed mutants in PSII. As in the case of green algae, the genes for PSII are located in both the nucleus and chloroplast. Since most of the polypeptides of PSII are coded for by the chloroplast (Gray 1987), mutagenesis of these components requires a chloroplast transformation procedure. The high velocity microprojectile procedure recently used successfully for chloroplast transformations in *Chlamydomonas reinhardtii* (Boynton *et al.* 1988) can also be used in higher plants (Svab *et al.* 1990) but at only 1% of the efficiency.

Another approach to the engineering of chloroplast-encoded components of the PSII complex in higher plants was recently described by Cheung *et al.* (1988). By fusing the promoter and transit peptide encoding regions of the nuclear gene encoding the small subunit of ribulose-1,5-bisphosphate carboxylase to the chloroplast *psbA* gene, they were able to relocate this gene to the nucleus and still be able to direct the *psbA* gene product to the thylakoid membrane of the chloroplast. In their particular case, they transformed tobacco plants with a *psbA* gene that had been isolated from *Amaranthus hybridus*, and which conferred resistance to the herbicide atrazine. However, the tobacco transformants still showed some sensitivity to atrazine, presumably because of the presence of a large number of endogenous herbicide-sensitive PSII centres (see Chapter 6).

If a favourable modification of PSII function is demonstrated by studies on model organisms, then this experimental strategy may provide a way in which a mutation in a chloroplast-encoded component could be transferred to higher plants.

There are a number of techniques for introducing DNA into the nucleus, such as: *Agrobacterium*-mediated transfer (Weising *et al.* 1988); electroporation (Fromm *et al.* 1986); calcium phosphate coprecipitation (Hain *et al.* 1985; Lörz *et al.* 1985) and microprojectiles (Klein *et al.* 1987). Therefore, a modified nuclear gene of PSII (such as *psbO*, *psbP*, *psbQ*, or *psbR*) could in theory be reintroduced into the plant. However, the second disadvantage with plants is that there is no method for the deletion or interruption of nuclear genes to generate the necessary genetic background. This poses a formidable barrier to using plants, although great advances are being made in the T-DNA interruption of nuclear genes in *Arabidopsis thaliana* (Feldmann *et al.* 1990). In addition, homologous recombination occurs infrequently in plants (Peterhans and Paszkowski 1990), thus preventing gene replacement.

## Green algae

Eukaryotic green algae, such as *Chlamydomonas*, have for many years been used as model organisms for the study of oxygenic photosynthesis (Bennoun and Delepelaire 1982; Harris 1989). The ability of strains such as *Chlamydomonas reinhardtii* to grow heterotrophically (in the absence of photosynthesis), in the dark on acetate-containing media, has enabled deletion and interruption mutants of PSII to be generated by random mutagenesis and their properties studied.

However, as in plants, many of the genes encoding PSII polypeptides are located in the chloroplast genome (Rochaix 1981) and so until recently *in vitro* manipulation of these genes was not possible because of the lack of a transformation procedure. However, Boynton *et al.* (1988) have now demonstrated that the large size of the chloroplast in *C. reinhardtii*, and its location close to the plasma membrane, make it a suitable target for transformation using the high velocity tungsten microprojectile method, originally developed by Sanford and co-workers for the introduction of DNA into onion epidermal cells (Klein *et al.* 1987). Boynton and co-workers (1988, 1990) were able to transform

several non-photosynthetic strains, with deletions in the chloroplast-encoded *atpB*, *psbA* and *rbcL* genes, to photoautotrophic growth by bombarding cells with beads coated with wild type DNA. Analysis of the chloroplast DNA from photosynthetic transformants showed that, in the case of the *atpB* gene, the wild type fragment had integrated at its normal location, presumably by homologous recombination.

Studies by Blowers *et al.* (1989) have confirmed these observations and have furthermore shown that circular plasmids, single-stranded DNA circles and linear duplex DNA molecules can all be used to transform *Chlamydomonas*. Although much work has still to be done to develop *C. reinhardtii* as a host system for analysing modified PSII genes, such as the creation of dominant markers to identify transformed cells carrying a modified gene, there is no denying the advantages of this system over cyanobacteria for the biochemical analysis of mutants.

There exists a large collection of nuclear mutants of *Chlamydomonas* that affect expression of the major membrane–protein complexes associated with the photosynthetic electron transport chain. These include mutants lacking any one of the following: PSI, PSII, cyt b6/f complex, ATP synthase or most of the light-harvesting chlorophyll proteins. As *C. reinhardtii* is a haploid organism and can undergo genetic crosses, mutants with backgrounds lacking one or more of these complexes can, in principle, be crossed into a strain containing a site-directed alteration in a nuclear or chloroplast gene. For example, PSII mutants could be analysed in a strain of *C. reinhardtii* that contains no PSI or ATPase complex, thus simplifying the isolation of PSII-enriched membranes or core complexes (Diner and Wollman, 1980). An additional advantage is that PSII reaction centres are also present at higher concentrations (as compared to PSI) in thylakoid membranes of *Chlamydomonas* than in those of cyanobacteria. Consequently, it should be easier to obtain (from *Chlamydomonas*) PSII-enriched thylakoid membranes that still evolve oxygen. The lack of such preparations from the cyanobacterium, *Synechocystis* 6803, has proven to be a major obstacle to the study of mutants affecting water oxidation in this organism. However, a recent publication by Noren *et al.* (1991) reports the isolation of core complexes from this organism showing high rates of oxygen evolution.

Clearly, *Chlamydomonas reinhardtii* is now a feasible system for studying engineered mutants of

PSII, particularly in the *psbA* gene (Boynton *et al.* 1990), for which deletion strains are already available.

## Cyanobacteria

For the last few years, cyanobacteria have been the most popular organisms for studying the effects of directed mutations in PSII genes. In particular, two strains—*Synechocystis* sp. PCC 6803 and *Synechococcus* sp. PCC 7002—appear well suited for experimentation for a number of reasons:

(1) Like all cyanobacteria, they possess photosystems that are highly homologous in both composition and function to those in higher plants (Bryant 1987; Rögner *et al.* 1990a). Indeed, two groups have shown that it is possible to replace the *psbA* gene family of *Synechocystis* 6803 by a single *psbA* gene from the chloroplast genome of a higher plant without impairing photoautotrophic growth (Nixon *et al.* 1990, 1991; Reiss *et al.* 1990). This ability to construct a functional interspecies hybrid PSII reaction centre not only demonstrates the conservation in structure and function between species, but provides strong support for using cyanobacteria as a model system for higher plants, in spite of obvious differences, such as their light-harvesting complexes.

(2) Unlike cyanobacterial strains that are obligate photoautotrophs, such as *Synechococcus* PCC 7942 (formerly known as *Anacystis nidulans* R2), these two strains can be grown photoheterotrophically in the absence of a functional PSII complex, when supplied with a suitable carbon substrate (Rippka *et al.* 1979). Such a characteristic enables mutants with impaired PSII complexes to be grown and studied (Astier *et al.* 1984). Although this limits the usefulness of *Synechococcus* 7942, mutants that grow photosynthetically can still be generated in this organism (by site-directed mutagenesis) and their properties studied. For example, herbicide resistant strains have been engineered (Ohad and Hirschberg 1990).

(3) They have a natural DNA uptake system that allows easy transformation by exogenously added DNA (Stevens Jr. and Porter 1980; Grigorieva and Shestakov 1982; Williams 1988). Coupled with this uptake mechanism, which is poorly understood, is an extremely active recombination pathway that allows stable

integration of added DNA into the chromosome by homologous recombination (Williams 1988; Labarre *et al.* 1989).

Efficient recombination occurs in *Synechocystis* 6803 when a piece of DNA (such as an antibiotic resistance gene) is flanked on both sides by DNA derived from the chromosome (Williams 1988; Labarre *et al.* 1989). However, it is still possible to obtain low rates of recombination even when foreign DNA is flanked only on one side by DNA from *Synechocystis* 6803 (Metz *et al.* 1989). Illegitimate recombination, in which heterologous DNA sequences are incorporated into the chromosome, has also been reported for *Synechocystis* 6803 (Dzelzkalns and Bogorad 1986) following pretreatment of the cells with low levels of ultra-violet radiation. The frequency of this type of recombination is much lower than that found with homologous recombination, which may explain why it has been difficult to reproduce this phenomenon in other laboratories (Labarre *et al.* 1989).

The mechanism behind homologous recombination is unclear. However, since Labarre *et al.* (1989) found that single integrative crossovers occurred rarely in *Synechocystis* 6803, they suggested that homologous recombination occurred via non-reciprocal exchange (gene conversion) rather than a double crossover mechanism. Williams and Szalay (1983) arrived at a similar conclusion from studies on *Synechococcus* 7942. The fidelity of this mechanism for introducing DNA back into the chromosome has not been fully investigated. Recent studies in our laboratory have indicated that 240 bp of flanking DNA is sufficient for accurate recombination and that with 75 bp > 65% of the recombination events were inexact and occurred either upstream or downstream of the target region (Nixon and Diner unpublished results).

Another avenue for introducing modified DNA into *Synechocystis* 6803 (Chauvat *et al.* 1986) and *Synechococcus* 7002 (Buzby *et al.* 1985) is through the use of shuttle vectors that can be propagated in both *Escherichia coli* and cyanobacteria (reviewed in Kuhlemeier and Van Arkel 1988). Such vectors have been constructed by ligating fragments of DNA from *E. coli* plasmids, such as pBR322 or pACYC184, to DNA fragments from one of the naturally occurring plasmids found in *Synechocystis* 6803 (Chauvat *et al.* 1986) or *Synechococcus* 7002 (Buzby *et al.* 1983). These vectors have not yet been exploited for the engineering of PSII genes in these organisms.

Despite these genetic advantages, there are several complicating features that need to be considered before site-directed mutagenesis is performed. For a start, the presence of multigene families in cyanobacteria increases the difficulty of creating deletion strains. However, a sufficient number of selectable markers for these organisms exist, such that once the appropriate deletions have been made for these genes, as is now the case in a number of laboratories, this disadvantage vanishes. Also, as each gene may be under quite different transcriptional control (see p. 111), a careful choice of gene is required for mutagenesis experiments. However, multigene families have only been identified for the *psbA*, *psbD* and *psbG* genes in cyanobacteria.

A second obstacle to using cyanobacteria, but one that is shared with the chloroplast genome of eukaryotic algae and higher plants, is that they may contain multiple copies of their genome in a single cell. For *Synechocystis* 6803, Labarre *et al.* (1989) have estimated a figure of 12 genomic copies per cell, close to the value of 10 estimated for *Synechococcus* 7942 (Golden *et al.* 1986). Therefore, in certain situations, such as the interruption of a PSII gene with an antibiotic-resistance gene, it may be necessary to restreak transformants many times on antibiotic-containing plates under photoheterotrophic growth conditions, until complete segregation of the mutant gene is achieved (Williams 1988). The presence of multiple copies also complicates the isolation of recessive mutants using a random mutagenesis approach, since homozygous mutant strains, in the absence of selectional pressure, will segregate only after many divisions of the original heterozygous mutant.

However, once complete segregation has occurred, cyanobacteria have the distinct advantage over green algae of being able to be stored frozen at $-80°C$. In our own experience, mutant strains of *Synechocystis* 6803 can be revived after 5 years when stored at $-80°C$ in 20% glycerol or 15% dimethylsulphoxide.

Even though a number of cyanobacteria possess properties amenable to the genetic manipulation of PSII genes through site-directed mutagenesis (e.g. the photoheterotrophic strains *Synechococcus* 7002, *Synechocystis* 6714 and *Synechocystis* 6803), much of the work so far published has focused on *Synechocystis* PCC 6803 as the organism of choice. The remainder of the review will therefore deal mainly with studies using this particular strain and, where possible, compare results to those from studies on other species, particularly the green alga *Chlamydomonas reinhardtii*.

## Mutagenesis of *Synechocystis* 6803

The strategies used to exploit *Synechocystis* 6803 as a model system for the analysis of PSII polypeptides have been reviewed in depth elsewhere (Vermaas 1988; Williams 1988; Carpenter and Vermaas 1989; Vermaas *et al.* 1990a). Methods for the mutagenesis of other cyanobacteria have been described by Golden *et al.* (1988).

The basic steps in the construction of directed mutants in our laboratory are summarized in Figure 5.4. The target gene is first cloned into *E. coli* along with as much flanking sequence as possible. This cloning step may not be trivial; many of the genes of PSII appear to be unclonable when present on multicopy plasmids such as pUC18 (Chisholm and Williams 1988), probably because the gene product is expressed in *E. coli* using the *Synechocystis* promoter (Philbrick and Zilinskas 1988) and is lethal to the cell because of its hydrophobic nature. In such cases, the genes are usually cloned as two fragments or recovered using other techniques such as inverted PCR (Triglia *et al.* 1988).

Once the cyanobacterial genes have been cloned into *E. coli* vectors, four objectives must be met before mutant genes can be established in *Synechocystis* 6803:

(1) The gene and flanking sequences should be sequenced. Knowledge of the nucleotide sequence is necessary for designing oligonucleotide primers needed for mutagenesis and sequencing, and helps immensely in subsequent cloning steps. In addition, open reading frames within the vicinity of the target gene, which could potentially be affected by modifications to the target gene, can be identified.

(2) To engineer the deletion strain, a version of the cloned gene should be constructed so that the coding region is replaced with a selectable marker surrounded by wild-type flanking sequences. Many of the resistance genes commonly used in *E. coli* can be used effectively in *Synechocystis* 6803, such as the genes encoding resistance to the antibiotics spectinomycin (Prentki and Krisch 1984), kanamycin (Taylor and Rose 1988), chloramphenicol (Chang and

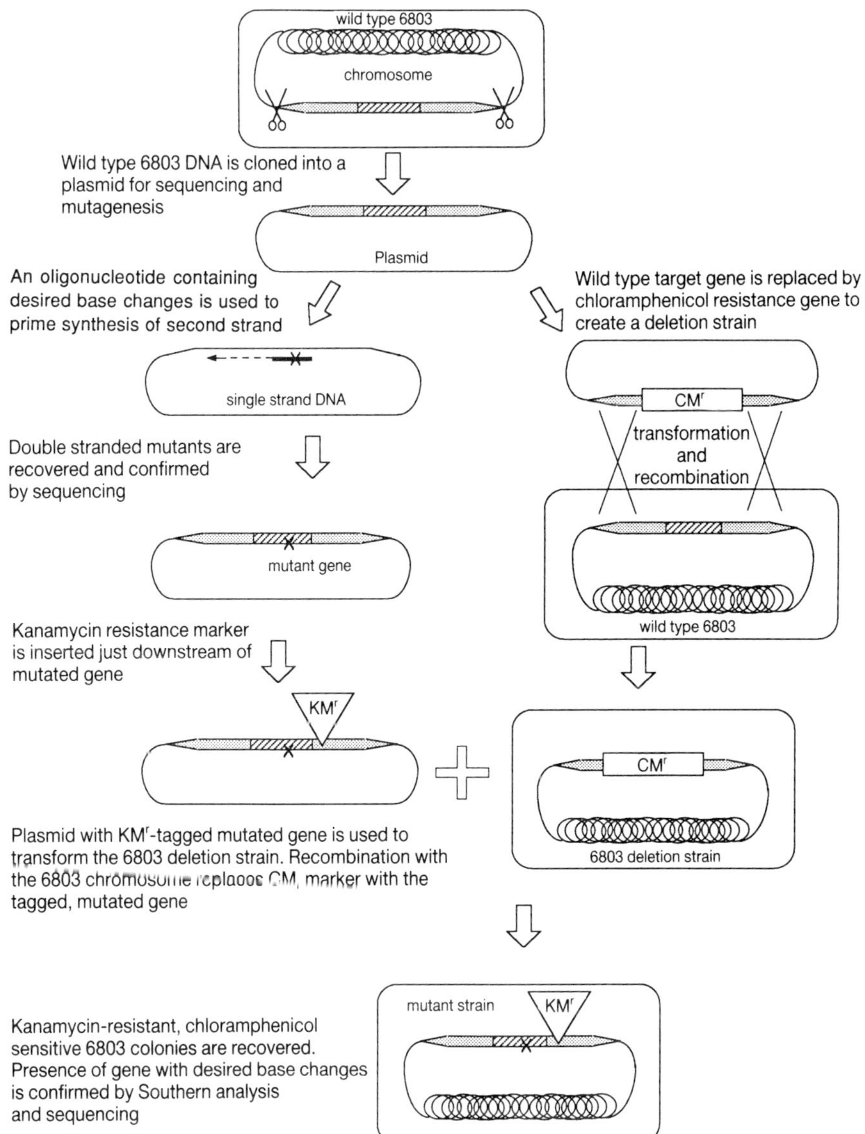



Wild type 6803 DNA is cloned into a plasmid for sequencing and mutagenesis

An oligonucleotide containing desired base changes is used to prime synthesis of second strand

Double stranded mutants are recovered and confirmed by sequencing

Kanamycin resistance marker is inserted just downstream of mutated gene

Plasmid with KM^r-tagged mutated gene is used to transform the 6803 deletion strain. Recombination with the 6803 chromosome replaces CM^r marker with the tagged, mutated gene

Kanamycin-resistant, chloramphenicol sensitive 6803 colonies are recovered. Presence of gene with desired base changes is confirmed by Southern analysis and sequencing

Wild type target gene is replaced by chloramphenicol resistance gene to create a deletion strain

**Fig. 5.4**   Overview of DNA manipulations used to create site-directed mutants in *Synechocystis* 6803. Two converging paths are depicted, one is followed to create specific mutations in cloned genes using plasmids in *E. coli*, the other is followed in creating *Synechocystis* 6803 hosts to receive the mutated genes.

Cohen 1978), erythromycin (Elhai and Wolk 1988), ampicillin and tetracycline (Rodriguez and Denhardt 1988). Since *E. coli* plasmids do not replicate in *Synechocystis* 6803, when cells are transformed with such a DNA construct, the vast majority of antibiotic-resistant colonies result from the replacement of the wild type coding region with the selectable marker through homologous recombination. The re-

sulting *Synechocystis* 6803 deletion mutant can then be used as the host for incoming mutated versions of the gene. In the case of multiple genes, such as *psbA* and *psbD*, every member of the family needs to be deleted to forestall recombination between homologous sequences that might result in reversion of the introduced mutations.

(3)   A version of the cloned gene should be

modified to contain the nucleotide changes desired in the final mutant. We use a mutagenic primer annealed to single-stranded template to initiate synthesis of a double-stranded molecule that is amplified in *E. coli*. For this step we have had good success with the *dut⁻-ung⁻* system described by Kunkel *et al.* (1987) and supplied in kit form by Bio-Rad (Richmond, California, USA). The frequency of mutation is sufficiently high (20–70%) to permit screening for mutants by DNA sequencing.

(4) To select for incorporation of the mutated gene into the chromosome of the deletion strain, the modified gene should be coupled to a different selectable marker to the one already present in the recipient deletion strain. It is important that the marker does not interfere with any promoter sequences, but otherwise can be placed upstream or downstream of the mutant gene.

Transformation of *Synechocystis* 6803 is accomplished simply by adding 1–10 μg/ml of DNA to cells at a density of approximately $10^9$ cells/ml. (Williams 1988). The form of DNA is unimportant: supercoiled plasmid, linear plasmid, circular concatamers and chromosomal fragments can all transform successfully. The cells are incubated in the light at 30°C for a few hours to allow DNA uptake, then plated onto filters placed on non-selective medium for 48 h to allow integration of DNA and expression of drug-resistance genes. The filters are then transferred to selective media for recovery of resistant cells. Resistant colonies are visible in 7 to 10 days and are restreaked once or twice on selective plates to ensure complete segregation of the mutant gene. Herbicide, such as DCMU at 10 μM, is included in the medium to ensure that transformants grow under photoheterotrophic conditions, to minimize the selectional pressure for photosynthetic revertants.

Genomic DNA isolated from resistant colonies, is analysed by Southern blotting to confirm the desired structure of the genes. The presence of the altered nucleotide sequence in the genomic DNA of the mutant is checked by sequencing the region of DNA containing the mutation, after amplification by PCR. If the mutant is non-photosynthetic, transforming the mutant back to photosynthetic growth, with small fragments of wild type DNA encompassing the site of mutation, is a useful demonstration that the engineered mutation does indeed cause the phenotype.

# Assay of PSII function

A range of techniques has been used to assay PSII function in mutant strains of algae and cyanobacteria. These fall into two categories, the biochemical and biophysical. Fortunately, many of these are applicable to whole cells and yield considerable functional information with little biochemical manipulation.

## Whole cells

### Photosynthetic growth

Certainly the simplest test of function is the ability to grow photoautotrophically on minimal medium. Growth rates are gross indicators of either reaction centre concentration or of photosynthetic quantum yield. These can occasionally be misleading, as strains grown on medium supplemented with glucose, with 30% of wild type rates for oxygen evolution, have been observed not to grow photosynthetically. Such strains may undergo photoinhibition in the light at rates that exceed repair in photoautotrophically grown cells. On the other hand, the ability to grow photoautotrophically necessarily requires oxygen evolution and indicates the non-essential nature of the site of mutation for PSII function.

### Oxygen evolution

Oxygen evolution has been measured in the steady state using a Clark-type concentration electrode. Typically, di-Cl-, dimethyl- or phenyl-*p*-benzoquinone plus ferricyanide are used as electron acceptors. These provide a large electron acceptor pool and simplify $O_2$ measurements by blocking respiration. These measurements are generally performed in saturating light and, as such, are the product of reaction centre concentration × the rate of the limiting step. Any change in rate of oxygen evolution, therefore, can be due to one or the other or both factors. Excitation with a train of saturating light flashes ($\leq$ 10 Hz) or with non-saturating light avoids the rate limitation.

Relative oxygen flash yields have also been measured in whole cells on a Joliot-type rate electrode (Joliot and Joliot 1968). Dark-adapted algal cells typically show a damped oscillation of period 4 of the oxygen yields, with maxima occurring on the 3rd and 7th flashes. This flash pattern, originally described by Joliot *et al.* (1969) was interpreted by Kok *et al.* (1970) as reflecting the need to accumulate four oxidizing equivalents before an oxygen molecule is released, consistent with the electrochemistry of water oxidation (see Joliot and Kok 1975 for review). Each state of the oxygen evolving complex (OEC) is represented by the letter S with a subscript indicating the number of stored oxidizing equivalents, each one contributed by a charge separation in its associated reaction centre. $S_4$ returns to $S_0$ with the release of $O_2$. An additional feature of the model is the stability of the singly oxidized state (state $S_1$) to explain the maximum yield on the third rather than on the fourth flash. Analysis of the damping of the oscillation of period 4 of the oxygen flash yields observed upon excitation of dark-adapted cells is an indication of the stability of the charge-separated state and of the equilibrium constants that govern equilibration of $Q_A/Q_B$ on the acceptor side and $Z/P680$ on the donor side. P680 must be present in the reduced form and $Q_A$ in the oxidized form to produce a stable charge separation. Lowered equilibrium constants will increase the concentration of $P680^+$ and $Q_A^-$, either of which will prevent stable charge separation and increase the damping of the oscillation. Once corrected for these 'misses', which lower the $O_2$ yield/flash, the amplitude of the oxygen flash yield with respect to WT is also an indicator of the relative number of centres/cell competent for $O_2$-evolution.

This oscillatory flash pattern can be used to determine the rate of $Q_A^-$ reoxidation by $Q_B$ (hundreds of $\mu$s time range; Kok *et al.* 1970; Bouges-Bocquet 1973) and the lifetime of the $S_2$ and $S_3$ states of the OEC (in seconds, Joliot *et al.* 1971; Forbush *et al.* 1971). A first flash given to dark-adapted cells generates $S_2Q_A^-$ in about 50 $\mu$s. A second flash is given a variable time later and $O_2$ is detected on the third flash. At short times $Q_A^-$ must be reoxidized by $Q_B$ before the second flash will produce a stable charge separation. At long times decay of $S_2$ will lower the third flash yield. Measurement of the $O_2$-yield of the third flash as a function of the time interval between the first and second flashes gives the relaxation rate for both phenomena.

Similarly, by varying the interval between the second and third flashes and detecting the $O_2$-yield on the third flash, the rate of $Q_A^-$ reoxidation by the secondary quinone can again be evaluated at short times and the lifetime of state $S_3$ of the OEC measured at long time intervals.

## Kinetic measurements using fluorescence

*Continuous light measurement of the evolution of the fluorescence yield*

One of the simplest kinetic techniques available for mutant characterization is that of the fluorescence induction curve. Here dark-adapted algal cells in suspension or on Petri plates (Bennoun and Delepelaire 1982) are exposed to continuous light ($\leq 640$ nm), which serves at the same time to drive electron transfer and to monitor the quantum yield of chlorophyll fluorescence. A high band pass filter transmits the chlorophyll emission at $> 670$ nm and blocks the actinic light protecting the detecting photomultiplier or photodiode. The variation with time in the quantum yield of fluorescence (see below) is recorded and stored on an analogue or digital memory oscilloscope or computer.

The quantum yield of chlorophyll fluorescence is a measure of the redox state of $Q_A$ (for review see Papageourgiou 1975). Where no energy transfer between reaction centres occurs, the variable yield of fluorescence is proportional to the concentration of $Q_A^-$. In many cases this is true in the cyanobacteria, where energy transfer between centres is poor. Where energy transfer occurs, the variable fluorescence yield depends non-linearly on $[Q_A^-]$, as described originally by Joliot and Joliot (1964).

A mutant with impaired electron transfer from $Q_A$ to $Q_B$ can be detected as a rapid rise in the fluorescence yield with kinetics comparable to those observed in the presence of inhibitors like 3-(3,4-dichlorophenyl)-1,1-dimethylurea (DCMU) or atrazine, which dissociate $Q_B$ from the reaction centre. Kinetic limitation on the electron donor side generally appears as only a very slight rise in the variable fluorescence, e.g. the LF1 mutant of *Scenedesmus obliquus* (Metz *et al.* 1980, 1985), which cannot process D1 (Diner *et al.* 1988; Taylor *et al.* 1988) and the BF25 mutant of *Chlamydomonas*, which cannot accumulate the 23 kDa extrinsic polypeptide (Bennoun *et al.* 1981; Mayfield *et al.* 1987b). Mutants such as these, which lack the tertiary electron donor (the OEC), will show an enhancement in their variable fluorescence yield upon the addition of DCMU (Bennoun *et al.* 1981) as a result of the formation of $Z^+Q_A^-$, which shows a high fluorescence yield. The variable fluorescence yield is stimulated further by the addition of 10 mM $NH_2OH$, an artificial electron donor (Bennoun *et al.* 1981; Metz *et al.* 1985). Under these conditions the BF25 and LF1 mutants are practically indistinguishable from their corresponding wild type (WT) strains, as $NH_2OH$ also extracts the Mn of the OEC in the WT. In mutants lacking functional Z (Debus *et al.* 1988b; Metz *et al.* 1989), the variable fluorescence yield is observed to rise only very slowly in the light, in the absence of an artificial electron donor, as $P680^+$ quenches the fluorescence yield despite the presence of $Q_A^-$. The kinetics of the rise of the variable fluorescence yield are particularly stimulated by hydroquinone (Debus *et al.* 1988b) and less so by $NH_2OH$ (Metz *et al.* 1989). Of the two, hydroquinone is the better donor to $P680^+$. In this way, donor side mutants can be detected. Those blocked at the level of the tertiary electron donor can be distinguished from those unable to oxidize Z. At the other extreme are mutants lacking PSII reaction centres altogether (e.g. Bennoun *et al.* 1981). These show no variable fluorescence but have an elevated invariant fluorescence yield, independent of the presence or absence of inhibitor.

In the case of *Synechocystis*, the cells can generally be pretreated with 0.2–0.3 mM benzoquinone and 0.2–0.3 mM ferricyanide to ensure complete oxidation of $Q_A$ and $Q_B$ prior to the addition of DCMU and an excess of artificial donor (10–20 mM $NH_2OH$ or 5 mM hydroquinone) followed by illumination. Measured in this way the total variable fluorescence is a fairly reliable measure of the concentration of PSII reaction centres. This measurement assumes a constant antenna size.

*Flash measurements of the evolution of the fluorescence yield*

Measurements similar to those described above can also be carried out using light flashes in an instrument described by Joliot and Joliot (1973). Here the actinic and detecting sources are physically separated. The actinic flash drives the electron transfer and the detecting flashes (hitting $\leq 1\%$ of the centres and delivered in a flash train at predetermined time intervals) monitor the relative quantum yield of fluorescence. Using such an instrument, the increase in the fluorescence yield produced by each of a series of saturating flashes in the presence of DCMU is an indicator of the stability of the charge separated state, i.e. the ability of the secondary electron donor to compete with the back reaction for the reduction of $P680^+$ (Metz *et al.* 1989).

Following each of a series of actinic light flashes given to dark-adapted preoxidized algae or cyanobacteria, the fluorescence yield rises as Z reduces $P680^+$ ($P680^+$ quenches the fluorescence, Butler *et*

*al.* 1973) and then falls as $Q_A^-$ is oxidized by $Q_B$ (or $Q_B^-$). $Q_A^-$ can also be oxidized by back reaction with the donor side if $Q_B$ is non-functional or displaced by an inhibitor. The rise in the fluorescence yield is, therefore, a measure of the rate of reduction of $P680^+$ by Z (Mauzerall 1972). The decay is the rate of oxidation of $Q_A^-$ by secondary quinone or by charge recombination with the donor side (Joliot 1974). The kinetics in both cases can be corrected for energy transfer (according to Joliot and Joliot 1964) as needed for greater accuracy (Robinson and Crofts 1984).

Charge recombination between state $S_2$ of the OEC and $Q_A^-$ can be measured by fluorescence in the presence of an inhibitor like DCMU and typically occurs with a $t_{\frac{1}{2}}$ of 0.5–1 s (Bennoun 1970). Where $Z^+$ cannot be reduced by Mn of the OEC, the fluorescence yield measurements show the recombination to occur in 20–80 ms ($t_{\frac{1}{2}}$, Dekker *et al.* 1984). Where $P680^+$ cannot be reduced by Z, recombination occurs with a $t_{\frac{1}{2}}$ of 0.1–1 ms (Reinman *et al.* 1981; Metz *et al.* 1989) and must be measured by other means (e.g. optical spectroscopy). There is little variable fluorescence yield observed at all as, following a light flash, $P680^+$ quenches the fluorescence yield.

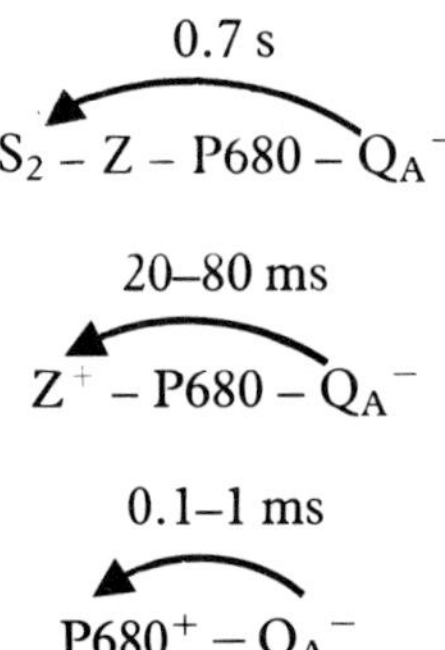



These measurements are therefore useful indicators of the blockage of electron flow on the oxidizing side of the reaction centre.

The flash detection fluorescence technique has also been used for monitoring electron transfer rates on the acceptor side of the reaction centre. The rate of electron transfer is normally faster for $Q_A^-$ to $Q_B$ than for $Q_A^-$ to $Q_B^-$ (Robinson and Crofts 1983):

$$Q_A^-Q_B \rightarrow Q_AQ_B^- \qquad t_{\frac{1}{2}} = 100\text{–}200\ \mu s$$

$$Q_A^-Q_B \xrightarrow{2H^+} Q_AQ_BH_2 \qquad t_{\frac{1}{2}} = 300\text{–}500\ \mu s$$

Therefore, during a series of saturating light flashes

given to dark-adapted preoxidized algae or cyanobacteria, the rate of $Q_A^-$ oxidation will be faster on odd flashes and slower on even flashes. Alternation of fast and slow rates following each of a series of light flashes is dependent on an intact electron donor side and elevated equilibrium constants favouring $Q_B$ reduction on the acceptor side and P680 reduction on the donor side. Mutants of D1 and D2 have been isolated from *Synechocystis* and higher plants in which the forward rates and the equilibrium constants have been altered (Erickson *et al.* 1989; Etienne *et al.* 1990; Taoka and Crofts 1990; work in this chapter). A measure of the equilibrium constant for $Q_A^-Q_B = Q_AQ_B^-$ can be obtained by giving a saturating light flash to cells incubated with 2 mM $NH_2OH$ following preoxidation by benzoquinone and ferricyanide. $NH_2OH$ prevents dark recombination following charge separation and the fluorescence yield indicates the distribution of the electron between $Q_A$ and $Q_B$ (Robinson and Crofts 1984). Such measurements have also been made in the absence of $NH_2OH$ by detecting the fluorescence yield at 0.5 s after the actinic flash (Etienne *et al.* 1990). An alternative method (Diner 1977; Robinson and Crofts 1983) involves measuring the rate of charge recombination between state $S_2$ and $Q_A^-$ and dividing by the rate of recombination between state $S_2$ and $Q_B^-$ (lifetime of $S_2$, see above). This ratio is equal to 1 plus the apparent equilibrium constant for reaction $Q_A^- + PQ = Q_A^-Q_B = Q_AQ_B^-$ ($PQ \equiv$ plastoquinone a).

By giving a single saturating flash followed by a variable time period, the rate of oxidation of $Q_B^-$ by S2 during the time interval can be measured using these fluorescence measurements (Robinson and Crofts 1983) and is equivalent to the method described above using oxygen flash yields. The phase of the oscillating pattern of fluorescence relaxation can be measured by giving a series of flashes at varying times after the single saturating flash. The time interval after which no oscillations are observed is the $t_\frac{1}{2}$ point for $Q_B^-$ oxidation, where half of the centres are in state $Q_AQ_B$ and the other half are in $Q_AQ_B^-$.

## Luminescence

Bennoun and Delepelaire (1982) have used the luminescence emission from chlorophyll following each of a series of light flashes to characterize mutants of *Chlamydomonas*. Although used by these authors to identify mutants affecting the chloroplast ATPase, any mutation that enhances the equilibrium concentration of $Q_A^-$ and/or $P680^+$ should also stimulate luminescence emission in the 100 ms time range, arising from charge recombination in the reaction centre, reforming P680* with low quantum yield (for review, see Lavorel 1975).

## Fluorescence emission spectra

The fluorescence emission spectra of whole cells and thylakoid membranes measured at 77 K have also been used to characterize mutants of PSII (Vermaas *et al.* 1987a; Dzelzkalns and Bogorad 1989). PSII normally shows emission bands at 685 and 695 nm, while PSI emits at ~720 nm. Such measurements should be treated with caution, as intact phycobilisomes also show an emission band at 685 nm (Elmorjani *et al.* 1986). The relative strength of the emission bands upon excitation of chlorophyll at 435 nm is an indicator of the relative concentrations of the two photosystems (Vermaas *et al.* 1987a; Burnap *et al.* 1989). Detection of emission at 695 nm upon excitation at 580 nm in cyanobacteria is an indicator of how well energy is transferred from the antenna phycobilin to PSII. As energy is normally preferentially transferred from the phycobilisome to PSII, such measurements also allow enhancement of the PSII emission spectrum relative to that of PSI.

## Optical spectroscopy

A number of investigators have measured the carotenoid band shift at 515 nm, an indicator of the transmembrane electrostatic potential in whole cells of *Chlorella* (Joliot and Delosme 1974) and in *Chlamydomonas* (Bennoun *et al.* 1981). This is generally the largest light-induced optical absorbance change and if measured at < 1 ms is an indicator of charge separation in both the PSI and PSII reaction centres. PSII can be blocked by preillumination in the presence of DCMU and $NH_2OH$. The remaining signals arising from PSI only can be compared with that obtained with both photosystems functioning to obtain the ratio of PSII/PSI reaction centres. Lavergne (1984, 1987) has used a PSI$^-$, light-harvesting antenna-depleted strain of *Chlorella sorokiniana* for measurements of optical changes associated with the S-state transitions of the OEC, $Q_A^-$–$Q_A$, $Q_B^-$–$Q_B$ and cyt b559 reduced–minus–oxidized difference spectra. Optical spectroscopy of PSII has proved to be difficult in whole cells of *Synechocystis* 6803, owing to the large excess of PSI over PSII. A two-fold improvement in the PSII/PSI ratio has been reported by growing cells under red illumination (Burnap *et al.* 1989) or by using a mutant deficient in synthesis of phycocyanin (Rögner *et al.* 1990a). The carotenoid band shift, which detects the transmembrane electrostatic potential, is particularly small in this organism compared to green algae and higher plants, probably owing to the extramembrane location of the light-harvesting antenna.

### EPR spectroscopy

EPR has been used in whole cells of *Synechocystis*, but only for the detection of the dark stable signal arising from the oxidized secondary electron donor, $D^+$ (Debus *et al.* 1988a; Vermaas *et al.* 1988a; Barry *et al.* 1990). Again, interference from the large excess of PSI reaction centres precludes the measurement of light-induced signals associated with PSII.

### Herbicide binding

Labelled diuron (5–100 nM) has been used to monitor the number of reaction centres per cell in *Synechocystis* mutants. The concentration dependence gives both the binding constant and the number of binding sites. Taking the difference in counts in the presence and absence of 20 μM unlabelled atrazine allowed Vermaas *et al.* (1990b) to correct for non-specific binding.

### Pulse labelling

A number of groups have characterized mutants of PSII by using pulse-labelling and pulse-chase experiments with $[^{14}C]$-acetate or $[^{35}S]$-methionine or tritiated amino acids to examine the synthesis and lifetime of the component polypeptides. It is possible to distinguish between nuclear and chloroplast-encoded gene products by using appropriate inhibitors: either anisomycin or cycloheximide to inhibit cytoplasmic protein synthesis on 80S ribosomes, or chloramphenicol to inhibit chloroplast encoded protein synthesis on 70S ribosomes (Delepelaire 1983, 1984; de Vitry *et al.* 1989). Labelling is often done in the light, which stimulates translation of PSII core polypeptides, particularly D1 (Fromm *et al.* 1985; Bennoun 1986). A number of core polypeptides have also been shown to be phosphorylated by labelling with $^{32}Pi$ or $\gamma$-$[^{32}P]$-ATP (Wettern *et al.* 1983; Delepelaire 1984; de Vitry *et al.* 1987). Either thylakoid membranes

(Delepelaire 1983; de Vitry *et al.* 1989) or whole cell lysates (Delepelaire 1984; Fromm *et al.* 1985, Diner *et al.* 1988) are prepared and the labelling of the constituent polypeptides is determined by SDS-PAGE and autoradiography. Yu and Vermaas (1990) have recently succeeded in doing pulse-labelling in the light in *Synechocystis* 6803. Unlike in plants, CP47, CP43 and D2 appear to be labelled to nearly the same extent as D1.

### Western blotting

The steady state accumulation of core polypeptides has been evaluated primarily by the use of Western blotting techniques (Vermaas *et al.* 1988b; de Vitry *et al.* 1989; Nilsson *et al.* 1990; Nixon *et al.* 1990 and see Jay 1987; Harlow and Lane 1988 for reviews). Antibodies are generally prepared either by immunization against polypeptides excised from gels or by using small synthetic oligopeptides, homologous to core polypeptides, which are cross-linked to particularly antigenic proteins, e.g. keyhole limpet haemocyanin, bovine serum albumin or bovine thyroglobulin, or by expressing the plant protein in *E. coli* (Harlow and Lane 1988). Antibodies directed against spinach CP47, CP43, D1, D2 and OEE 1 (Vermaas *et al.* 1988b; Burnap *et al.* 1989) and *Chlamydomonas* CP47 and CP43 (Nixon *et al.* 1990) have been successfully used for detecting their homologues in *Synechocystis* 6803. The quantitation of the relative levels of detected polypeptides is rather approximate.

The pulse-labelling, pulse-chase and Western blotting techniques have all been heavily used to evaluate the pleiotropic effects of nuclear and chloroplast mutations on the expression of core subunits of PSII. Many of these results are described later in the chapter.

## Membranes

Chloroplast and thylakoid membrane fractions have been used for many years for both biochemical and spectroscopic studies of higher plant photosynthesis. Membranes isolated from *Chlamydomonas* have been used for detailed characterization by SDS-PAGE of the polypeptide composition of photosynthesis mutants (Chua and Bennoun 1975). Gels run at 4°C (Delepelaire and Chua 1979) and at 25°C with (Piccioni *et al.* 1981) and without 8 M urea (Delepelaire and Chua 1979) have shown, through staining techniques, Western blotting and autoradiography of pulse-labelling and pulse-chase experiments, the steady-state concentrations and the rates of synthesis and degradation of reaction centre polypeptides. SDS-PAGE of thylakoid membranes isolated from cyanobacterial membranes has been difficult to characterize through simple staining techniques. The enormous number of components, the low concentration of PSII and the much higher concentration of PSI have necessitated the use of Western blotting and, more recently, pulse-labelling to ascertain the fate of PSII reaction centre polypetides in mutant strains.

Functional studies have been limited in *Chlamydomonas* and cyanobacterial membranes. Broken cell preparations have been isolated from mutant and wild type strains of *Chlamydomonas* and have been used for studies of state transitions controlled by the redox state of the plastoquinone pool (Wollman and Delepelaire 1984) and by varying $Mg^{2+}$ concentrations (Wollman and Diner 1980). They have also been used for studies of the iron–quinone complex of PSII by Mössbauer and EPR spectroscopy of mutant strains, including those labelled using the Mössbauer isotope $^{57}$Fe (Petrouleas and Diner 1986). Little optical spectroscopy has been done on cyanobacterial membranes, although oxidation-induced reduction of cyt b559 has been measured in the presence of tetraphenylboron in mutant and wild type strains of *Synechocystis* (Pakrasi *et al.* 1989). Attempts have also been made to detect $Q_A$ reduction in the ultra-violet (Pakrasi *et al.* 1989). Other absorbance changes interfere. Cyanobacterial membranes have also been used for oxygen rate measurements (Burnap *et al.* 1989; Vermaas *et al.* 1990c).

## Core complexes and reaction centres

Oxygen-evolving core complexes have been isolated from the thermophilic cyanobacteria with the greatest success (Dekker *et al.* 1988; Koike *et al.* 1989). These show high flash yields and light-saturated rates of oxygen evolution. A version of this preparation (Schatz and Witt 1984) gives the only known detergent solubilized core complex to retain electron transfer between $Q_A$ and $Q_B$. Optical studies of $Q_A^-$–$Q_A$ and $Q_B^-$–$Q_B$ (Schatz and van Gorkom 1985) have been carried out on this material, as have fluorescence lifetime measurements (Schatz *et al.* 1987). Oxygen-evolving core complexes have now been isolated from mesophilic cyanobacteria (Burnap *et al.* 1989; Noren *et al.*

1991) and these are just beginning to be used in spectroscopic studies.

Purified core complexes (~50 chl/reaction centre), that no longer evolve oxygen and are free of PSI contamination, have been isolated from both *Chlamydomonas* (Diner and Wollman 1980) and *Synechocystis* 6803 (Rögner *et al.* 1990). These have been used for spectroscopic and kinetic studies involving optical difference spectroscopy in the ultra-violet and visible and EPR spectroscopy (Metz *et al.* 1989). These core complexes isolated from *Synechocystis* 6803 are currently finding extensive use in the optical and EPR characterization of site-directed mutants. A core complex lacking CP43 (~33 chl/reaction centre) has been isolated from a mutant of *Synechocystis* 6803 unable to express this polypeptide (Rögner *et al.* 1991). This complex, like that containing CP43, shows light-driven electron transfer from Z through to $Q_A$ (Rögner *et al.* 1991).

Reaction centre complexes containing only polypeptides D1, D2, cyt b559 and the *psbI* gene product have been isolated from cyanobacterial sources (Ikeuchi *et al.* 1989b; Gounaris *et al.* 1989). These, like their higher plant counterparts (Nanba and Satoh 1987; Barber *et al.* 1987) have lost $Q_A$ and are only able to do charge separation and recombination between P680 and pheophytin. No such complex has been isolated from *Chlamydomonas*.

## Identification of PSII genes

Much effort has been directed recently at identifying, cloning and sequencing all the genes that encode the polypeptides that copurify with the PSII complex (Gray 1987; Gray *et al.* 1990). Currently there are thought to be at least 20 polypeptides coded for by genes in both the nucleus and chloroplast of eukaryotes. In cyanobacteria, of course, no such compartmentalization of DNA exists.

Unfortunately, the function of the majority of these polypeptides has not yet been determined (particularly the many low molecular weight polypeptides of less than 10 kDa), and the possibility that some may have been misassigned roles in PSII cannot yet be discounted. Hopefully, mutagenesis experiments in which the putative PSII gene is deleted or interrupted will help establish more fully its function in PSII.

Polypeptide components of PSII have been identified from the analysis by SDS-PAGE of PSII extracts ranging in complexity from PSII-enriched membranes (Berthold *et al.* 1981) to detergent-solubilized $O_2$-evolving (Ikeuchi and Inoue 1986) and non-$O_2$-evolving PSII core complexes (Diner and Wollman 1980; Satoh *et al.* 1983) to the smallest complex so far studied, the PSII reaction centre complex (Nanba and Satoh 1987). Originally, oxygen-evolving core complexes could be resolved into approximately six polypeptide bands (Ikeuchi *et al.* 1985). However, by using better gel systems with improved resolution in the low molecular weight region (below 10 kDa), it has been possible to detect many more low molecular weight polypeptides than previously thought (Ljungberg *et al.* 1986).

The genes for the more easily identifiable 'high molecular weight' polypeptides have been cloned from higher plants using relatively standard procedures (Gray 1987). The equivalent genes from cyanobacteria have usually been isolated using the chloroplast gene as a heterologous probe (for a review see Houmard and Tandeau de Marsac 1988), although antibody probes have been used (Philbrick and Zilinskas 1988).

Characterization of the low molecular weight polypeptides is now under way. Inoue and co-workers have conducted an intensive investigation into the number of such polypeptides in PSII preparations from the higher plants, wheat and spinach, and the thermophilic cyanobacterium *Synechococcus vulcanus* (Ikeuchi *et al.* 1990). They have also been able to obtain partial N-terminal sequence information for many of these polypeptides which, together with the complete nucleotide sequence of the chloroplast genomes from tobacco (Shinozaki *et al.* 1986a, b), liverwort (Ohyama *et al.* 1986a, b) and rice (Hiratsuka *et al.* 1989), have allowed the identification of several polypeptides as products of small, previously unassigned open reading frames within the chloroplast genome. Such comparative studies are helped by the high degree of conservation in the primary structures of PSII polypeptides between chloroplasts and cyanobacteria. For example, two polypeptides that so far have only been identified in *Synechococcus vulcanus*, were partially sequenced and found to show an identity of 74% and 42% with open reading frames of 34 and 43 codons, respectively, present in the chloroplast genomes of tobacco and liverwort. On this basis they have been designated the *psbM* and *psbN* genes, respectively (Ikeuchi *et al.* 1989a).

Table 5.1 lists those genes that have so far been designated PSII genes (Hallick 1989) and includes their size in codons, their genome location and in what type of PSII preparation their product can be found. It should be noted that many more genes may yet be added to this list in the coming years. For comparison, the table also lists the current status of these genes from *Synechocystis* 6803 and, where possible, the growth characteristics of mutants possessing specifically inactivated genes. Further details of this work will be described in the next section.

## Mutagenesis of PSII genes

### The *psbA* gene

The highly-conserved *psbA* gene has been sequenced from over 15 organisms (Erickson *et al.* 1985) and codes for a component of the PSII reaction centre complex with an apparent molecular weight of 30–34 kDa upon denaturing polyacrylamide gel electrophoresis (Nanba and Satoh 1987; Marder *et al.* 1987). It is now clear that the *psbA* gene product, usually called the D1 polypeptide, is equivalent to the 32 kDa herbicide-binding protein identified by photoaffinity labelling (Pfister *et al.* 1981), the $Q_B$-protein that binds the secondary quinone acceptor $Q_B$ (reviewed by Kyle 1985), and the rapidly-turning-over 32 kDa polypeptide (reviewed by Mattoo *et al.* 1989). Originally thought of as a polypeptide whose role lay primarily on the acceptor side of PSII, it is widely speculated that its function may extend to the binding of the manganese cluster, as well as to the binding of the cofactors required for transmembrane charge separation. Other possible roles include the binding of calcium and chloride ions, necessary for optimal oxygen-evolving activity (Coleman and Govindjee 1987). Studies on higher plants have shown that the D1 polypeptide undergoes extensive post-translational modification, including removal of the initiating formyl methionine residue; acetylation and phosphorylation of the N-terminal threonine residue (Michel *et al.* 1988); removal of nine amino acid residues from the carboxy terminus (Takahashi *et al.* 1988) and palmitoylation (Mattoo and Edelman 1987). The reasons behind these modifications are unclear and are currently under investigation.

Cyanobacteria have generally been found to contain three copies of the *psbA* gene, as shown in

*Synechocystis* 6803 (Jansson *et al.* 1987), *Synechococcus* 6714 (Ajlani *et al.* 1989a), *Synechococcus* 7942 (Golden *et al.* 1986), and *Synechococcus* 7002 (Gingrich *et al.* 1988). One exception to this is *Anabaena* 7120, where four copies have been identified (Vrba and Curtis 1989).

In those species where all the *psbA* genes have been sequenced (i.e. *Synechococcus* 7942, *Anabaena* 7120 and *Synechocystis* 6803), two different forms of the D1 polypeptide can be synthesized. In *Synechococcus* 7942, these two forms, designated FORM I (the product of the *psbA1* gene) and FORM II (the product of the *psbA2* and *psbA3* genes) differ from each other by 25 residues, 12 of which are located in the first 16 amino acids. In some extremely elegant work, Schaefer and Golden have studied the effect of light intensity on the expression of each of the three genes. Using a reporter gene, they found that the expression of each gene at low light intensity differed markedly with *psbA1* $\gg$ *psbA3* > *psbA2*; and that with increasing light intensity, expression of *psbA1* declined while that of *psbA2* and *psbA3* increased (Schaefer and Golden 1989a). Although the total amount of D1 polypeptide remained constant, these variations were also reflected in the amount of FORM I and FORM II polypeptide detected in the membrane (Schaefer and Golden 1989b). The reason for such regulation has yet to be determined, but may reflect differences in the biophysical properties of PSII reaction centres containing each form of the D1 polypeptide.

In the case of *Synechocystis* 6803 the nucleotide sequences of *psbA1* (Osiewacz and McIntosh 1987), *psbA2* (Ravnikar *et al.* 1989) and *psbA3* (Metz *et al.* 1990) have been published. The *psbA3* gene encodes an identical product to that of the *psbA2* gene. Analysis of steady-state levels of mRNA from strains in which two of the three genes were insertionally inactivated, has indicated that only *psbA2* and *psbA3* can accumulate transcripts and allow *Synechocystis* 6803 to grow photoautotrophically (Mohamed and Jansson 1989). In agreement with these observations we have found that under normal laboratory growth conditions, strains containing *psbA1*, but lacking *psbA2* and *psbA3*, lack functional PSII reaction centres (Nixon and Diner, unpublished data). In this regard, *psbA1* is useless as a gene in which to construct mutations for the study of D1 function.

## Inactivation of the *psbA* gene family

The effect of the loss of the D1 polypeptide on PSII function has been analysed by the construction of strains in which the *psbA* gene family has been inactivated either by insertional mutagenesis (Jansson *et al.* 1987) or by deletion (Nixon *et al.* 1990; Debus *et al.* 1990). In all cases, photoautotrophic growth and oxygen-evolving activity are lost. Fluorescence induction experiments have indicated that the variable fluorescence arising from PSII is absent in a triple-deletion strain, consistent with loss of the ability to perform stable charge separation (Nixon *et al.* 1990).

Steady-state levels of several of the PSII polypeptides have been analysed in such strains through immunoblotting experiments. Analysis of a *psbA* triple deletion strain showed that the CP43 apoprotein was present at normal wild type levels, whereas D2 and the CP47 apoprotein were present at vastly reduced levels (Nixon *et al.* 1990). However, Nilsson *et al.* (1990), who performed a similar study on an insertion mutant, did not detect any CP47 apoprotein or D2. The reason for this difference probably lies in differences in the sensitivity of the antisera used in the two studies. Despite the lack of D1 and the apparent loss of D2 in this strain, the cyanobacterial equivalent of the 33 kDa extrinsic polypeptide of higher plants is present at normal levels in a membrane-bound form (Jansson *et al.* 1987). The nature of this binding and its differences to that observed in wild type thylakoids has not yet been determined. The only reaction centre polypeptides so far analysed that are present at substantial levels in the *psbA*-interruption mutant are the cyt b559 apopolypeptides (Nilsson *et al.* 1990).

Random mutagenesis has also been used to generate *psbA* deletion mutants. Some 22 chloroplast mutants of *Chlamydomonas* have now been isolated. These have lost all or nearly all of the two copies of the *psbA* gene present in the inverted repeat region of the chloroplast genome (Bennoun *et al.* 1986). No *psbA* transcripts are, therefore, detectable in these mutants. In one of the best characterized of these mutants, FUD7, and in another isolated by Spreitzer and Mets (1981), 8-36C, there is no synthesis of polypeptide D1 (de Vitry *et al.* 1989; Jensen *et al.* 1986, respectively). At the same time, there is a low rate of synthesis of apoCP47, but only minor slowing of the synthesis of apoCP43, D2 and the $\alpha$-subunit of cyt b559. There is very little accumulation of CP47 and of

D2. However, higher levels of CP43 (20% of WT, Rochaix *et al.* 1989) are observed. The lack of accumulation of D2 is attributed to its rapid proteolysis in the absence of assembled complex ($t_{\frac{1}{2}}$ = 30–60 min, de Vitry *et al.* 1989). The accumulation of the extrinsic polypeptides OEE 1–3 is not affected in FUD7, although the polypeptides are no longer tightly bound to the thylakoid membrane. Poor binding of the OEE polypeptides to the thylakoid membrane is observed in all cases where they accumulate in the absence of assembled PSII core complex (de Vitry *et al.* 1989).

Taken together, analyses of *psbA* mutants from both *Synechocystis* 6803 and *C. reinhardtii* suggest that D1 has a role in stabilizing the other core polypeptides in the membrane, in addition to its other roles in electron transport. One minor difference appears to lie in the levels of CP43 that accumulate in the absence of D1; higher levels appear in *Synechocystis* 6803 than in *Chlamydomonas*. A major difference is the binding affinity of OEE1 in the *psbA* mutants, tighter binding occurs in *Synechocystis* than in *Chlamydomonas*.

## Herbicide-resistant mutants

There is now a large collection of point mutations in D1 that have arisen spontaneously or through random mutagenesis, which provide some resistance to PSII-directed herbicides. These herbicides all act by displacing the secondary quinone electron acceptor, $Q_B$, from its binding site on the reaction centre. Consequently, these mutations, all of which involve codon changes in *psbA*, map out the $Q_B$ binding pocket. Furthermore, some show greater resistance to one class of inhibitor than to another. For example, Ser264Gly shows a greater enhancement of resistance to atrazine than to DCMU, while the reverse is true of the double mutant Ser264Ala, Phe255Leu (Table 5.2), indicating that DCMU and atrazine interact with different residues lining the $Q_B$ pocket. These mutations are not without consequence for $Q_B$ binding, as some show slowed electron transfer from $Q_A^-$ to $Q_B$ and a lowered apparent equilibrium constant for the reaction $Q_A^- + PQ = Q_A^- Q_B = Q_A Q_B^-$ (Taoka and Crofts 1990; Etienne *et al.* 1990). The latter class of mutants shows an elevated quantum yield of fluorescence at 0.5 s following flash excitation of the centre. This is caused by a shift to the left of the above-mentioned equilibrium, resulting in a more elevated equilibrium concentration of $Q_A^-$ (see pp. 105–7). The decrease in the apparent equilibrium constant can arise either from a decrease in the free energy difference between the $Q_A/Q_A^-$ and $Q_B/Q_B^-$ redox couples or a lowered binding affinity of plastoquinone for the $Q_B$ binding site (i.e. an increased dissociation constant). Taoka and Crofts (1990) have argued in favour of the second mechanism in the case of the Ser264Gly mutant from *Amaranthus hybridus*.

Some of the mutants mentioned in Table 5.2

**Table 5.2**  Herbicide-resistant mutants. Single and double-site herbicide resistance mutations located in PSII and found in sources ranging from higher plants to cyanobacteria. In the second column, the relative dissociation constant, $k_d$, in each of the mutants, of DCMU and atrazine is indicated relative to the wild type of the same organism. Also indicated in the same column is the effect of the mutation on the overall rate of electron transfer for the reaction $Q_A^- Q_B \rightarrow Q_A Q_B^-$. The last column shows the remaining variable fluorescence measured 0.5 s after a saturating light flash (Etienne *et al.* 1990). This measurement is a qualitative indicator of the equilibrium constant for $Q_A^- + PQ = Q_A^- Q_B = Q_A Q_B^-$. The higher the value, the lower the apparent equilibrium constant.

| Sites of PSII mutation | $t_{1/2}$ of $Q_A^-$ to $Q_B$ electron transfer (rel. to WT) and resistance to atrazine and DCMU (both relative to $k_d$ of WT) | %$F_{var}$ remaining at 0.5 s after saturating flash (Mut., WT) (from ref. 4) |
|---|---|---|
| D1-phe211ser[4,5] | unchanged, (7–10A,1–2D) | 25, 20 |
| D1-phe211ser,ala251val[4] | slowed, [100A,2D] | 35, 20 |
| D1-val219ile[3,5] | 1–2x, (2A,10–20D) | |
| D1-ala251val[7] | not measured, (25A,5D) | |
| D1-phe255tyr[3] | 1–2x, (15A,0.5D) | |
| D1-gly256asp[3] | >3x, (15A,3D) | |
| D1-ser264ala[3,4] | >3x, (100A,10–20D) | 40, 20 |
| D1-ser264ala,phe255leu[4] | slowed, [1A,1000D] | 36, 20 |
| D1-ser264gly[4,6] | ≤3x[2], (500–1000A,1D) | 24, 12 |
| D1-asn266thr[1,4] | unchanged, (1A,6D) | |
| D1-leu275phe[3] | 1–2x, (1A,5D) | |

Key to references: [1] Ajlani *et al.* 1989b; [2] Taoka and Crofts 1990; [3] Erickson *et al.* 1989; [4] Etienne *et al.* 1990; [5] Gingrich *et al.* 1988; [6] Hirschberg and McIntosh 1983; [7] Johanningmeier *et al.* 1987.

have been isolated from cyanobacteria as well as from green algae and higher plants. Indeed, Ohad and Hirschberg (1990) have engineered the Ser264Gly mutant into *Synechococcus* PCC 7942. This mutant shows a phenotype similar to that observed in higher plants, affirming the use of the cyanobacterial system as a model for higher plant photosynthesis.

## Oligonucleotide-directed mutagenesis

Two groups have now developed systems for the rapid generation of *psbA* mutants in *Synechocystis* 6803. Site-directed mutations are created in cloned fragments of either the *psbA2* (Debus *et al.* 1990) or *psbA3* gene (Nixon and Diner unpublished data), which are then used to transform deletion strains lacking most of the coding region of *psbA*. Such an approach is a distinct improvement on the previous methods that used a double deletion strain as a host for the transformation (Debus *et al.* 1988b; Metz *et al.* 1989; Ravnikar *et al.* 1990), since every transformant now carries the mutant gene. Table 5.3 is a list of some of the mutants generated by oligonucleotide-directed mutagenesis of the *psbA* gene.

*Mutants affecting the acceptor side of the reaction centre*
A mutant in which His215 was replaced by leucine (Metz and Diner unpublished results) does not grow photosynthetically. D1-His215 is thought to be one of the ligands to the non-haem iron and may hydrogen bond to $Q_B$. Loss of the ability to transfer an electron to $Q_B$ would explain its inability to grow photosynthetically. Apparently not all the ligands to the non-haem iron are required for assembly of the reaction centre, as this mutant accumulates PSII centres.

D1-His252 is probably located in a position homologous to that of L-Glu212 or L-Asp213, both of which have been implicated (Paddock *et al.* 1989; Takahashi and Wraight 1990, respectively) in protonation reactions coupled to the reduction of $Q_B$ in *Rb. sphaeroides* reaction centres. We have replaced D1-His252 with leucine (Metz *et al.* unpublished data, see Table 5.3) and found a greatly reduced ($\leq 1$) equilibrium constant for the reaction $Q_A^- + PQ = Q_A Q_B^-$ compared to the value of 20–30 measured for wild type. His252 is probably located close to Ser264 (H. Robinson and A. Crofts, personal communication and Diner *et al.* 1991), which is thought to hydrogen-bond to $Q_B$ (Michel and Deisenhofer 1988). Loss of the pro-

tonateable histidine residue appears to destabilize the $Q_B^-$ semiquinone anion. The mutant grows photoautotrophically, but slowly.

*Mutants affecting the electron-donor side of the reaction centre*
Barry and Babcock (1987) have shown that secondary electron donor D to the PSII reaction centre is a tyrosine. Debus *et al.* (1988a) suggested that D1-Tyr161 might be the secondary donor, Z, responsible for linking the OEC to the primary electron donor, P680. This suggestion was based on the location of this residue in a position symmetric with respect to that of D2-Tyr160 shown by Debus *et al.* (1988a) and Vermaas *et al.* (1988a) to be secondary donor, D.

Replacement of Tyr161 with phenylalanine (Debus *et al.* 1988b; Metz *et al.* 1989) and with tryptophan, histidine, methionine and cysteine (Nixon and Diner unpublished results) all resulted in non-photosynthetic strains that no longer showed oxidation of the secondary electron donor, Z. In core complexes, isolated from each of these mutants, the oxidized primary donor, P680$^+$, is reduced by charge recombination with $Q_A^-$ in 1 ms ($t_{\frac{1}{2}}$; pH 5.9). In wild type core complex, P680$^+$ is normally reduced by Z in 17 μs ($t_{\frac{1}{2}}$; pH 5.9). $NH_2OH$, which normally donates to Z$^+$ in isolated core complexes, is a poor donor to P680$^+$ in this system. None of the replacement residues appeared to act as efficient electron donors, either because of their elevated midpoint potentials or because of an irreversible oxidation that had occurred on an earlier charge separation. These observations provide strong evidence in favour of attributing D1-Tyr161 to Z.

*Role of D1 in ligating the manganese cluster*
One of the least understood aspects of water oxidation is how the managanese is bound to the protein components of PSII. Four manganese are bound in a cluster and probably undergo oxidation from an oxidation state of +2 through to +4 (Hansson and Wydrzynski 1990.) All three oxidation states can form complexes with hard Lewis bases such as the oxyanions. Possible amino acid ligands are therefore the carboxylic acids, glutamate and aspartate; the alcohols, serine and threonine; and the phenol, tyrosine. Other potential ligands are the primary amides, asparagine and glutamine (for a more detailed discussion see Pecoraro 1988). Nitrogen atoms from histidine residues may also act as ligands, although there are thought to be a greater

**Table 5.3**   Directed mutants, all constructed in *Synechocystis* PCC 6803 with the exception of the D1-ser264gly mutant of *Synechococcus* PCC 7942 (Ohad and Hirschberg 1990). The mutants listed, unless otherwise indicated, were isolated in our laboratory. Indicated under 'Photosynthesis' is the ability of the mutant to grow photoautotrophically together with an estimate of the doubling time in comparison to wild type. The oxygen measurements are light-saturated rates indicated relative to those obtained in wild type in the presence of 0.3 mM 2,5-dichloro-*p*-benzoquinone and 1 mM ferricyanide. The relative concentration of reaction centres was determined, unless otherwise indicated, by measuring the variable fluorescence yield in the presence of 40 $\mu$M DCMU, 20 mM $NH_2OH$ or 5 mM hydroquinone following preoxidation of the cyanobacterial cells with 0.2 mM *p*-benzoquinone and 0.2 mM ferricyanide. The letter codes before and after the numbers indicate the wild type and mutant residues, respectively, at that position.

| Mutant | Photosynthesis | $O_2$ evolution (% of WT) | [RC] (% of WT) | Comments |
|---|---|---|---|---|
| *psb*A | | | | |
| D1-D59N | + 1.4x | | 100 | donor-side-$Q_A^-$ recombination rate = WT |
| D1-D59V | + 3.3x | ~10 | 30 | donor-side-$Q_A^-$ recombination rate 2xWT |
| D1-D61N | + 1.9x | | 100 | donor-side-$Q_A^-$ recombination rate = WT |
| D1-D61V | + 2.2x | ~5 | ~75 | donor-side-$Q_A^-$ recombination rate 2xWT |
| D1-E65Q | + 1.9x | ~20 | 100 | donor-side-$Q_A^-$ recombination rate 1.3xWT |
| D1-E65L | + 4.0x | ~10 | 100 | donor-side-$Q_A^-$ recombination rate 2xWT |
| D1-H92L | + | | | |
| D1-E98A | + 3.3x | ~10 | 100 | donor-side-$Q_A^-$ recombination rate 2xWT |
| D1-D103V | + | ~30 | ~70 | |
| D1-Y112F | + | | | |
| D1-Y161C,F[2,3],H,M,W | − | 0 | 10–50[3] | no Z, 1ms $P^+Q_A^-$ recombination[3] |
| D1-D170E | + | 60–100 | 100 | donor-side-$Q_A^-$ recombination rate = WT |
| D1-D170F | − | | 10 | |
| D1-D170S | − | 0 | 100 | no tertiary donor, 20ms $Z^+Q_A^-$ recombination |
| D1-E189Q | + | | | |
| D1-H190L | − | | 0 | |
| D1-H195L | + | | 100 | donor-side-$Q_A^-$ recombination = WT |
| D1-H198L | − | | 0 | |
| D1-H215L | − | | >50 | |
| D1-H252L | + | ≪WT | 100 | low Kapp. for $Q_A^-Q_B = Q_AQ_B^-$ |
| D1-S264G[4] | + | | | herbicide resistant, 1000X WT for atrazine, 2X WT for DCMU |
| D1-D308A | + | | | |
| D1-D319A | + | | | |
| D1-E329Q | + | | | |
| D1-E333Q | + 3.3x | ~20 | 100 | |
| D1-N335stop | − | | ≤10 | |
| D1-H337L | − | ~20 | 35 | $O_2$ rate slows in bright light |
| D1-H337Y | − | | 10 | |
| D1-S345A | + | | | |
| D1-S345P | − | 0 | 20–30 | |
| D1-S345R | + | | | |
| D1-deletion beyond A344 | + | | 100 | D1 C-terminal extension not required for photosynthesis |
| D1-full replacement with *Poa annua psb*A[5] | + 1.3x | 100[5] | 100[5] | *Poa annua* D1 is processed at C-terminus[5] |
| *psb*C | | | | |
| CP43-M1A[1,6,7] | − | 0 | 10[6] | Assembles photoactive core complexes lacking CP43[6] |
| CP43-M1V[7] | − | | | Reported to have no PSII activity |
| *psb*D | | | | |
| D2-E69Q[8] | − | 30 | 50* | $O_2$-rate slows in bright light |
| D2-E69V[8] | − | 0 | 0 | |
| D2-M159R[9] | + | | 75[9] | Alteration in shape of $D^+$ EPR signal[9] |
| D2-Y160F[9,10] | + | | 75[10] | $D^+$ signal lost, donor D is tyr160[7,8] |
| D2-Y160W[11] | + | | | Replacement of stable $D^+$ tyrosine free radical signal with tryptophan free radical[11] |
| D2-H197Q[12] | + | ~40 | 00 | $E_mP680^+/P680$ shifted negative by 20–30 mV. Enhanced misses in oxygen flash yields. Consistent with H197 as ligand to P680 |

| Mutant | Photosynthesis | $O_2$ evolution (% of WT) | [RC] (% of WT) | Comments |
|---|---|---|---|---|
| D2-H197Y[13] | − | | 0 | |
| D2-H214N[13], K[1],M[1] | − | | 0[1,13] | |
| D2-G215W[14] | poor | ~30, initially | 40* | $O_2$-evolution inactivated in saturating light |
| D2-E224F | + | | 100 | $Q_A/Q_B$ electron transfer like WT |
| D2-D225F | + | | 100 | $Q_A/Q_B$ electron transfer like WT |
| D2-D225Y | + | | 100 | $Q_A/Q_B$ electron transfer near normal ($t_{1/2} \sim 1$ ms) |
| D2-E227D,F,Y | + | | | $Q_A/Q_B$ electron transfer near normal ($t_{1/2} \sim 1$ ms) |
| D2-W253L[14] | − | | 0* | |
| D2-K264E,L,Q,R | + | | 100 | $Q_A/Q_B$ electron transfer has large slow phase |
| D2-R265H,N,M,Q | + | | 100 | $Q_A/Q_B$ electron transfer has large slow phase |
| D2-H268D | − | | 0 | |
| *psb*F | | | | |
| cytb559-H23L[15] | − | | 0 | |

* Centre quantitation based on DCMU binding.
References: [1]This work; [2] Debus *et al.* (1988b); [3]Metz *et al.* (1989); [4]Ohad and Hirschberg (1990); [5]Nixon *et al.* (1991); [6]Rögner *et al.* (1991); [7]Carpenter *et al.* (1990b); [8]Vermaas *et al.* (1990c); [9]Vermas *et al.* (1988a); [10]Debus *et al.* (1988a); [11]Barry *et al.* (1990); [12]W.F.J. Vermaas and B.A. Diner, unpublished data; [13]Vermaas *et al.* (1987a); [14]Vermaas *et al.* (1990b); [15]Nyhus *et al.* (1990).

number of oxygen ligands, at least to those manganese species responsible for the $S_2$ multiline EPR signal (Andréasson 1989; Britt *et al.* 1990; De Rose *et al.* 1991).

Chemical modification experiments using diethylpyrocarbonate (a relatively specific modifier of histidine residues), and 1-ethyl-3-(3-dimethyl-aminopropyl) carbodiimide, a probe for carboxy late residues, have confirmed that these types of residues are crucial for the photoactivation of manganese into a functional cluster (Tamura *et al.* 1989). Whether this is because of a direct role in binding manganese, or merely the result of a secondary effect, has not been determined.

The polypeptides involved in manganese ligation are unknown (Babcock *et al.* 1989; Rutherford 1989), although D1 and D2 represent the most likely candidates. Two extreme models can be considered. In the first, residues in the carboxy-terminal regions of both polypeptides participate in binding the cluster, so maintaining the pseudo-$C_2$ symmetry within the complex (Dismukes 1988). In the second type of model the manganese cluster is displaced, so that residues linking the first two transmembrane helices of the D1 polypeptide have primary responsibility in binding manganese (e.g. Babcock *et al.* 1989). This model attempts to reflect the fact that the manganese appears to be more intimately linked to donor Z (Tyr161 of D1) than to donor D (Tyr160 of D2) on the basis of the EPR power saturation characteristics of $Z^+$ and $D^+$ (Babcock *et al.* 1989) and the known role of Z as the sole electron transfer component between manganese and P680$^+$ (Gerken *et al.* 1988; Debus *et al.* 1988a, b; Vermaas *et al.* 1988a; Metz *et al.* 1989).

Since there may be many amino acid ligands to manganese (up to 14), it is not possible at this stage to rule out any of the components of the PSII oxygen-evolving complex in binding manganese. Of particular interest is the possible role of the chlorophyll-binding proteins, CP47 and CP43, in such a function. So far it has not been possible to purify a PSII complex with a functional manganese cluster in their absence. This raises the possibility that they may play additional roles other than light-harvesting (Bricker 1990), especially as hydropathy plots (Frankel and Bricker 1989) suggest the presence in the lumen of large hydrophilic loops of at least 130 amino acids. Equivalent regions are not conserved in a homologous protein that is thought to be a chlorophyll-binding protein induced by iron deficiency (Laudenbach and Straus 1988). These observations would suggest that CP47 and CP43 may be needed for an additional function specific to PSII, such as binding the 33 kDa extrinsic polypeptide (Bricker *et al.* 1988) or manganese. In addition, it is interesting to note that engineered strains of *Synechocystis* 6803 that fail to synthesize CP43 are incapable of oxidizing water despite the presence of intact reaction centres that can generate

$Z^+$ (Rögner *et al.* 1991). These strains need to be investigated further, to see whether this inability to evolve oxygen is as a result of a lesion on the oxidizing side of PSII or perhaps on the acceptor side. Evidence in favour of the latter comes from herbicide-binding experiments that have indicated that PSII complexes lacking CP43 cannot bind DCMU (Carpenter *et al.* 1990a).

The engineering of site-directed mutants offers an elegant way to examine the role of specific residues in oxygen evolution. Our approach to investigate the importance of the carboxylate and histidine residues of the D1 polypeptide in water oxidation is first to screen for the ability of the mutant *psbA* gene, carrying the site-directed change, to transform to photosynthetic growth the non-photosynthetic triple-deletion strain.

*Carboxylate residues*
Table 5.3 summarizes the results of the initial screen of carboxylate residues. Mutated residues that restore photosynthetic growth are assumed at this stage to have a minor role in binding manganese; the involvement of these residues in binding the other ion cofactors necessary for water oxidation, such as calcium, cannot yet be ruled out. In this way, carboxylate residues that have been proposed to ligate manganese, such as Asp308 (Dismukes 1988), Asp61, Glu65, Glu98, Asp103, Asp319, Glu329 and Glu333 (Coleman and Govindjee 1987) have all been tested. None of the changes analysed have abolished oxygen evolution and the mutants can grow photosynthetically under appropriate conditions. For instance, E333Q is unable to grow photoautotrophically on solid media, but can do so in liquid culture at low rates.

However, we have found two carboxylic acid residues that do appear critical to oxygen evolution. Significantly, one of these is the carboxylic acid residue that is the closest in the primary structure to Tyr161, donor Z, the immediate oxidant of the Mn complex. Replacement of Asp170 with serine gave a non-photosynthetic strain that did not evolve oxygen, whereas replacement with glutamate showed wild type rates of $O_2$-evolution and growth. The serine mutant showed the absence of electron donation to $Z^+$ from the Mn cluster of the OEC. Charge recombination between $Q_A{}^-$ and the oxidized donor side was 40-fold faster ($t_{\frac{1}{2}} = 15$–20 ms compared with the value for wild type of 0.7 s). Core complexes isolated from the serine mutant showed a 60-fold higher $K_m$ than wild type

for the reduction of $Z^+$ by exogenous $Mn^{2+}$, indicating the loss in the mutant of a high affinity binding site or access route for $Mn^{2+}$ to $Z^+$. D170 is, therefore, the most promising candidate yet described for the ligation of OEC-Mn (Nixon and Diner, 1990, 1992). Replacement of D170 by phenylalanine showed very little accumulation of reaction centre.

Another strong candidate for Mn ligation is Asp342. Its replacement by valine results in a complete loss of oxygen evolution and an inhibition of electron donation to $Z^+$.

*Histidine residues*
Histidine residues in D1 that have been speculated to be possible manganese ligands include His92, His190, His332 and His337 (Tamura *et al.* 1989; Seibert *et al.* 1989). Mutants of *Synechocystis* 6803 have been constructed in which three of these histidine residues have been changed to the similarly-sized leucine residue. Mutant His92Leu grows photosynthetically; whereas His190Leu and His337Leu are non-photosynthetic. However, mutant His337Leu does evolve oxygen, but at approximately 20–40% the level of wild type. Replacement of His190 by glutamine results in greatly slowed oxidation of Z(Tyr161). Such an observation is consistent with model building by Svensson *et al.* (1990) in which His190 was proposed to hydrogen bond to Tyr161, stabilizing the tyrosine free radical.

Replacement of His332 by leucine gives a phenotype similar to the Asp342 mutant mentioned above, making His332 also a strong candidate for new ligation.

*Processing of the* psbA *gene product*
A low fluorescence mutant of the green alga *Scenedesmus obliquus*, known as LF1, was originally analysed by Metz *et al.* (1980). It was found to be unable to evolve oxygen and to contain thylakoid membranes with less than half of the wild type levels of manganese on a per chlorophyll basis. Apart from this, LF1 contains a normally functional PSII reaction centre present at wild type levels in thylakoids. Subsequent studies showed that LF1 was unable to process the precursor D1 polypeptide at its carboxy terminus (Diner *et al.* 1988; Taylor *et al.* 1988b), due to the absence or inactivity of the processing protease (Taylor *et al.* 1988a). These observations indicated that processing of the D1 was required for stable assembly of the manganese cluster and, importantly, provided

the first real demonstration that the D1 polypeptide played a role on the donor side of PSII.

Direct sequencing of the carboxy terminal amino acid of the mature D1 polypeptide from spinach (Takahashi *et al.* 1988; Takahashi *et al.* 1990) and *Chlamydomonas reinhardtii* (B.A. Diner, unpublished data), and *Synechocystis* 6803 (B.A. Nixon, unpublished data) has indicated that Ala344 is the terminal amino acid in all three cases; nine amino acids are therefore removed during maturation in spinach, eight in *C. reinhardtii* and sixteen in *Synechocystis*. An immunochemical analysis of strain KWPAS, in which the *psbA* gene product of the higher plant, *Poa annua*, is incorporated into a functional PSII complex in *Synechocystis* 6803, also supports the idea that processing of D1 does occur in *Synechocystis*. Using antibodies specific for the carboxy terminal extension (Diner *et al.* 1988) we have found that the D1 polypeptide in PSII complexes from KWPAS no longer cross-reacts (Nixon *et al.* 1991). Not only does this result support the idea of processing in cyanobacteria, but it would also suggest that the processing protease from *Synechocystis* can recognize a substrate from a higher plant, despite the considerable differences in primary structure at the carboxy terminus (Morden and Golden 1989). Presumably, recognition of the D1 precursor polypeptide by the protease depends more upon sequence elements within the mature polypeptide than in the carboxy extension.

Our laboratory has also begun to investigate the possible function of the carboxy terminal region of the D1 polypeptide. A mutant, SΔ, was constructed in which the last 16 codons of the *psbA* gene were deleted to leave Ala344 as the final residue. This mutant is still capable of growing photoautotrophically, suggesting that the last 16 amino acid residues are neither required for incorporation of the D1 polypeptide into the membrane nor for its correct assembly into an $O_2$-evolving complex. In contrast, mutant Asn335stop, in which a stop codon was placed at codon 335 to produce a truncated polypeptide some 25 residues smaller than normal, is unable to accumulate any functional PSII reaction centre.

The serine residue after the proposed cleavage site, at position 345, has also been changed to see if it were possible to block the proposed cleavage of the D1 precursor into its mature form. When replaced by the amino acids alanine and arginine, no effect was observed; the mutants could still grow photoautotrophically. However, when the serine residue was changed to proline (Ser345Pro),

a mutant with a phenotype similar to the LF1 mutant was obtained, i.e. Ser345Pro is incapable of growing photoautotrophically, cannot evolve oxygen but can still perform charge separation within PSII. This mutant, therefore, shows functional characteristics consistent with the non-processing of D1. Western blotting also supports the absence of processing in this mutant.

## The *psbB* gene

The *psbB* gene codes for the apopolypeptide of one of the two chlorophyll–protein antenna complexes of the PSII core complex, CP47 (Camm and Green 1980; Bricker 1990). Until recently, CP47 was also thought to contain the reaction centre components P680 and $Q_A$ (Nakatani *et al.* 1984), but isolation of photoactive reaction centres lacking CP47 (Barber *et al.* 1987; Nanba and Satoh 1987) has discounted this hypothesis. In *Synechocystis* 6803, the coding region is 1521 base pairs in length and is present as a single copy (Vermaas *et al.* 1987b). The predicted amino acid sequence lacks the last residue found in higher plants but is otherwise 72% identical (Ohyama *et al.* 1986b; Shinozaki *et al.* 1986a; Morris and Herrmann, 1984). It is of interest to note that 40% of the non-conservative amino acid replacements occur in four regions representing only 12% of the protein. According to the folding model of Bricker (1990), these regions all lie on hydrophilic loops that are lumenally exposed. On the basis of hydropathy plots, CP47 is predicted to consist of six transmembrane alpha helices, which contain a number of histidine residues located toward the stromal and lumenal surfaces of the membrane that may act as ligands to chlorophyll molecules (Bricker 1990).

Vermaas and collaborators have engineered two strains of *Synechocystis* 6803 that contain an inactivated *psbB* gene, either by deleting a portion of the gene (mutant *psbB⁻*, Yu and Vermaas 1990) or by interrupting the 5′ end of the gene with a kanamycin cartridge (mutant *psbB*-int, Vermaas *et al.* 1986; Vermaas *et al.* 1988b). Both mutants are non-photosynthetic. Yu and Vermaas (1990) have shown that in both mutants there is no accumulation of the *psbB* transcript. Pulse-labelling experiments with [³⁵S] methionine have demonstrated that apoCP43, D1 and D2 are all synthesized at close to wild type rates and, as expected, there is no detectable synthesis of apoCP47. Western blotting experiments indicated that only apoCP43 accumulated to a significant level, although this was still below that of wild type. D2 was present at considerably reduced levels and D1 could not be detected.

Several mutants that contain a hybrid *psbB* gene consisting of sequences from both *Synechocystis* 6803 and spinach have also been constructed. Chimeric mutants that were non-photosynthetic did not accumulate CP47, but now contained near wild type levels of CP43 and moderate levels of D1 and D2 (Vermaas 1988b). The enhanced accumulation of these subunits in these particular strains may reflect an increased but unstable presence of CP47 in the chimeric mutants, as opposed to mutants *psbB*⁻ and *psbB-int*.

Jensen *et al.* (1986) have identified a nuclear mutant of *Chlamydomonas*, GE2.10, which does not accumulate transcripts of *psbB* and which cannot synthesize apoCP47. This mutant shows slowed synthesis of D1 while apoCP43, D2 and α-subunit of cyt b559 are all translated at wild type rates. However, there is no accumulation of CP43, D1 and D2, indicating that they are quickly degraded if not assembled into a complex that probably also contains CP47. D2 is degraded at the fastest rate. CP43 and D1 are degraded more slowly, but still more quickly than in wild type.

Jensen *et al.* (1986) proposed that translation of apoCP47 and D1 is coupled in *Chlamydomonas*, with the inability to express one reflected in impaired synthesis of the other. This proposal appears to be borne out in mutants of *Chlamydomonas* that do not synthesize D1 (Jensen *et al.* 1986; de Vitry *et al.* 1989), where synthesis of apoCP47 is indeed slowed. However, as mentioned above, deletion of *psbB* in *Synechocystis* has no effect on the rate of translation of D1. Synthesis of apoCP47 has not been examined in mutants of *Synechocystis* lacking D1.

No mutants containing specific amino acid changes in CP47 have yet been described.

## The *psbC* gene

The *psbC* gene codes for the apopolypeptide of the second longest chlorophyll–protein complex of the PSII core, CP43 (Camm and Green 1980; Bricker 1990). In *Synechocystis* 6803, the coding region is 1380 base pairs in length and is present as a single copy that overlaps the *psbD1* gene by 17 bases (Chisholm and Williams 1988; Carpenter *et al.*

1990b). As in the case of CP47, the CP43 apopolypeptide has six predicted transmembrane helices that contain histidine residues in positions that are remarkably similar to those in CP47 (Chisholm and Williams 1988). A comparison of the deduced amino acid sequences from ten plant and cyanobacterial genes indicates that 67% of the residues are identical, with the most dissimilar region occurring at the amino terminal end, where only five out of the first eighteen residues are conserved (Bricker 1990). This region is also where apoCP43 from *Synechocystis* 6803 is missing one amino acid when compared to the other species.

A number of non-photosynthetic *psbC* mutants have been engineered in *Synechocystis* 6803. In one of these, psbD1/C⁻, the operon that contains *psbD1* and *psbC* has been entirely deleted. This mutant has been reported by Yu and Vermaas (1990) to show near normal rates of synthesis of apoCP47, D1 and D2. However, only reduced levels of apoCP47, D1 and D2 accumulate, indicating an enhanced proteolysis of these subunits in the absence of CP43. The second copy of the *psbD* gene, *psbD2*, functions in place of the deleted *psbD1* gene and encodes an identical form of polypeptide D2 (see below).

One similar and two additional mutants have been characterized by Rögner et al. (1991). These are the deletion mutant Tol 1357, which is similar to mutant psbD1/C⁻ described above, mutant T 1292, which contains an insertion of a kanamycin resistance gene about half-way into the coding region of *psbC*, and mutant Tol 107, which contains no translational start codon for *psbC*. While all are incapable of photoautotrophic growth, analysis of cells and membranes has shown that PSII core complexes lacking CP43 are still assembled and accumulate to about 10% of wild type levels (Rögner *et al.* 1991). They are capable of light-driven electron transfer from Z to $Q_A$ with a quantum yield similar to that of wild type, but do not evolve oxygen. It is clear, therefore, that CP43 is not required for incorporation and function of the primary quinone electron acceptor $Q_A$, contrary to a recent proposal based on biochemical extraction of CP43 (Petersen *et al.* 1990). These core complexes contain approximately 33 chlorophylls per photoreducible $Q_A$ (i.e. per reaction centre), as opposed to ~48 when CP43 is bound, indicating a chlorophyll content for CP43 of about 15 chlorophyll molecules per monomer.

Vermaas *et al.* (1988b) have also reported some evidence of photoactivity in a non-photoautotrophic mutant (psbC-I1), which contains a kanamycin resistance gene inserted about 90% of the way into the coding region. psbC-I1, like the mutants described above, shows low levels of D1, D2 and CP47.

A chemically-induced mutant of *Synechocystis* 6803 (PS3) with a 202 bp deletion in the coding region of *psbC* has also been isolated (Dzelzkalns and Bogorad 1989). It does not contain CP43, but it does contain a 'trace' amount of D2, 80% of wild type levels of CP47 and normal levels of the 33 kDa extrinsic polypeptide. The ability to perform charge separation was not analysed in this mutant. Normal phycobilisome composition was confirmed by comparing fluorescence maxima at 668 nm for mutant and wild type samples.

Although only a few *psbC* mutants have been constructed in which specific codons have been altered by site-directed mutagenesis, they amply demonstrate the power of this approach. For example, the true translation initiation codon of *psbC* in cyanobacteria has been uncertain ever since the discovery that the presumptive AUG start codon assigned to chloroplasts and *Synechocystis* 6803 (Chisholm and Williams 1988) was not present in *Synechococcus* 7942 (Golden and Stearns 1988), *Synechococcus* 7002 (Gingrich *et al.* 1990) or *Chlamydomonas reinhardtii* (Rochaix *et al.* 1989). Instead, a GUG codon 36 bp downstream from the AUG codon was suggested as an alternative start site (Golden and Stearns 1988). Indirect evidence supporting this proposal came from a sequence comparison of a number of *psbC* genes, which indicated the conservation of a Shine–Dalgarno sequence preceding the GUG codon, but not the AUG codon (Chisholm and Williams 1988). To test the efficacy of each codon, Carpenter *et al.* (1990b) generated four mutants in *Synechocystis* 6803 that contained altered codons; the GUG codon was altered to GCG or GUA and AUG was changed to ACG or AAG. The mutants in which AUG was altered were still capable of photoautotrophic growth, whereas the GUG mutants were neither able to grow photoautotrophically nor accumulate apoCP43. These mutants, therefore, provide overwhelming evidence in favour of assigning the GUG triplet as the initiation codon of translation.

A number of chloroplast and nuclear mutants that affect the expression of CP43 have been isolated from *Chlamydomonas*. In the chloroplast mutants MA16 and FUD34 (Rochaix *et al.* 1989), there is, respectively, a 6 bp duplication in the middle of the coding region and a modification in the 5′ non-coding region of the *psbC* gene. In these mutants, apoCP47, D1, D2 and the α-subunit of cyt b559 are synthesized at wild type rates. There is no synthesis of apoCP43 in FUD34 but there is some low level synthesis of this polypeptide in MA16. In neither case is there accumulation of CP43. Contrary to Rochaix *et al.* (1989), de Vitry *et al.* (1989) report in FUD34 some low level accumulation of CP47, D1, D2 and cyt b559, which can form a complex as judged by cosedimentation during sucrose density gradient centrifugation. The oxygen evolving enhancer (OEE) extrinsic polypeptides are also synthesized and accumulate at wild type levels. However, de Vitry and colleagues (1989) have reported that OEE2 is no longer restricted to the granal region of the thylakoid membrane in FUD34. This observation, and the instability of CP43 in the absence of OEE2 (see below), suggests binding of OEE2 to CP43.

Two nuclear mutants F34 and F64, which affect expression of CP43 (Chua and Bennoun 1975; Bennoun *et al.* 1980) have also been isolated. These accumulate normal levels of *psbC* transcript but do not synthesize apoCP43 (Rochaix *et al.* 1989). Rochaix *et al.* (1989) and de Vitry *et al.* (1989) report near wild type rates of synthesis of apoCP47, D1, D2 and the α-subunit of cyt b559. The former study reports some low level accumulation of apoCP47 but almost no D1 polypeptide in F34 and F64 and the latter paper reports somewhat higher levels of accumulation of CP47, D1 and D2 in F34, as in FUD34. The small differences in these reports may arise from differences in antibody sensitivity. It is clear that loss of CP43 is less consequential for the stability of CP47, D1 and D2 than are any of these for each other.

## The *psbD* gene

### Description of the gene

The D2 polypeptide of the PSII reaction centre in *Synechocystis* 6803 is encoded by two *psbD* genes, designated *psbD1* and *psbD2*, each of which is 1059 basepairs in length. As observed with higher plants, the *psbD1* gene overlaps the *psbC* gene by 17 nucleotides. This overlap occurs in all the cyanobacterial and chloroplast genomes examined to date (Alt *et al.* 1984; Holschuh *et al.* 1984; Bookjans *et al.* 1986; Ohyama *et al.* 1986b; Shinozaka *et al.* 1986a; Chisholm and Williams 1988; Golden and Stearns 1988; Neumann 1988; Gingrich *et al.* 1990), with the exception of two eukaryotic algae, *Chlamydomonas reinhardtii* (Erickson *et al.* 1986) and *Euglena gracilis* (Schantz 1985). The two genes are cotranscribed (Mohamed and Jansson 1989; Golden *et al.* 1989; Gingrich *et al.* 1990). In contrast to chloroplasts, *Synechocystis* 6803 and the other species of cyanobacteria that have been studied contain a second copy of the *psbD* gene designated *psbD2* (Williams and Chisholm 1987; Golden and Stearns 1988; Gingrich *et al.* 1990). Although there are slight nucleotide differences between the genes, in *Synechococcus* they both code for identical products. (Golden *et al.* 1989; Gingrich *et al.* 1990). There were originally thought to be three amino acid differences between the two forms of D2 in *Synechocystis* 6803 (Williams and Chisholm 1987). However, following a re-evaluation of the nucleotide sequence data, it now appears that there are no differences between the *psbD1* gene product and the *psbD2* gene products. Transcript (Mohamed and Jansson 1989; Golden *et al.* 1989; Gingrich *et al.* 1990) and functional (Yu and Vermaas 1990) analyses indicate that both gene copies are active in all three species of cyanobacteria.

Overall homology between *psbD* genes is high, with about 85% of the amino acid residues conserved between all known genes (Williams and Chisholm 1987; Gingrich *et al.* 1990). Many of the non-conserved residues occur in patches that are confined to hydrophilic loops. One extensive region occurs between residues 5 and 19 at the amino terminus of the protein, and it is within this region that the cyanobacterial and *Chlamydomonas reinhardtii* D2 proteins have lost a residue, causing them to be one amino acid shorter than their higher plant counterparts (Williams and Chisholm 1987; Gingrich *et al.* 1990).

## Random mutagenesis

There exist a number of *Chlamydomonas* mutants of chloroplast and nuclear origin that affect the synthesis of polypeptide D2. Chloroplast mutant FUD47 contains a 46 bp repeat in the coding region of the *psbD* gene and produces a transcript that accumulates at greatly reduced levels and that codes for only 186 out of the 352 amino acids in D2 (Erickson *et al.* 1986). There is no detectable synthesis of D2. However, there is some disagreement on rates of synthesis of the other polypeptides, with Kuchka *et al.* (1988) reporting wild type rates for both D1 and apoCP47 and

Erickson *et al.* (1986) and de Vitry *et al.* (1989) reporting depressed rates for D1 and for D1 and apoCP47, respectively. According to de Vitry *et al.* (1989), there is little effect on the rates of synthesis of apoCP43 and the α-subunit of cyt b559. Only very low levels of CP47 and D1 accumulate in the thylakoid membrane, indicating that the rate of breakdown exceeds the rate of synthesis. CP43 is less depleted. The three OEE extrinsic polypeptides accumulate to normal levels but are not tightly bound to the membrane.

Several nuclear mutants, in which the synthesis of D2 is most affected, have been isolated. Mutants NAC-1-18 and NAC-1-11 (Kuchka *et al.* 1988) have elevated levels of the *psbD* message, possibly through a negative feedback mechanism, but show very little synthesis of D2. Synthesis of the other subunits—apoCP47, apoCP43 and D1—are near normal. The steady-state levels of D1 and D2 are undetectable but CP47 and CP43 are clearly observed on western blots. The OEE polypeptides are present at wild type levels.

Another nuclear mutant, NAC-2-26 (Kuchka *et al.* 1989), will synthesize but not accumulate *psbD* message. No D2 is detected in a pulse labelling experiment. Rates of synthesis of apoCP47, apoCP43 and D1 are normal but there appears to be little accumulation of any of these. Oddly, this mutant does not accumulate the OEE extrinsic polypeptides either, despite normal rates of synthesis. This is the first mutant affecting expression of a core subunit that also loses the ability to accumulate the OEE extrinsic polypeptides. It is possible that some gene product, either altered in this mutant or not expressed, plays a structural role in the stabilization of these polypeptides.

## Inactivation of the *psbD* gene in *Synechocystis* 6803

Specific deletion or interruption of the *psbA*, *psbB* and *psbC* genes has enabled an assessment to be made of the relative importance of each gene to the assembly of the reaction centre complex. However, similar experiments with *psbD* have been difficult because of the added complication that *psbD1* overlaps the *psbC* gene. Only the independent *psbD2* copy has been replaced by antibiotic resistance genes (Debus *et al.* 1988a; Vermaas *et al.* 1988a; Vermaas *et al.* 1990a). In the case of *psbD1*, the overlap with *psbC* has precluded any inactivation scheme that may affect transcription of *psbC*, such as the interruption of *psbD1* with a resistance marker. Possible ways to circumvent this problem may be to engineer an unrecognizable *psbD1* start codon or, alternatively, to make a precise excision of *psbD1* that would place the *psbC* start codon exactly where the *psbD1* start codon used to be.

Although no simple insertion or deletion mutants of the *psbD* gene family have thus far been characterized, there are single amino acid mutants that have caused a total loss of PSII reaction centre (Trp253Leu, Vermaas *et al.* 1990b; and His214Asn, Vermaas *et al.* 1987a). Such results suggest that loss of the entire D2 protein would have equally dramatic consequences. The amino acid numbering used here for polypeptide D2 is that of *Synechocystis* 6803.

## Oligonucleotide-directed mutagenesis

### *Mutants affecting the acceptor side of the reaction centre (Table 5.3)*

*Glu224, Asp225 and Glu227*   The non-haem iron in bacterial reaction centres is coordinated by four histidines and by Glu232 contributed by the M-subunit (*R. viridis* numbering; Michel and Deisenhofer 1988; Komiya *et al.* 1989). We have looked for the functional homologue of Glu232 among carboxylate residues of polypeptide D2 located at nearly homologous positions. Consequently, we have targeted for mutagenesis Glu224, Asp225 and Glu227. These were replaced by phenylalanine and tyrosine (Chisholm and Diner unpublished results). The rationale was that loss of a ligand to iron (by substitution with phenylalanine) might perturb the $Q_A/Q_B$ electron transfer rate or that replacement of a carboxylate residue with tyrosine would facilitate the oxidation of the non-haem iron to the Fe(III) form by lowering its midpoint potential (Pryz *et al.* 1985), giving rise to a fast component of electron transfer from $Q_A^-$ to Fe(III) (Petrouleas and Diner 1987). None of these mutants gave a very significant alteration in the rates of $Q_A/Q_B$ electron transfer. Therefore, it appears unlikely that these residues are involved in Fe ligation.

*His214 and His268*   Both of these residues are likely candidates for ligation of the non-haem iron (Michel and Deisenhofer 1988; Komiya *et al.* 1989). Replacement of His214 with asparagine resulted in complete loss of the PSII reaction centre (Vermaas *et al.* 1987a). Replacement of His214 with lysine and methionine, and His268 with aspar-

tate, gave no PSII photoactivity (Chisholm and Diner unpublished results). It is not known if the centres are still physically present but are inactive.

*Trp253*   This residue is thought to be involved in binding of $Q_A$ and is the homologue to M-Trp250 (*R. viridis* numbering; Michel and Deisenhofer 1988; Komiya *et al.* 1989). Its replacement by leucine resulted in loss of the reaction centre, arguing for a role of $Q_A$ binding in centre stability (Vermaas *et al.* 1990b).

*Lys264 and Arg265*   Lys264 and Arg265 are located four and three residues away from His268, which is thought to coordinate the iron. All the mutations that have been made at these sites show a considerable slowing of $Q_A/Q_B$ electron transfer, reminiscent of that observed following bicarbonate depletion of PSII using formate (Eaton-Rye and Govindjee 1988; Cao and Govindjee 1990). Increases in the $t_{\frac{1}{2}}$ by as much as a factor of 10 are observed, more so on the second than on the first flash. A large slow phase is also present.

Diner and Petrouleas (1990) have shown that $HCO_3^-$ and NO compete for binding to the reaction centre, with NO binding directly to the iron. It is likely, therefore, that $HCO_3^-$ also binds to the iron. This observation agrees with an earlier suggestion of Michel and Deisenhofer (1988) that bicarbonate could be the PSII homologue of Glu232.

It is possible that Lys264 and Arg265 increase the binding affinity of the reaction centre for $HCO_3^-$ (Diner *et al.* 1990) in a manner similar to that observed in serum transferrin (Bailey *et al.* 1988) and human lactoferrin (Anderson *et al.* 1989), each of which has an arginine (124 and 121, respectively) that is implicated in the binding of iron and of $HCO_3^-(CO_3^-)$. Blubaugh and Govindjee (1988) have also suggested a role for cationic residues in $HCO_3^-$ binding. The residues they have proposed, Arg257 and His252, reside on polypeptide D1.

### *Mutants affecting the donor side of the reaction centre*

*Glu69*   Vermaas and co-workers have been attempting to locate residues that are likely to be involved in manganese ligation in polypeptide D2. These include carboxylate residues. As a result, Vermaas *et al.* (1990c) have replaced Glu69 by glutamine and by valine. In both cases the mutant strains are non-photosynthetic. Thylakoid membranes isolated from the glutamine mutant are able to evolve oxygen transiently in bright light, becoming progressively inhibited with a $t_{\frac{1}{2}}$ of 25 s, but less so in the presence of $Mn^{2+}$. The valine mutant does not accumulate PSII reaction centres. These authors consider Glu69 to be a potential ligand to Mn with its replacement leading to a looser binding of Mn, though an effect on binding of $Ca^{2+}$ or of OEE1 should probably not be excluded.

*Tyr160*   This residue was thought to be a good candidate for donor D, the slow electron donor to $P680^+$. Tyr160 was mutated to phenylalanine (Debus *et al.* 1988a; Vermaas *et al.* 1988a). The EPR signal arising from the oxidized secondary donor, $D^+$, was lost, implicating Tyr160 as donor D. The mutant was still able to grow photosynthetically, albeit more slowly, indicating the non-essential nature of this species. In a more positive identification, Barry *et al.* (1990) replaced Tyr160 with tryptophan. The typical $D^+$ signal II-type EPR resonance was replaced by a narrower signal with a g-value of 2.0035, which these authors attributed to the oxidized free radical of tryptophan.

*His197*   This residue has been considered a likely ligand to P680. It has been replaced by tyrosine (Vermaas *et al.* 1987b) leading to a loss of PSII reaction centres. His197 has also been replaced by glutamine (Vermaas and Diner unpublished results). This mutant shows a shift of about 20–30 mV negative in the midpoint potential of $P680^+$/P680. There is also an enhancement in the 'misses' observed in the oscillatory pattern of oxygen flash yields upon flash illumination of the dark-adapted mutant (see p. 104), indicating that one of the oxidation steps of the OEC has a midpoint potential close to that of $P680^+$/P680. These observations support a role for His197 in the coordination of P680.

## The *psbE* gene

The *psbE* gene codes for the $\alpha$-subunit of cyt b559, which is present in the minimal PSII reaction centre complex (Nanba and Satoh 1987). The gene has been sequenced from a wide variety of organisms (for a comparison see Cantrell and Bryant 1988) including *Synechocystis* PCC 6803 (Pakrasi *et al.* 1988). The conservation in primary structure for the gene product between species is not as high as that found for the *psbA* and *psbD* genes with, for example, only a 70% identity between *Synechocystis*

and spinach. In comparison, the *psbA* gene products show greater than 85% identity. The greatest sequence conservation is found in the region of the single transmembrane helix, which contains the histidine residue that is thought to bind the haem in conjunction with a histidine residue from the *psbF* gene product. This $\alpha\beta$ heterodimer model for the structure is more likely than the existence of two homodimers ($\alpha_2$ and $\beta_2$), because the one form of cyt b559 that has been isolated biochemically (Widger *et al.* 1984) contains both the $\alpha$- and $\beta$-subunits.

The *psbE* gene product is one of the few core polypeptides of PSII for which a study of membrane orientation has been performed (Tae *et al.* 1988; Vallon *et al.* 1989). Antibodies specific for a carboxy-terminal peptide were exploited to show that the $\alpha$-subunit was orientated with its carboxy-terminus in the lumen. The presence of the haem-coordinating histidine residue towards the N-terminus suggests that the haem is therefore located towards the stromal side of the membrane.

The function of the cytochrome within PSII is uncertain (Cramer *et al.* 1986). The most current of the many hypotheses that have been proposed is that the cytochrome may have a protective role against photoinhibition within PSII (Thompson and Brudvig 1988).

In all the photosynthetic organisms so far examined, *psbE* is located in an operon (usually covering about 700 bp) of three or possibly four genes, in the following order: *psbE-psbF-psbL* and ORF40. It is important to note that in some earlier publications these genes were referred to as the *psbE-psbF-psbI and psbJ* cluster (Cantrell and Bryant 1988). This nomenclature is now no longer used.

The presence of the *psbE* gene in an operon complicates its manipulation through *in vitro* mutagenesis techniques because changes in the *psbE* gene may exert polar effects on the steady-state levels of the downstream gene products (Lewin 1990). Extreme caution is thus required in the interpretation of experimental results.

No strain of *Synechocystis* 6803 has yet been constructed specifically with an interrupted or deleted *psbE* gene. A deletion strain, T1297, has been constructed in which the entire *psbE-psbF-psbL-*ORF40 operon was deleted from the chromosome and replaced by an antibiotic resistance cassette conferring resistance to kanamycin (Pakrasi *et al.* 1988). The mutant is incapable of evolving oxygen and was shown by immunochemical studies to have undetectable levels of the reaction centre polypeptides D1 and D2, reduced levels of the CP47 apoprotein but normal levels of OEE1 and the CP43 apoprotein (Pakrasi *et al.* 1987; Pakrasi *et al.* 1989). Therefore, at least one product from this operon is required for the accumulation of the D1 and D2 polypeptides in the membrane. Originally, the α- and β-subunits of b559 (the *psbE* gene and *psbF* gene products respectively) were assigned such a role (Pakrasi *et al.* 1989). However, the inadvertent deletion of *psbL* and ORF40 from strain T1297, necessitates a re-evaluation of this conclusion until more precise mutants are constructed and analysed.

## The *psbF* gene

The *psbF* gene is located downstream of the *psbE* gene and in *Synechocystis* 6803 codes for a small polypeptide of 44 residues. As explained on page 122, the gene product is thought to span the membrane once and cooperate with the *psbE* gene product in binding the haem group of cyt b559.

Recently, a mutant of *Synechocystis* 6803 was constructed in which 40 out of the 44 codons of the *psbF* gene were deleted, leaving the rest of the operon intact (Pakrasi *et al.* 1990). The resulting strain contains near normal levels of the CP43 and CP47 apoproteins, but is unable to accumulate D1 and D2 and, therefore, shows no PSII activity. The *psbF* gene product would appear to be crucial for the accumulation of the PSII reaction centre. However, pleiotropic effects of this deletion, such as the destabilization of the mRNA derived from this operon, have yet to be investigated.

The histidine residue that is thought to ligate the haem group has been changed to a leucine residue (Nyhus *et al.* 1990). In this particular case, the PSII core complex does not appear to accumulate. Whether this is due to a structural requirement for cyt b559, or is the result of damage from photoinhibition because of the absence of functional cytochrome, has yet to be investigated.

## The *psbG* gene

Although originally identified as a component of PSII from immunochemical studies of maize (Steinmetz *et al.* 1986), current ideas suggest that the *psbG* gene is more likely to be a component of a NADH or NADPH dehydrogenase, which acts to reduce plastoquinone (Nixon *et al.* 1989). In all the chloroplast genomes so far studied, the *psbG* gene is located between the *ndhC* gene, which shows homology to a mitochondrial gene that encodes a subunit of the mitochondrial NADH dehydrogenase; and an open reading frame of approximately 158 residues. In tobacco this cluster is cotranscribed (Matsubayashi *et al.* 1987). Steinmüller *et al.* (1989) showed that a homologous cluster of *ndhC*-*psbG*-'ORF157' exists in *Synechocystis* 6803 and that they are again cotranscribed. This copy of *psbG*, now termed *psbG1*, is approximately 52% identical to the maize *psbG* gene. More recently, a second copy of the *psbG* gene, designated *psbG2*, has been cloned and sequenced (Mayes *et al.* 1990; Steinmüller and Bogorad 1990). Steinmüller and Bogorad (1990) have concluded that this copy is not present in the chromosome but is found instead on a large megaplasmid. There is as yet no evidence to indicate that *psbG2* is expressed.

Although there is uncertainty as to the role of the *psbG* gene product, studies using *Synechocystis* 6803 should indicate if that role lies exclusively within PSII because, in such a case, a deletion strain could be constructed and maintained on glucose if necessary. On the other hand, if the gene product functions as a component of a NAD(P)H dehydrogenase, then it may be crucial for cell growth and be impossible to delete.

## The *psbH* gene

In higher plants, the *psbH* gene codes for the so-called 9 kDa or 10 kDa phosphoprotein (Farchaus and Dilley 1986), a protein that undergoes reversible light-regulated phosphorylation (Bennett 1977) at a threonine residue near its N-terminus (Michel and Bennett 1987). Its role in PSII is currently unknown. The *psbH* gene has recently been sequenced from *Synechocystis* 6803 (Mayes and Barber 1990; Abdel-Mawgood and Dilley 1990). No work has yet shown if this gene is expressed, but this would seem likely because a highly homologous polypeptide has been sequenced from a PSII core complex from the cyanobacterium *Synechococcus vulcanus* (Koike *et al.* 1989). Interestingly, the cyanobacterial *psbH* gene product does not contain the conserved threonine residue that is phosphorylated in higher plants. This feature, together with the lack of low molecular weight phosphoproteins in the cyanobacterium

*Synechococcus* 6301 (Allen *et al.* 1985), has led to the conclusion that the *psbH* gene product is not phosphorylated in cyanobacteria (Koike *et al.* 1989) and therefore should be referred to as the H-polypeptide, and not the 10 kDa phosphoprotein.

No attempt has yet been made to investigate its function in PSII through mutagenesis.

## The *psbI* gene

The *psbI* gene product is a component of PSII reaction centre complexes isolated from both higher plants (Ikeuchi and Inoue 1988; Webber *et al.* 1989a) and the cyanobacterium *Synechococcus vulcanus* (Ikeuchi *et al.* 1989b). The function of this polypeptide is unknown but it has been speculated that it may fulfil a similar function to that of the H subunit of reaction centres from purple non-sulphur photosynthetic bacteria (Webber *et al.* 1989a). The *psbI* gene has not yet been cloned from any cyanobacterium.

## The *psbJ* gene

In all the photosynthetic organisms so far examined, the genes for the $\alpha$- and $\beta$-subunits of cyt b559 (*psbE* and *psbF*, respectively) are co-transcribed with two downstream open reading frames of approximately 38 (*psbL*) and 40 codons. The high degree of conservation in primary structure for the product of ORF40 between species (Cushman *et al.* 1988) together with its transcription with known PSII genes has led to the assumption that it must also encode a low molecular weight polypeptide of PSII. However, the gene product has so far not been detected. Therefore, the gene designation *psbJ* has been reserved for ORF40, to be used only when the gene product has been identified in PSII extracts (Hallick 1989). In *Synechocystis* 6803, the equivalent gene to ORF40 has been sequenced and found to encode a product of 39 amino acids (Nyhus and Pakrasi 1989), which shows a 64% identity to ORF40 from pea.

The possible role of ORF39 in *Synechocystis* 6803 has been examined by constructing a strain in which the fourth codon was replaced by a stop codon, so that a full length gene product could no longer be synthesized. The mutant still evolves oxygen at 40% wild type rates and grows photo-autotrophically at 55% wild type rates (Nyhus and Pakrasi 1989). It would, therefore, appear that ORF39 is not essential for PSII activity. The reason for the reduced oxygen-evolving activity in this strain is as yet unclear, and may either be a direct consequence of the absence of the gene product or a pleiotropic effect on the expression of the *psbE*, *psbF* and *psbL* genes.

## The *psbK* gene

The *psbK* gene product is another low molecular weight component predicted to contain a single transmembrane helix. Originally identified in crude $O_2$-evolving preparations from the comparison of N-terminal sequence data to an open reading frame found in the chloroplast genome (Murata *et al.* 1988), this gene has subsequently been identified in the cyanobacterium *Synechococcus* 6301 (Fukuda *et al.* 1989) as well as in *Synechocystis* 6803 (Zhang *et al.* 1990). The gene product has also been identified in $O_2$-evolving complexes from *Synechococcus vulcanus* (Koike *et al.* 1989) and partially sequenced. Comparison of these data to the sequence deduced from the *psbK* gene, suggests that the cyanobacterial gene product is synthesized as a precursor of 45 amino acids and that it is processed to its mature form by removal of the first eight amino acids. The length of this presequence in higher plants is uncertain, ranging from a possible 61 or 24 amino acids in tobacco to 18 amino acids in liverwort (Murata *et al.* 1988).

Again the function of this polypeptide is unknown, but its absence in purified $O_2$-evolving complexes of spinach (Murata *et al.* 1988) strongly suggests a peripheral involvement in PSII function. Interestingly, there appears to be some sequence homology with the small light-harvesting polypeptides of photosynthetic bacteria (Umesono *et al.* 1988). No *psbK* mutants have yet been described.

## The *psbL* gene

This gene, which is cotranscribed with the *psbE* and *psbF* genes (Cushman *et al.* 1988) codes for a gene product of 37 residues after removal of the initiating N-formyl methionine residue, and is found in oxygen-evolving core complexes of higher plants (Webber *et al.* 1989b; Ikeuchi *et al.* 1989) and cyanobacteria (Ikeuchi *et al.* 1989b). Again its function is unknown, but preliminary evidence suggests that it may be more closely associated with the PSII reaction centre complex than other core polypeptides such as the *psbH* and *psbK* gene products (Ikeuchi *et al.* 1989b). This gene has been sequenced from *Synechocystis* 6803 (Nyhus and Pakrasi 1989) and found to encode a gene product of 39 amino acids with a 70% identity to the equivalent gene product from pea. Mutants of the *psbL* gene have yet to be described.

## The *psbM* and *psbN* genes

As described on page 110, these two genes were originally identified in the chloroplast genome by comparison with the partial N-terminal sequence for two small polypeptides found in $O_2$-evolving complexes from *Synechococcus vulcanus* (Ikeuchi *et al.* 1989a). The equivalent genes have not been located in cyanobacteria; nor have the gene products been identified in higher plants, possibly because their N-termini are blocked. The *psbN* gene product is, however, likely to be present since the *psbN* gene, located between the *psbB* and *psbH* genes but on the opposite strand, is transcribed in both liverwort and pea (Kohchi *et al.* 1988).

## The *psbO* gene

The *psbO* gene product is often referred to in higher plants as the 33 kDa extrinsic polypeptide because of its apparent molecular weight upon SDS-polyacrylamide electrophoresis and its ability to be removed from the thylakoid membrane by high salt concentrations. Other designations include the oxygen evolving enhancer protein 1 (OEE1) and the manganese stabilizing protein (MSP) (for review, see Murata and Miyao 1985).

The *psbO* gene is located in the thylakoid lumen and is considered to be more directly associated with the manganese cluster than any of the other extrinsic polypeptides. At first, the protein was thought to have a catalytic role in water oxidation (Barber 1984) because under oxidizing conditions it could be isolated with two bound manganese (Abramowicz and Dismukes 1984). However, later work showed that it was still possible to obtain water oxidation in the absence of the *psbO* gene product, if suitable ions were added (Ono and Inoue 1983, 1984; Miyao and Murata 1984). The gradual loss of manganese in the absence of the *psbO* gene product suggested more of a role in stabilization than catalysis. A role in binding cal-

cium has also been suggested from sequence homology with known calcium-binding proteins (Wales *et al.* 1989; Burnap *et al.* 1990).

The *psbO* gene in eukaryotes is located in the nucleus and includes the sequence for a large transit peptide to enable transport of the gene product into the thylakoid lumen of the chloroplast (Keegstra 1989). In cyanobacteria, such as *Synechocystis* 6803, there is a much smaller signal sequence because the gene product now only has to cross one membrane to reach the lumen (Philbrick and Zilinskas 1988). In *Synechocystis* 6803, the mature product shows an apparent molecular weight of 32 kDa, rather than 33 kDa observed in higher plants (Philbrick and Zilinskas 1988). However, for convenience we shall still refer to the *psbO* gene product as the 33 kDa extrinsic polypeptide, or OEE1.

## Inactivation of the *psbO* gene

The first reported *psbO* mutant was obtained in *C. reinhardtii* when Mayfield *et al.* (1987a) described the isolation of a nuclear mutant, FUD44, which was found to contain a 5 kb insert in the 5′ region of the *psbO* gene. This mutant accumulates no stable transcript of *psbO* and is unable to synthesize its gene product, OEE1. The core polypeptides (apoCP47, apoCP43, D1, D2 and the α-subunit of cyt 559) are all translated at wild type rates but accumulate to levels ranging from 10 to 30% of wild type (Mayfield *et al.* 1987a; de Vitry *et al.* 1989). Extrinsic polypeptides OEE2 and 3 are synthesized at wild type rates and accumulate at levels comparable to those of wild type. Thus, the absence of OEE1 destabilizes the core complex, increasing the rate of turnover of the core subunits. As this mutant is unable to evolve oxygen, it would appear that OEE1 is crucial for oxygen evolution and important for stabilization of the oxygen evolving complex (however, see below). De Vitry *et al.* (1989) reported that the absence of OEE1 has no influence on the binding affinity of OEE2 and 3 to the thylakoid membrane, indicating that OEE1 and OEE2 bind at separate sites. This observation is consistent with the report of Bricker *et al.* (1988) of cross-linking of OEE3 to CP47 and evidence from de Vitry and co-workers (1989) of possible binding of OEE2 to CP43. In a contrary opinion, however, Murata and Miyao (1985) describe cooperative binding of OEE1 and OEE2.

The importance of the *psbO* gene product for oxygen evolution was initially supported by an analysis of an interposon mutant of *Synechocystis* 6803, in which *psbO* was interrupted by a kanamycin resistance cartridge (Philbrick and Zilinskas 1988). As expected, this mutant did not synthesize the *psbO* gene product and was also incapable of evolving oxygen. However, subsequent to this report, it has been found that this particular mutant contained some other undetermined lesion within PSII and that its phenotype was unrelated to the interruption of the *psbO* gene (J. Philbrick personal communication).

Instead, when a different mutant was created by deletion of the *psbO* gene, it was found that this strain could grow photoautotrophically, although at a somewhat reduced rate than wild type (Burnap and Sherman 1991; Philbrick *et al.* 1991). These studies would therefore support a stabilizing role for the *psbO* gene product in oxygen evolution, rather than the crucial one suggested by the *Chlamydomonas* mutant. The deletion strain does show an increased susceptibility to photoinhibition and an enhanced dependence on $Ca^{2+}$ (Philbrick *et al.* 1991). No site-directed mutants have so far been reported using *Synechocystis* 6803 but clearly, with the isolation of a deletion strain, many may well be reported in the future.

Recent work by Mayfield and Kindle (1990) has demonstrated that it may also be possible to use *Chlamydomonas reinhardtii* for the construction of *psbO* mutants. Using the deletion strain FUD44, described above, they were able to transform cells at low frequency back to photoautotrophic growth by bombardment with microprojectiles covered with DNA containing the wild type *psbO* gene. Unlike the chloroplast transformation procedure, the wild type DNA now incorporated itself randomly into the nuclear genome. However, one drawback with this procedure is that the frequency of transformation is extremely low; it is actually less than the reversion rate for FUD44.

An alternative method for analysing the effects of mutations in the *psbO* gene may be through the use of an *in vitro* system. Seidler and Michel (1990) have demonstrated that it is possible to overexpress the mature form of the *psbO* gene product from spinach in *E. coli* and to rebind it to washed thylakoid membranes with restoration of oxygen evolution. Purification of the polypeptide from *E. coli* was simplified by using the last 18 amino acids of the 84-amino-acid-long transit peptide to act as a signal sequence to direct the protein into the periplasm. The endogenous signal peptidase of *E. coli* correctly recognized and processed

the precursor form into the mature form.

With this elegant expression system it should be possible to analyse the ability of site-directed mutant polypeptides to (i) bind to the PSII complex and (ii) restore oxygen evolution. The use of the *psbO* gene from cyanobacteria for such studies appears more limited because the gene product is far more hydrophobic than the higher plant form and is relatively insoluble (Kuwabara *et al.* 1989).

## The *psbP* gene

The *psbP* gene product, also known as the 23 kDa extrinsic polypeptide and the oxygen-evolving enhancer protein 2 (OEE2) is thought to be a regulatory factor involved in water oxidation, exerting its influence by somehow enhancing the local concentrations of chloride and calcium needed for efficient water splitting (for the effects of calcium and chloride on PSII function, see Babcock, 1987). Oddly, this gene product appears to be absent in cyanobacteria. However, these conclusions are based on immunochemical data that showed no cross-reactivity between PSII-enriched preparations and antibodies specific for the *psbP* gene product of spinach (Stewart *et al.* 1985). It is still possible that a protein analogous to the *psbP* gene product is present in cyanobacteria, but that the sequence homology in the antigenic regions is insufficient to cause a cross-reaction.

Bennoun has isolated two nuclear mutants, BF25 (Bennoun *et al.* 1981) and FUD39 (de Vitry *et al.* 1989), which inhibit expression of OEE2. One of these, BF25, has been shown by Mayfield *et al.* (1987b) to be unable to accumulate *psbP* transcript. This mutant and FUD39 are unable to synthesize OEE2. Both mutants synthesize the core subunits apoCP47, apoCP43, D1, D2 and the α-subunit of cyt b559 as well as OEE1 and OEE3 at wild type rates. The core subunits in BF25 accumulate to 65% (de Vitry *et al.* 1989) of wild type levels, indicating some destabilization of the core complex. The OEE1 and OEE3 polypeptides accumulate to near wild type levels. Upon isolation of the thylakoid membranes, OEE3 is lost in both mutants (de Vitry *et al.* 1989), consistent with the requirement of OEE2 for binding of OEE3 (Miyao and Murata 1983). BF25 shows only 5% of wild type $O_2$-evolving activity (Mayfield *et al.* 1987b) probably owing to a loss, with the dissociation of OEE2 and OEE3, of tight binding of $Ca^{2+}$ and $Cl^-$, cofactors for oxygen evolution (Babcock

1987). Both mutants also show enhanced proteolytic degradation of CP43 (de Vitry *et al.* 1989), implying a stabilization of CP43 by OEE2 and a binding of OEE2 to CP43 (see above).

## The *psbQ* gene

The *psbQ* gene product is referred to as the 16 kDa to 18 kDa extrinsic polypeptide in higher plants, and the oxygen-evolving enhancer protein 3 (OEE3) in *C. reinhardtii*. It is thought to function in water oxidation by enhancing the local concentration of chloride ions to an optimal level (Murata and Miyao 1985). No equivalent protein has been found in cyanobacteria and no mutants have been obtained in this gene.

## The *psbR* gene

The *psbR* gene, previously called the ST-LSI gene (Eckes *et al.* 1986), codes for a 10 kDa polypeptide that is located in the lumen of higher plants (Ljungberg *et al.* 1986). Extraction studies using high salt and detergent have suggested that it is an extrinsic polypeptide, but of a more hydrophobic nature than the other three extrinsic polypeptides involved in water oxidation (Ljungberg *et al.* 1986). In fact, the possibility that the protein is anchored in the membrane by a single transmembrane helix, located towards its carboxy-terminus, cannot be totally discounted (Ljungberg *et al.* 1986; Lautner *et al.* 1988), particularly in the light of a recent analysis that has identified a hydrophobic segment that is thought to act as a non-cleavable thylakoid transfer domain in this region (Webber *et al.* 1989c).

The function of this gene product is unknown, but immunoprecipitation experiments have suggested that it is involved in the binding of both the *psbO* and *psbP* gene products (Ljungberg *et al.* 1984). The gene product could not be identified immunochemically in either extracts of the cyanobacteria *Phormidium laminosum* or the red alga *Porphyridium purpureum* (Ljungberg *et al.* 1986) using antibodies specific for the spinach protein, and so the ubiquity of this gene product has yet to be demonstrated.

By using an anti-sense RNA approach, Stockhaus and co-workers (Stockhaus *et al.* 1990) have recently succeeded in generating transgenic potato plants that do not accumulate the *psbR* gene product. The depletion of this polypeptide to 1–3% of WT levels does not have a drastic effect on growth rates and so does not appear to be crucial for PS2 activity.

# Conclusions

## Random mutagenesis

### *Chlamydomonas*

The results derived from random mutagenesis in *Chlamydomonas* have led to several conclusions concerning the synthesis and accumulation of core polypeptides (see also Rochaix and Erickson, 1988):

(1) CP47, D1 and D2 are capable of forming a complex and can only accumulate in a concerted manner. Loss of any one polypeptide strongly affects the stability of the others. These subunits probably associate at an early stage of centre assembly.
(2) Synthesis of D1 and CP47 is coupled with an inability to translate one that impairs translation of the other.
(3) There is no consistent picture governing the dependence of D1 and CP47 translation on D2. Two out of three reports claim impairment of D1 synthesis in the absence of D2.
(4) Synthesis and accumulation of apoCP43 are more independent of the other three. Its absence will partially destabilize a CP47-, D1- or D2-containing complex.
(5) The extrinsic polypeptides do not bind tightly to the thylakoid membrane in the absence of core complex.
   (i) OEE1 and OEE2 bind at separate sites, possibly CP47 and CP43, respectively. OEE1 has another binding site on the reaction centre;
   (ii) OEE3 binds to OEE2.

### *Synechocystis*

Points (1) and (4) also apply to *Synechocystis*, although the destabilization of the three remaining large subunits in the absence of CP43 or in the absence of D1 is less marked than in *Chlamydomonas*. Destabilization of CP47 in the absence of D2 and CP43 is also less marked. In the absence of CP43, a fully active non-oxygen evolving core complex is assembled at 10% of wild type levels, showing that, while the complex is destabilized, CP43 is not required for light-driven oxidation of Z and reduction of $Q_A$.

In contrast to point (2), loss of CP47 in *Synechocystis* has little effect on the rates of synthesis of CP43, D1 and D2. In contrast to point (3), absence of D2 and CP43 has little effect on the synthesis of CP47 and D1. Based on a more limited analysis of mutants, translational regulation of the large core subunits in *Synechocystis* appears to occur more independently (less coupled) than in *Chlamydomonas*. Synthesis of the other large subunits has not been investigated in deletion mutants unable to express D1 alone or D2 alone.

## Site-directed mutagenesis

With more and more researchers applying molecular biological techniques to the study of photosynthesis, the power of site-directed mutagenesis for the mapping of the binding sites of redox components within the PSII core complex is becoming increasingly evident. These techniques have, for the most part, confirmed the existence of localized homologies that were suggested to exist between the PSII and the bacterial reaction centres concerning the binding sites of the primary donor and of the quinone electron acceptors.

Homology with bacterial reaction centres is not a prerequisite for the successful application of directed mutagenesis. No homology exists for the secondary and tertiary electron donors, yet the construction of directed mutants at D1-Tyr161 and D2-Tyr160 have permitted the localization of the secondary donors, Z and D. An active effort is also underway in several groups to locate the ligands to the manganese cluster of the OEC. One likely candidate has already been located in D1-Asp170 (Nixon and Diner 1992).

Site-directed mutagenesis has also made a contribution towards resolving the role of individual amino acids in mechanisms of electron transfer, particularly as concerns coupled protonation steps.

The choice of mutational sites for the location of ligands to Mn and $Ca^{2+}$ has been largely based on educated guesses as to the chemical nature of the ligand and its probable location within the complex. While this approach has been successful, the number of mutants required to pin down a co-ordination site can be considerable. There is something to be said for a more random mutagenesis

approach, but within a restricted region of the complex (e.g. an interhelical loop) coupled with suitable screening techniques.

However, one should not overlook the limitations of site-directed mutagenesis. It is, after all, limited to a replacement with any one of only 19 other residues. There is also the risk of the appearance of second-order structural perturbations that could obscure the interpretation of the primary mutation. Where a crystal structure of a mutant can be obtained, such fears can be allayed. In the case of PSII, where there is no check of this sort, one can only try and make conservative substitutions with respect to amino acid volume and make control substitutions in adjacent but non-functional residues. Where charges are created or lost one must resort to functional and spectroscopic assays at nearby sites to check for such secondary effects and the use of multiple amino acid replacements to arrive at a consensus image of what is occurring at the targeted site.

These are not overwhelming objections. As more and more research groups enter the field, using not only *Synechocystis* and *Synechococcus* but also *Chlamydomonas*, a consensus picture will emerge that will increase the confidence with which functional roles are assigned to individual residues. The mapping of the reaction centre and the understanding of protein regulation of electron transfer is going to have to be a collective effort.

# References

Abdel-Mawgood, A.L. and Dilley, R.A. (1990). Cloning and nucleotide sequence of the *psbH* gene from the cyanobacterium *Synechocystis* 6803. *Plant Molecular Biology*. **14**: 445–6.

Abramowicz, D.A. and Dismukes, G.C. (1984). Manganese proteins isolated from spinach thylakoid membranes and their role in $O_2$ evolution. *Biochimica et Biophysica Acta*. **765**: 318–28.

Ajlani, G., Kirilovsky, D., Picaud, M. and Astier, C. (1989a). Molecular analysis of *psbA* mutations responsible for various herbicide resistance phenotypes in *Synechocystis* 6714. *Plant Molecular Biology*. **13**: 469–79.

Ajlani, G., Meyer, I., Vernotte, C. and Astier, C. (1989b). Mutation in phenol-type herbicide resistance maps within the *psbA* gene in *Synechocystis* 6714. *FEBS Letters*. **246**: 207–10.

Allen, J.F., Sanders, C.E. and Holmes, N.G. (1985). Correlation of membrane protein phosphorylation with excitation energy distribution in the cyanobacterium *Synechococcus* 6301. *FEBS Letters*. **193**: 271–5.

Allen, J.P., Feher, G., Yeates, T.O., Komiya, H. and Rees, D.C. (1987). Structure of the reaction center from *Rhodobacter sphaeroides* R-26: the protein subunits. *Proceedings of the National Academy of Sciences, USA.* **84:** 6162–6.

Alt, J., Morris, J., Westhoff, P. and Herrmann, R. (1984). Nucleotide sequence of the clustered genes for the 44 kd chlorophyll *a* apoprotein and the '32 kd'-like protein of the Photosystem II reaction center in the spinach plastid chromosome. *Current Genetics* **8:** 597–606.

Anderson, B.F., Baker, H.M., Norris, G.E., Rice, D.W. and Baker, E.N. (1989). Structure of human lactoferrin: crystallographic structure analysis and refinement at 2.8Å resolution. *Journal of Molecular Biology.* **209:** 711–34.

Andersson, B. and Åkerlund, H.-E. (1987). Proteins of the oxygen-evolving complex. In *Topics in Photosynthesis, The Light Reactions*, Vol. 8 (ed. J. Barber), pp. 379–420. Elsevier, Amsterdam.

Andréasson, L.-E. (1989). Is nitrogen liganded to manganese in the photosynthetic oxygen-evolving system? EPR studies after isotopic replacement with $^{15}$N. *Biochimica et Biophysica Acta.* **973:** 465–7.

Astier, C., Elmorjani, K., Meyer, I., Joset, F. and Herdman, M. (1984). Photosynthetic mutants of the cyanobacteria *Synechocystis* sp. strains PCC 6714 and PCC 6803: sodium *p*-hydroxymercuribenzoate as a selective agent. *Journal of Bacteriology.* **158:** 659–64.

Babcock, G.T. (1987). The photosynthetic oxygen evolving process. In *New Comprehensive Biochemistry: Photosynthesis* (ed. J. Amesz), pp. 125–58. Elsevier, Amsterdam.

Babcock, G.T., Barry, B.A., Debus, R.J., Hoganson, C.W., Atamian, M., McIntosh, L., Sithole, I. and Yocum, C.F. (1989). Water oxidation in photosystem II: from radical chemistry to multielectron chemistry. *Biochemistry.* **28:** 9557–65.

Bailey, S., Evans, R.W., Garratt, R.C., Gorinsky, B., Hasnain, S., Horsburgh, C., Jhoti, H., Lindley, P.F., Mydin, A., Sarra, R. and Watson, J.L. (1988). Molecular structure of serum transferrin at 3.3Å resolution. *Biochemistry.* **27:** 5804–12.

Barber, J. (1984). Has the mangano-protein of the water splitting reaction of photosynthesis been isolated? *Trends in Biochemical Sciences.* **9:** 79–80.

Barber, J., Chapman, D.J. and Telfer, A. (1987) Characterisation of a PSII reaction centre isolated from the chloroplasts of *Pisum sativum*. *FEBS Letters.* **220:** 67–73.

Barry, B.A. and Babcock, G.T. (1987). Tyrosine radicals are involved in the photosynthetic oxygen-evolving system. *Proceedings of the National Academy of Sciences, USA* **84:** 7099–103.

Barry, B.A., El-Deeb, M., Sithole, I., Debus, R., McIntosh, L. and Babcock, G.T. (1990). Structural studies of the stable tyrosine radical, $Y_D^+$, in photosystem II. In *Current Research In Photosynthesis*, Vol. I

(ed. M. Baltcheffsky), pp. 483–6. Kluwer Academic Publishers, Dordrecht, The Netherlands.

Bennett, J. (1977). Phosphorylation of chloroplast membrane polypeptides. *Nature.* **269:** 344–6.

Bennoun, P. (1970). Réoxydation du quencher de fluorescence 'Q' en présence de 3-(3,4-dichlorophényl)-1,1-dimethylurée. *Biochimica et Biophysica Acta.* **216:** 357.

Bennoun, P. (1986). Le contrôle par la lumière des synthèses protéiques chloroplastiques chez *Chlamydomonas reinhardtii* fait intervenir les ribosomes attachés aux thylakoïdes. *Comptes Rendus de l'Académie des Sciences de Paris.* **303:** 645–50.

Bennoun, P. and Delepelaire, P. (1982). Isolation of photosynthesis mutants in *Chlamydomonas*. In *Methods in Chloroplast Molecular Biology*, (ed. M. Edelman, R.B. Hallick and N.-H. Chua), pp. 25–38. Elsevier, Amsterdam.

Bennoun, P., Masson, A. and Delosme, N. (1980). A method for complementation analysis of nuclear and chloroplast mutants of photosynthesis in *Chlamydomonas*. *Genetics.* **95:** 39–47.

Bennoun, P., Diner, B.A., Wollman, F.-A., Schmidt, G. and Chua, N.-H. (1981). Thylakoid polypeptides associated with Photosystem II in *Chlamydomonas reinhardtii*: comparison of system II mutants and particles. In *Photosynthesis III. Structure and Molecular Organization of the Photosynthetic Apparatus*, (ed. G. Akoyunoglou), pp. 839–49.

Bennoun, P., Spierer-Herz, M., Erickson, J., Girard-Bascou, J., Pierre, Y., Delosme, M. and Rochaix, J.-D. (1986). Characterization of photosystem II mutants of *Chlamydomonas reinhardtii* lacking the *psbA* gene. *Plant Molecular Biology.* **6:** 151–60.

Berthold, D.A., Babcock, G.T. and Yocum, C.F. (1981). A highly resolved, oxygen-evolving photosystem II preparation from spinach thylakoid membranes. EPR and electron-transport properties. *FEBS Letters.* **134:** 231–4.

Blowers, A.D., Bogorad, L., Shark, K.B. and Sanford, J.C. (1989). Studies on *Chlamydomonas* chloroplast transformation: foreign DNA can be stably maintained in the chromosome. *The Plant Cell.* **1:** 123–32.

Blubaugh, D. and Govindjee (1988). The molecular mechanism of the bicarbonate effect at the plastoquinone reductase site of photosynthesis. *Photosynthesis Research.* **19:** 85–128.

Bookjans, G., Stummann, B.M., Rasmussen, O.F. and Henningsen, K.W. (1986). Structure of a 3.2 kb region of pea chloroplast DNA containing the gene for the 44kd photosystem II polypeptide. *Plant Molecular Biology.* **6:** 359–66.

Bouges-Bocquet, B. (1973). Limiting steps in photosystem II and water decomposition in *Chlorella* and spinach chloroplasts. *Biochimica et Biophysica Acta.* **292:** 772–85.

Boynton, J.E., Gillham, N.W., Harris, E.H., Hosler, J.P., Johnson, A.M., Jones, A.R., Randolph-

Anderson, B.L., Robertson, D., Klein, T.M., Shark, K.B. and Sanford, J.C. (1988). Chloroplast transformation in *Chlamydomonas* with high velocity microprojectiles. *Science.* 240: 1534–8.

Boynton, J.E., Gillham, N.W., Harris, E.H., Newman, S.M., Randolph-Anderson, B.L., Johnson, A.M. and Jones, A.R. (1990). Manipulating the chloroplast genome of *Chlamydomonas*. Molecular genetics and transformation. In *Current Research in Photosynthesis*, Vol. III, (ed. M. Baltscheffsky), pp. 509–16. Kluwer Academic Publishers, Dordrecht, The Netherlands.

Brettel, K., Schlodder, E. and Witt, H.T. (1984). Nanosecond reduction kinetics of photooxidized chlorophyll-a$_{II}$ (P680) in single flashes as a probe for the electron pathway, H$^+$-release and charge accumulation in the O$_2$-evolving complex. *Biochimica et Biophysica Acta.* 766: 403–15.

Bricker, T.M. (1990). The structure and function of CPa-1 and CPa-2 in photosystem II. *Photosynthesis Research.* 24: 1–13.

Bricker, T.M., Odom, W.R. and Queirolo, C.B. (1988). Close association of the 33 kDa extrinsic protein with the apoprotein of CPa1 in photosystem II. *FEBS Letters.* 231: 111–17.

Britt, R.D., Derose, V.J., Yachandra, V.K., Kim, D.H., Sauer, K. and Klein, M.P. (1990). Pulsed EPR studies of the manganese center of the oxygen-evolving complex of photosystem II. In *Current Research in Photosynthesis*, Vol. I (ed. M. Baltscheffsky, pp. 769–72. Kluwer Academic Publishers, Dordrecht, The Netherlands.

Brusslan, J. and Haselkorn, R. (1989). Resistance to the photosystem II herbicide diuron is dominant to sensitivity in the cyanobacterium *Synechococcus* sp. PCC7942. *The EMBO Journal.* 8: 1237–45.

Bryant, D. (1987). The cyanobacterial photosynthetic apparatus: comparison of those of higher plants and photosynthetic bacteria. In *Canadian Bulletin of Fisheries and Aquatic Sciences 214, Photosynthetic Picoplankton* (ed. T. Platt and W. Li), pp. 423–500. Fisheries and Oceans, Canada.

Burnap, R., Koike, H., Sotiropoulou, G., Sherman, L.A. and Inoue, Y. (1989). Oxygen evolving membranes and particles from the transformable cyanobacterium *Synechocystis* 6803. *Photosynthesis Research.* 22: 123:30.

Burnap, R.L. and Sherman, L.A. (1991). Deletion mutagenesis in *Synechocystis* sp. PCC 6803 indicates that the Mn-stabilizing protein of Photosystem II is not essential for O$_2$ evolution. *Biochemistry.* 30: 440–6.

Burnap, R., Webb, R. and Sherman, L.A. (1990). Progress towards the production and analysis of H$_2$O splitting mutants of the cyanobacterium *Synechocystis* sp. PCC6803. In *Current Research in Photosynthesis*, Vol. I (ed. M. Baltscheffsky), pp. 255–8. Kluwer Academic Publishers, Dordrecht, The Netherlands.

Butler, W.L., Visser, J.W.M. and Symons, H.L. (1973). The kinetics of light-induced changes of C-550, cytochrome b559 and fluorescence yield in chloroplasts at low temperature. *Biochimica et Biophysica Acta.* 292: 140–51.

Buzby, J.S., Porter, R.D. and Stevens Jr., S.E. (1983). Plasmid transformation in *Agmenellum quadruplicatum* PR-6: construction of biphasic plasmids and characterization of their transformation properties. *Journal of Bacteriology.* 154: 1446–50.

Buzby, J.S., Porter, R.D. and Stevens Jr., S.E. (1985). Expression of the *Escherichia coli lacZ* gene on a plasmid vector in a cyanobacterium. *Science.* 230: 805–7.

Camm, E.L. and Green, B.R. (1980). Fractionation of thylakoid membranes with the non-ionic detergent octyl-β-D-glucopyranoside. *Plant Physiology.* 66: 428–32.

Cantrell, A. and Bryant, D.A. (1988). Nucleotide sequence of the genes encoding cytochrome b-559 from the cyanelle genome of *Cyanophora paradoxa*. In *Molecular Biology of Photosynthesis* (ed. Govindjee, H.J. Bohnert, W. Bottomley, D.A. Bryant, J.E. Mullet, W.L. Ogren, H. Pakrasi and C.R. Somerville), pp. 371–87. Kluwer Academic Publishers, Dordrecht, The Netherlands.

Cao, J. and Govindjee (1990). Anion effects on the electron acceptor side of photosystem II in a transformable cyanobacterium *Synechocystis* 6803. In *Current Research in Photosynthesis*, Vol. I (ed. M. Baltscheffsky), pp. 515–18. Kluwer Academic Publishers, Dordrecht, The Netherlands.

Carpenter, S.D. and Vermaas, W.F.J. (1989). Directed mutagenesis to probe the structure and function of photosystem II. *Physiologia Plantarum.* 77: 436–43.

Carpenter, S.D., Charité, J., Eggers, B. and Vermaas, W. (1990a). Characterization of site-directed and hybrid *psbC* mutants of *Synechocystis* 6803. In *Current Research In Photosynthesis*, Vol. I (ed. M. Baltscheffsky), pp. 359–62. Kluwer Academic Publishers, Dordrecht, The Netherlands.

Carpenter, S.D., Charité, J., Eggers, B. and Vermaas, W.F.J. (1990b). The *psbC* start codon in *Synechocystis* sp PCC 6803. *FEBS Letters.* 260: 135–7.

Chang, A.C.Y. and Cohen, S. (1978). Construction and characterization of amplifiable multicopy DNA cloning vehicles derived from the P15A cryptic miniplasmid. *Journal of Bacteriology.* 134: 1141–56.

Chang, C.H., Tiede, D., Tang, J., Smith, U., Norris, J. and Schiffer, M. (1986). Structure of *Rhodopseudomonas sphaeroides* R-26 reaction center. *FEBS Letters.* 205: 82–6.

Chapman, D.J., Gounaris, K., Vass, I. and Barber, J. (1990). Properties and stability of the isolated photosystem two reaction centre. In *Current Research in Photosynthesis*, Vol. I (ed. M. Baltscheffsky), pp. 223–30. Kluwer Academic Publishers, Dordrecht, The Netherlands.

Chauvat, F., De Vries, L., Van der Ende, A. and Van Arkel, G.A. (1986). A host–vector system for gene cloning in the cyanobacterium *Synechocystis* PCC 6803. *Molecular and General Genetics*. **204**: 185–91.

Cheung, A.Y., Bogorad, L., Van Montagu, M. and Schell, J. (1988). Relocating a gene for herbicide tolerance: a chloroplast gene is converted into a nuclear gene. *Proceedings of the National Academy of Sciences, USA*. **85**: 391–5.

Chisholm, D. and Williams, J.G.K. (1988). Nucleotide sequence of *psbC*, the gene encoding the CP-43 chlorophyll *a*-binding protein of Photosystem II, in the cyanobacterium *Synechocystis* 6803. *Plant Molecular Biology*. **10**: 293–301.

Chua, N.-H. and Bennoun, P. (1975). Thylakoid membrane polypeptides of *Chlamydomonas reinhardtii*: wild-type and mutant strains deficient in Photosystem II reaction centers. *Proceedings of the National Academy of Sciences, USA*. **72**: 2175–9.

Clackson, T. and Winter, G. (1989). 'Sticky feet'-directed mutagenesis and its application to swapping antibody domains. *Nucleic Acids Research*. **17**: 10163–70.

Coleman, W.J. (1990). Chloride binding proteins: mechanistic implications for the oxygen-evolving complex of Photosystem II. *Photosynthesis Research*. **23**: 1–27.

Coleman, W.J. and Govindjee (1987). A model for the mechanism of chloride activation in photosystem II. *Photosynthesis Research*. **13**: 199–223.

Cramer, W.A., Theg, S.M. and Widger, W.R. (1986). On the structure and function of cytochrome *b*-559. *Photosynthesis Research*. **10**: 393–403.

Cushman, J.C., Christopher, D.A., Little, M.C., Hallick, R.B. and Price, C.A. (1988). Organization of the *psbE*, *psbF*, orf38, and orf42 gene loci on the *Euglena gracilis* chloroplast genome. *Current Genetics*. **13**: 173–80.

Danielius, R.V., Satoh, K., van Kan, P.J.M., Plijter, J.J., Nuijs, A.M. and van Gorkom, H. (1987). The primary reaction of Photosystem II in the D1-D2-cyt b559 complex. *FEBS Letters*. **213**: 241–4.

de Paula, J.C., Innes, J.B. and Brudvig, G.W. (1985). Electron transfer in Photosystem II at cryogenic temperatures. *Biochemistry*. **24**: 8114–20.

de Vitry, C., Diner, B.A., Lemoine, Y. (1987). Chemical composition of Photosystem II reaction centers (PSII): phosphorylation of PSII polypeptides. In *Progress in Photosynthesis Research, Vol II* (ed. J. Biggins), pp. 105–8. Martinus Nijhoff Publishers, Dordrecht, The Netherlands.

de Vitry, C., Olive, J., Drapier, D., Recouvreur, M. and Wollman, F.-A. (1989). Posttranslational events leading to the assembly of Photosystem II protein complex: a study using photosynthesis mutants from *Chlamydomonas reinhardtii*. *Journal of Cell Biology*. **109**: 991–1006.

Debus, R.J., Barry, B.A., Babcock, G.T. and McIntosh, L. (1988a). Site-directed mutagenesis identifies a tyrosine radical involved in the photosynthetic oxygen-evolving system. *Proceedings of the National Academy of Sciences, USA*. **85**: 427–30.

Debus, R.J., Barry, B.A., Sithole, I., Babcock, G.T. and McIntosh, L. (1988b). Directed mutagenesis indicates that the donor to $P_{680}^+$ in photosystem II is tyrosine-161 of the D1 polypeptide. *Biochemistry*. **27**: 9071–4.

Debus, R.J., Nguyen, A.P. and Conway, A.B. (1990). Identification of ligands to manganese and calcium in photosystem II by site-directed mutagenesis. In *Current Research in Photosynthesis*, Vol. I (ed. M. Baltscheffsky), pp. 829–32. Kluwer Academic Publishers, Dordrecht, The Netherlands.

Deisenhofer, J. and Michel, H. (1989). The photosynthetic reaction center from the purple bacterium *Rhodopseudonmonas viridis*. *Science*. **245**: 1463–73.

Dekker, J.P., Van Gorkom, H.J., Brok, M. and Ouwehand, L. (1984). Optical characterization of Photosystem II electron donors. *Biochimica et Biophysica Acta*. **764**: 301–9.

Dekker, J.P., Boekema, E.J., Witt, H.T. and Rögner, M. (1988). Refined purification and further characterization of oxygen-evolving and Tris-treated Photosystem II particles from the thermophilic cyanobacterium, *Synechococcus* sp. *Biochimica et Biophysica Acta*. **936**: 307–18.

Dekker, J.P., Bowlby, N.R. and Yocum, C.F. (1989). Chlorophyll and cytochrome b559 content of the photochemical reaction center of Photosystem II. *FEBS Letters*. **254**: 150–4.

Delepelaire, P. (1983). Characterization of additional thylakoid membrane polypeptides synthesized inside the chloroplast in *Chlamydomonas reinhardtii*. *Photobiochemistry and Photobiophysics*. **6**: 279–91.

Delepelaire, P. (1984). Partial characterization of the biosynthesis and integration of the Photosystem II reaction centers in the thylakoid membrane of *Chlamydomonas reinhardtii*. *The EMBO Journal*. **3**: 701–6.

Delepelaire, P. and Chua, N.-H. (1979). Lithium dodecyl sulfate/polyacrylamide gel electrophoresis of thylakoid membranes at 4°C: characterizations of two additional chlorophyll a-protein complexes. *Proceedings of the National Academy of Sciences, USA*. **76**: 111–15.

De Rose, V.J., Yachandra, V.K., McDermott, A.E., Britt, R.D., Sauer, K. and Klein, M.P (1991). Nitrogen ligation to manganese in the photosynthetic oxygen-evolving complex: continuous-wave and pulsed EPR studies of Photosystem II particles containing $^{14}$N or $^{15}$N. *Biochemistry*. **30**: 1335–41.

Diner, B.A. (1977). Dependence of the deactivation reactions of Photosystem II on the redox state of the plastoquinone pool A varied under anaerobic conditions. Equilibria on the acceptor side of Photosystem II. *Biochimica et Biophysica Acta*. **460**: 247–58.

Diner, B.A. and Petrouleas, V. (1987). Light-induced oxidation of the acceptor-side Fe(II) of Photosystem II by exogenous quinones acting through the $Q_B$ binding site. I. Quinones, kinetics and pH-dependence. *Biochimica Biophysica Acta.* **895**: 107–25.

Diner, B. and Petrouleas, V. (1990). Formation by NO of nitrosyl adducts of redox components of the Photosystem II reaction center. II. Evidence that $HCO_3^-$/$CO_2$ binds to the acceptor-side non-heme iron. *Biochimica Biophysica Acta.* **1015**: 141–9.

Diner, B.A. and Wollman, F.-A. (1980). Isolation of highly active Photosystem II particles from a mutant of *Chlamydomonas reinhardtii*. *European Journal of Biochemistry.* **110**: 521–6.

Diner, B.A., Schenck, C.C. and de Vitry, C. (1984). Effect of inhibitors, redox state, and isoprenoid chain length on the affinity of ubiquinone for the secondary acceptor binding site in the reaction centers of photosynthetic bacteria. *Biochimica et Biophysica Acta.* **766**: 9–20.

Diner, B.A., Ries, D.F., Cohen, B.N. and Metz, J.G. (1988). COOH-terminal processing of polypeptide D1 of the Photosystem II reaction center of *Scenedesmus obliquus* is necessary for the assembly of the oxygen-evolving complex. *The Journal of Biological Chemistry.* **263**: 8972–80.

Diner, B.A., Petrouleas, V. and Wendoloski, J.J. (1991). The iron–quinone complex of Photosystem II. *Physiologia Plantarum.* **81**: 423–36.

Dismukes, G.C. (1988). The spectroscopically derived structure of the manganese site for photosynthetic water oxidation and a proposal for the protein-binding sites for calcium and manganese. *Chemica Scripta.* **28A**: 99–104.

Dzelzkalns, V.A. and Bogorad, L. (1986). Stable transformation of the cyanobacterium *Synechocystis* sp. PCC 6803 induced by UV irradiation. *Journal of Bacteriology.* **165**: 964–71.

Dzelzkalns, V.A. and Bogorad, L. (1989). Spectral properties and composition of reaction center and ancillary polypeptide complexes of photosystem II deficient mutants of *Synechocystis* 6803. *Plant Physiology.* **90**: 617–23.

Eaton-Rye, J.J. and Govindjee (1988). Electron transfer through the quinone acceptor complex of Photosystem II after one or two actinic flashes in bicarbonate-depleted spinach thylakoid membranes. *Biochimica et Biophysica Acta.* **935**: 248–57.

Eckert, H.J., Weise, N., Bernarding, J. Eichler, H.J. and Renger, G. (1988). Analysis of the electron transfer from Pheo to $Q_A$ in PSII membrane fragments from spinach by time resolved 325nm absorption changes in the picosecond domain. *FEBS Letters.* **240**: 153–8.

Eckes, P., Rosahl, S., Schell, J. and Willmitzer, L. (1986). Isolation and characterization of a light-inducible, organ-specific gene from potato and analysis of its expression after tagging and transfer into tobacco and potato shoots. *Molecular and General Genetics.* **205**: 14–22.

Edelman, M., Hallick, R.B. and Chua, N.-H. (1982). *Methods in chloroplast molecular biology.* Elsevier, Amsterdam.

Elhai, J. and Wolk, C.P. (1988). A versatile class of positive-selection vectors based on the nonviability of palindrome-containing plasmids that allows cloning into long polylinkers. *Gene.* **68**: 119–38.

Elmorjani, K., Thomas, J.-C. and Sebban, P. (1986). Phycobilisomes of wild type and pigment mutants of the cyanobacterium *Synechocystis* PCC 6803. *Archives of Microbiology.* **146**: 186–91.

Erickson, J.M., Rochaix, J.-D. and Delepelaire, P. (1985). Analysis of genes encoding two Photosystem II proteins of the 30–34-kD size class. In *Molecular Biology Of The Photosynthetic Apparatus* (ed. K.E. Steinback, S. Bonitz, C.J. Arntzen and L. Bogorad), pp. 53–65. Cold Spring Harbor Laboratory, Cold Spring Harbor, USA.

Erickson, J.M., Rahire, M., Malnoe, P., Girard-Bascou, J., Pierre, Y., Bennoun, P. and Rochaix, J.-D. (1986). Lack of the D2 protein in a *Chlamydomonas reinhardtii psbD* mutant affects Photosystem II stability and D1 expression. *The EMBO Journal.* **5**: 1745–54.

Erickson, J.M., Pfister, K., Rahier, M., Togasaki, R.K., Mets, L. and Rochaix, J.-D. (1989). Molecular and biophysical analysis of herbicide-resistant mutants of *Chlamydomonas reinhardtii*: structure–function relationship of the Photosystem II D1 polypeptide. *The Plant Cell.* **1**: 361–71.

Etienne, A.-L., Ducruet, J.-M., Ajlani, G. and Vernotte, C. (1990). Comparative studies on electron transfer in Photosystem II of herbicide-resistance mutants from different organisms. *Biochimica et Biophysica Acta.* **1015**: 435–40.

Farchaus, J. and Dilley, R.A. (1986). Purification and partial sequence of the $M_r$ 10,000 phosphoprotein from spinach thylakoids. *Archives of Biochemistry and Biophysics.* **244**: 94–101.

Feldmann, K.A., Carlson, T.J., Coomber, S.A., Farrance, C.E., Mandel, M.A. and Wierzbicki, A.M. (1990). T-DNA insertional mutagenesis in *Arabidopsis thaliana*. In *Horticulture Biotechnology, Plant Biology*, Vol. 11 (ed. A.B. Bennett and S.D. O'Neill), pp. 109–20. John Wiley & Sons Inc., New York.

Forbush, B., Kok, B. and McGloin, M. (1971). Cooperation of charges in photosynthetic $O_2$ evolution – II. Damping of flash yield oscillation, deactivation. *Photochemistry and Photobiology.* **14**: 307–21.

Frankel, L.K. and Bricker, T.M. (1989). Epitope mapping of the monoclonal antibody FAC2 on the apoprotein of CPa-1 in Photosystem II. *FEBS Letters.* **257**: 279–82.

Fromm, H., Devic, M., Fluhr, R. and Edelman, M. (1985). Control of *psbA* gene expression: in mature *Spirodela* chloroplasts light regulation of 32-kd protein synthesis is independent of transcript level. *The EMBO Journal.* **4**: 291–5.

Fromm, M.E., Taylor, L.P. and Walbot, V. (1986).

Stable transformation of maize after gene transfer by electroporation. *Nature*. **319**: 791–3.

Fukuda, M., Meng, B.Y., Hayashida, N. and Sugiura, M. (1989). Nucleotide sequence of the *psbK* gene of the cyanobacterium, *Anacystis nidulans* 6301. *Nucleic Acids Research*. **17**: 7521.

Gerken, S., Brettel, K., Schlodder, E. and Witt, H.T. (1988). Optical characterization of the immediate electron donor to chlorophyll $a_{II}^+$ in $O_2$-evolving photosystem II complexes. Tyrosine as possible electron carrier between chlorophyll $a_{II}$ and the water-oxidizing complex, *FEBS Letters*. **237**: 69–75.

Ghanotakis, D.F., Demetriou, D.M. and Yocum, C.F. (1987). Isolation and characterization of an oxygen-evolving Photosystem II reaction center core preparation and a 28 kDa Chl-a-binding protein. *Biochimica et Biophysica Acta*. **891**: 15–21.

Gingrich, J.C., Buzby, J.S., Stirewalt, V.L. and Bryant, D.A. (1988). Genetic analysis of two new mutations resulting in herbicide resistance in the cyanobacterium *Synechococcus* sp. PCC 7002. *Photosynthesis Research*. **16**: 83–99.

Gingrich, J.C., Gasparich, G.E., Sauer, K. and Bryant, D.A. (1990). Nucleotide sequence and expression of the two genes encoding D2 protein and the single gene encoding the CP43 protein of photosystem II in the cyanobacterium *Synechococcus* sp. PCC 7002. *Photosynthesis Research*. **24**: 137–50.

Golbeck, J.H. and Bryant, D.A. (1990). Photosystem I. In *Current Topics in Bioenergetics*, Vol. 16 (ed. C.P. Lee), pp. 83–177. Academic Press, New York.

Golden, S. and Stearns, G.W. (1988). Nucleotide sequence and transcript analysis of three photosystem II genes from the cyanobacterium *Synechococcus* sp. PCC7942. *Gene*. **67**: 85–96.

Golden, S.S., Brusslan, J. and Haselkorn, R. (1986). Expression of a family of *psbA* genes encoding a photosystem II polypeptide in the cyanobacterium *Anacystis nidulans* R2. *The EMBO Journal*. **5**: 2789–98.

Golden, S.S., Brusslan, J. and Haselkorn, R. (1988). Genetic engineering of the cyanobacterial chromosome. *Methods In Enzymology*. **153**: 215–31.

Golden, S.S., Cho, D.S.C. and Nalty, M.S. (1989). Two functional *psbD* genes in the cyanobacterium *Synechococcus* sp strain PCC 7942. *Journal of Bacteriology*. **171**: 4707–13.

Gounaris, K., Pick, H and Barber, J. (1987). Stoichiometry and turnover of photosystem two polypeptides. *FEBS Letters*. **211**: 94–8.

Gounaris, K., Chapman, D.J. and Barber, J. (1988). The interaction between the 33kDa manganese-stabilising protein and the D1/D2 cytochrome b-559 complex. *FEBS Letters*. **234**: 374–8.

Gounaris, K., Chapman, D.J. and Barber, J. (1989). Isolation and characterization of a D1/D2 cytochrome b-559 complex from *Synechocystis* 6803. *Biochimica et Biophysica Acta*. **973**: 296–301.

Gray, J.C. (1987). Genetics and synthesis of chloroplast membrane proteins. In *Photosynthesis* (ed. J. Amesz), pp. 319–41. Elsevier Science Publishers B.V. (Biomedical Division), Amsterdam.

Gray, J.C., Webber, A.N., Hird, S.M., Willey, D.L. and Dyer, T.A. (1990). Genes for photosystem II polypeptides. In *Current Research in Photosynthesis*, Vol. III (ed. M. Baltscheffsky), pp. 461–8. Kluwer Academic Publishers, Dordrecht, The Netherlands.

Grigorieva, G. and Shestakov, S. (1982). Transformation in the cyanobacterium *Synechocystis* sp. 6803. *FEMS Microbiology Letters*. **13**: 367–70.

Hain, R., Stabel, P., Czernilofsky, A.P., Steinbiss, H.H., Herrera-Estrella, L. and Schell, J. (1985). Uptake, integration, expression and genetic transmission of a selectable chimaeric gene by plant protoplasts. *Molecular and General Genetics*. **199**: 161–8.

Hallick, R.B. (1989). Prposals for the naming of chloroplast genes. II. Update to the nomenclature of genes for thylakoid membrane polypeptides. *Plant Molecular Biology Reporter*. **7**: 266–75.

Hansson, Ö. and Wydrzynski, T. (1990). Current perceptions of Photosystem II. *Photosynthesis Research*. **23**: 131–62.

Harlow, E. and Lane, D. (1988). *Antibodies. A Laboratory Manual*. Cold Spring Harbor Laboratory, Cold Spring Harbor, USA.

Harris, E.H., (1989). *The Chlamydomonas Sourcebook— A Comprehensive Guide to Biology and Laboratory Use*. Academic Press, USA.

Hill, D.E., Hope, I.A., Macke, J.P. and Struhl, K. (1986). Saturation mutagenesis of the yeast *his3*. Regulatory site: requirements for transcriptional induction and for binding by GCN4 activator protein. *Science*. **234**: 451–7.

Hiratsuka, J., Shimada, H., Whittier, R., Ishibashi, T., Sakamoto, M., Mori, M., Kondo, C., Honji, T., Sun, S.R., Meng, B.Y., Li, Y.Q., Kanno, A., Nishizawa, Y., Hirai, A., Shinozaki, K. and Sugiura M. (1989). The complete sequence of the rice (*Oryza sativa*) chloroplast genome: intermolecular recombination between distinct tRNA genes accounts for a major plastid DNA inversion during the evolution of the cereals. *Molecular and General Genetics*. **217**: 185–94.

Hirschberg, J. and McIntosh, L. (1983). Molecular basis of herbicide resistance in *Amaranthus hybridus*. *Science*. **222**: 1346–9.

Ho, S.N., Hunt, H.D., Horton, R.M., Pullen, J.K. and Pease, L.R. (1989). Site-directed mutagenesis by overlap extension using the polymerase chain reaction. *Gene*. **77**. 51–9.

Holschuh, K., Bottomley, W. and Whitfeld, P.R. (1984). Structure of the spinach chloroplast genes for the D2 and 44kDa reaction center proteins of Photosystem II and for tRNA$^{Ser}$ (UGA). *Nucleic Acids Research*. **12**: 8819–34.

Houmard, J. and Tandeau de Marsac, N. (1988). Cyanobacterial genetic tools: current status. *Methods In Enzymology*. **167**: 808–47.

Ikeuchi, M. and Inoue, Y. (1986). Characterization of $O_2$ evolution by a wheat Photosystem II reaction center complex isolated by a simplified method: disjunction of secondary acceptor quinone and enhanced $Ca^{2+}$ demand. *Archives of Biochemistry and Biophysics*. **247**: 97–107.

Ikeuchi, M. and Inoue, Y. (1988). A new photosystem II reaction center component (4.8 kDa protein) encoded by chloroplast genome. *FEBS Letters*. **241**: 99–104.

Ikeuchi, M., Yuasa, M. and Inoue, Y. (1985). Simple and discrete isolation of an $O_2$-evolving PSII reaction center complex retaining Mn and the 33 kDa protein. *FEBS Letters*. **185**: 316–22.

Ikeuchi, M., Koike, H. and Inoue, Y. (1989a). N-terminal sequencing of low-molecular-mass components in cyanobacterial photosystem II core complex. Two components correspond to unidentified open reading frames of plant chloroplast DNA. *FEBS Letters*. **253**: 178–82.

Ikeuchi, M., Koike, H. and Inoue, Y. (1989b). Identification of *psb*I and *psb*L gene products in cyanobacterial photosystem II reaction center preparation. *FEBS Letters*. **252**: 155–60.

Ikeuchi, M., Takio, K. and Inoue, Y. (1989c). N-terminal sequencing of photosystem II low-molecular-mass proteins 5 and 4.1 kDa components of the $O_2$-evolving core complex from higher plants. *FEBS Letters*. **242**: 263–9.

Ikeuchi, M., Koike, H., Mamada, K., Takio, K. and Inoue, Y. (1990). Comparative study of PSII low-molecular-mass components between *Synechococcus vulcanus* and higher plants. In *Current Research in Photosynthesis*, Vol. I (ed. M. Baltscheffsky), pp. 347–50. Kluwer Academic Publishers, Dordrecht, The Netherlands.

Jansen, T., Rother, C., Steppuhn, J., Reinke, H., Beyreuther, K., Jansson, C., Andersson, B. and Herrmann, R.G. (1987). Nucleotide sequence of cDNA clones encoding the complete '23 kDa' and '16 kDa' precursor proteins associated with the photosynthetic oxygen-evolving complex from spinach. *FEBS Letters*. **216**: 234–40.

Jansson, C., Debus, R.J., Osiewacz, H.D., Gurevitz, M., and McIntosh, L. (1987). Construction of an obligate photoheterotrophic mutant of the cyanobacterium *Synechocystis* 6803. *Plant Physiology*. **85**: 1021–5.

Jay, F.A. (1987). Techniques for the biochemical and immunological characterization of bacterial thylakoid proteins. In *The Light Reactions* (ed. J. Barber), pp. 261–303. Elsevier, Amsterdam.

Jensen, K., Herrin, D.L., Plumley, F.G. and Schmidt, G.W. (1986). Biogenesis of Photosystem II complexes: transcriptional, translational, and posttranslational regulation. *The Journal of Cell Biology*. **103**: 1315–25.

Johanningmeier, U., Bodner, U. and Wildner, G.F. (1987). A new mutation in the gene coding for the herbicide-binding protein in *Chlamydomonas*. *FEBS Letters*. **211**: 221–4.

Joliot, A. (1974). Effect of low temperature ($-30$ to $-60°C$) on the reoxidation of the Photosystem II primary electron acceptor in the presence and absence of 3-(3,4-dichlorophenyl)-1,1-dimethylurea. *Biochimica et Biophysica Acta*. **357**: 439–48.

Joliot, P. and Delosme, R. (1974). Flash-induced 519 nm absorption change in green algae. *Biochimica et Biophysica Acta*. **357**: 267–84.

Joliot, A. and Joliot, P. (1964). Étude cinétique de la réaction photochimique libérant l'oxygène au cours de la photosynthèse. *Comptes Rendus de l'Académie des Sciences, Paris*. **258**: 4622–5.

Joliot, P. and Joliot, A. (1968). A polarographic method for detection of oxygen production and reduction of Hill reagent by isolated chloroplasts. *Biochimica et Biophysica Acta*. **153**: 625–34.

Joliot, P. and Joliot, A. (1973). Different types of quenching involved in Photosystem II centers. *Biochimica et Biophysica Acta*. **305**: 302–16.

Joliot, P. and Kok, B. (1975). Oxygen evolution in photosynthesis. In *Bioenergetics of Photosynthesis* (ed. Govindjee), pp. 387–412, Academic Press, New York.

Joliot, P., Barbieri, G., and Chabaud, R. (1969). Un nouveau modèle des centres photochimique du système II. *Photochemistry and Photobiology*. **10**: 309–29.

Joliot, P., Joliot, A., Bouges, B. and Barbieri, G. (1971). Studies of System II photocenters by comparative measurements of luminescence, fluorescence and oxygen emission. *Photochemistry and Photobiology*. **14**: 287–305.

Keegstra, K. (1989). Transport and routing of proteins into chloroplasts. *Cell*. **56**: 247–53.

Klein, T.M., Wolf, E.D., Wu, R. and Sanford, J.C. (1987). High-velocity microprojectiles for delivering nucleic acids into living cells. *Nature*. **327**: 70–3.

Klimov, V.V. and Krasnovskii, A.A. (1981). Pheophytin as the primary electron acceptor in Photosystem 2 reaction centres. *Photosynthetica*. **15**: 592–609.

Koboyashi, M., Maeda, H., Watanabe, T., Nakane, H. and Satoh, K. (1990). Chlorophyll *a* and β-carotene content in the D1/D2 cytochrome *b*-559 reaction center complex from spinach. *FEBS Letters*. **260**: 138–40.

Kohchi, T., Yoshida, T., Komano, T. and Ohyama, K. (1988). Divergent mRNA transcription in the chloroplast *psb*B operon. *The EMBO Journal*. **7**: 885–91.

Koike, H., Mamada, K., Ikeuchi, M. and Inoue, Y. (1989). Low-molecular-mass proteins in cyanobacterial photosystem II: identification of *psb*H and *psb*K gene products by N-terminal sequencing. *FEBS Letters*. **244**: 391–6.

Koike, H., Ikeuchi, M., Hiyama, T., Mamada, K. and Inoue, Y. (1990). Identification of PSI and PSII components from the cyanobacterium *Synechococcus* by N-terminal sequencing. In *Current Research in Photosynthesis*, Vol. I (ed. M. Baltscheffsky), pp. 351–4. Kluwer Academic Publishers, Dodrecht, The Netherlands.

Kok, B., Forbush, B. and McGloin, M.P. (1970). Cooperation of charges in photosynthetic $O_2$ evolution—I. A linear four step mechanism. *Photochemistry and Photobiology*. **11**: 457–75.

Komiya, H., Yeates, T.O., Rees, D.C., Allen, J.P. and Feher, G. (1988). Structure of the reaction center from *Rhodobacter sphaeroides* R-26 and 2.4.1: symmetry relations and sequence comparison between different species. *Proceedings of the National Academy of Sciences, USA*. **85**: 9012–16.

Kuchka, M.R., Mayfield, S.P. and Rochaix, J.-D. (1988). Nuclear mutations specifically alter the synthesis and/or degradation of the chloroplast-encoded D2 polypeptide of photosystem II in *Chlamydomonas reinhardtii*. *The EMBO Journal*. **7**: 319–24.

Kuchka, M.R., Goldschmidt-Clermont, M., van Dillewijn, J. and Rochaix, J.-D. (1989). Mutation at the *Clamydomonas* nuclear NAC2 locus specifically affects stability of the chloroplast *psb*D transcript encoding polypeptide D2 of PSII. *Cell*. **58**: 869–76.

Kuhlemeier, C.J. and Van Arkel, G.A. (1988). Host–vector systems for gene cloning in cyanobacteria. *Methods in Enzymology*. **153**: 199–215.

Kunkel, T.A., Roberts, J.D. and Zakour, R.A. (1987). Rapid and efficient site-specific mutagenesis without phenotypic selection. *Methods In Enzymology*. **154**: 367–82.

Kuwabara, T., Nagata, R. and Shinohara, K. (1989). Expression and processing of cyanobacterial Mn-stabilizing protein in *Escherichia coli*. *European Journal of Biochemistry*. **186**: 227–32.

Kyle, D.J. (1985). The 32000 dalton $Q_B$ protein of Photosystem II. *Photochemistry and Photobiology*. **41**: 107–16.

Labarre, J., Chauvat, F. and Thuriaux, P. (1989). Insertional mutagenesis by random cloning of antibiotic resistance genes into the genome of the cyanobacterium *Synechocystis* strain PCC 6803. *Journal of Bacteriology*. **171**: 3449–57.

Laudenbach, D.E. and Straus, N.A. (1988). Characterization of a cyanobacterial iron stress induced gene similar to *psb*C. *Journal of Bacteriology*. **170**: 5018–26.

Lautner, A., Klein, R., Ljungberg, U., Reiländer, H., Bartling, D., Andersson, B., Reinke, H., Beyreuther, K. and Herrmann, R.G. (1988). Nucleotide sequence of cDNA clones encoding the complete precursor for the '10-kDa' polypeptide of photosystem II from spinach. *The Journal of Biological Chemistry*. **263**: 10077–81.

Lavergne, J. (1984). Absorption changes of Photosystem II donors and acceptors in algal cells. *FEBS Letters*. **173**: 9–14.

Lavergne, J. (1987). Optical-difference spectra of the S-state transitions in the photosynthetic oxygen-evolving complex. *Biochimica et Biophysica Acta*. **894**: 91–107.

Lavorel, J. (1975). Luminescence. In *Bioenergetics of Photosynthesis* (ed. Govindjee), pp. 223–317. Academic Press, New York.

Leto, K.J., Bell, E. and McIntosh, L. (1985). Nuclear mutation leads to an accelerated turnover of chloroplast-encoded 48 kd and 34.5 kd polypeptides in thylakoids lacking photosystem II. *The EMBO Journal*. **4**: 1645–53.

Lewin, B. (1990). *Genes IV*. Oxford University Press, Oxford, England.

Ljungberg, U., Åkerlund, H.-E., Larsson, C. and Andersson, B. (1984). Identification of polypeptides associated with the 23 and 33 kDa proteins of photosynthetic oxygen evolution. *Biochimica et Biophysica Acta*. **767**: 145–52.

Ljungberg, U., Åkerlund, H.-E. and Andersson, B. (1986). Isolation and characterization of the 10-kDa and 22-kDa polypeptides of higher plant photosystem 2. *European Journal of Biochemistry*. **158**: 477–82.

Ljungberg, U., Henrysson, T., Rochester, C.P., Åkerlund, H.-E. and Andersson, B. (1986). The presence of low-molecular-weight polypeptides in spinach Photosystem II core preparations. Isolation of a 5 kDa hydrophilic polypeptide. *Biochimica et Biophysica Acta*. **849**: 112–20.

Lörz, H., Baker, B. and Schell, J. (1985). Gene transfer to cereal cells mediated by protoplast transformation. *Molecular and General Genetics*. **199**: 178–82.

Marder, J.B., Chapman, D.J., Telfer, A., Nixon, P.J. and Barber, J. (1987). Identification of *psb*A and *psb*D gene products, D1 and D2, as reaction centre proteins of photosystem 2. *Plant Molecular Biology*. **9**: 325–33.

Matsubayashi, T., Wakasugi, T., Shinozaki, K., Yamaguchi-Shinozaki, K., Zaita, N., Hidaka, T., Meng, B.Y., Ohto, C., Tanaka, M., Kato, A., Maruyama, T. and Sugiura, M. (1987). Six chloroplast genes (*ndh*A-F) homologous to human mitochondrial genes encoding components of the respiratory chain NADH dehydrogenase are actively expressed: determination of the splice sites in *ndh*A and *ndh*B pre-mRNAs. *Molecular and General Genetics*. **210**: 385–93.

Mattoo, A.K. and Edelman, M. (1987). Intramembrane translocation and posttranslational palmitoylation of the chloroplast 32-kDa herbicide-binding protein. *Proceedings of the National Academy of Sciences, USA*. **84**: 1497–1501.

Mattoo, A.K., Marder, J.B. and Edelman, M. (1989). Dynamics of the Photosystem II reaction center. *Cell*. **56**: 241–6.

Mauzerall, D. (1972). Light-induced fluorescence changes in *Chlorella*, and the primary photoreactions for the production of oxygen. *Proceedings of the National Academy of Sciences, USA*. **69**: 1358–62.

Mayes, S.R. and Barber, J. (1989). Nucleotide sequence of the *psb*H gene of the cyanobacterium *Synechocystis* 6803. *Nucleic Acids Research*. **18**: 194.

Mayes, S.R., Cook, K.M. and Barber, J. (1990). Nucleotide sequence of the second *psb*G gene in *Synechocystis* 6803. Possible implications for *psb*G function as a

NAD(P)H dehydrogenase subunit gene. *FEBS Letters*. 262: 49–54.

Mayfield, S.P. and Kindle, K.L. (1990). Stable nuclear transformation of *Chlamydomonas reinhardtii* by using a *C. reinhardtii* gene as the selectable marker. *Proceedings of the National Academy of Sciences, USA*. 87: 2087–91.

Mayfield, S.P., Bennoun, P. and Rochaix, J.D. (1987a). Expression of the nuclear encoded OEE1 protein is required for oxygen evolution and stability of photosystem II particles in *Chlamydomonas reinhardtii*. *The EMBO Journal*. 6: 313–18.

Mayfield, S.P., Rahire, M., Frank, G., Zuber, H. and Rochaix, J.-D. (1987b). Expression of the nuclear gene encoding oxygen-evolving enhancer protein 2 is required for high levels of photosynthetic oxygen evolution in *Chlamydomonas reinhardtii*. *Proceedings of the National Academy of Sciences, USA*. 84: 749–53.

McPherson, P.H., Okamura, M.Y. and Feher, G. (1990). Electron transfer from the reaction center of *Rb. sphaeroides* to the quinone pool: doubly reduced $Q_B$ leaves the reaction center. *Biochimica et Biophysica Acta*. 1016: 289–92.

Metz, J.G., Wong, J. and Bishop, N.I. (1980). Changes in the electrophoretic mobility of a chloroplasts membrane polypeptide associated with the loss of the oxidizing side of Photosystem II in low fluorescent mutants of *Scenedesmus*. *FEBS Letters*. 114: 61–6.

Metz, J.G., Bricker, T.M. and Seibert, M. (1985). The azido-[$^{14}$C]-atrazine photoaffinity technique labels a 34-kDa protein in *Scenedesmus* which functions on the oxidizing side of Photosystem II. *FEBS Letters*. 185: 191–6.

Metz., J.G., Nixon, P.J., Rögner, M., Brudvig, G.W. and Diner, B.A. (1989). Directed alteration of the D1 polypeptide of Photosystem II: Evidence that tyrosine-161 is the redox component, Z, connecting the oxygen-evolving complex to the primary electron donor, P680. *Biochemistry*. 28: 6960–9.

Metz, J.G., Nixon, P.J. and Diner, B.A. (1990). Nucleotide sequence of the *psbA3* gene from the cynobacterium *Synechocystis* PCC 6803. *Nucleic Acids Research*. 18: 6715.

Michel, H.P. and Bennett, J. (1987). Identification of the phosphorylation site of an 8.3 kDa protein from photosystem II of spinach. *FEBS Letters*. 212: 103–8.

Michel, H. and Deisenhofer, J. (1988). Relevance of the photosynthetic reaction center from purple bacteria to the structure of Photosystem II. *Biochemistry*. 27: 1–7.

Michel, H., Hunt, D.F., Shabanowitz, J. and Bennett, J. (1988). Tandem mass spectrometry reveals that three Photosystem II proteins of spinach chloroplasts contain N-acetyl-O-phosphothreonine at their $NH_2$ termini. *The Journal of Biological Chemistry*. 263: 1123–30.

Miles, D., Leto, K.J., Neuffer, M.G., Polacco, M., Hanks, J.F. and Hunt, M.A. (1985). Chromosome arm location of photosynthesis mutants in *Zea mays* L.

using B–A translocations. In *Molecular Biology of the Photosynthetic Apparatus* (ed. K.E. Steinback, S. Bonitz, C.J. Arntzen and L. Bogorad), pp. 361–5. Cold Spring Harbor Laboratory, Cold Spring Harbor, USA.

Miyao, M. and Murata, N. (1983). Partial disintegration and reconstitution of the photosynthetic oxygen evolution system. Binding of 24 kilodalton and 18 kilodalton polypeptides. *Biochimica et Biophysica Acta*. 725: 87–93.

Miyao, M. and Murata, N. (1984). Role of the 33-kDa polypeptide in preserving Mn in the photosynthetic oxygen-evolution system and its replacement by chloride ions. *FEBS Letters*. 170: 350–4.

Miyazaki, A., Shina, T., Toyoshima, Y., Gounaris, K. and Barber, J. (1989). Stoichiometry of cytochrome b-559 in Photosystem II. *Biochimica et Biophysica Acta*. 975: 142–7.

Mohamed, A. and Jansson, C. (1989). Influence of light on accumulation of photosynthesis-specific transcripts in the cyanobacterium *Synechocystis* 6803. *Plant Molecular Biology*. 13: 693–700.

Morden, C.W. and Golden, S.S. (1989). *psbA* genes indicate common ancestry of prochlorophytes and chloroplasts. *Nature*. 337: 382–5. [Corrigendum: *Nature* 339: 400.]

Morris, J. and Herrmann, R.G. (1984). Nucleotide sequence of the gene for the $P_{680}$ chlorophyll and apoprotein in the photosystem II reaction center from spinach. *Nucleic Acids Research*. 12: 2837–50.

Murata, N. and Miyao, M. (1985). Extrinsic membrane proteins in the photosynthetic oxygen-evolving complex. *Trends in Biochemical Sciences*. 10: 122–4.

Murata, N., Miyao, M., Hayashida, N., Hidaka, T. and Sugiura, M. (1988). Identification of a new gene in the chloroplast genome encoding a low-molecular-mass polypeptide of photosystem II complex. *FEBS Letters*. 235: 283–8.

Nakatani, H.Y., Ke, B., Dolan, E. and Arntzen, C.J. (1984). Identity of the photosystem II reaction center polypeptide. *Biochimica et Biophysica Acta*. 765: 347–52.

Nanba, O. and Satoh, K. (1987). Isolation of a Photosystem II reaction center consisting of D-1 and D-2 polypeptides and cytochrome b-559. *Proceedings of the National Academy of Sciences, USA*. 84: 109–12.

Neumann, E.M. (1988). Primary structure of barley genes encoding quinone and chlorophyll a binding proteins of Photosystem II. *Carlsberg Research Communications*. 53: 259–75.

Nilsson, F., Simpson, D.J., Andersson, B. and Jansson, C. (1990). Structural and ultrastructural organization of thylakoids in a constructed photosystem II mutant of the cyanobacterium *Synechocystis* 6803. In *Current Research in Photosynthesis*, Vol. I (ed. M. Baltscheffsky), pp. 299–302. Kluwer Academic Publishers, Dordrecht, The Netherlands.

Nixon, P.J., Gounaris, K., Coomber, S.A., Hunter,

C.N., Dyer, T.A. and Barber, J. (1989). *psbG* is not a Photosystem Two gene but may be an *ndh* gene. *Journal of Biological Chemistry*. **264**: 14129–35.

Nixon, P.J., Metz, J.G., Rögner, M. and Diner, B.A. (1990). A *Synechocystis* PCC 6803 *psbA* deletion mutant and its transformation with a *psbA* gene from a higher plant. In *Current Research in Photosynthesis*, Vol. I (ed. M. Baltcheffsky), pp. 471–4. Kluwer Academic Publishers, Dordrecht, The Netherlands.

Nixon, P.J. and Diner, B.A. (1990). Protein coordination of the photosynthetic oxygen-evolving complex, a quaternary electron counter. In *Proceedings of the Twelfth Annual International Conference of the IEEE Engineering in Medicine and Biology Society* (ed. P.C. Pedersen and B. Onaral), pp. 1732–4. IEEE, New York.

Nixon, P.J., Rögner, M. and Diner, B.A. (1991). Expression of a higher plant *psbA* gene in *Synechocystis* 6803 yields a functional hybrid photosystem II reaction center complex. *The Plant Cell*. **3**: 383–95.

Nixon, P.J. and Diner, B.A. (1992). Aspartate 170 of the Photosystem II reaction center polypeptide D1 is involved in the assembly of the oxygen-evolving manganese cluster. *Biochemistry*. (In press.)

Noren, G.H., Boerner, R.J. and Barry, B.A. (1991). EPR characterization of an oxygen-evolving photosystem II preparation from the transformable cyanobacterium *Synechocystis* 6803. *Biochemistry*. **30**: 3943–50.

Nuijs, A.M., van Gorkem, H.J., Plijter, J.J. and Duysens, L.N.M. (1986). Primary charge separation and excitation of chlorophyll a in Photosystem II particles from spinach as studied by picosecond absorbance-difference spectroscopy. *Biochimica et Biophysica Acta*. **848**: 167–75

Nyhus, K.J. and Pakrasi, H.B. (1989). Possible role of *psbJ* in Photosystem II: site-directed mutagenesis in the cyanobacterium, *Synechocystis* 6803. In *Techniques and New Developments in Photosynthesis Research* (ed. J. Barber and R. Malkin), pp. 469–72. Plenum Publishing Corporation, New York and London.

Nyhus, K.J., Granok, H. and Pakrasi, H.B. (1990). Deletion mutagenesis of *psbF*, the gene for the β-subunit of cytochrome $b_{559}$ in *Synechocystis* PCC 6803. In *Current Research in Photosynthesis*, Vol. I (ed. M. Baltscheffsky), pp. 367–70. Kluwer Academic Publishers, Dordrecht, The Netherlands.

Ohad, N. and Hirschberg, J. (1990). A similar structure of the herbicide binding site in Photosystem II of plants and cyanobacteria is demonstrated by site specific mutagenesis of the *psbA* gene. *Photosynthesis Research*. **23**: 73–9.

Ohyama, K., Fukuzawa, H., Kohchi, T., Shirai, H., Sano, T., Sano, S., Umesono, K., Shiki, Y., Takeuchi, M., Chang, Z., Aota, S.I., Inokuchi, H. and Ozeki, H. (1986a). Chloroplast gene organization deduced from complete sequence of liverwort *Marchantia polymorpha* chloroplast DNA. *Nature*. **322**: 572–4.

Ohyama, K., Fukazawa, H., Kohchi, T., Shirai, H., Sano, T., Sano, S., Umesono, K., Shiki, Y., Takeuchi, M., Chang, Z., Aota, S., Inokuchi, H. and Ozeki, H. (1986b). Complete nucleotide sequence of liverwort *Marchantia polymorpha* chloroplast DNA. *Plant Molecular Biology Reporter*. **4**: 149–75.

Ono, T. and Inoue, Y. (1983). Mn-preserving extraction of 33-, 24- and 16-kDa proteins from $O_2$-evolving PS II particles by divalent salt-washing. *FEBS Letters*. **164**: 255–60.

Ono, T. and Inoue, Y. (1984). $Ca^{2+}$-dependent restoration of $O_2$-evolving activity in $CaCl_2$-washed PS II particles depleted of 33, 24 and 16 kDa proteins. *FEBS Letters*. **168**: 281–6.

Osiewacz, H.D. and McIntosh, L. (1987). Nucleotide sequence of a member of the *psbA* multigene family from the unicellular cyanobacterium *Synechocystis* 6803. *Nucleic Acids Research*. **15**: 10585.

Paddock, M.L., Rongey, S.H., Feher, G. and Okamura, M.Y. (1989). Pathway of proton transfer in bacterial reaction centers: replacement of glutamic acid 212 in the L subunit by glutamine inhibits quinone (secondary acceptor) turnover. *Proceedings of the National Academy of Sciences, USA*. **86**: 6602–6.

Pakrasi, H.B., Williams, J.G.K. and Arntzen, C.J. (1987). Genetically engineered cytochrome b559 mutants of the cyanobacterium *Synechocystis* 6803. In *Progress in Photosynthesis Research*, Vol. IV (ed. J. Biggins), pp. 813–16. Martinus Nijhoff Publishers, Dordrecht, The Netherlands.

Pakrasi, H.B., Williams, J.G.K. and Arntzen, C.J. (1988). Targeted mutagenesis of the *psbE* and *psbF* genes blocks photosynthetic electron transport: evidence for a functional role of cytochrome $b_{559}$ in photosystem II. *The EMBO Journal*. **7**: 325–32.

Pakrasi, H.B., Diner, B.A., Williams, J.G.K. and Arntzen, C.J. (1989). Deletion mutagenesis of the cytochrome $b_{559}$ protein inactivates the reaction center of photosystem II. *The Plant Cell*. **1**: 591–7.

Pakrasi, H.B., Nyhus, K.J. and Granok, H. (1990). Targeted deletion mutagenesis of the β subunit of cytochrome b559 protein destabilises the reaction center of photosystem II. *Zeitschrift für Naturforschung*. **45c**: 67–73.

Papageorgiou, G. (1975). Chlorophyll fluorescence: an intrinsic probe of photosynthesis. In *Bioenergetics of Photosynthesis*, (ed. Govindjee), pp. 319–71. Academic Press, New York.

Pecoraro, V.L. (1988). Structural proposals for the manganese centers of the oxygen evolving complex: an inorganic chemist's perspective. *Photochemistry and Photobiology*. **48**: 249–64.

Peterhans, A. and Paszkowski, J. (1990). Prospects for homologous recombination in plants. In *Horticulture Biotechnology, Plant Biology*, Vol. 11 (ed. A.B. Bennett and S.D. O'Neill), pp 155–70. John Wiley & Sons Inc., New York.

Petersen, J., Dekker, J.P., Bowlby, N.R., Ghanotakis,

D.F., Yocum, C.F. and Babcock, G.T. (1990). EPR characterization of the CP47-D1-D2-cytochrome b559 complex of Photosystem II. *Biochemistry.* **29:** 3226–31.

Petrouleas, V. and Diner, B.A. (1986). Identification of $Q_{400}$, a high-potential electron acceptor of Photosystem II, with the iron of the quinone-iron acceptor complex. *Biochimica et Biophysica Acta.* **849:** 264–75.

Petrouleas, V. and Diner, B.A. (1987). Light-induced oxidation of the acceptor-side Fe(II) of Photosystem II by exogenous quinones acting through the $Q_B$ binding site. I. Quinones, kinetics and pH-dependence. *Biochimica et Biophysica Acta.* **893:** 126–37.

Pfister, K., Steinback, K.E., Gardner, G. and Arntzen, C.J. (1981). Photoaffinity labeling of an herbicide receptor protein in chloroplast membranes. *Proceedings of the National Academy of Sciences, USA.* **78:** 981–5.

Philbrick, J.B. and Zilinskas, B.A. (1988). Cloning, nucleotide sequence and mutational analysis of the gene encoding the Photosystem II manganese-stabilising polypeptide of *Synechocystis* 6803. *Molecular and General Genetics.* **212:** 418–25.

Philbrick, J.B., Diner, B.A. and Zilinskas, B.A. (1991). Construction and characterization of cyanobacterial mutants lacking the manganese-stabilizing polypeptide of Photosystem II. *The Journal of Biological Chemistry.* **266:** 13370–6.

Piccioni, R.G.P., Bennoun, P. and Chua, N.-H. (1981). A nuclear mutant of *Chlamydmonas reinhardtii* defective in photosynthetic photophosphorylation. *European Journal of Biochemistry.* **117:** 93–102.

Prentki, P. and Krisch, H.M. (1984). *In vitro* insertional mutagenesis with a selectable DNA fragment. *Gene.* **29:** 303–13.

Pryz, J.W., Roe, A.L., Stern, L.J. and Que, L. (1985). Model studies of iron-tyrosinate proteins. *Journal of the American Chemical Society.* **107:** 614–20.

Ravnikar, P.D., Debus, R., Sevrinck, J., Saetaert, P. and McIntosh, L. (1989). Nucleotide sequence of a second *psbA* gene from the unicellular cyanobacterium *Synechocystis* 6803. *Nucleic Acids Research.* **17:** 3991.

Ravnikar, P., Sithole, I., Debus, R., Babcock, G. and McIntosh, L. (1990). The molecular biology of site-directed modification of photosystem II in the unicellular cyanobacterium *Synechocystis* sp. PCC 6803. In *Current Research in Photosynthesis*, Vol. III, (ed. M. Baltscheffsky), pp. 499–507. Kluwer Academic Publishers, Dordrecht, The Netherlands.

Reinman, S., Mathis, P., Conjeaud, H. and Stewart, A. (1981). Kinetics of the reduction of the primary donor of Photosystem II. Influence of pH in various preparations. *Biochimica et Biophysica Acta.* **635:** 429–33.

Reiss, T., Jansson, C. and McIntosh, L. (1990). Insertion of a chloroplast *psbA* gene into the chromosome of the cyanobacterium *Synechocystis* 6803. In *Current Research in Photosynthesis*, Vol. III (ed. M. Baltscheffsky), pp. 649–52. Kluwer Academic Publishers, Dordrecht, The Netherlands.

Rippka, R., Deruelles, J., Waterbury, J.B., Herdman, M. and Stanier, R.Y. (1979). Generic assignments, strain histories and properties of pure cultures of cyanobacteria. *Journal of General Microbiology.* **111:** 1–61.

Robinson, H.H. and Crofts, A.R. (1983). Kinetics of the oxidation–reduction reactions of the Photosystem II quinone acceptor complex, and the pathway for de-activation. *FEBS Letters.* **153:** 221–6.

Robinson, H.H. and Crofts, A.R. (1984). Kinetics of proton uptake and the oxidation–reduction relations of the quinone acceptor complex of PSII from pea chloroplasts. In *Advances in Photosynthesis Research*, Vol. I (ed. C. Sybesma), pp. 477–80. Martinus Nijhoff/Dr. W. Junk, The Hague, The Netherlands.

Rochaix, J.D. (1981). Organization, function and expression of the chloroplast DNA of *Chlamydomonas reinhardii.* *Experientia.* **371:** 323–440.

Rochaix, J.D. and Erickson, J. (1988). Function and assembly of Photosystem II. genetic and molecular analysis. *Trends In Biochemical Sciences.* **13:** 56–9.

Rochaix, J.-D., Kuchka, M., Mayfield, S., Schirmer-Rahire, M., Girard-Bascou, J. and Bennoun, P. (1989). Nuclear and chloroplast mutations affect the synthesis or stability of the chloroplast *psbC* gene product in *Chlamydomonas reinhardtii.* *The EMBO Journal.* **8:** 1013–21.

Rodriguez, R.L. and Denhardt, D.T. (1988). In *Vectors: a survey of molecular cloning vectors and their uses.* Butterworths, Stoneham, Massachusetts.

Rögner, M., Nixon, P.J. and Diner, B.A. (1990). Purification and characterization of Photosystem I and Photosystem II core complexes from wild-type and phycocyanin-deficient strains of the cyanobacterium *Synechocystis* PCC 6803. *The Journal of Biological Chemistry.* **265:** 6189–96.

Rögner, M., Chisholm, D. and Diner, B.A. (1991). Site-directed mutagenesis of the *psbC* gene of photosystem II: isolation and functional characterization of CP43-less Photosystem II core complexes. *Biochemistry.* **30:** 5387–95.

Rossi, J. and Zoller, M. (1987). Site-specific and regionally directed mutagenesis of protein-encoding sequences. In *Protein Engineering* (ed. D.L. Oxender and C.F. Fox), pp. 51–63. Alan R. Liss Inc., New York.

Rutherford, A.W. (1989). Photosystem II, the water-splitting enzyme. *Trends In Biochemical Sciences.* **14:** 227–32.

Saiki, R.K., Gelfand, D.H., Stoffel, S., Scharf, S.J., Higuchi, R., Horn, G.T., Mullis, K.B. and Erlich, H.A. (1988). Primer-directed enzymatic amplification of DNA with a thermostable DNA polymerase. *Science.* **239:** 487–91.

Satoh, K. and Nakane, H. (1990). Refined purification and characterization of the D1-D2 reaction center of Photosystem II. In *Current Research in Photosynthesis*, Vol. I (ed. M. Baltcheffsky), pp. 271–4,

Kluwer Academic Publishers, Dordrecht, The Netherlands.

Satoh, K., Nakatam, H.Y., Steinback, K.E., Watson, J. and Arntzen, C.J. (1983). Polypeptide Composition of a Photosystem II Core Complex. *Biochimica et Biophysica Acta.* **724**: 142–50.

Sayre, R.T., Andersson, B. and Bogorad, L. (1986). The topology of a membrane protein: the orientation of the 32kd Qb-binding chloroplast thylakoid membrane protein. *Cell.* **47**: 601–8.

Schaefer, M.R. and Golden, S.S. (1989a). Differential expression of members of a cyanobacterial *psbA* gene family in response to light. *Journal of Bacteriology.* **171**: 3973–81.

Schaefer, M.R. and Golden, S.S. (1989b). Light availability influences the ratio of two forms of D1 in cyanobacterial thylakoids. *The Journal of Biological Chemistry.* **264**: 7412–17.

Schantz, R. (1985). Mapping of the chloroplast genes coding for the chlorophyll a-binding proteins in *Euglena gracilis. Plant Science Letters.* **40**: 43–39.

Schatz, G.H. and van Gorkom, H.J. (1985). Absorbance difference spectra upon charge transfer to secondary donors and acceptors in Photosystem II. *Biochimica et Biophysica Acta.* **810**: 283–94.

Schatz, G.H. and Witt, H.T. (1984). Extraction and characterization of oxygen-evolving Photosystem II complexes from a thermophilic cyanobacterium *Synechococcus. Photobiochemistry and Photobiophysics.* **7**: 1–14.

Schatz, G.H., Brock, H. and Holzwarth, A.R. (1987). Picosecond kinetics of fluorescence and absorbance changes in Photosystem II particles excited at low photon density. *Proceedings of the National Academy of Sciences, USA.* **84**: 8414–18.

Seibert, M., Tanura, N. and Inoue, Y. (1989). Lack of photoactivation capacity in *Scenedesmus obliquus* LF-1 results from loss of half the high-affinity manganese binding site. Relationship to the unprocessed D1 protein. *Biochimica et Biophysica Acta.* **974**: 185–91.

Seidler, A. and Michel, H. (1990). Expression in *Escherichia coli* of the *psbO* gene encoding the 33 kd protein of the oxygen-evolving complex from spinach. *The EMBO Journal.* **9**: 1743–8.

Shinozaki, K., Ohme, M., Tanaka, M., Wakasugi, T., Hayshida, N., Matsubayasha, T., Zaita, M., Chunwongse, J., Obokata, J., Yamaguchi-Shinozaki, K., Ohto, C., Torazawa, K., Meng, B.Y., Sugita, M., Deno, H., Kamogashira, T., Yamada, K., Kusuda, J., Takaiwa, F., Kata, A., Tohdoh, N., Shimada, H. and Sugiura, M. (1986a). The complete nucleotide sequence of the tobacco chloroplast genome. *Plant Molecular Biology Reporter.* **4**: 111–47.

Shinozaki, K., Ohme, M., Tanaka, M., Wakasugi, T., Hayashida, N., Matsubayashi, T., Zaita, N., Chunwongse, J., Obokata, J., Yamaguchi-Shinozaki, K., Ohto, C., Torazawa, K., Meng, B.Y., Sugita, M., Deno, H., Kamogashira, T., Yamada, K., Kusuda, J., Takaiwa, F., Kato, A., Tohdoh, N., Shimada, H. and Sugiura, M. (1986b). The complete nucleotide sequence of the tobacco chloroplast genome: its gene organization and expression. *The EMBO Journal.* **5**: 2043–9.

Spreitzer, R.J. and Mets, L. (1981). Photosynthesis-deficient mutants of *Chlamydomonas reinhardtii* with associated light-sensitive phenotypes. *Plant Physiology.* **67**: 565–9.

Steinmetz, A.A., Castroviejo, M., Sayre, R.T. and Bogorad, L. (1986). Protein PSII-G. An additional component of photosystem II identified through its plastid gene in maize. *Journal of Biological Chemistry.* **261**: 2485–8.

Steinmüller, K. and Bogorad, L. (1990). Identification of a *psbG*-homologous gene in *Synechocystis* sp. PCC6803. In *Current Research in Photosynthesis*, Vol. III (ed. M. Baltscheffsky), pp. 557–60. Kluwer Academic Publishers, Dordrecht, The Netherlands.

Steinmüller, K., Ley, A.C., Steinmetz, A.A., Sayre, R.T. and Bogorad, L. (1989). Characterization of the *ndhC-psbG*-ORF157/159 operon of maize plastid DNA and of the cyanobacterium *Synechocystis* sp. PCC 6803. *Molecular and General Genetics.* **216**: 60–9.

Stevens Jr., S.E. and Porter, R.D. (1980). Transformation in *Agmenellum quadruplicatum. Proceedings of the National Academy of Sciences, USA.* **77**: 6052–6.

Stewart, A.C., Ljungberg, U., Åkerlund, H.-E. and Andersson, B. (1985). Studies on the polypeptide composition of the cyanobacterial oxygen-evolving complex. *Biochimica et Biophysica Acta.* **808**: 353–62.

Stockhaus, J., Höfer, M., Renger, G., Westhoff, P., Wydrzynski, T. and Willmitzer, L. (1990). Anti-sense RNA efficiently inhibits formation of the 10 kd polypeptide of photosystem II in transgenic potato plants: analysis of the role of 10 kd protein. *The EMBO Journal.* **9**: 3013–21.

Sugiura, M. (1990). Organisation of monocot and dicot chloroplast genomes. In *Current Research in Photosynthesis*, Vol. III (ed. M. Baltscheffsky), pp. 469–74. Kluwer Academic Publishers, Dordrecht, The Netherlands.

Svensson, B., Vass, I., Cedergren, E. and Styring, S. (1990). Structure of donor side components in photosystem II predicted by computer modelling. *The EMBO Journal* **9**: 2051–9.

Svab, Z., Hajdukiewicz, P. and Maliga, P. (1990). Stable transformation of plastids in higher plants. *Proceedings of the National Academy of Sciences, USA.* **87**: 8526–30.

Tae, G.-S., Black, M.T., Cramer, W.A., Vallon, O. and Bogorad, L. (1988). Thylakoid membrane protein topography: transmembrane orientation of the chloroplasts cytochrome b559 *psbE* gene product. *Biochemistry.* **27**: 9075–80.

Takahashi, E. and Wraight, C.A. (1990). A crucial role for Asp$^{L213}$ in the proton transfer pathway to the secondary quinone of reaction centers from *Rhodobacter sphaeroides*. *Biochimica et Biophysica Acta.* **1020**: 107–11.

Takahashi, Y., Hansson, O., Mathis, P. and Satoh, K. (1987). Primary radical pair in Photosystem II reaction center. *Biochimica et Biophysica Acta.* **893**: 49–59.

Takahashi, M., Shiraishi, T. and Asada, K. (1988). COOH-terminal residues of D1 and the 44 kDa CPa-2 at spinach Photosystem II core complex, *FEBS Letters.* **240**: 6–8.

Takahashi, Y., Nakane, H., Kojima, H. and Satoh, K. (1990). Chromatographic purification and determination of the carboxy-terminal sequences of photosystem II reaction center proteins, D1 and D2. *Plant Cell Physiology.* **31**: 273–80.

Tamura, N., Ikeuchi, M. and Inoue, Y. (1989). Assignment of histidine residues in D1 protein as possible ligands for functional manganese in photosynthetic water-oxidizing complex. *Biochimica et Biophysica Acta.* **973**: 281–9.

Taoka, S. and Crofts, A.R. (1990). Two-electron gate in triazine resistant and susceptible *Amaranthus hybridus*. In *Current Research in Photosynthesis*, Vol. I (ed. M. Baltscheffsky), pp. 547–50. Kluwer Academic Publishers, Dordrecht, The Netherlands.

Taylor, L.A. and Rose, R.E. (1988). A correction in the nucleotide sequence of the Tn903 kanamycin resistance determinant in pUC4K. *Nucleic Acids Research.* **16**: 358.

Taylor, M.A., Packer, J.C.L. and Bowyer, J.R. (1988a). Processing of the D1 polypeptide of the photosystem II reaction centre and photoactivation of a low fluorescence mutant (LF-1) of *Scenedesmus obliquus*. *FEBS Letters.* **237**: 229–33.

Taylor, M.A., Nixon, P.J., Todd, C.M., Barber, J. and Bowyer, J.R. (1988b). Characterisation of the D1 protein in a Photosystem II mutant (LF-1) of *Scenedesmus obliquus* blocked on the oxidising side. Evidence supporting non-processing of D1 as the cause of the lesion. *FEBS Letters.* **235**: 109–16.

Thompson, L.K. and Brudvig, G.W. (1988). Cytochrome *b*-559 may function to protect Photosystem II from photoinhibition. *Biochemistry.* **27**: 6653–8.

Trebst, A. (1986). The topology of the plastoquinone and herbicide binding peptides of Photosystem II in the thylakoid membrane. *Zeitschrift für Naturforschung.* **41c**: 240–5.

Triglia, T., Peterson, M.G. and Kemp, D.J. (1988). A procedure for *in vitro* amplification of DNA segments that lie outside the boundaries of known sequences. *Nucleic Acids Research.* **16**: 8186.

Trissl, H.W. and Leibl, W. (1989). Primary charge separation in Photosystem II involves two electrogenic steps. *FEBS Letters.* **244**: 85–8.

Tyagi, A., Hermans, J., Steppuhn, J., Jansson, Ch., Vater, F. and Herrmann, R.G. (1987). Nucleotide sequence of cDNA clones encoding the complete "33 kDa" precursor protein associated with the photosynthetic oxygen-evolving complex from spinach. *Molecular and General Genetics.* **207**: 288–93.

Umesono, K., Inokuchi, H., Shiki, Y., Takeuchi, M., Chang, Z., Fukuzawa, H., Kohchi, T., Shirai, H., Ohyama, K. and Ozeki, H. (1988). Structure and organization of *Marchantia polymorpha* chloroplast genome. II. Gene organization of the large single copy region from *rps'* 12 to *atpB*. *Journal of Molecular Biology.* **203**: 299–331.

Vallon, O., Tae, G.S., Cramer, W.A., Simpson, D., Hoyer-Hansen, G. and Bogorad, L. (1989). Visualization of antibody binding to the photosynthetic membrane: the transmembrane orientation of cytochrome *b*-559. *Biochimica et Biophysica Acta.* **975**: 132–41.

Vandeyar, M.A., Weiner, M.P., Hutton, C.J. and Batt, C.A. (1988). A simple and rapid method for the selection of oligodeoxynucleotide-directed mutants. *Gene.* **65**: 129–33.

Vermaas, W.F.J. (1988). Photosystem II function as probed by mutagenesis. In *Light-Energy Transduction in Photosynthesis: Higher Plant and Bacterial Models* (ed. S.E. Stevens, Jr. and D.A. Bryant), pp. 197–214. The American Society of Plant Physiologists, USA.

Vermaas, W.F.J., Williams, J.G.K., Rutherford, A.W., Mathis, P. and Arntzen, C.J. (1986). Genetically engineered mutant of the cyanobacterium *Synechocystis* 6803 lacks the photosystem II chlorophyll-binding protein CP47. *Proceedings of the National Academy of Science, USA.* **83**: 9474–7.

Vermaas, W.F.J., Williams, J.G.K. and Arntzen, C.J. (1987a). Site-directed mutations of two histidine residues in the D2 protein inactivate and destabilize photosystem II in the cyanobacterium *Synechocystis* 6803. *Zeitschrift für Naturforschung.* **42c**: 762–8.

Vermaas, W.F.J., Williams, J.G.K. and Arntzen, C.J. (1987b). Sequencing and modification of *psbB* the gene encoding the CP47 protein of photosystem II in the cyanobacterium *Synechocystis* 6803. *Plant Molecular Biology.* **8**: 317–26.

Vermaas, W., Rutherford, A.W. and Hansson, Ö. (1988a). Site-directed mutagenesis in photosystem II of the cyanobacterium *Synechocystis* sp. PCC 6803: donor D is a tyrosine residue in the D2 protein. *Proceedings of the National Academy of Sciences, USA.* **85**: 8477–81.

Vermaas, W.F.J., Ikeuchi, M. and Inoue, Y. (1988b). Protein composition of the Photosystem II core complex in genetically engineered mutants of the cyanobacterium *Synechocystis* sp. PCC 6803. *Photosynthesis Research.* **17**: 97–113.

Vermaas, W.F.J., Charité, J. and Eggers, B. (1990a). System for site-directed mutagenesis in the psbD1/C operon of *Synechocystis* sp PCC 6803. In *Current Research in Photosynthesis*, Vol. I, (ed. M. Baltscheffsky), pp. 231–8. Kluwer Academic Publishers, Dordrecht, The Netherlands.

Vermaas, W., Charité, J. and Shen, G. (1990b). $Q_A$ binding to D2 contributes to the functional and structural integrity of Photosystem II. *Zeitschrift für Naturforschung*. **45c**: 359–65.

Vermaas, W., Charité, J. and Shen, G. (1990c). Glu-69 of the D2 protein of Photosystem II is a potential ligand to Mn involved in photosynthetic oxygen evolution. *Biochemistry*. **29**: 5325–32.

Vrba, J.M. and Curtis, S.E. (1989). Characterization of a four-member *psbA* gene family from the cyanobacterium *Anabaena* PCC 7120. *Plant Molecular Biology*. **14**: 81–92.

Wales, R., Newman, B.J., Pappin, D. and Gray, J.C. (1989). The extrinsic 33-kDa polypeptide of the oxygen-evolving complex of Photosystem II is a putative calcium-binding protein and is encoded by a multigene family in pea. *Plant Molecular Biology*. **12**: 439–52.

Wasielewski, M.R., Johnson, D.G., Seibert, M. and Govindjee (1989). Determination of the primary charge separation rate in isolated Photosystem II reaction centers with 500-fs time resolution. *Proceedings of the National Academy of Sciences, USA*. **86**: 524–8.

Webber, A.N., Packman, L., Chapman, D.J., Barber, J. and Gray, J.C. (1989a). A fifth chloroplast-encoded polypeptide is present in the photosystem II reaction centre complex. *FEBS Letters*. **242**: 259–62.

Webber, A.N., Hird, S.M., Packman, L.C., Dyer, T.A. and Gray, J.C. (1989b). A Photosystem II polypeptide is encoded by an open reading frame co-transcribed with genes for cytochrome b-559 in wheat chloroplast DNA. *Plant Molecular Biology*. **12**: 141–52.

Webber, A.N., Packman, L.C. and Gray, J.C. (1989c). A 10 kDa polypeptide associated with the oxygen-evolving complex of photosystem II has a putative C-terminal non-cleavable thylakoid transfer domain. *FEBS Letters*. **242**: 435–8.

Weising, K., Schell, J. and Kahl, G. (1988). Foreign genes in plants: transfer, structure, expression and applications. *Annual Review of Genetics*. **22**: 421–77.

Wells, J.A., Vasser, M. and Powers, D.B. (1985). Cassette mutagenesis: an efficient method for generation of multiple mutations at defined sites. *Gene*. **34**: 315–23.

Wettern, M., Owens, J.C. and Ohad, I. (1983). Role of thylakoid polypeptide phosphorylation and turnover in the assembly and function of Photosystem II. *Methods in Enzymology*. **97**: 554–67.

Widger, W.R., Cramer, W.A., Hermodson, M., Meyer, D. and Gullifor, M. (1984). Purification and partial amino acid sequence of the chloroplast cytochrome b-559. *The Journal of Biological Chemistry*. **259**: 3870–6.

Williams, J.G.K. (1988). Construction of specific mutations in the photosystem II photosynthetic reaction center by genetic engineering methods in the cyanobacterium *Synechocystis 6803*. *Methods in Enzymology*. **167**: 766–78.

Williams, J.G.K. and Chisholm, D.A. (1987). Nucleotide sequences of both *psbD* genes from the cyanobacterium *Synechocystis 6803*. In *Progress in Photosynthesis Research*, Vol. IV (ed. J. Biggins), pp. 809–12. Martinus Nijhoff Publishers, Dordrecht, The Netherlands.

Williams, J.G.K. and Szalay, A.A. (1983). Stable integration of foreign DNA into the chromosome of the cyanobacterium *Synechococcus* R2. *Gene*. **24**: 37–51.

Wollman, F.-A. and Delepelaire, P. (1984). Correlation between changes in light energy distribution and changes in thylakoid membrane polypeptide phosphorylation in *Chlamydomonas reinhardtii*. *The Journal of Cell Biology*. **98**: 1–7.

Wollman, F.-A. and Diner, B.A. (1980). Cation control of fluorescence emission, light scatter, and membrane stacking in pigment mutants of *Chlamydomonas reinhardtii*. *Archives of Biochemistry and Biophysics*. **201**: 646–59.

Yu, J. and Vermaas, W.F.J. (1990). Transcript levels and synthesis of photosystem II components in cyanobacterial mutants with inactivated photosystem II genes. *The Plant Cell*. **2**: 315–22.

Zhang, Z.H., Mayes, S.R. and Barber, J. (1990). Nucleotide sequence of the *psbK* gene of the cyanobacterium *Synechocystis 6803*. *Nucleic Acids Research*. **18**: 1284.

# 6
# Organellar Targeting in Plants

Newell F. Bascomb, Steven Gutteridge and Thomas H. Lubben

## Introduction

Plant cells contain many intracellular compartments. Two of these, chloroplasts and mitochondria, contain the machinery necessary for protein synthesis. However, only about 10% of the proteins required by these organelles are synthesized using this machinery. The remaining 90%, and all the proteins destined for all other intracellular compartments, originate from nuclear-encoded genes that are translated by cytosolic polysomes. These organellar, cytoplasmically synthesized proteins reach their correct location via several different protein transport mechanisms (Fig. 6.1).

The various protein import and targeting mechanisms present within the plant cell share some common features and display important differences. Proteins that are destined for chloroplasts, mitochondria or distribution via the endoplasmic reticulum (e.r.) are usually synthesized as longer precursor (pr) proteins, with extra amino acids at the amino-terminal end. This N-terminal region is known as a signal peptide for proteins that are sorted via the e.r., or as a transit peptide for proteins destined for mitochondria and chloroplasts. The protein that is translocated is commonly referred to as a passenger protein, or mature protein. There are differences that extend beyond the merely semantic between e.r.-mediated transport and protein transport to chloroplasts and mitochondria.

Although most of what we know of e.r.-mediated translocation has been learned from non-plant systems, the basic mechanisms have presumably been conserved during evolution. E.r.-mediated protein transport is normally cotranslational, although in some cases translocation can proceed post-translationally (for review see Walter and Lingappa 1986). Translocation of the polypeptide begins while the nascent chain is attached to the ribosome. Indeed, the recognition of the precursor protein, by various cellular factors involved in the initial steps of translocation, is dependent upon the polypeptide being attached to the ribosome (Perara et al. 1986).

In contrast, proteins destined for chloroplasts (for review see Keegstra 1989; Keegstra et al. 1989; Smeekens et al. 1990) or mitochondria (for review see Pfanner et al. 1988a) are present as soluble precursors prior to translocation, and association with ribosomes is not necessary for import. There are many other differences between e.r.-based and post-translational protein translocation; these are discussed in more detail in the body of this chapter. For peroxisomal targeting, the targeting signal is within the protein itself, and no post-translocational modification takes place (for review see Lazarow and Fujiki 1985). However, the basic mechanisms may be conserved for peroxisomal targeting across plant peroxisomes (Gould et al. 1990), although the study of these mechanisms, and of those for glyoxysomal and nuclear targeting in plants, has not advanced as far.

The study of protein translocation in plant cells has mainly involved those pathways that are unique to plants, rather than those pathways common to all eukaryotes. There are several reasons for this. An important factor has been the relative ease of isolating intact organelles from soft animal tissues as opposed to plant cells, with their robust cell walls. Furthermore, genetic studies involving mutants are more easily performed with microorganisms than on whole plants or plant cells in tissue culture. Therefore, for phenomena such as

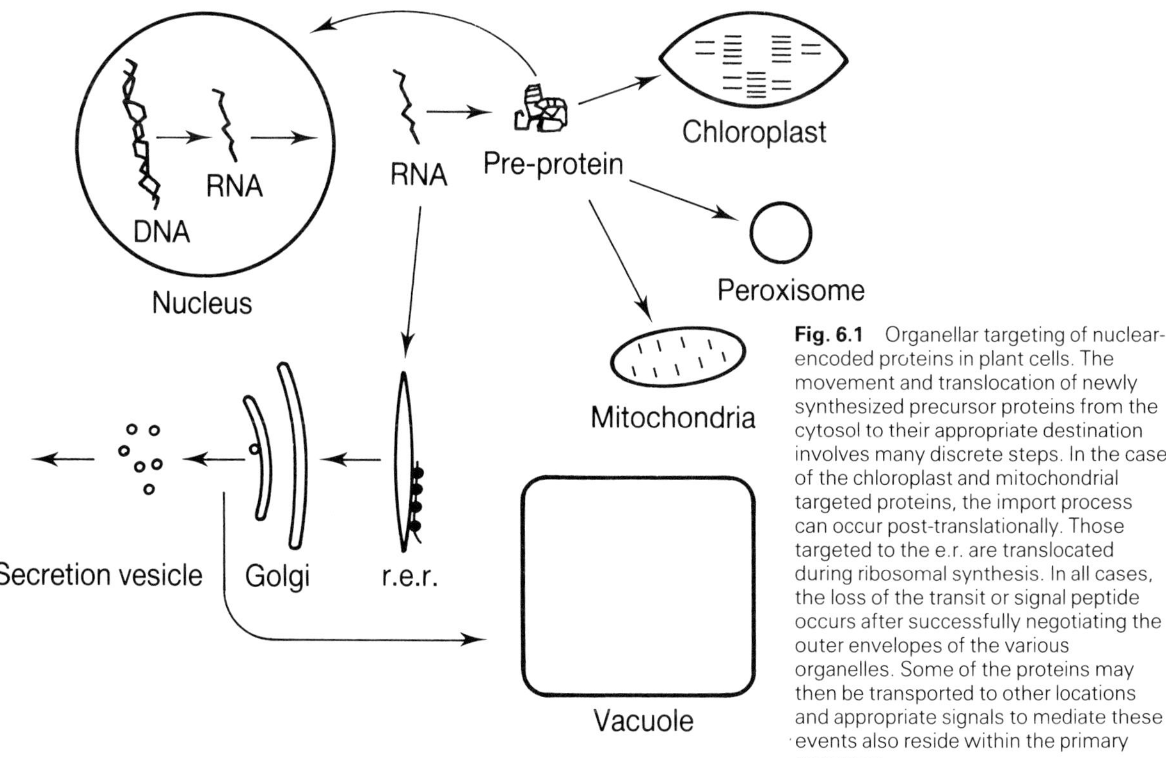



**Fig. 6.1** Organellar targeting of nuclear-encoded proteins in plant cells. The movement and translocation of newly synthesized precursor proteins from the cytosol to their appropriate destination involves many discrete steps. In the case of the chloroplast and mitochondrial targeted proteins, the import process can occur post-translationally. Those targeted to the e.r. are translocated during ribosomal synthesis. In all cases, the loss of the transit or signal peptide occurs after successfully negotiating the outer envelopes of the various organelles. Some of the proteins may then be transported to other locations and appropriate signals to mediate these events also reside within the primary sequence.

protein secretion, nuclear targeting, or mitochondrial protein import, the tendency has been to study animal or microbial systems and to assume that, during evolution, the basic structure of the targeting pathways have been conserved across species and even across kingdom lines.

However, there are targeting systems that exist solely in plants, and study of these systems provides an opportunity to address questions relating directly to plants. Plants have many unique structures and the study of targeting to these is both scientifically interesting and of economic interest for those who wish to manipulate plant properties. Several examples include the e.r.-based pathways for seed storage proteins and vacuolar targeting, and the pathway for post-translational chloroplast protein import. Of these, chloroplast protein import has probably been the most widely studied.

There are several reasons for this. Historically, the first description of post-translation protein import, in any system, was described for chloroplasts (Highfield and Ellis 1978); post-translational import into yeast mitochondria was described later (Maccecchini *et al.* 1979). It is relatively easy to isolate chloroplasts from species such as spinach and pea, especially when chloroplast isolation is compared to the isolation of other organelles from plants, such as e.r. and mitochondria. Another important reason for the study of chloroplast targeting, especially with regard to this chapter, is because of the vital role played by chloroplasts—not only in plant metabolism but generally within the biosphere—as the major source of reduced carbon and essential amino acids. Therefore, modification of chloroplast function is especially relevant with regard to genetic modification of plant traits. Because most chloroplast proteins are nuclear-encoded, and post-translationally translocated to the chloroplast, an understanding of protein import is an essential prerequisite for attempts to alter most chloroplast functions.

## Post-translational protein targeting

This discussion will focus primarily on import into mitochondria and chloroplasts. Of these two, chloroplast protein import has been more studied in plants and mitochondrial import has been exten sively studied in yeast and mammalian systems. Import of proteins into plastids other than chloroplasts has been less extensively studied. What is known about import into leucoplasts (Boyle *et al.* 1986) and amyloplasts (Strzalka *et al.* 1987; Klosgen *et al.* 1989) indicates that transport into these and other plastids (Boyle *et al.* 1990) is probably very similar to import into chloroplasts.

Because chloroplasts are unique to plants, the targeting of proteins to these organelles is likely to have unique features and thus offer insights into the different targeting mechanisms that have evolved in plants. One particularly intriguing aspect of this is the way in which mitochondrial protein import differs from chloroplast protein import. It is tempting to conclude that mitochondrial import and chloroplast protein import differ only to the extent necessary to ensure that mistargeting does not occur (Singer *et al.* 1987). However, this view is probably naive, and ignores several important physiological differences between these two organelles, as well as the requirements and known features of import into each.

One question that is relevant to both chloroplasts and mitochondria may affect efforts to genetically modify plants. This question is 'Why do these organelles retain certain proteins in their genome?' Already, studies of chloroplast import and yeast mitochondria import have eliminated one possible answer. It has been shown that at least some of the proteins encoded within chloroplasts can be post-translationally imported into chloroplasts (Cheung *et al.* 1988; Gatenby *et al.* 1988) or even into mitochondria (Hurt *et al.* 1986a), if the genes are modified to contain a region encoding an appropriate transit peptide. This eliminates the possibility that proteins are encoded by the organellar genome because organelle-encoded proteins might be incompatible with post-translational protein import.

Two other possibilities for organellar encoding of genes need to be considered, both are relevant for those who might wish to modify organelle function through nuclear expression of genes normally present in the organellar genome. One is that from a viewpoint of regulation, the gene expression of certain subunits of proteins is best controlled within the organelle. Another possibility (von Heijne 1986a) is that the organellar-encoded proteins contain sequences that would interfere with or be inadvertently misdirected by the e.r.-based translocation system, and that these proteins could not peacefully coexist with the cotranslational protein transport system in the cytoplasm. In fact, second-

ary structure prediction algorithms support this latter hypothesis (von Heijne 1986a, b), suggesting that such sequences may compete for those other factors that are involved in mediating protein transport and processing. If either of these possibilities is true, then it may not be possible, or at least practicable, to delete an organellar-encoded protein and replace it with a nuclear-encoded protein, of modified function, with an organellar transit peptide. It may be that alteration of at least some organellar functions will have to await technologies that allow transformation of organellar genomes. Another potential problem is that an organellar gene, transferred to the nucleus, may not have the optimum codon usage (Oliver *et al.* 1989). However, until there is strong evidence against using a nuclear-encoded gene to substitute for a particular organelle gene, this approach seems to be the most worthwhile (but see below). The reality may be that, in practice, most organellar genes can be substituted by nuclear genes. However, it may be necessary to engineer altered structure or regulation as well as the altered properties originally sought after.

Irrespective of the problems that still exist and require further investigation, enough is now known about the process to exploit it for practical purposes. For example, photosynthetic $CO_2$ fixation is dependent on the catalytic properties of ribulose bisphosphate carboxylase (Rubisco) but this enzyme is inefficient in terms of turnover rate and carboxylation specificity. It is now clear that the construction of a more efficient Rubisco will require changes in the large subunit of the enzyme, a protein that is synthesized in the chloroplast stroma. It is also now clear that the large subunit, endowed with the transit peptide from the small subunit (prSS), can be transported relatively efficiently to the chloroplast stroma, processed correctly and be assembled into holoenzyme (Gatenby *et al.* 1988). This suggests that a suitable plant might be transformed to enable nuclear-encoding of the Rubisco large subunit. The presence of pre-existing Rubisco in the chloroplast might be suppressed by overexpressing the nuclear encoded protein or by generating plant cells deficient in Rubisco. Alternatively, a means of organellar mutagenesis might be devised to introduce the changes into the indigenous Rubisco large subunit. However, this will probably require chloroplast transformation procedures, something that is only now feasible, for higher plants. It is worth noting that DNA import into mitochondria has been demonstrated using DNA covalently linked to a mitochondrial precursor protein (Vestweber and Schatz 1989). If this is possible for chloroplasts, then it may provide one means of chloroplast transformation.

## Mitochondrial import in plants

Only a few studies have addressed the issue of mitochondrial protein import in plants. Two of these (White and Scandalios 1987; Whelan *et al.* 1988) examined protein import *in vitro*. In these studies, precursor proteins were transcribed and translated and presented to mitochondria *in vitro*. The precursors were imported and processed to a size that corresponded to that expected for the mature protein. One study (White and Scandalios 1987) used a maize RNA preparation to prepare the precursor for Mn-superoxide dismutase; this was then imported into maize mitochondria. Another study (Whelan *et al.* 1988) used mitochondria from *Vicia faba*, the ATP/ADP translocator of this species, and the *Neurospora crassa* β-F1-ATPase precursor, were imported. Import of the *N. crassa* protein is significant in that it demonstrates that mitochondrial transit peptides are functionally conserved between fungi and plants. It is tempting to conclude that such conservation of function is widespread between species.

In a different approach, Boutry *et al.* (1987) used transgenic plants to study the targeting of a foreign protein into chloroplasts and mitochondria in *Nicotiana plumbaginifolia*. The bacterial protein chloramphenicol acetyl transferase (CAT) was used. Two different transit peptides from *N. plumbaginifolia* were used, one from a mitochondrial protein, pre-β-F1-ATPase subunit and one from a chloroplast protein, prSS. The researchers found that the location of CAT depended solely on which transit peptide was used, and no incorrect targeting could be detected. While this finding may not be surprising to some, it did contradict some *in vitro* studies that had found a chloroplast transit peptide to target polypeptides to yeast mitochondria (Hurt *et al.* 1986b). With regard to these studies, several groups have shown that not only a chloroplast transit peptide, but a variety of other peptides can act as mitochondrial transit peptides under certain conditions (Allison and Schatz 1986; Baker and Schatz 1987; Banroques *et al.* 1987; Hurt and Schatz 1987; Vassarotti *et al.* 1987; Lemire *et al.* 1989; van Loon and Brandll 1986). However, it

seems likely that this represents a non-specific event that does not occur significantly *in vivo* (Pfanner *et al.* 1988b) and, despite earlier reports to the contrary, organellar transit peptides are specific *in vivo*. Significantly, it also appears that plants can be genetically engineered to express chimeric proteins with a mitochondrial transit peptide that results in the foreign protein being transported to mitochondria in a specific manner.

## Transit peptides

This subject has been extensively reviewed by Keegstra *et al.* (1989). It is interesting that there is remarkably little primary sequence homology between different transit peptides. This is true whether one examines mitochondrial or chloroplast transit peptides. It seems apparent for mitochondrial transit peptides that, rather than conservation of primary structure, the conserved feature is secondary structure in the form of an amphipathic helix (von Heijne 1986b; Roise *et al.* 1986; Gavel *et al.* 1988). A diversity of sequences exists among chloroplast transit peptides and the information that must be common to all transit peptides remains obscure. If one compares only transit peptides from one protein, such as prSS, from different species, the sequences appear to be fairly well conserved (Keegstra *et al.* 1989). All are about 57 amino acids long, with a similar distribution of charged and hydrophobic residues and a conserved processing site. However, the lack of conservation, both for length and composition, is striking if one compares sequences from different precursors (Keegstra *et al.* 1989). This is consistent with the proposal (Keegstra *et al.* 1989; von Heijne *et al.* 1989) that secondary structural features, rather than primary sequence, are the conserved elements among chloroplast transit peptides. One such feature may be an amphipathic β-stand near the transit peptide-imported protein junction (Keegstra *et al.* 1989; von Heijne *et al.* 1989), although this feature is not as prominent in hydrophobic moment analysis, or as easy to model, as the amphipathic helices of mitochondria transit peptides. Length of transit sequence is another recognizable common feature, i.e. the transit peptides are composed of at least 30 amino acids; they nearly all start MetAla at the N-terminus.

Alternative hypotheses for the sequence diversity of transit peptides are possible. These include the concept that the different transit peptides interact with different receptors on the organelle, and also the idea that the differences reflect some sort of coevolution of transit peptides and targeted proteins, so that transit peptides have evolved to react with their respective mature peptides. However, the enormous number of different proteins imported into either chloroplasts or mitochondria would argue against the former view, and the wide variety of foreign proteins that have been imported using different transit peptides would argue against the latter. Therefore, the idea that the conserved feature of transit peptides may be one of secondary structure, rather than primary sequence, seems to be the best explanation at this time.

It has been proposed by several groups (Karlin-Neumann and Tobin 1986; Schmidt and Mishkind 1986) that chloroplast transit peptides contain three domains. However, re-examination of this hypothesis in the light of sequence data beyond that available from small subunit precursors proves the existence of transit peptides that lack some or all of these conserved regions. The results of several deletion studies (Reiss *et al.* 1987; Szabo and Cashmore 1987; see also Russ *et al.* 1989) can be combined to show that all three domains can be removed and yet the transit peptide can still support import. Although reduction in import occurs when deletions are made, no region appears to be absolutely essential. This supports the view that transit peptide structure, rather than primary sequence, is the essential conserved feature.

Figure 6.2 shows three transit peptides to indicate the complete absence of homology and also the apparent lack of any relationship between the length of the transit peptide and the size of the mature protein. Thus, the wheat Rubisco small subunit (120 amino acids) is readily transported utilizing a 60 amino acid transit peptide Acetolactate synthase (ALS), from tobacco or *Arabidopsis*, is a 64 kDa mature protein with one of the longest transit sequences—about 100 amino acids. In contrast, plastocyanin (8 kDa) utilizes a transit peptide of 6 kDa.

In the case of ALS, a comparison of the mature and precursor proteins on SDS gels indicated that the transit peptide must be at least 10 kDa. The approximate position of the processing site was first located by exploiting the unusual composition of the N-terminal sequence of the precursor, which includes 33% of all the His residues and 25% of the Ser. A comparison of the Ser and His compositions of the unprocessed and mature proteins using tritiated amino acids during translation confirmed the size of the transit peptide. The unique pattern of Val and Ser amino acids within this region of the N-terminus of the precursor enzyme established the exact position of the cleavage site and the composition of the transit peptide. The precursor, labelled with either one of these amino acids, was imported into intact chloroplasts and the mature enzyme was isolated. From the pattern of radioactivity emerging from the sequenator, the N-terminus of the mature protein was located and thus the composition of the transit peptide was revealed. Various deletion mutants generated at the C-terminus of the mature part of the sequence did not affect import *in vitro*, indicating that translocation into the stroma is not dependent on the mature sequence (Bascomb *et al.* unpublished work).

The function of the transit peptides discussed above is to target proteins to the stroma. The final destination for a number of nuclear-encoded proteins, is the thylakoids; these proteins require a bipartite transit peptide. The first portion is a stromal targeting domain; the second is a domain whose function is to direct the protein to the thylakoids. This thylakoid targeting domain appears to be similar to an e.r. signal peptide or a bacterial signal peptide (for reviews, see Keegstra 1989; Keegstra *et al.* 1989; Smeekens *et al.* 1990). The role of this second domain in thylakoid targeting has been confirmed by import of several foreign passenger proteins into thylakoids. Thus, the transit peptide from the 33 kDa oxygen evolving protein has been used to target dihydrofolate reductase (Meadows *et al.* 1989; Ko and Cashmore 1989) and Rubisco small subunit (Ko and Cashmore 1989) to thylakoids. However, as described on p. 150, certain passenger proteins either cannot be transported to thylakoids or contain internal, noncleaved signals for thylakoid localization.

## Chloroplast protein import

The transit peptide on chloroplast protein precursors is somehow recognized by the organelle. Traditionally, a recognition event such as this would be expected to involve a receptor on the organelle. Several reports of the identification of such a protein receptor have appeared (Cornwell and Keegstra 1987; Pain *et al.* 1988; Hinz and Flugge 1988; see also Kloppstech and Bitsch 1986). However, each has identified a different protein. In view of this disparity, assignment of a single receptor must be postponed, assuming that not all of

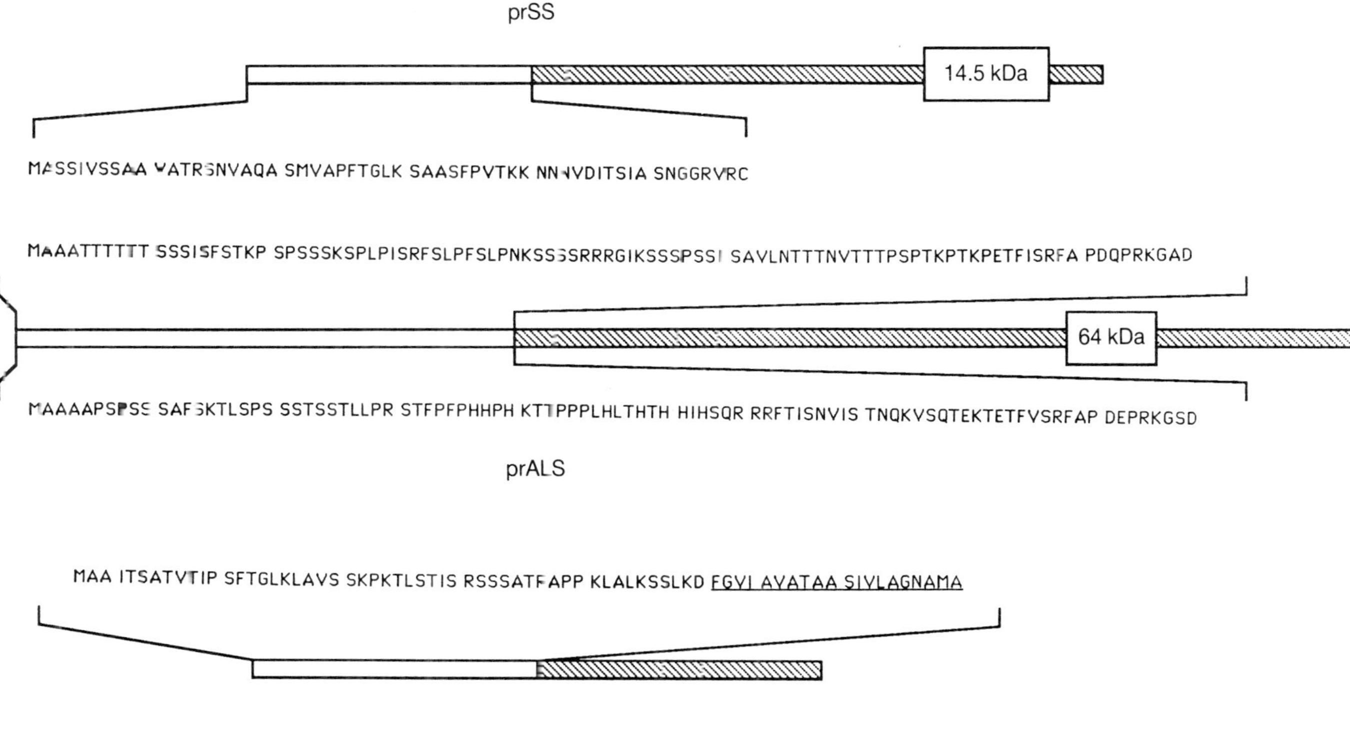


**Fig. 6.2** Chloroplast transit sequences show little homology when different precursor protein sequences are compared. More consensus is often found within the same family of proteins, e.g. the precursor of Rubisco small subunit (prSS), although in the case of prALS this really only extends to length rather than composition. The upper sequence is that for the *Arabidopsis* ALS, the lower for the tobacco enzyme. The third example is the transit peptide for plastocyanin, which has a transit and intrachloroplastid signal (underlined) that accounts for almost 50% of the total protein composition.

these proteins are involved. It seems likely that the requisite specificity for chloroplast protein import would only be met if one invoked the involvement of some sort of protein–protein interaction. Furthermore, binding to chloroplasts is saturable (Friedman and Keegstra 1989; Pfisterer *et al.* 1982), and somewhat sensitive to pre-treatment with protease (Cline *et al.* 1985; Friedman and Keegstra 1989). As proposed by Keegstra *et al.* (1989), both protein–protein interactions and protein–lipid interactions may be involved in chloroplast import. In their model, specificity is ensured by protein–protein interactions at the putative receptor, whereas translocation across the envelope membrane also involves protein–lipid interactions between the envelope lipids and secondary structural elements of the precursor protein.

Although chloroplast transit peptides do not fit into predicted amphipathic helices, amphipathic β structures have been proposed (Tyagi *et al.* 1987; von Heijne *et al.* 1989). If such an amphipathic structure is a functional part of chloroplast transit peptides, then import may proceed by a mechanism that combines protein–protein interactions and protein–lipid interactions as described above. The identification, characterization and cloning of the proteins involved in chloroplast import is important with respect to commercial applications. One way that this could be manifested is in the development of herbicides that might specifically inhibit chloroplast, but not mitochondrial, import. Such compounds would have total specificity for plants, while being harmless to non-plant life forms.

## Energy requirements for chloroplast import

The difference in energy requirements is one of the ways in which chloroplast import can be distinguished from mitochondrial import (Keegstra *et al.* 1989). The latter depends on an electrochemical potential (Gasser *et al.* 1982), with ATP essential for maintenance of many precursors in an import competent form (Hwang and Schatz 1989). Ten years ago, Grossman *et al.* (1980) reported that ATP, rather than an electrochemical potential was required for chloroplast protein import. This view is still held to be true. However, more is known today about the details of this requirement. After much debate (Flügge and Hinz 1986; Pain and Blobef 1987), several studies which conclusively establish that millimolar internal ATP concentra-tions are required for translocation, have appeared (Theg *et al.* 1989), whereas only micromolar external ATP concentrations are adequate for binding of precursors (Olsen *et al.* 1989). This requirement for external ATP for binding eluded previous investigators because enough residual ATP from *in vitro* translation of precursors usually remained to fulfil this micromolar requirement.

In view of the different morphology of mitochondria and chloroplasts, the different energy requirements are not at all suprising, and are perhaps to be expected. Chloroplasts are morphologically distinct from mitochondria, particularly with regard to their internal organization. In mitochondria, the ATP-producing electrochemical potential exists across the mitochondrial inner membrane. In chloroplasts, the equivalent ATP-producing membrane is not the inner envelope membrane, but rather the thylakoid membrane. A pH gradient exists across the chloroplast envelope membranes, but this is not large. Furthermore, ATP, produced from photosynthesis, is present in the stroma. Chloroplasts have a readily available supply of ATP, but do not have a strong gradient across the envelopes. However, mitochondria have a strong electrochemical gradient and it is not surprising that the protein import mechanisms for these two organelles utilize different sources of energy. It may be that because of this, they also utilize fundamentally different mechanisms of translocation, although this remains to be established. If so, then this may provide another target for those wishing to modify plant metabolism without affecting animal processes.

## Transport of foreign proteins to chloroplasts

Many essential metabolic processes are localized within the chloroplasts. Apart from photosynthetic carbon reduction, the essential amino acids, oils and lipids of commercial interest rely on the metabolism of the chloroplast. Some of these enzymes are the site of action of herbicides (see chapter 15). Therefore, the ability to understand and exploit the translocation process to import proteins of choice into the chloroplast is attractive. Clearly, unravelling the rules of the mechanism would be helpful in designing appropriate constructs that give efficient import. Furthermore, the ability to generate a chimeric precursor construct might also be ex-

**Table 6.1** Foreign proteins imported into chloroplasts

| Protein | Reference |
| --- | --- |
| Bacterial neomycin | Van den Broek *et al.* 1985 |
| phosphotransferase | Schrier *et al.* 1985 |
| Soybean heatshock protein | Lubben and Keegstra 1986 |
| Yeast Mn-superoxide | Smeekens *et al.* 1987 |
| dismutase | |
| Bacterial chloramphenicol | Boutry *et al.* 1987 |
| acetyl transferase | Lubben *et al.* 1989b |
| *Synechococcus* Rubisco large | Gatenby *et al.* 1988 |
| subunit | |
| Brome mosaic virus coat | Lubben *et al.* 1989a |
| protein | |
| Bacterial pre-β-lactamase | Lubben *et al.* 1989b |
| Alfalfa glutamine synthetase | Lubben *et al.* 1989b |

ploited to establish some of the steps involved in the translocation.

Many foreign proteins have been imported into chloroplasts using various transit peptides from stromal or thylakoid proteins. These are shown in Table 6.1. The first report detailed the successful targeting of neomycin phosphotransferase (NPTII) precursor (Van den Broek *et al.* 1985) in transgenic plants. Subsequently, reports have established that a wide variety of foreign proteins can be imported into the organelle (della-Cioppa *et al.* 1987; Boutry *et al.* 1987; Lubben and Keegstra 1986; Lubben *et al.* 1988, 1989a, 1989b). From the point of view of economic exploitation, it has to be shown that similar chimeras function in transgenic plants and that the foreign protein is ultimately located in the chloroplast. Experiments involving two of the enzymes, normally located in the chloroplast, that are targets of potent herbicides have been reported in detail. EPSP synthase, an enzyme essential for aromatic amino acid biosynthesis in the chloroplast, is inhibited by glyphosate. However, glyphosate-resistant plants have been obtained by transformation with synthases unaffected by the herbicide. The enzymes from both plant (Shah *et al.* 1986) and bacteria fused to the plant transit peptide (Thompson *et al.* 1987; della-Cioppa 1987; Fillati *et al.* 1987) have successfully conferred glyphosate resistance to various crop plants. Similar studies with the genes coding for plant ALS have provided a means of introducing resistance to sulphonylurea based herbicides. (Haughn *et al.* 1988). Both plant and yeast genes coding for resistant ALS have been used to introduce herbicide tolerance into sensitive plants. In many cases the mutations have been identified and reintroduced into sensitive genes by site specific mutagenesis to fine tune enzyme resistance to

different sulphonylurea based herbicides. The mutated genes have then in turn been used to increase crop resistance to selected herbicides (Hartnett *et al.* 1990). Undoubtedly, other examples of the import of economically significant foreign proteins exist, e.g. the *psbA* gene product (see pp. 110–17). However, because of their economic value and, therefore, because of proprietary concerns, such examples may not be as readily reported as studies of model proteins.

Because chloroplasts seem receptive to many transit-foreign protein chimeras and removal of the transit peptide occurs normally, several important facets of chloroplast import now seem apparent. First, there is no essential message within the mature protein sequence required for translocation. Secondly, no information resides within the mature sequence necessary to direct processing. Presumably, the transit peptide carries this information as well as the targeting information. Finally, the number of precursor chimeras that have been successfully imported argues that, if there is a common mechanism involved, then it is largely unaffected by the foreign protein. Furthermore, once inside the chloroplasts, imported foreign proteins appear to function normally (Herrara-Estrella *et al.* 1984; della-Cioppa *et al.* 1986, 1987; Gatenby *et al.* 1988).

In a limited number of cases, foreign polypeptides appear to be non-importable. Certain chimeric constructs of the brome mosaic virus (BMV) coat protein and Rubisco small subunit precursor could not be imported into chloroplasts. Import of a transit peptide-coat protein chimera proceeded at a rate only 2.5% that of prSS, but when most of the small subunit mature peptide sequences were included along with the transit peptide, no import could be detected (Lubben *et al.* 1989a). None of the chimeric precursors were found associated with chloroplast envelope or inner membrane, suggesting that the presence of the foreign protein interfered in the first step of translocation, namely the interaction of the transit peptide with the putative receptor.

Several other experiments have addressed the question of what effect the imported protein has on import. Lubben *et al.* (1987) showed that protein import does not proceed by a process involving threading an unfolded protein directly through a lipid bilayer. They used chimeras containing small subunit and hydrophobic stop-transfer regions from two different e.r.-translocated proteins, vesicular stomatitis viral G-protein and IgM. Both chimeras successfully crossed both envelope membranes and were found in the stroma. It has been suggested that unfolding of proteins is required for import of proteins into chloroplasts (della-Cioppa and Kishore 1988) as well as in mitochondria (Eilers and Schatz 1988; Eilers *et al.* 1988). For chloroplasts, the primary evidence is that binding of the inhibitor herbicide glyphosate to the precursor of EPSP synthase prevented import (della-Cioppa and Kishore 1988). In mitochondria, dihydrofolate reductase is not imported in the presence of the tight-binding inhibitor methotrexate. Presumably, such binding prevents unfolding of the protein. However, an alternative explanation is that inhibitor binding causes the precursor to assume a conformation incompatible with import.

Recent evidence from mitochondrial import has suggested that a variety of ligands may be covalently attached to a precursor and still allow import. These include stilbene disulphonate and branched polypeptides (Vestweber and Schatz 1988a, 1988b) as well as DNA conjugates (Vestweber and Schatz 1989). These observations are consistent with the hypothesis of Keegstra (1989) that import involves protein conformational changes (but not complete unfolding) and rearrangements in the lipid bilayer (de Kruiff *et al.* 1985) to accommodate protein structural elements during translocation. Such a model would allow for all the observations to date. Thus, inhibition of import by binding of ligands might prevent conformational changes (rather than prevent unfolding) and passage of stop-transfer regions would be allowed if translocation proceeded through a non-bilayer structure. If such a model proves true, then mitochondrial and chloroplast import proceed by mechanisms distinct from e.r.-mediated transport, which appears to involve threading (for a review, see Walter and Lingappa 1986).

### Import of proteins normally plastid-encoded

The inability to manipulate the plastid genome and thus introduce mutations into plastid-encoded proteins has meant that alternative strategies have had to be devised to alter phenotype through these proteins. One way is to locate the gene for the protein in the nucleus. Several groups have reported that it is possible to import proteins that are normally plastid-encoded into chloroplasts. The β-subunit of chloroplast CF1 coupling factor and

*Synechococcus* Rubisco large subunit have been imported *in vitro* (Gatenby *et al.* 1988), and the herbicide binding protein (psbA) of photosystem II has been imported *in vivo* in transformed plants (Cheung *et al.* 1988). In the case of the CF1-β-subunit, assembly into CF1–CF0 complex was not observed but imported Rubisco large subunit was assembled into Rubisco holoenzyme.

The experiments with the herbicide-binding protein were done by transformation of tobacco with a chimera that contained the transit peptide and regulatory regions of Rubisco small subunit precursor and a mutant *psbA* gene. The fact that a resistant phenotype was observed in transgenic plants indicated that the precursor protein, now nuclear-encoded, was import-competent and at least some of the imported resistant psbA was assembled and functioned in the photosynthetic apparatus. Thus, it appears that it is possible to modify some chloroplast functions by substituting imported nuclear gene products for chloroplast-encoded proteins, albeit superimposed on and presumably in competition with the plastid-encoded protein (see also Rodermel *et al.* 1988). This is significant in that transformation of chloroplast genomes is not yet routine, particularly with respect to important crop plants.

## Intraorganellar targeting

Some of the precursor proteins do not reside in the stroma but are destined for other locations within the organelle. One protein that has been studied in some detail is the light-harvesting complex protein (LHCP). This protein is present in the thylakoid membrane associated with chlorophyll and carotenoid pigments. The protein not only becomes integrated into the membrane but spans the bilayer three times. The precursor is processed in the stroma to a mature form, which still carries the information necessary for thylakoid integration (Cline 1988), and insertion into thylakoids appears to be independent of the presence or absence of the transit peptide (Viitanen *et al.* 1988; Lampaa 1988), which can act as a stromal targeting transit peptide (Hand *et al.* 1989). Deletion experiments that generate C-terminal truncated forms of LHCP indicate that it is this region of the mature protein that is essential for integration (Kohorn and Tobin 1989).

Another protein destined for the thylakoid membrane is the herbicide binding 32 kDa protein (or *psbA* gene product described above) of photosystem II. The protein is synthesized in the chloroplast on stromal lamellae as a 34 kDa precursor protein that eventually becomes integrated into the photosystem II complex of the granal lamellae. However, in this case processing removes a C-terminal peptide from the protein before integration into the granal complex.

Both LHCP (also known as cab, chlorophyll a/b binding protein) and the 33 kDa oxygen-evolving protein have been the subject of studies where they have been used in chimeric constructs. For both of these proteins, the thylakoid targeting region appears to act independently of stromal targeting. Recently, with the 33 kDa protein transit peptide, successful targeting of proteins to thylakoids has been reported. These foreign proteins include dihydrofolate reductase (Meadows *et al.* 1989; Ko and Cashmore 1989) and Rubisco small subunit (Ko and Cashmore 1989). However, apparently not all proteins can be directed to thylakoids. The same transit sequence was unable to direct glycolate oxidase to thylakoids (Ko and Cashmore 1989), and the transit sequence of plastocyanin is unable to direct ferredoxin (Smeekens *et al.* 1986) or mitochondrial superoxide dismutase (Smeekens *et al.* 1987) to thylakoids.

## Molecular chaperones

One important facet of import that has been appreciated only recently is the involvement of proteins known as molecular chaperones or chaperonins. The involvement of the Rubisco binding protein in Rubisco assembly was proposed about ten years ago (Barraclough and Ellis 1980). Early studies involved incubation of chloroplasts with radioactively labelled amino acids. Because Rubisco large subunit is the major labelled protein under such conditions, it was possible to follow the time course of its synthesis and assembly (see also Roy and Cannon 1988; Roy *et al.* 1988). Under these conditions, large subunit first appeared complexed to the binding protein (Barraclough and Ellis 1980), a high molecular oligomeric protein now known to be related to the groEL protein of *Escherichia coli* (Hemmingsen *et al.* 1988), and was subsequently termed a chloroplast chaperonin (Hemmingsen *et al.* 1988). When *Synechococcus* Rubisco large subunit was imported into chloroplasts it was complexed with this chaperonin (Gatenby *et al.* 1988). Imported small subunit also

associates with the chaperonin (Gatenby *et al.* 1988; Ellis and van der Vies 1988). Lubben *et al.* (1989b) report that imported large and small subunits of Rubisco, CF1-β-subunit, CAT, β-lactamase, LHCP and glutamine synthetase all bind to chloroplast chaperonin upon import. In the same study, imported superoxide dismutase and ferredoxin were not found to be associated. Thus, it appears that association with chaperonin upon import is widespread, although perhaps not universal. Furthermore, because newly synthesized proteins such as large subunit also associate with chaperonin, similar structural features may be exposed in newly imported proteins as in newly synthesized proteins.

A chaperonin system homologous to the groE protein of *E. coli* is also present in mitochondria (McMullin and Hallberg 1988). Imported proteins have been shown to bind to this mitochondrial chaperonin (Ostermann *et al.* 1989). One of these imported proteins, the Reiske Fe/S protein, may require association with chaperonin for its correct intraorganellar targeting to the inner membrane (Cheng *et al.* 1989; Ostermann *et al.* 1989). If chloroplast chaperonin is similar, then it may also be involved in targeting to the inner membrane and to the thylakoids.

Because the chloroplast and mitochondrial chaperonins are homologous to the groEL protein of *E. coli*, it is tempting to extrapolate studies with the bacterial proteins to the organellar situation. If this is valid, then we can expect the existence of organellar equivalents of the other member of the groE family, the groES protein. Although no organellar groES-like protein has yet been described, it seems likely that such proteins exist. When they are discovered, a comparison of the organellar and bacterial groE proteins may be useful. A certain degree of coevolution of chaperonins and their substrates is possible, and certain chaperones may function better with some proteins than with others. If so, then expression of foreign or retargeted genes may require expression of suitable chaperonins.

Another class of molecular chaperones, the 70 kDa class, also exists in chloroplasts (Amir-Shapira *et al.* 1990; Marshall *et al.* 1990) and mitochondria (Amir-Shapira *et al.* 1990). These proteins are ubiquitous and are found in endosymbiotic organelles, e.r. systems and cytoplasm (Murakami *et al.* 1988). Apparently, there are both constitutive and heat-induced members of this family, equivalent to the dnaK protein in *E. coli*,

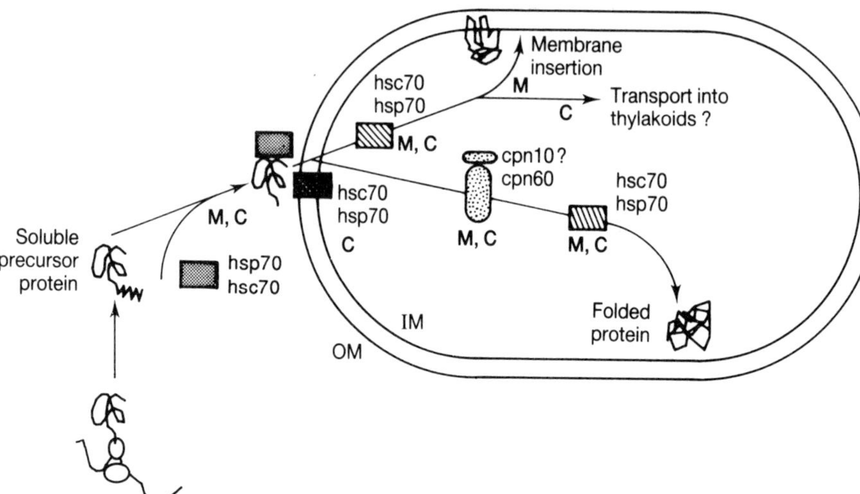



**Fig. 6.3** The proposed mediation of translocation folding and assembly by organellar and cytosolic chaperonins. The cytosolic group, hsp70 or hsc70, retains the precursor in an 'unfolded' import-competent state. There is some experimental evidence that suggests import is more efficient into isolated chloroplasts when cocktails are supplemented with chaperonin. Similar components exist within the stroma to mediate internal targeting, e.g. to membrane, or to assist in correct assembly as with cpn60 and cpn10 species.

The constitutive hsp70s (hsc70s) have been implicated in maintaining precursor proteins in an import-competent form for mitochondrial import and also in folding inside mitochondria (Deshaies *et al.* 1988; Chirico *et al.* 1988). Recently, hsp70-like proteins have also been found in chloroplasts (Marshall *et al.* 1990; Amir-Shapira *et al.* 1990) by using antibodies against hsp70 protein (Marshall *et al.* 1990) or against dnaK (Amir-Shapira *et al.* 1990). The studies by Marshall *et al.* utilized pea chloroplasts and determined the intraorganellar location. Three hsc70s were found: two were stromal (78 kDa and 75 kDa) and one was tightly associated with the outer membrane (75 kDa), but not susceptible to external proteolysis. Although no functional studies were done, the authors suggested that the outer membrane hsc70 may be involved in protein import in a manner similar to the mitochondrial hsc70s, to maintain or establish import competence or to participate in folding reactions related to import. Recently, it has been reported that some soluble factor is required for import of LHCP into chloroplasts, and hsc70 can only partially substitute for this activity (Waegemann *et al.* 1990). Figures 6.3 and 6.4 depict protein transport into e.r., chloroplasts and mitochondria, incorporating molecular chaperones where they are known to assist or where there is good evidence for such a role. A significant difference is that the cystosolic chaperones may maintain

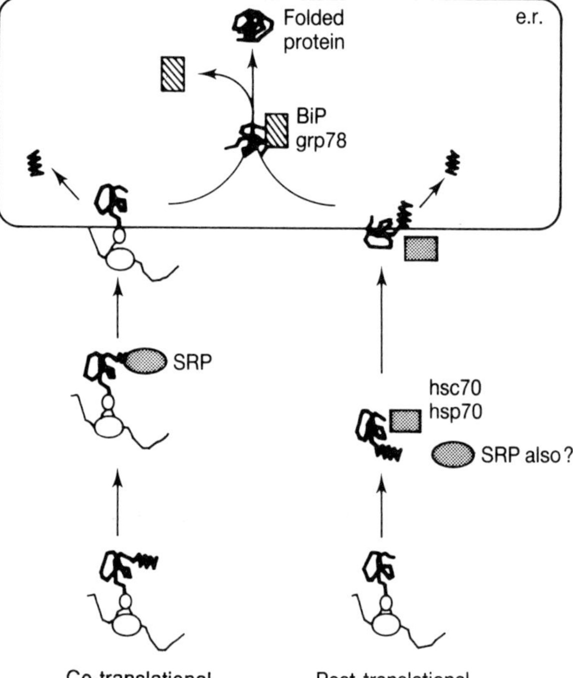



**Fig. 6.4** Molecular chaperone-mediated e.r. transport. Two processes are recognized in the initial events of e.r. targeting. One that has been established is cotranslational involving a signal recognition particle (SRP). Alternatively, the cytosolic chaperonins may also assist precursors post-translationally. Those proteins retained in the e.r. lumen, such as BiP, harbour a C-terminal peptide KDEL that is recognized by a salvage receptor.

proteins in an unfolded state, whereas the organellar species may enhance the folding and assembly of imported proteins.

# Endomembrane targeting

Targeting of proteins through the endomembrane system of plants is markedly different from targeting to the chloroplast. Targeting to the semiautonomous organelles (chloroplast and mitochondria) occurs via a post-translational mechanism as described in the previous section. However, targeting to the endomembrane system (e.r., Golgi apparatus, vacuole, protein bodies, plasma membrane and nuclear envelope) occurs by cotranslational transport of a preprotein to the e.r. followed by vesicular flow to subsequent membrane systems (see Fig. 6.1).

Proteins targeted to the endomembrane system are nuclear-encoded and synthesized on ribosomes bound to the e.r. Once a protein enters the e.r. it can follow a simplified vectorial flow from the e.r. to the Golgi apparatus and from there to the vacuole, protein body or plasma membrane. The final location of the protein can be in any membrane system in the pathway and is dependent on the presence or absence of particular sorting sequences.

The structure of e.r.-targeted proteins is similar to chloroplast-targeted proteins in that the protein is made as a longer precursor that is processed to produce the mature protein once it has entered the e.r. lumen. However, the structure of the N-terminal extension, termed a signal peptide, is significantly different from the chloroplast transit sequence. Although there is little, if any, sequence homology between signal sequences they all have common structural components (von Heijne 1983, 1984, 1985). The sequence can be in the range of 15–30 amino acids with three identifiable domains. The first is usually 1–9 amino acids and contains at least one positively charged residue (R or K). The second domain varies in length between 7 and 15 amino acids and is defined by a high content of hydrophobic amino acids (F, L, I, M, V, W). The third and final domain is 5–9 amino acids in length and is characterized by a higher number of polar amino acids (Q, N, R, K, H, D, E, S, T, Y). The presequence shown in Figure 6.5 is the N-terminal extension (presequence) of carrot extensin protein and provides an example of the distribution of amino acids in the three domains.

**Fig. 6.5** The localization of various amino acid types within three regions of the e.r. signal peptide. The example shown is that for the signal of extensin from carrot with a basic N-terminus, hydrophobic mid-sequence and more polar C-terminus.

Our understanding of the mechanisms of targeting of proteins to the endomembrane systems of animals and yeast has increased considerably over the past several years. However, the study of endomembrane sorting of plant proteins has only been examined fairly recently. In yeast the genetic analysis of mutants has allowed the evaluation of the movement of targeted proteins in intricate detail (see, for example, Supplement 14C of the *Journal of Cellular Biochemistry*, 1990 and the references contained therein). Scores of mutants have been isolated and used to elucidate mechanisms of transfer at all points within the sorting process. With the aid of recent advances in genetic engineering, the ability to answer the questions of protein sorting in plants has been addressed using chimeric proteins. The information from yeast will serve as an invaluable roadmap to assist in recognizing the similarities and differences between these systems.

## Targeting to the endoplasmic reticulum

It has been clearly demonstrated in animal and yeast that proteins destined for the e.r. are transported by a cotranslational process across the membrane to the e.r. lumen. During synthesis of the precursor protein, the signal peptide is recognized by a cytosolic ribonucleoprotein called the Signal Recognition Particle (SRP) as it emerges from the ribosome (Walter and Blobel 1986). The SRP is an 11S particle that contains six proteins and a 7S RNA component (Walter and Blobel 1980). The SRP binds to both the signal sequence and the ribosome and results in an arrested or a decreased rate of translation. The primary function of the SRP is to bring the precursor and ribosome to the e.r. The SRP binds to a specific receptor, at which point it is released for recycling and the precursor signal sequence is recognized by a signal sequence receptor. Loss of the SRP allows translation to resume and the growing nascent chain is transported across the e.r. membrane. The signal sequence is proteolytically removed by a signal peptidase.

Although SRP has been identified in wheat germ and maize endosperm it did not arrest translation when tested in an *in vitro* assay with the wheat germ cell-free translation system. The putative SRP of maize endosperm is a 12S particle with a complex RNA composition that can be resolved into three species by electrophoresis. The RNA sediments as an 8S particle. The protein fraction of the particle has not been investigated because of the low yield from endosperm (Campos *et al.* 1988). The wheat germ SRP has been characterized as an 11S particle composed of one 7S RNA and uncharacterized protein (Prehn *et al.* 1987). The inability of either plant SRP to cause translational arrest might be a characteristic of the heterologous nature of a plant translation system being used with animal SRP and precursor proteins (Prehn *et al.* 1988).

The mechanism for translocation outlined above has been elucidated in animal and yeast systems. Experiments involving the targeting of plant proteins indicate that a similar process occurs in plants. Plant mRNA can direct the targeting of proteins to canine microsomes (Burr and Burr 1981; Torrent *et al.* 1986; Hattori *et al.* 1987; DellaPenna and Bennet 1988) and the expression of plant genes in heterologous systems (Rothstein *et al.* 1984; Tague and Chrispeels 1987) shows correct localization of the protein in the e.r.

Signal sequences from several sources have recently been used to construct chimeric genes to examine the extent of the sequence information required for endomembrane targeting. The 68 kDa bacterial reporter protein, β-glucuronidase (GUS), has been fused to a putative signal peptide of the potato tuber storage protein, patatin (Itturiaga *et al.* 1989). In this series of experiments the GUS gene was fused with increasing lengths of the gene coding for the signal peptide including part of the patatin mature gene. The chimeric constructs were then put into tobacco and the location of GUS was determined. Unfortunately, GUS has two potential glycosylation sites that are apparently recognized by the transport processes. Glycosylation at these sites results in the inactivation of GUS. All the enzyme activity could be recovered from the transgenic plants if they were treated with tunicamycin to inhibit glycosylation during development. A similar series of experiments, performed in another laboratory (Denecke *et al.* 1990a) examined the fate of three bacterial proteins, phosphinothricin acetyl transferase (PAT, 21 kDa), neomycin phospho-

transferase (NPTII 33 kDa) and β-glucuronidase fused to the signal peptide of the 18 kDa 'pathogenesis-related protein' (PR protein) PR-1b or the prepro-peptide of cecropin B (a peptide secreted into the haemolymph of the insect *Hyalophora cecropia*). In these experiments the DNA constructs were electroporated into tobacco protoplasts that were then investigated for transient activity. After 36 h the enzyme could be recovered from the cells and the media and the amount of secreted protein was measured. In all cases secretion of the protein was dependent on the presence of the signal peptide.

This example of protein targeting has potential agronomic importance. In the case of transformation to introduce disease resistance, the best defence would involve secretion of the protein outside the cell where it would interact with the invading pathogen before the pathogen became established. For example, the secretion of a bacterial chitinase to the extracellular space may afford just this type of protection. When the 59 kDa chitinase ChiA is expressed in plants along with its own targeting sequence that directs its secretion in bacteria, the enzyme is secreted efficiently and glycosylated correctly (Lund *et al.* 1989). Another successful example of the exploitation of this form of targeting has involved the expression of both light and heavy chains of an antibody in transgenic tobacco. The functional antibody was obtained only when both polypeptides were directed out of the cell with their own signal peptide (Hiatt *et al.* 1989).

## Glycosylation of targeted proteins

As the protein is transported through the e.r. membrane into the lumen it is modified by the addition of oligosaccharide side chains. The glycosyl moiety is first built on a membrane-bound dolichol phosphate carrier and transferred to the amide nitrogen of asparagine in the protein. The glycan that is transferred is composed of nine mannose, two N-acetylglucosamine and three glucose residues ($GlcNAc_2Man_9Glc_3$, see Fig. 6.6). The dolichol carrier is located on the lumenal side of the e.r. and is composed of 16–20 unsaturated isoprenoid alcohols. In its phosphorylated form, the glycan becomes attached through a phosphate ester linkage. The transfer of the entire glycan moiety to the growing protein then occurs as it is transported through the membrane (Kornfeld and Kornfeld 1985; Hirschber and Snider 1987). The

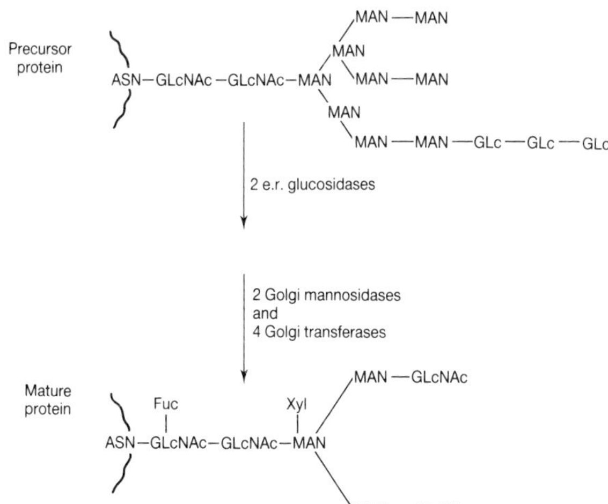


**Fig. 6.6**   The complicated glycosylation process of e.r.-targeted proteins. The core glycan moiety composed of GlcNAc$_2$Man$_9$Glc$_3$ is transferred to the Asn residue of the lumenal emerging protein from phosphorylated dolichol. The glycan is then further processed by other endomembrane localized enzymes (see the text). The final form of the glycan appendage is dependent on the distribution and relative quantities of the hydrolytic and transferase enzymes. At some time after emerging into the lumen the signal is removed by proteolytic enzymes.

specific sequence that is recognized as a site for glycan attached is N-X-S/T, in which X can be any amino acid except P or D. In addition to the presence of the specific sequence, one further restriction is that the N residue must be accessible to the oligosaccharide transferring enzyme.

The core glycan transferred to the protein is processed as it passes through the various associated endomembrane systems (see Fig. 6.6). In the e.r. the core glycan is modified by glucosidase I and II to remove the three terminal glucose residues giving Man$_9$GlcNAc$_3$ (Szumillo *et al.* 1986a). As the protein passes from the e.r. into the cis Golgi it is further acted on by α-mannosidase I to remove as many as four mannose residues (Szumillo *et al.* 1986b). A glycan core with at least five of the original nine mannose residues is a 'high mannose' glycan. Removal of the next two mannose residues occurs in the medial Golgi and is catalysed by α-mannosidase II (Johnson and Chrispeels 1987). This enzyme results in the formation of the core structure found in 'complex' glycans, i.e. less than five of the original nine mannose residues remain. The final steps of glycan processing include the addition of fucose, xylose and GlcNac in the trans Golgi (Chrispeels 1985; Sturm and Kindl 1983;

Johnson and Chrispeels 1987). The enzymes responsible for the transfer of the respective residues have been characterized in plants and shown to be localized in the Golgi as predicted (Sturm *et al.* 1987).

Localization of the specific enzymes to different subcompartments of plant Golgi has not been very successful. The standard procedure of density gradient fractionation, as used for yeast, does not give demonstrable separation of the Golgi compartments of plants. One report illustrated the separation of galactosyltransferase from xylosyltransferase and fucosyltransferase by density gradient centrifugation of membranes from sycamore cells (Ali *et al.* 1986) but other attempts have not been successful. The actual processing that any particular glycan may undergo on a particular protein is dependent on the presence of the various processing enzymes and their relative amounts. In addition, the local structure around a glycan might effect the form of the processed glycan (Voelker *et al.* 1989).

The function of the glycan moiety attached to proteins traversing the endomembrane system is unknown. The presence of the glycan, which can constitute 50% of the mass of the protein, can drastically effect the chemical and biological properties of the protein. In animal cells the presence of a mannose-6-phosphate moiety is the signal responsible for the transport of proteins to the lysosome. Since the animal cell lysosome is similar to the plant vacuole, the glycan was hypothesized to be involved in targeting in plant cells. However, when glycosylation was either inhibited by treatment of cells with tunicamycin (which prevents assembly of the glycan-dolichol precursor) or eliminated by site-directed mutagenesis of glycosylation sites (e.g. the N residue changed to S or the T to A) the protein (phytohaemagglutinin) was still correctly targeted to the vacuole (Voelker *et al.* 1989). In addition, the assembly of phytohaemagglutinin subunits into the native tetrameric structure was also unaffected by elimination of the glycan moieties. Further understanding of the importance of this complex derivatization process in transport will require the study of many other precursor proteins and the effect of removing glycosylation sites.

## E.r. retention

Most of the proteins that enter the e.r. are transported further into the Golgi and, from there, are usually transported to the plasma membrane to be secreted or to protein bodies or the vacuole. However, those proteins that are maintained in the e.r. must have a unique property that functions either to retain the protein in the e.r. or causes the protein to be recovered from the post-e.r. compartment.

E.r.-retained proteins of yeast have been examined to identify a sequence responsible for e.r. retention. It was observed that the sequence KDEL or the related sequences RDEL, HTEL, HIEL or HEEL were present at the C-terminus of 28 of the proteins examined (Pelham 1989). In plants, three proteins that are known to be in the e.r. lumen—immunoglobulin binding protein, protein disulphide isomerase and putative auxin receptor—have the C-terminal sequence KDEL (Hesse *et al.* 1989).

To determine whether the C-terminal sequence is sufficient for retaining proteins, chimeric proteins were produced with the sequence HDEL, KDEL or RDEL fused to the C-terminus of bacterial reporter proteins that have also been modified to include a signal sequence at the N-terminus. Each of the sequences were found to be necessary and sufficient for e.r. retention (Denecke *et al.* 1990b). In yeast the sequence KDEL was shown to be required for e.r. retention by deleting the C-terminal sequence from the e.r.-retained proteins or by adding it to proteins that are usually secreted (Munro and Pelham 1987). In addition, proteins retained in the e.r. were also found to be in the nuclear envelope (Rose *et al.* 1989). These results suggest that not only is the KDEL sequence sufficient for e.r. retention, it is a further indication of the continuity of the nuclear envelope with the e.r. and the sufficiency of this sequence to retain proteins in the e.r. as well as the nuclear envelope.

From the results described above, in which proteins targeted to the e.r. can diffuse to the nuclear envelope, it is difficult to understand how a protein could cause retention on the one hand and yet allow free diffusion of the protein within the e.r. This apparent discrepancy was addressed in an experiment with yeast, in which a protein normally destined for the lysosome was altered to contain the KDEL sequence on the C-terminus (Pelham 1988). The protein was found in the e.r., as anticipated, and was subsequently analysed for composition of the glycan moiety attached. The analysis showed the presence of mannose-6-phosphate. Since the transferase that catalyses this reaction is not located

in the e.r. it was postulated that e.r. 'retained' proteins, e.g. BiP (see Fig. 6.4), actually move to a post-e.r. compartment (either the cis Golgi or a 'salvage compartment') and are then transported back to the e.r. (Pelham 1988).

Experimental support for a salvage process that recognizes the C-terminal KDEL signal of those proteins that should be retained in the lumen, has recently gained experimental support. Anti-idiotypic antibodies raised to monoclonals against the C-terminal sequences of two resident lumenal proteins have led to the identification of a receptor that recognizes this signal. The receptor functions in a distal compartment and captures escaping e.r.-resident proteins with the signal and returns them to the e.r. (Vaux *et al.* 1990).

## Vacuole/protein bodies

Protein bodies develop primarily in seeds and act as a store of reserve protein to supply amino acids and during the initial stages of germination. In dicotyledons, particularly legumes, the protein bodies are formed from the division of vacuoles and subsequently enlarge by fusion with Golgi-derived membrane vesicles (Robinson 1985). It has been proposed that targeting of proteins to the plant vacuole is similar to vacuolar targeting in yeast (Marty 1978). When the plant vacuolar protein phytohaemagglutinin was expressed in yeast, it was correctly localized in the vacuole. However, unlike yeast proteins normally destined for the vacuole, that have a recognizable N-terminal sequence, no such sequence has been identified for plant proteins.

Production of protein bodies may also occur in monocotyledons by retention and assembly of the storage proteins in the e.r. and formation of the protein body by budding from the e.r. (Robinson 1985). Since the first step was apparent retention in the e.r., it was anticipated that the KDEL type retention sequence would be present at the C-terminus of these proteins. Examination of the zein amino acid sequences did not show the expected sequence (Larkins *et al.* 1989). Instead an alternative mechanism must be used to retain the proteins and promote aggregation and formation of the protein body. Microinjection of zein mRNA into *Xenopus* oöcytes caused the accumulation of structures similar to the plant protein body, but the specific interactions and structure were not the same as in authentic protein bodies (Lending and

Larkins 1989; Wallace *et al.* 1988, see chapter 10). One possible mechanism proposed for the formation of protein bodies is that specific interactions occur between different subunits of zein. The aggregate formed is presumed to be insoluble in the e.r. lumen, thereby preventing the secretion or forward movement within the endomembrane system (Lending and Larkins 1989).

## Protein secretion

Secretion of proteins from plant cells occurs by fusion of Golgi-derived secretory vesicles with the plasma membrane (Akazawa and Hara-Nishimura 1985). A second mechanism may involve a mechanism that bypasses the Golgi and allows fusion of e.r.-derived vesicles with the plasma membrane. The pathway for secretion through the Golgi has been clearly demonstrated, unlike the route from the e.r. to the plasma membrane.

Secretion is considered to be the default fate of those proteins that lack any other targeting signal. Their movement to the plasma membrane is a result of bulk flow through the endomembrane system (Pfeffer and Rothman 1987). Although this default pathway has been demonstrated in animals and yeast, the situation is not as well defined in plants. Several studies that have addressed the process of bulk flow in plants have used bacterial proteins expressed with their own targeting sequence, as in the case of chitinase discussed above (Lund *et al.* 1989) or chimeric constructs of non-secretory proteins and plant signal sequence (Denecke *et al.* 1990a and b). Since there is no additional targeting information within the protein, their efficient secretion suggests that this is also a default pathway in plants.

The default secretion of proteins that lack additional sequences does not always operate. As with other targeting processes, the protein to be transported across membranes has to be 'competent'. Unfortunately, 'transport competence' is not well defined. One example is the incorrect localization of the cytosolic albumin that carried the signal sequence of phytohaemagglutinin when expressed in transgenic tobacco (Dorel *et al.* 1989).

The glycan moiety is involved in the targeting of some proteins in animals (Pfeffer and Rothman 1987) but there is no example of its involvement in the limited number of proteins studied in plant cells. The evidence is based on the absence of glycosylation of some secreted proteins, e.g. α-amylases of barley aleurone (Jones *et al.* 1987), nor does tunicamycin inhibit secretion of those proteins that are normally glycosylated, e.g. rice scutellum α-amylase (Miyata and Akazawa 1982) or β-fructosidase (Faye *et al.* 1986). One report does indicate that the secretion of β-fructosidase was reduced from carrot cells treated with tunicamycin, but it was determined that the glycan moiety in this case was important for stabilizing the protein. The unglycosylated form was degraded late in the secretory process (Faye and Chrispeels 1989).

## Peroxisomal targeting

Peroxisomes are not part of the endomembrane system but are derived from pre-existing peroxisomes through budding and division (Trelase 1984; Lazarow and Fujiki 1985). The organelle is unique because it is bound by a single membrane. Transport of proteins into the peroxisome is post-translational after the proteins are synthesized on free polysomes (Lazarow and Fujiki 1985). The imported protein does not appear to be modified in any detectable way.

Firefly luciferase is located in the peroxisomes of transgenic insects, mammals, yeast and plants (Gould *et al.* 1990). The targeting sequence was initially located within the C-terminal 12 amino acids of the protein. Most recently, it has been shown that the terminal tripeptide SKL is sufficient to direct targeting and a comparison of many peroxisomal proteins shows that they have this sequence, or one very similar.

Further proof that the terminal tripeptide is adequate was obtained using CAT or DHFR with the tripeptide at their C-termini. Both enzymes were localized in the peroxisome (Gould *et al.* 1990). Mutagenesis of the tripeptide in luciferase indicated the degeneracy of the tripeptide signal as S (A or C)K (H or R)L, from those that targeted correctly. Some plant peroxisomal proteins have a C-terminal methionine in place of L but this has yet to be examined as an alternative substitute in the signal of the chimeras used.

## Summary

Protein transport in plants, as well as in other eukaryotes, involves both post-translational and cotranslational import. Import to the semiautonomous organelles is post-translational and transport through the endomembrane system is primarily cotranslational. Import to chloroplasts and mitochondria is similar with respect to the overall features. Precursor proteins are either partially unfolded by, or maintained in, a partially unfolded state by molecular chaperones.

Translocation through the envelopes depends upon energy and may involve conformational changes in the protein, the lipid bilayer or both. The transit peptide is removed during or after import, and chaperones assist at least some proteins in refolding and assembly, or in maintaining a conformation compatible with translocation to other intraorganellar locations, such as thylakoids or inner envelope membranes. However, import into chloroplasts and mitochondria differs from in important details, such as the energy required for import and the secondary structure of the transit peptides. The translocation mechanisms are possibly different.

It is possible to transport a variety of foreign proteins to chloroplasts and mitochondria, and this has proved useful in terms of altering plant characteristics, particularly herbicide tolerance. Interference with chloroplast import may be another route for manipulation of plant traits. Another avenue for modification of crop plants may be through the replacement of organellar genes with nuclear genes containing an organellar transit peptide.

Proteins destined for the endomembrane system, including secreted proteins and storage proteins, are translocated differently from proteins targeted for chloroplasts or mitochondria. Transport usually occurs cotranslationally and sorting is dependent upon the signal present in the sorted protein sequence. Glycosylation is an important yet little understood part of this pathway. In an example of utilizing the e.r. pathway for agronomic purposes, bacterial chitinase has been engineered into plants to provide insect resistance. For those wishing to introduce storage proteins with altered sequences into selected crop plants, it is this system that will need to be recruited to produce storage vesicles. In this case, the modification in protein structure will have to minimally affect the accumulation process (see Chapter 10). An understanding of protein trafficking in plants, as well as homologous trafficking in non-plant species, will be essential in efficient manipulation of plant traits for agronomic benefit.

# References

Akazawa, T. and Hara-Nishimura, I. (1985). Topographic aspects of biosynthesis, extracellular secretion, and intracellular storage of proteins in plant cells. *Annual Reviews Plant Physiology*. **36**: 441–72.

Ali, S., Mitsui, T. and Akazawa (1986). Golgi specific localization of transglycosylases engaged in glycoprotein biosynthesis in suspension-cultured cells of Sycamore (*Acer pseudoplatanus* L.) *Archives Biochemistry Biophysics*. **251**: 421–31.

Allison, D.S. and Schatz, G. (1986). Artificial mitochondrial presequences. *Proceedings of the National Academy of Science, USA*. **83**: 9011–15.

Amir-Shapira, D., Leustek, T. and Dalie, B. (1990). Hsp70 proteins, similar to *Escherichia coli* DnaK, in chloroplasts and mitochondria of *Euglena gracilis*. *Proceedings of the National Academy of Science, USA*. **87**: 1749–52.

Baker, A. and Schatz, G. (1987). Sequences from a prokaryotic genome or the mouse dihydrofolate reductase gene can restore the import of a truncated precursor protein into yeast mitochondria. *Proceedings of the National Academy of Science, USA*. **84**: 3117–21.

Banroques, J., Perea, J. and Jacq, C. (1987). Efficient splicing of two yeast mitochondrial introns controlled by a nuclear-encoded maturase. *EMBO Journal*. **6**: 1085–91.

Barraclough, R. and Ellis, R.J. (1980). Protein synthesis in chloroplasts. IX. Assembly of newly-synthesized large subunits into ribulose bisphosphate carboxylase in isolated pea chloroplasts. *Biochimica Biophysica Acta*. **608**: 19–31.

Boutry, M., Nagy, F. and Poulsen, C. (1987). Targeting of bacterial chloramphenicol acetyltransferase to mitochondria in transgenic plants. *Nature*. **328**: 340–2.

Boyle, S.A., Hemmingsen, S.M. and Dennis, D.T. (1986). Uptake and processing of the precursor to the small subunit of ribulose 1,5-bisphosphate carboxylase by leucoplasts from the endosperm of developing castor oil seeds. *Plant Physiology*. **81**: 817–22.

Boyle, S.A., Hemmingsen, S.M. and Dennis D.T. (1990). Energy requirement for the import of protein into plastids from developing endosperm of *Ricinus communis* L. *Plant Physiology*. **92**: 151–4.

Burr, F.A. and Burr, B. (1981). *In vitro* uptake and processing of prezein and other maize preproteins by maize membranes. *Journal Cell Biology*. **90**: 427–34.

Campos, N., Palau, J., Torrent, M. and Ludevid, D. (1988). Signal recognition-like particles are present in maize. *Journal Biological Chemistry*. **263**: 9646–50.

Cheng, M.Y., Hartl, F.-U., Martin, J., Pollock, R.A. and Kalousek, F. (1989). Mitochondrial heat-shock protein hsp60 is essential for assembly of proteins imported into yeast mitochondria. *Nature*. **337**: 620–5.

Cheung, A.Y., Bogorad, L. and van Montagu, M. (1988). Relocating a gene for herbicide tolerance: a chloroplast gene is converted into a nuclear gene.

*Proceedings of the National Academy of Science, USA.* **85:** 391–5.

Chirico, W.J., Waters, M.G. and Blobel, G. (1988). 70k heat shock related proteins stimulate protein translocation into microsomes. *Nature.* **332:** 805–10.

Chrispeels, M.J. (1985). UDP-GlcNAc: glycoprotein GlcNAc-transferase is located in the Golgi apparatus of developing bean cotyledons. *Plant Physiology.* **78:** 835.

Cline, K. (1988). Light-harvesting chlorophyll a/b protein. Membrane insertion, proteolytic processing, assembly into LHC II, and localization to appressed membranes occurs in chloroplast lysates. *Plant Physiology.* **86:** 1120–6.

Cline, K., Werner-Washburne, M., Lubben, T.H. and Keegstra, K. (1985). Precursors to two nuclear-encoded chloroplast proteins bind to the outer envelope membrane before being imported into chloroplasts. *Journal of Biological Chemistry.* **260:** 3691–6.

Cornwell, K.L. and Keegstra, K. (1987). Evidence that a chloroplast surface protein is associated with a specific binding site for the precursor to the small subunit of ribulose-1,5-bisphosphate carboxylase. *Plant Physiology.* **85:** 780–5.

de Kruijff, B., Cullis, P.R., Verkleij, A.J., Hope, M.J., van Echteld, C.J.A. and Taraschi, T.F. (1985). Lipid polymorphism and membrane function. In *The Enzymes of Biological Membranes* (ed. A. Martinosi), pp. 131–204. Plenum Press, New York.

della-Cioppa, G. and Kishore, G.M. (1988). Import of a precursor protein into chloroplasts is inhibited by the herbicide glyphosphate. *EMBO Journal.* **7:** 1299–305.

della-Cioppa, G., Bauer, S.C. and Klein, B.K. (1986). Translocation of the precursor of 5-enol pyruvylshikimate 3 phosphate synthase into chloroplasts of higher plants *in vitro. Proceedings of the National Academy of Science, USA.* **83:** 6873–7.

della-Cioppa, G., Kishore, G.M., Beachy, R.N. and Fraley, R.T. (1987). Targeting a herbicide-resistant enzyme from *Escherichia coli* to chloroplasts of higher plants. *Biotechnology.* **5:** 579–83.

DellaPenna, D. and Bennet, A.B. (1988). *In vitro* synthesis and processing of tomato fruit polygalacturonase. *Plant Physiology.* **86:** 1057.

Denecke, J., Botterman, J. and Deblaere, R. (1990a). Protein secretion in plant cells can occur via a default pathway. *Plant Cell.* **2:** 51–9.

Denecke, J., Hellemans, B. and Botterman, J. (1990b). Functional analysis of signals for protein retention in the plant endoplasmic reticulum. *Journal of Cellular Biochemistry.* **14C:** H223.

Deshaies, R.J., Koch, B.D. and Werner-Washburne, M. (1988). A subfamily of stress proteins facilitates translocation of secretory and mitochondrial precursor polypeptides. *Nature.* **232:** 800–5.

Dorel, C., Voelker, T.A., Herman, E.M. and Chrispeels, M.J. (1989). Transport of proteins to the plant vacuole is not by bulk flow through the secretory system, and requires positive sorting information.

*Journal of Cell Biology.* **108:** 327–32.

Eilers, M. and Schatz, G. (1988). Protein unfolding and the energetics of protein translocation across biological membranes. *Cell.* **52:** 481–3.

Eilers, M., Hwang, S. and Schatz, G. (1988). Unfolding and refolding of a purified precursor protein during import into isolated mitochondria. *EMBO Journal.* **7:** 1139–45.

Ellis, R.J. and van der Vies, S.M. (1988). The Rubisco subunit binding protein. *Photosynthesis Research.* **16:** 101–15.

Faye, L. and Chrispeels, M.J. (1989). Apparent inhibition of β-fructosidase secretion by tunicamycin may be explained by breakdown of the unglycosylated protein during secretion. *Plant Physiology.* **89:** 206.

Faye, L., Mouatassim, B. and Ghorbel, A. (1986). Cell wall and cytoplasmic isozymes of radish β-fructosidase have different N-linked oligosaccarides. *Plant Physiology.* **80:** 27–33.

Fillatti, J., Kiser, J., Rose, R. and Comai, L. (1987). Efficient transfer of a glyphosate tolerance gene into tomato using a binary *Agrobacterium tumefaciens* vector. *Biotechnology.* **5:** 726–30.

Flügge, U.K. and Hinz, G. (1986). Energy dependence of protein translocation into chloroplasts. *Biochemistry.* **160:** 563–70.

Friedman, A.L. and Keegstra, K. (1989). Chloroplast protein import: quantitative analysis of precursor binding. *Plant Physiology.* **89:** 993–9.

Gasser, S.M., Daum, G. and Schatz, G. (1982). Import of proteins into mitochondria. *Journal of Biological Chemistry.* **257:** 13034–41.

Gatenby, A.A., Lubben, T.H., Ahlquist, P. and Keegstra, K. (1988). Imported large subunits of ribulose bisphosphate carboxylase/oxygenase, but not imported β-ATP synthase subunits, are assembled into holoenzyme in isolated chloroplasts. *EMBO Journal.* **7:** 1307–14.

Gavel, Y., Nilsson, L. and von Heijne, G. (1988). Mitochondrial targeting sequences: why 'non-amphiphilic' peptides may still be amphiphilic. *FEBS Letters.* **235:** 173–7.

Gould, S.J., Keller, G.-A., Schneider, M., Howell, S.H., Garrard, L.J., Goodman, J.M., Distel, B., Tabak, H. and Subramani, S. (1990). Peroxisomal protein import is conserved between yeast, plants, insects and mammals. *EMBO Journal.* **9:** 85–90.

Grossman, A., Bartlett, S. and Chua, N.-H. (1980). Energy dependent uptake of cytoplasmically synthesized polypeptides by chloroplasts. *Nature.* **285:** 625–8.

Hand, J.M., Szabo, L.J. and Vasconcelos, A.C. (1989). The transit peptide of a chloroplast thylakoid membrane protein is functionally equivalent to a stromal-targeting sequence. *EMBO Journal.* **8:** 3195–206.

Hartnett, M.E., Chui, C.-F., Mauvais, C.J., McDevitt, R.E., Knowlton, S., Smith, J.K., Falco, S.C. and Mazur, B.J. (1990). Herbicide resistant plants carrying mutated acetolactate synthase genes. In *Managing Resistance to Agrochemicals: from fundamental research to practical strategies* (ed. M.B. Green, W.K. Moberg and H. LeBaron), pp. 459–73. American Chemical Society, Washington D.C.

Hattori, T., Ichihara, S. and Nakamura, K. (1987). Processing of a plant vacuolar protein precursor *in vitro*. *European Journal Biochemistry*. **166**: 533–8.

Haughn, G.W., Smith, J., Mazur, B. and Somerville, C. (1988). Transformation with a mutant Arabidopsis acetolactate synthase gene renders tobacco resistant to sulfonylurea herbicides. *Molecular General Genetics*. **211**: 266–71.

Hemmingsen, S.M., Woolford, C., van der Vies, S.M., Tilley, K. and Dennis, D.T. (1988). Homologous plant and bacterial proteins chaperone oligomeric protein assembly. *Nature*. **333**: 330–4.

Herrara-Estrella, L., Van den Broeck, G., Maenhaut, R., Van Montague, M., Schell, J., Timko, M. and Cashmore, A. (1984). Light-inducible and chloroplast-associated expression of a chimaeric gene introduced into *Nicotiana tabacum* using a Ti plasmid vector. *Nature*. **310**: 115–20.

Hesse, T., Feldwisch, J., Balshuserman, D., Bauw, G., Puype, M., Vandekerckhove, J., Lobler, M., Klambt, D., Schell, J. and Palme, K. (1989). Molecular cloning and structural analysis of a gene from *Zea mays* (1.) coding for a putative receptor for the plant hormone auxin. *EMBO Journal*. **8**: 2453–8.

Hiatt, A., Cafferkey, R. and Bowdish, K. (1989). Production of antibodies in transgenic plants. *Nature*. **342**: 76–8.

Highfield, P.E. and Ellis, R.J. (1978). Synthesis and transport of the small subunit of chloroplast ribulose bisphosphate carboxylase. *Nature*. **271**: 420–4.

Hinz, G. and Flügge, U.I. (1988). Phosphorylation of a 51-kDA envelope membrane polypeptide involved in protein translocation into chloroplasts. *European Journal of Biochemistry*. **175**: 649–59.

Hirschber, C.B. and Snider, M.D. (1987). Topography of glycosylation in the rough endoplasmic reticulum and Golgi apparatus. *Annual Review of Biochemistry*. **56**: 63.

Hurt, E.C. and Schatz, G. (1987). A cytosolic protein contains a cryptic mitochondrial targeting signal. *Nature*. **325**: 499–503.

Hurt, E.C., Goldschmidt-Clermont, M. and Pebol-Hurt, B. (1986a). A mitochondrial presequence can transport a chloroplast-encoded protein into yeast mitochondria. *Journal of Biological Chemistry*. **261**: 11440–3.

Hurt, E.C., Soltanifar, N. and Goldschmidt-Clermont, M. (1986b). The cleavable pre-sequence of an imported chloroplast protein directs attached polypeptides into yeast mitochondria. *EMBO Journal*. **5**: 1343–50.

Hwang, S.T. and Schatz, G. (1989). Translocation of proteins across the mitochondrial inner membrane, but not into the outer membrane, requires nucleoside triphosphates in the matrix. *Proceedings of the National Academy of Science, USA*. **86**: 8432–6.

Itturiaga, G., Jefferson, R.A. and Bevan, M.W. (1989). Endoplasmic reticulum targeting and glycosylation of hybrid proteins in transgenic tobacco. *Plant Cell*. **1**: 381–90.

Johnson, K.D. and Chrispeels, M.J. (1987). Substrate specificities of N-acetylglucosamine, fucosyl, and xylosyltransferases that modify glycoproteins in the Golgi apparatus of bean cotyledons. *Plant Physiology*. **84**: 1301.

Jones, R.L., Bush, D.S., Sticher, L., Simon, P. and Jacobsen, N. (1987). Intracellular transport and secretion of barley aleurone alpha-amylase. In: *Plant Membranes: structure, function, biogenesis* (ed. C. Leaver, H. Sze), pp. 325–40. Alan R. Liss Inc., New York.

Karlin-Neumann, G.A. and Tobin, E.M. (1986). Transit peptides of nuclear-encoded chloroplast proteins share a common amino acid framework. *EMBO Journal*. **5**: 9–13.

Keegstra, K. (1989). Transport and routing of proteins into chloroplasts. *Cell*. **56**: 247–53.

Keegstra, K., Olsen, L.J. and Theg, S.M. (1989). Chloroplastic precursors and their transport across the envelope membranes. *Annual Reviews of Plant Physiology*. **40**: 471–501.

Kloppstech, K. and Bitsch, A. (1986). Crosslinking of envelope proteins presumably involved in the binding of nuclear coded chloroplast precursor proteins. In *Regulation of Chloroplast Differentiation*, pp. 235–40. Alan R. Liss Inc., New York.

Klösgen, R.B., Saedler, H., and Weil, J.-H. (1989). The amyloplast-targeting transit peptide of the *waxy* protein of maize also mediates protein transport *in vitro* into chloroplasts. *Molecular and General Genetics*. **217**: 155–61.

Ko, K. and Cashmore, A.R. (1989) Targeting of proteins to the thylakoid lumen by the bipartitite transit peptide of the 33 kd oxygen-evolving protein. *EMBO Journal*. **8**: 3187–94.

Kohorn, B. and Tobin, E.M. (1989). A hydrophobic, carboxy-proximal region of a light-harvesting chlorophyll a/b protein is necessary for stable integration into thylakoid membranes. *Plant Cell*. **1**: 159–66.

Kornfeld, R. and Kornfeld, S. (1985). Assembly of asparagine linked oligosaccharides. *Annual Reviews of Biochemistry*. **54**: 631–54.

Lamppa, G.K. (1988). The chlorophyll a/b-binding protein inserts into the thylakoids independent of its cognate transit peptide. *Journal of Biological Chemistry*. **263**: 14996–9.

Larkins, B.A. *et al.* (1989). Characterization and modifications of maize storage proteins. In: *Biotechnology and Food Quality* (ed. S.-D. Kung, D.D. Bills and R. Quatrano), pp. 67–75. Butterworths, Guildford.

Lazarow, P.B. and Fujiki, Y. (1985). Biogenesis of peroxisomes. *Annual Reviews of Cell Biology*. **1**: 489.

Lemire, B.D., Fankhauser, C. and Baker, A. (1989).

The mitochondrial targeting function of randomly generated peptide sequences correlates with predicted helical amphiphilicity. *Journal of Biological Chemistry.* **264:** 20206–15.

Lending, C.R. and Larkins, B.A. (1989). Changes in the zein composition of protein bodies during maize endosperm development. *Plant Cell.* **1:** 1011.

Lubben, T.H. and Keegstra, K. (1986). Efficient *in vitro* import of a cytosolic heat shock protein into pea chloroplasts. *Proceedings of the National Academy of Science, USA.* **83:** 5502–6.

Lubben, T.H., Bansberg, J. and Keegstra, K. (1987). Stop-transfer regions do not half translocation of proteins into chloroplasts. *Science.* **283:** 1112–14.

Lubben, T.H., Theg, S.M. and Keegstra, K. (1988). Transport of proteins into chloroplasts. *Photosynthesis Research.* **17:** 173–94.

Lubben, T.H., Gatenby, A.A., Ahlquist, P. and Keegstra, K. (1989a). Chloroplast import characteristics of chimeric proteins. *Plant Molecular Biology.* **12:** 13–18.

Lubben, T.H., Donaldson, G.K., Viitanen, P. and Gatenby, A.A. (1989b). Several proteins imported into chloroplasts form stable complexes with the GroEL-related chloroplast molecular chaperone. *Plant Cell.* **1:** 1223–30.

Lund, P., Rino, Y.L. and Dunsmuir, P. (1989). Bacterial chitinase is modified and secreted in transgenic tobacco. *Plant Physiology* **91:** 130–5.

Maccecchini, M., Rubin, Y., Blobel, G. and Schatz, G. (1979). Import of proteins into mitochondria: Precursor forms of the extramitochondrially made $F_1$-ATPase subunits in yeast. *Proceedings of the National Academy of Science, USA.* **76:** 343–7.

Marshall, J.S., Derocher, A.E., Keegstra, K. and Fielding, E. (1990). Identification of heat shock protein hsp70 homologues in chloroplasts. *Proceedings of the National Academy of Science, USA.* **87:** 374–8.

Marty, F. (1978). Cytochemical studies on GERL, provacuoles, and vacuoles in root meristematic cells of Euphorbia. *Proceedings of the National Academy of Science, USA.* **75:** 852.

McMullin, T.W. and Hallberg, R.L. (1988). A highly evolutionarily conserved mitochondrial protein is structurally related to the protein encoded by the *Escherichia coli groEL* gene. *Molecular and Cellular Biology.* **8:** 371–80.

Meadows, J.W., Shackleton, J.B and Hurlford, A. (1989). Targeting of a foreign protein into the thylakoid lumen of pea chloroplasts. *FEBS Letters.* **253:** 244–6.

Miyata, S. and Akazawa, T. (1982). Enzymic mechanism of starch breakdown in germinating rice seeds. 12. Biosynthesis of alpha-amylase in relation to protein glycosylation. *Plant Physiology.* **70:** 147–53.

Munro, S. and Pelham, H.R.B. (1987). A C-terminal signal prevents secretion of lumenal ER proteins. *Cell.* **48:** 899.

Murakami, H., Pain, D. and Blobel, G. (1988). 70-kD heat shock-related protein is one of at least two distinct cytosolic factors stimulating protein import into mitochondria. *Journal of Cell Biology.* **107:** 2051–7.

Oliver, J.L., Marin, A. and Martinez-Zapater, J.M. (1989). Chloroplast genes transferred to the nuclear plant genome have adjusted to nuclear base composition and codon usage. *Nucleic Acids Research.* **18:** 65–73.

Olsen, L.J., Theg, S.M. and Selman, B.R. (1989). ATP is required for the binding of precursor proteins to chloroplasts. *Journal of Biological Chemistry.* **264:** 6724–9.

Ostermann, J., Horwich, A.L. and Neupert, W. (1989). Protein folding in mitochondria requires complex formation with hsp60 and ATP hydrolysis. *Nature.* **341:** 125–30.

Pain, D. and Blobel, G. (1987). Protein import into chloroplasts requires a chloroplast ATPase. *Proceedings of the National Academy of Science, USA.* **84:** 3288–92.

Pain, D., Kanwar, Y.S. and Blobel, G. (1988). Identification of a receptor for protein import into chloroplasts and its localization to envelope contact zones. *Nature.* **331:** 232–7.

Perara, E., Rothman, R.E. and Lingappa, V.R. (1986). Uncoupling translocation from translation: implications for transport of proteins across membranes. *Science.* **232:** 348–52.

Pelham, H.R.B. (1988). Evidence that the luminal ER proteins are sorted from secreted proteins in a post-ER compartment. *EMBO Journal.* **7:** 913–18.

Pelham, H.R.B. (1989). Control of protein exit from the endoplasmic reticulum. *Annual Reviews of Cell Biology.* **5:** 1.

Pfanner, N., Hartl, F.U. and Neupert, W. (1988a). Import of proteins into mitochondria: a multi-step process. *European Journal of Biochemistry.* **175:** 205–12.

Pfanner, N., Pfaller, R. and Neupert, W. (1988b). How finicky is mitochondrial protein import? *TIBS.* **13:** 165–7.

Pfeffer, S.R. and Rothman, J.E. (1987). Biosynthetic protein transport and sorting by the endoplasmic reticulum and Golgi. *Annual Review of Biochemistry.* **56:** 829.

Pfisterer, J., Lachmann, P. and Kloppstech, K. (1982). Transport of proteins into chloroplasts. Binding of nuclear coded chloroplast proteins to the chloroplast envelope. *European Journal of Biochemistry.* **126:** 143–8.

Prehn, S., Weidmann, M., Rapoport, T.A. and Zweib, C. (1987). *EMBO Journal.* **6:** 2093–7.

Reiss, B., Wassmann, C.C. and Bohnert, H.J. (1987). Regions in the transit peptide of SSU essential for transport into chloroplasts. *Molecular and General Genetics.* **209:** 116–21.

Reiss, B., Wasmann, C.C., Schell, J. and Bohnert, H.J. (1989). Effect of mutations on the binding and translocation functions of a chloroplast transit peptide. *Proceedings of the National Academy of Science, USA.* **86**: 886–90.

Robinson, D.G. (1985). *Endo- and plasma membranes of plant cells.* John Wiley and Sons Inc., New York.

Rodermel, S.R., Abbott, M.S. and Bogorad, L. (1988). Nuclear-organelle interactions: nuclear antisense gene inhibits ribulose bisphosphate carboxylase enzyme levels in transformed tobacco plants. *Cell.* **55**: 673–81.

Roise, D., Horvath, S.J. and Tomich, J.M. (1986). A chemically synthesized pre-sequence of an imported mitochondrial protein can form an amphiphilic helix and perturb natural and artificial phospholipid bilayers. *EMBO Journal.* **5**: 1327–34.

Rose, M.D., Misra, L.M. and Vogel, J.P. (1989) KAR2, a karyogamy gene, is the yeast homolog of the mammalian BiP/GRP78 gene. *Cell.* **57**: 1211.

Rothstein, S.J., Lazarus, C.M., Smith, W.E., Baulcombe, D.C. and Gatenby, A.A. (1984). Secretion of wheat alpha amylase expressed in yeast. *Nature.* **308**: 662.

Roy, H. and Cannon, S. (1988). Ribulose bisphosphate carboxylase assembly: what is the role of the large subunit binding protein? *TIBS.* **13**: 163–5.

Roy, H., Chaudhari, P. and Cannon, S. (1988). Incorporation of large subunits into ribulose bisphosphate carboxylase in chloroplast extracts: influence of added small subunits and of conditions during synthesis. *Plant Physiology.* **86**: 44–9.

Schmidt, G.W. and Mishkind, M.L. (1986). The transport of proteins into chloroplasts. *Annual Review of Biochemistry.* **55**: 879–912.

Schmidt, R.J., Gillham, N.W. and Boynton, J.E. (1985). Processing of the precursor to a chloroplast ribosomal protein made in the cytosol occurs in two steps, one of which depends on a protein made in the chloroplast. *Molecular and Cellular Biology.* 1093–9.

Shah, D.M., Horsch, R.B. and Klee, H.J. (1986). Engineering herbicide tolerance in transgenic plants. *Science.* **233**: 478–81.

Singer, S.J., Maher, P.A. and Yaffe, M.P. (1987). On the translocation of proteins across membranes. *Proceedings of the National Academy of Science, USA.* **84**: 1015–19.

Smeekens, S., Bauerle, C., Hageman, J., Keegstra, K. and Weisbeck, P. (1986). The role of the transit peptide in the routing of precursors toward different chloroplast compartments. *Cell.* **46**: 365–75.

Smeekens, S., van Steeg, H. and Bauerle, C. (1987). Import into chloroplasts of a yeast mitochondrial protein directed by ferredoxin and plastocyanin transit peptides. *Plant Molecular Biology.* **9**: 377–88.

Smeekens, S., Weisbeek, P. and Robinson, C. (1990). Protein transport into and within chloroplasts. *TIBS.* **15**: 73–6.

Steppuhn, J. *et al.* (1987). The complete amino-acid sequence of the Rieske FeS-precursor protein from spinach chloroplasts deduced from cDNA analysis. *Molecular and General Genetics.* **210**: 171–7.

Strzalka, K., Ngernrrasirtsiri, J. and Watanabe, A. (1987). Sycamore amyloplasts can import and process precursors of nuclear encoded chloroplast proteins. *Biochemical and Biophysical Research Communications.* **149**: 799–806.

Sturm, A. and Kindl, H. (1983). Fucosyltransferase and fucose incorporation *in vivo* as markers for subfractionating cucumber microsomes. *Zeit. Physiol. Chem.* **160**: 165–8.

Sturm, A., Johnson, K.D., Szumillo, T., Elbein, A.D. and Chrispeels, M.J. (1987). Subcellular localization and glycosyltransferases involved in the processing of N-linked oligosaccharides. *Plant Physiology.* **85**: 741.

Szabo, L.J. and Cashmore, A.R. (1987). Targeting nuclear gene products into chloroplasts. In *Plant DNA Infectious Agents* (ed. T. Hohn, J. Schell), pp. 321–39. Springer-Verlag, Vienna.

Szumillo, T., Kaushal, G.P. and Elbein, A.D. (1986a). Purification and properties of glucosidase I from mung bean seedlings. *Archives of Biochemistry and Biophysics.* **247**: 261.

Szumillo, T., Kaushal, G.P., Hidetaka, H. and Elbein, A.D. (1986b). Purification and properties of a glycoprotein processing alpha mannosidase from mung bean seedlings. *Plant Physiology.* **81**: 383–9.

Tague, B.W. and Chrispeels, M.J. (1987). The plant vacuolar protein, phytohemagglutinin is transported to the vacuole of transgenic yeast. *Journal of Cell Biology.* **105**: 1971.

Theg, S.M., Bauerle, C., Olsen, L.J., Selman, B.R. and Keegstra, K. (1989). Internal ATP is the only energy requirement for the translocation of precursor proteins across chloroplastic membranes. *Journal of Biological Chemistry.* **264**: 6730–6.

Thompson, G.A., Hialt, N.R., Facciotti, D., Stalker, D.M. and Comai, L. (1987). Expression in plants of a bacterial gene coding for glyphosate resistance. *Weed Science.* **35**: 19–23.

Torrent, M., Poca, E., Campos, N. Ludevid, M.D. and Palau, J. (1986). In maize, glutelin-2 and low molecular weight zeins are synthesized by membrane bound polyribosomes and translocated into microsomal membranes. *Plant Molecular Biology.* **7**: 393.

Trelase, R.N. (1984). Biogenesis of glyoxysomes. *Annual Review of Plant Physiology.* **35**: 321.

Tyagi, A., Hermans, J., Steppuhn, J., Jansson, C., Vater, F. and Herrmann, R.G. (1987). Nucleotide sequence of cDNA clones encoding the complete "33 kDa" precursor protein associated with the photosynthetic oxygen-evolving complex from spinach. *Molecular and General Genetics.* **207**: 288–93.

Van den Broeck, G., Timko, M.P., Kausch, A.P., Cashmore, A.R., Van Montagu, M. and Herrera-Estrella, L. (1985). Targeting of a foreign protein to chloroplasts by fusion to the transit peptide from the small subunit of ribulose 1.5-bisphosphate carboxylase. *Nature.* **313**: 358–63.

van Loon, A.P.G.M. and Brandll, A.W. (1986). The presequences of two imported mitochondrial proteins contain information for intracellular and intramitochondrial sorting. *Cell*. **44**: 801–12.

Vassarotti, A., Stroud, R., Douglas, M. (1987). Independent mutations at the amino terminus of a protein act as surrogate signals for mitochondrial import. *EMBO Journal*. **6**: 705–11.

Vaux, D., Tooze, J. and Fuller, S. (1990). Identification by anti-idiotype antibodies of an intracellular membrane protein that recognises a mammalian endoplasmic reticulum retention signal. *Nature*. **345**: 495–8.

Vestweber, D. and Schatz, G. (1988a). A chimeric mitochondrial precursor protein with internal disulfide bridges blocks import of authentic precursors into mitochondria and allows quantitation of import sites. *Journal of Cell Biology*. **107**: 2037–43.

Vestweber, D. and Schatz, G. (1988b). Mitochondria can import artificial precursor proteins containing a branched polypeptide chain or a carboxy-terminal stilbene disulfonate. *Journal of Cell Biology*. **107**: 2045–9.

Vestweber, D. and Schatz, G. (1989). DNA-protein conjugates can enter mitochondria via the protein import pathway. *Nature*. **338**: 170–3.

Vitanen, P.V., Doran, E.R. and Dunsmuir, P. (1988). What is the role of the transit peptide in thylakoid integration of the light-harvesting chlorophyll a/b protein? *Journal of Biological Chemistry*. **263**: 15 000–7.

Voelker, T.A., Flokiewicz, R.Z. and Chrispeels, M.J. (1989). *In vitro* mutated phytohemagglutinin genes expressed in tobacco seed: role of glycans in protein targeting and stability. *Plant Cell*. **1**: 95.

von Heijne, G. (1983). Patterns of amino acids near signal-sequence cleavage sites. *European Journal of Biochemistry*. **133**: 17.

von Heijne, G. (1984). How signal sequences maintain cleavage specificity. *Journal of Molecular Biology*. **173**: 243.

von Heijne, G. (1985). Signal sequences: the limits of variation. *Journal of Molecular Biology*. **184**: 99–108.

von Heijne, G. (1986a). Why mitochondria need a genome. *FEBS Letters*. **198**: 1–4.

von Heijne, G. (1986b). Mitochondrial targeting sequences may form amphiphilic helices. *EMBO Journal*. **5**:1335–42.

von Heijne, G., Steppuhn, J. and Herrmann, R.G. (1989). Domain structure of mitochondrial and chloroplast targeting peptides. *European Journal of Biochemistry*. **180**: 535–45.

Waegemann, K., Paulsen, H. and Soll, J. (1990). Translocation of proteins into isolated chloroplasts requires cytosolic factors to obtain import competence. *FEBS Letters*. **261**: 89–92.

Wallace, J.C., Galili, G., Kawata, E.E., Cuellar, R.E., Shotwell, M.A. and Larkins, B.A. (1988). Aggregation of lysine-containing zeins in proteins bodies in *Xenopus* oocytes. *Science*. **240**: 662.

Walter, P. and Blobel, G. (1980). *Proceedings of the National Academy of Science, USA*. **77**: 7112–16.

Walter, P. and Blobel, G. (1986). Mechanism of protein translocation across the endoplasmic reticulum. *Biochemical Society Transactions*. **47**: 183.

Walter, P. and Lingappa, V.R. (1986). Mechanism of protein translocation across the endoplasmic reticulum membrane. *Annual Review of Cell Biology*. **2**: 499–516.

Whelan, J., Dolan, L. and Harmey, M.A. (1988). Import of precursor proteins into *Vicia faba* mitochondria. *FEBS Letters*. **236**: 217–20.

White, J.A. and Scandalios, J.G. (1987). *In vitro* synthesis, importation and processing of Mn-superoxide dismutase (SOD-3) into maize mitochondria. *Biochimica Biophysica Acta*. **926**: 16–25.

# Part III  PLANT PROTEINS USED IN THE FOOD AND FEED INDUSTRIES

# 7

# Engineering Legume Seed Storage Proteins

Nigel Lambert and Jenny N. Yarwood

## Introduction

After cereals, legume seeds are the most important source of food protein in the world, and consequently are of immense nutritional and economic significance. This review concentrates solely upon the seed storage proteins that constitute the bulk of total seed protein. Much has been written about these proteins in various contexts, particularly over the last 10 years, which have seen an explosion of information in this area. It is not the intention here to reiterate these studies but rather to draw together various facets and focus them towards strategies that are/or could be employed in tailoring these important food proteins to the needs of the food industry, the agricultural community and the consumer.

To do justice to the subject of 'engineering legume seed proteins' we feel obliged to devote the first part of the text to a discussion of the structure of legume seed proteins, their commercial utilization and the nutritional and food functional qualities that are or that could be targets for engineering. This, we believe, 'sets the scene' for the second half of the chapter, which describes the central molecular biological practicalities of producing designer legume seed proteins via genetic engineering. Unlike more traditional genetic/protein engineering programmes, progress towards transgenic legumes and engineered globulins is still very much in its infancy, primarily as a consequence of the intrinsic difficulties associated with legumes and the complex nature of their seed proteins. Nevertheless, it will become apparent on reading the second half of this chapter that major strides have been made in this area and, perhaps more importantly, the direction for the future is much clearer.

## What are legume seed storage proteins?

Leguminosae is an immense botanical family in which the seeds show considerable variation in protein content. For example, Earle and Jones (1962) found a range of 12–55% protein among 320 different species. The genotype and agronomic conditions also influence protein levels within species. Seeds of the most widely cultivated legumes display a slightly narrower range of 20–40% protein (Table 7.1). These values are high compared to other plant food sources; two to four times greater than cereal grains, for example (Mossé and Baudet 1983). The protein content and quality of legumes are prime factors in making these crops an important food source. Up to 80% of the legume seed protein can be accounted for by seed storage proteins, reflecting their central role in seed viability (Coates *et al.* 1985; Croy and Gatehouse 1985). The remaining seed proteins (of which there are over 1000 types) are involved chiefly in cell structure and in basic metabolic processes. Some also have commercial significance because they possess anti-nutritional properties, e.g. lectins and protease inhibitors, and have been reviewed extensively (Liener 1983, 1989).

The sole physiological function of seed storage proteins is to provide a source of amino acids and nitrogen for seed germination (Spencer 1984). A protein may be classified as a seed storage protein if it:

**Table 7.1**   Plant seed globulins

| Plant[p] | % Seed protein[e] | 11S | 7S | 11S/7S ratio |
|---|---|---|---|---|
| Soyabean[o] (*Glycine max*) | 34.8–40.5 (40) | Glycinin | Conglycinin | 0.62–3.0[g,i] |
| Pea (*Pisum satiuum*) | 21.2–37.5 (25) | Legumin | Vicilin | 0.5–4.2[h,i] |
| French bean[b] (*Phaseolus vulgaris*) | 21.1–39.4 (28) | — | Phaseolin | Mainly 7S[i] |
| Broad bean[c] (*Vicia faba*) | 22.9–38.5 (30) | Legumin | Vicilin | 1.6–3.7[i] |
| Lupin (*Lupinus albus*) | 17.0–50.0 (40) | Conglutinα | Conglutinβ | 0.77[i] |
| Peanut[o] (*Arachis hypogaea*) | 23.5–33.3 (25) | Arachin | Conarchin | 3.0[j] |
| Chickpea[d] (*Cicer arietinum*) | 14.9–29.6 (21) | Legumin | Vicilin | 1.1[k,n] |
| Cow pea (*Vigna unguiculata*) | 22.9–34.6 (27) | Legumin | Vicilin | — |
| Mungbean (*Vigna radiata*) | — (27) | — | — | Mainly 7S[n] |
| Lentil (*Lens culinaris*) | 19.5–31.0 (28)[f] | — | — | 0.7[f,n] |
| Pigeon pea (*Cajanus cajan*) | — | — | — | — |
| Coconut[a] (*Cocos nucifera*) | — | Cocosin | Concocosin | — |
| Sunflower[a] (*Helianthus annus*) | — | Helianthin | — | Mainly 11S[l] |
| Pumpkin[a] (Cucurbitaceae spp.) | — | Cucurbitin | — | Mainly 11S[m] |

[a] non-legume plants; [b] also known as haricot bean, common bean, dry bean, field bean, kidney bean; [c] also known as field bean, faba bean, horse bean; [d] also known as Bengal gram; [e] unless stated otherwise values taken from Mossé and Baudet (1983) and Mossé and Pernollet (1983), values in parentheses are typical values; [f] Bhatty (1988); [g] Nielsen (1984); [h] Croy and Gatehouse (1985); [i] Gueguen (1983); [j] Jayarama-Setty and Narasinga Rao (1973); [k] Ganeth-Kumar and Venkataraman (1980); [p] Schwenke *et al.* (1985); [m] Hara-Nishimura *et al.* (1985); [n] Chakraborty *et al.* (1979); [o] oilseed legumes; [p] all species are pulses unless stated otherwise.

(i)   accumulates in the seed in large amounts;
(ii)   occurs only in seeds;
(iii)   is hydrolysed to constituent amino acids during germination and early seedling growth;
(iv)   possesses high levels of nitrogen-rich amino acids, e.g. amides and arginine.

Legume seed storage proteins are commonly referred to as 'globulins' (after the famous classification of Osborne (1924)) because they are salt-soluble at neutral pH. This distinguishes them clearly from the major seed storage proteins of cereals such as wheat, barley and rye, which are insoluble in salt solutions and are termed 'prolamins' (Croy and Gatehouse 1985; Dieckert and Dieckert 1985). Legume seed globulins can be divided into two distinct classes, termed 7S and 11S on the basis of their sedimentation coefficients (Derbyshire *et al.* 1976; Casey *et al.* 1986; Wright 1987). Other seed proteins have also been accredited with a storage function, e.g. pea albumin (Murray 1979) and jackbean urease (Bailey and Boulter 1971), but for the purposes of simplicity the discussion herein will be centred on the major seed globulins. 7S- and 11S-like storage globulins are found in the seeds of many angiosperms, not just in those of the Leguminosae family (Wright 1987). However, in several non-legume seeds the globulins play only a minimal physiological role, for example, in wheat where the major storage proteins are prolamins. The ratio of 7S to 11S globulins can vary quite significantly within a given species and also across species, although in some cases one globulin predominates, e.g. French beans possess little 11S globulin. Table 7.1 gives reported ratios for many species in addition to the trivial names frequently ascribed to the globulins. The evolutionary significance of the botanical distribution of seed globulins has been discussed in depth elsewhere (Borroto and Dure 1987; Alexenko *et al.* 1988).

11S and 7S globulins are synthesized on membrane-bound ribosomes in the developing seed. The mature proteins are packaged into organelles called protein bodies, where they remain throughout seed development and dormancy until they are hydrolysed by proteases during germination (Croy and Gatehouse 1985; Pernollet 1985; Casey *et al.* 1986). The chemical and biophysical structure of the globulins is undoubtedly an evolutionary consequence of these basic physiological roles. For example, their large multimeric nature and low solubility at physiological pH is ideally suited for packaging and deposition. The following section describes more fully the structural properties of isolated seed globulins.

# Chemical and biophysical structure of isolated legume seed storage proteins

A vast literature, stretching back to the early 19th century, exists on this subject (see Wright 1987) although much of what we know now has arisen out of work published in the last 20 years. This recent work has focused largely upon soya, pea, faba bean and French bean, primarily because they represent the major legume crops of the world's most powerful nations. In addition to studies with legume seed proteins, data from many non-legume seed globulins, e.g. oats, coconut, rapeseed, etc. are highly relevant because of their close structural similarities. The general structures of 11S and 7S globulins will now be considered briefly in turn.

## 11S globulins

### Quaternary structure

Data concerning the overall molecular shape and dimensions are not clear cut, varying with the experimental approach, research laboratories and for different 11S molecules (Plietz and Damaschun 1986). Generally, however, they indicate that 11S proteins are slightly ellipsoid with a maximum dimension of 10–13 nm, and molecular dimensions of approximately $11.0 \times 11.0 \times 8.0$ nm. 11S globulins are typically hexameric oligomers, with a subunit mol. wt of approximately 50–60 000 depending on the botanical source (Plietz *et al.* 1987; Wright 1987). The subunit arrangement within the molecule has been investigated, principally by X-ray scattering and electron microscopy (Tulloch and Blagrove 1985; Plietz and Damaschun 1986; I'Anson *et al.* 1987) in the absence of any available detailed crystallographic structures. A favoured model is a trigonal antiprism arrangement, i.e. two trimeric rings superimposed and twisted about 60°. However, much uncertainty remains as to the true structure.

The subunit association–dissociation properties have physiological relevance in terms of how the globulins pack inside protein bodies (see p. 168) and also with respect to their behaviour during food processing (see p. 173). Under dissociating conditions 11S globulins generally break down into monomers via a half molecule, trimer intermediate (Wright 1987; Gueguen *et al.* 1988). The susceptibility of different 11S globulins to dissociation and association is highly dependent upon ionic strength, pH and protein concentration (Wright 1987). For example sesame 11S hexamers are > 75% dissociated by 0.5 M NaBr, whereas no dissociation of pea hexamers occurs under similar conditions (Chambers *et al.* 1990a). For some 11S

globulins denaturation occurs simultaneously with dissociation, in others it occurs *after* dissociation (Prakash and Nandi 1976; Suresh Chandra *et al.* 1985). Dissociation constants and free energies of dissociation were recently deduced for pea 11S globulin and for the approximate number of amino acids involved in the trimer–trimer contact area (Chambers *et al.* 1990a). Workers have also shown that it is possible, under defined conditions, to reassociate denatured subunits from different 11S species into 'pseudo-11S globulins'. (Hasegawa *et al.* 1981; Nakamura *et al.* 1985).

### Subunit structure

As mentioned above, no detailed crystal structures have been solved for any 11S globulin, and this is *the* major limitation to furthering our understanding of these proteins. The 11S subunits of any given species are highly homologous but are not identical (Croy and Gatehouse 1985; Casey *et al.* 1986; Wright 1987). The subunits are the products of a multigene family comprising about 10 genes; the precise number varying with the species. Whether individual 11S globulin molecules are composed of identical or different subunits is not known. Subunit homology is also high *across* species, as revealed by sequence data (see p. 170). The similarity in structure between subunits has presented problems in isolating truly homogeneous material (Bacon *et al.* 1987), another fundamental limitation in this research area.

The typical 11S subunit chemical structure is shown in Figure 7.1. Each subunit is synthesized as a single chain (mol. wt $\sim 55\,000$ plus signal sequence) and is then nicked specifically post-translationally to produce an interdisulphide bonded A-chain (mol. wt $\sim 35\,000$) and B-chain (mol. wt $\sim 20\,000$) (Croy and Gatehouse 1985; Pernollet 1985). For certain soya subunits the A-chain is further cleaved at another specific site (Momma *et al.* 1985; Kagawa and Hirano 1988). The A-chains are generally acidic and hydrophilic and the B-chains are generally basic and hydrophobic. It has been proposed that the basic polypeptides are buried within the 11S complex and that the acidic polypeptides are exposed on the surface. This was based not only upon their hydrophobic/hydrophilic character, but also from limited proteolysis data, which showed the acidic chains to be much more susceptible to hydrolysis. Recent studies have pinpointed the initial trypsin cleavage sites of pea 11S globulins to be the C-termini of the

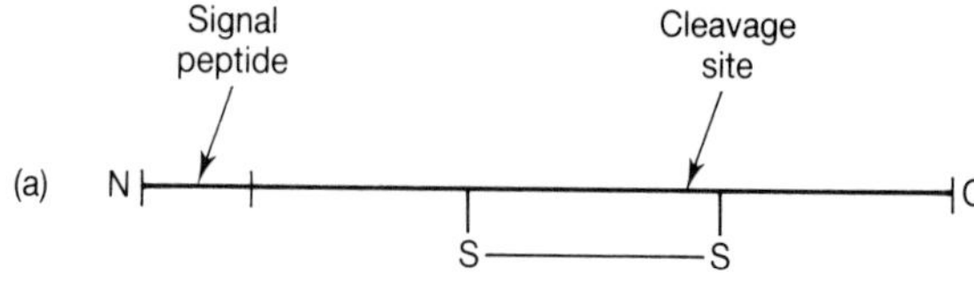

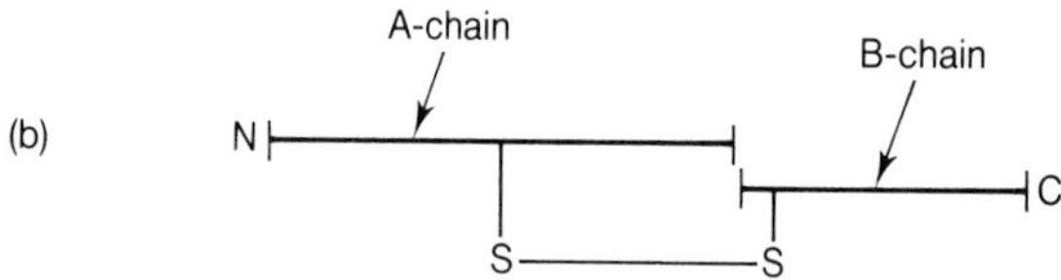

**Fig. 7.1**   Line structure of a typical 11S globulin subunit. (a), nascent subunit polypeptide (~55 000 mol. wt) with signal peptide and intradisulphide bond; (b), mature subunit after post-translational processing to remove signal peptide, and specific proteolytic nicking to generate an interdisulphide bond.

A-chains (Plumb *et al.* 1989). A panel of monoclonal antibodies is currently being utilized to epitope-map the surface of pea 11S; the data so far are consistent with the basic chains being buried (E. N. C. Mills personal communication).

### Secondary structure

Estimates of the proportions of α-helix, β-sheet, β-turns and random coil have been deduced from circular dichroism (CD) spectroscopy and computerized sequence analysis (Plietz *et al.* 1987; Wright 1987). Unfortunately both approaches have significant limitations. Secondary structure predictions from sequence data for any protein still have a level of accuracy of only 60–65%; and CD spectroscopy of 11S globulins gives different results depending on the method of data analysis used. Despite these limitations the data indicate that 11S proteins typically possess low levels of α-helix (10% or less) and approximately 40–50% each of β-sheet and random coil.

### Sequence data

Whole or partial sequences of many 11S globulin subunits have now been deduced (see Wright 1988 for a comprehensive catalogue). Much of this work arose during the 1980s, principally as a result of DNA sequencing of isolated cDNAs and genes. Such data has led to the following observations:

(1)  The existence of close homologies both within and across species.

(2)  Mol. wt determinations of A-chains by SDS-PAGE are frequently too high, by as much as 25% in some instances.

(3)  Sequence alignment reveals a series of alternating conserved and variable regions. Furthermore, two regions in the B-chain are similar to regions in the A-chain, perhaps lending support to the two-domain subunit structure theory.

(4)  The most conserved region is the A/B processing site, presumably reflecting protease specificity.

(5)  The presence of large glutamate/aspartate inserts (up to 12 consecutive residues in one subunit) at the C-terminal ends of some A-chains.

Such interpretations will clearly become more focused as more sequences emerge and other facets will undoubtedly appear.

## 7S globulins

### Quaternary structure

Such information for 7S globulins derives almost exclusively from proteins isolated from French bean (Blagrove *et al.* 1984), jackbean (McPherson 1980) and pea (I'Anson *et al.* 1988). Surprisingly little biophysical work has been performed with soya 7S. 7S molecules have overall dimensions of approximately $12.5 \times 12.5 \times 3.75$ nm and are trimeric in nature, being composed of three touching subunits in a planar arrangement. The subunit molecular weight is approximately 50 000. Evidence of two structural domains within each subunit has been reported for both jackbean 7S (from low resolution crystal structures) and pea 7S (from synchrotron X-ray scattering). The roughly spherical domains in the former are equal in size, in the latter they are unequal; however, by both methods the 7S presents a pseudohexagonal appearance. This hexagonal nature was confirmed recently for pea 7S using scanning tunnelling microscopy (Fig. 7.2); the first globular protein ever to be imaged by this exciting new technique (Welland *et al.* 1989).

In common with the 11S globulins, association–dissociation properties of 7S proteins are highly dependent upon protein concentration, ionic strength, pH and botanical source (Wright 1987). One form of soya 7S (β-conglycinin), for example,

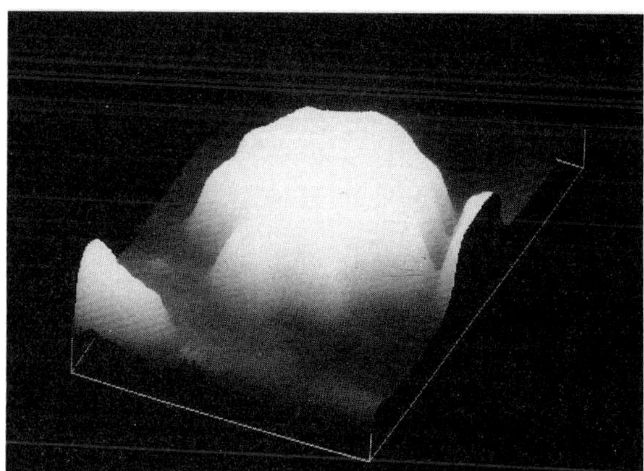

**Fig. 7.2**  Structure of vicilin. High magnification image of an individual vicilin molecule in air as revealed by scanning tunnelling microscopy. Molecular width is approximately 10 nm. (Taken from Welland *et al.* 1989.)

occurs in certain 7S subunits (e.g. those from pea, mungbean and broad bean), namely proteolytic processing at specific sites. In pea, the best studied example, 7S subunits are divided into three structural regions, $\alpha$, $\beta$ and $\gamma$, with some subunits possessing processing sites at the $\alpha/\beta$ and $\beta/\gamma$ junctions, some only one processing site, and others none at all (Gatehouse *et al.* 1982, 1984). The resulting fragments do not dissociate in aqueous media, but remain associated by non-covalent forces.

A complex spectrum of polypeptides is generated in the presence of dissociating agents such as SDS or urea, as revealed by the SDS-PAGE analysis of pea 7S shown in Figure 7.3. Size heterogeneity is

dimerizes at low ionic strength and neutral pH, whereas another form ($\gamma$-conglycinin) remains 7S. French bean 7S is relatively stable to changes of ionic strength, but forms a tetramer (18S) near its isoelectric point.

## Subunit structure

A low resolution computer representation of the polypeptide chain of jackbean 7S has been determined from X-ray crystallographic studies (McPherson 1980) but, as for 11S proteins, the paucity of detailed structural information severely limits our understanding. The subunits of a given 7S globulin are homologous but chemically distinct, deriving from a multigene family of about 10 members, depending upon the species. Subunits from soya 7S (Coates *et al.* 1985; Hirano *et al.* 1985) and French bean 7S (Paaren *et al.* 1987) have been the most extensively studied. The soya 7S fraction contains $\beta$-conglycinin (the major form), $\gamma$-conglycinin and other minor forms.

$\beta$-conglycinin is made up of *at least* four different types of subunit: $\alpha$, $\alpha'$, $\beta$ and $\beta'$, and at least seven distinct combinations of subunits have been recognized. For phaseolin (French bean 7S), three classes of subunit have been identified: $\alpha$, $\beta$ and $\gamma$ In addition to heterogeneity resulting from minor sequence differences, some 7S subunits are glycosylated to varying degrees, in marked contrast to 11S subunits (Wright 1987). Carbohydrate levels for *total 7S fractions* are usually 2–5%, although for pea the value is significantly less than 2%.

Another form of post-translational modification

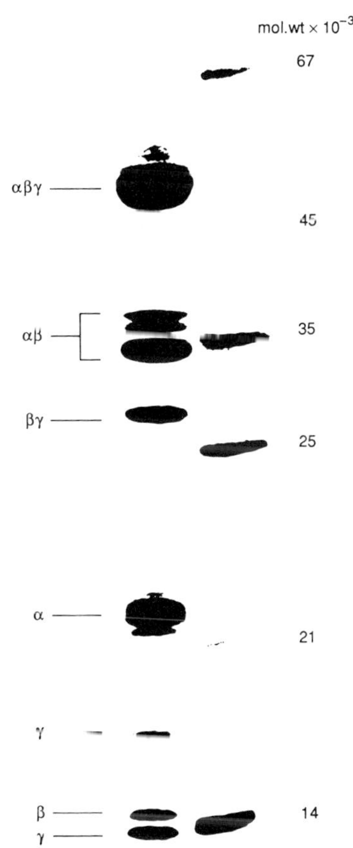


**Fig. 7.3**  SDS-PAGE of pea 7S globulin. The left-hand track reveals the complex banding pattern obtained for purified pea 7S globulin, resulting from specific post-translational proteolytic nicking. Right-hand track is mol. wt standards

increased further by glycosylation in the γ region of some subunits.

Another type of 7S globulin has been identified in pea and broad bean and has been termed convicilin (Croy *et al.* 1980: Casey *et al.* 1986; Wright 1988). This is a trimer of subunit mol. wt ~ 70 000; the higher mol. wt arising chiefly from an N-terminal extension (Bown *et al.* 1988). Both immunologically and by comparing available sequence data, concivilin units show close homology to the other 7S subunits.

### Secondary structure

As with 11S globulins, CD spectroscopy and structure predictions from sequence data suggest that 7S globulins have low contents of α-helix (<10%) and higher contents of β-sheet and random coil (Zirwer *et al.* 1985; I'Anson *et al.* 1988).

### Sequence data

Complete or partial sequences of many 7S subunits have now been reported, mainly as a result of sequencing cDNAs and genes. Such data have led to the following observations (Wright 1988):

(1) Close homologies exist within and across species.
(2) Glycosylation may influence molecular weight determinations.
(3) As with 11S subunits, sequence alignments reveal a series of alternating variable and conserved regions.
(4) Internal homology is present, which lends support to the two-domain theory. Comparison of 7S and 11S sequences reveals four homologous regions, which supports the hypothesis that these proteins are descended from a common ancestral storage globulin (Wright 1988).

# Current uses of legume seed proteins

Because the globulins are such vital and major components of legume seeds, the utilization of legume seed globulins is inextricably linked to the utilization of the seed as a whole. Totally refined seed globulins *per se* are not used in any commercial sense. Legume seeds are a major source of dietary protein for mankind, being consumed either directly, e.g. peas and beans, usually in a cooked form, or indirectly as a feed for livestock and poultry. In addition to their value as food, legumes are also important in cropping systems because of their ability to fix nitrogen, hence they increase the overall fertility of the soil and reduce the use of expensive nitrogen fertilizers.

Food legumes can be divided into oilseeds (peanut and soyabean) and grain legumes (or pulses). A digest of food legumes listed 27 different pulses grown worldwide (Kay 1979); the most significant of these are shown in Table 7.2. This gives data for recent world pulse and soya harvests and also the major world regions of production. Commercially, the world legume market continues to be dominated by the soyabean. In 1987 over 100 million metric tonnes (MMT) were grown worldwide, nearly *twice* the total production of all grain legumes! The USA is the major soya producer, their 1987 harvest alone being equivalent to the rest of the world pulse harvest. The attractions of soybeans are:

(i) 'cost', both in terms of production *and* marketable products;
(ii) they grow readily in the USA, a major economic power;
(iii) in addition to having a high protein content, they also contain substantial levels of oil, which is the major economic commodity.

Many developed countries unable to grow soyabeans are heavily dependent upon the US harvest, the EEC being the major importer with a staggering 12 MMT in 1987/88, representing approximately 45% of the world trade (Mackie 1988). Significantly, the EEC production of soyabeans has increased over the last 5 years, notably in Italy and in Spain; this trend is likely to continue into the next decade (1987/88 World Fats and Oils Report). Furthermore, indigenous crops are being grown as alternatives to soya, for example, rape and sunflower for edible oil and the legumes peas, lupins and beans for protein. Interestingly, although world soyabean production has increased sharply over the last 20 years (three-fold since 1965), the US percentage share continues to fall as other countries (most notably Brazil and Argentina) increase their production (Haumann 1988).

Unlike pulses, soyabeans are rarely consumed whole, but are fractionated into an edible oil and a protein-rich meal, most of which is used as an animal feed. The storage globulins form the main component of such animal feeds and are implicated

**Table 7.2**　World legume production

| Legume | World harvest (MMT)[a] | Major countries | Amounts (MMT) |
|---|---|---|---|
| Total pulses | 53.3[b] | India | 11.2 |
|  |  | USSR | 8.5 |
|  |  | 'Africa' | 6.5 |
|  |  | China | 5.8 |
|  |  | 'EEC' | 4.2 |
| Soyabean | 102.0 | USA | 51.2 |
|  |  | Brazil | 16.9 |
| Peanut | 20.1 | China | 6.1 |
|  |  | India | 4.5 |
| Pea | 14.5[c] | USSR | 6.8 |
|  |  | 'EEC' | 2.2 |
| French bean | 14.0[c] | India | 2.5 |
|  |  | Brazil | 2.0 |
| Chickpea | 6.9 | India | 4.4 |
| Broad bean | 4.5 | China | 2.3 |
| Lentil | 2.6 | Turkey | 0.9 |
|  |  | India | 0.7 |
| Cowpea | — | Nigeria | — |
| Mungbean | — | India | — |
|  |  | Thailand | — |
| Pigeon pea | — | India | — |

[a]Values are for 1987 and are expressed as million metric tonnes (MMT), and were obtained from the 1987 FAO production yearbook (FAO statistics series No. 82); [b]values are believed to be underestimated by up to 20% as in many Third World countries pulses are grown and consumed at a rural level and do not enter official statistics; [c]value is for the 'dry' legume only and not the 'green' form.

in key nutritional attributes, e.g. digestibility, amino acid balance, palatability and bioavailability of micronutrients (Waggle *et al.* 1989).

Another market for legume protein (almost solely derived from soya) that has arisen in the last 20 years is as a food ingredient in developed countries (Wright and Bumstead 1984; Uzzan 1988). It is used in this form either to increase protein content (a major selling factor) or to impart functionality and texture. Soya ingredients of this kind are currently available in three forms:

(1) *Soya protein flour* (at least 50% protein). Most manufacturing processes are 'wet' and involve solvent extraction of soya 'flakes' to remove the oil, followed by solvent removal and toasting to destroy residual antinutritional proteins. Flours produced from pulses are obtained by 'dry' mechanical processing, e.g. dehulling, milling and air-classification.

(2) *Soya protein concentrate* (at least 65% protein). This is the main vegetable protein produced to date on an industrial scale. The above material is purified further by aqueous alcohol or acid extraction. This material has improved colour, flavour and odour characteristics and, furthermore, sugars causing flatulence are removed. Concentrates can be obtained from pulse flours using similar methods or by further mechanical air-classification.

(3) *Soya protein isolate* (at least 90% protein). This is the purest form of vegetable protein available industrially. Intensive research has been performed over the last 15 years into improving the production and the nutritional and functional qualities of this material. It is manufactured by extracting the protein from flour using aqueous alkali followed by acid precipitation and drying. Similar regimes have proved effective for other legumes.

The storage globulins are again the major component of these soya ingredients and are responsible for conferring most of their nutritional and functional properties. The largest market for soya protein in human foods continues to be as a meat substitute (Uzzan 1988; Waggle *et al.* 1989) in products such as burgers, meat balls, sausages, patés, cooked pork meats, etc., where the chief advantages are reduced lipid content and increased protein levels, but above all reduction in cost—soya is substantially cheaper than meat. Burger products usually contain 20–30% soya protein in the form of a texturized isolate or concentrate suitably coloured and flavoured. In recent years totally meat-free products for vegetarians have become increasingly fashionable. Other markets include bread and bakery products (mainly for increasing protein levels), dietetic and health foods, pet foods and dairy analogues (e.g. soya milk, soya cheese, whipped toppings and ice-creams). Soya isolates are also used as whipping agents, emulsifiers and binding (lipid or water) agents in many general foodstuffs as a replacement for the more expensive animal-based protein ingredients from eggs and milk (Uzzan 1988; Waggle *et al.* 1989). In Oriental countries speciality foods are made from soyabeans (Haytowitz and Matthews 1989) such as the fermented products Miso, Natto and Tempeh, and Tofu which is formed by coagulating soya milk.

The expansion in soya protein utilization predicted in the mid-1970s did not occur although current signs are that the market is on the upturn. In 1986 there were approximately 10 firms in the USA manufacturing soya protein material and several thousand companies using their products. Table 7.3 gives production details for soya protein.

**Table 7.3**  Soya protein production[a]

| Soya protein | USA production (tonnes) | | EEC production |
| --- | --- | --- | --- |
| | 1980 | 1985 | 1985 |
| Flour | 261 000 | 150 000 | 35–45 000 |
| Concentrate | 35 000 | 50 000 | — |
| Isolate | 33 000 | 70 000 | 15 000 |
| Textured flour and concentrate | 41 000 | 75 000 | 25–30 000 |

[a] Values taken from Uzzan (1988).

The processing of pulses for human food ingredients has never 'taken off' in the way soya has. Several attempts in the 1970s and early 1980s by various small firms (in France and Canada, for example) met with little economic success. However, in 1988 a large Danish company launched fibre, starch and protein food ingredients obtained from the fractionation of peas using a novel ultra-filtration technique. The firm is currently processing 14 000 tonnes of peas per year. The success of this Danish venture (£16 million over 6 years) will no doubt determine for the forseeable future whether pulse processing is viable commercially (Anon 1988, 1989a; Bilidt 1990). Some considerations that will have a large bearing include:

(1) The US reaction to reduced exports of soya from the USA.
(2) The price of imported soya protein compared with home-grown legume protein.
(3) Competition from animal ingredients, especially caseinates and egg-white, which are supported by a strong political lobby.
(4) The unpredictability of the EEC Common Agricultural Policy.

A major factor in the whole issue of legume protein for human consumption is consumer appeal, which is difficult to assess, often contradictory and frequently voiced by minority groups. On the one hand there is a desire for food that is cheap, easy to prepare, wholesome and non-animal, all of which favour legume protein-based products. On the other hand vegetable protein, especially soya, has a poor image and there is a general suspicion of additives and ingredients.

# Approaches for improving the quality of legume seed storage proteins

The previous sections have outlined briefly the structure of these important proteins, and how they are used commercially. Clearly the ability to improve the quality of these (or any other) food proteins would be of enormous benefit to the food industry, agricultural community and, through consumers, the population as a whole. There are several characteristics of seed globulins that could be targeted for improvement, which can be separated broadly into two categories: 'nutrition' and 'functionality' (Table 7.4).

Before one can contemplate any *rational* strategy for improving various quality parameters one must have a cogent knowledge base at a molecular level of the processes/factors involved. Needless to say, much effort has gone into understanding the mechanisms behind the functions listed in Table 7.4 (Wright and Bumstead 1984) and ways for amelioration. The last 20 years have been especially fruitful for research work in this area, particularly with respect to soya. Major limitations to such work have been the heterogeneity of the seed globulins and the complexity of the systems in which they are used commercially. If we consider animal feeds, for which most of the nutritional considerations apply, biochemically crude flours are blended with other components and heat-treated to destroy antinutrional factors. Human food systems are even more complex: consider for example the structure and chemistry of a beefburger or a doughnut! The 'crude' soya protein isolate is but one of many ingredients, and frequently much 'processing' is performed with numerous biophysical consequences. Such a complexity is reflected in the multiplicity of approaches scientists have used to unravel the mechanisms involved (Cherry 1981; Blanshard and Mitchell 1988). As well as understanding the underlying biophysics (e.g. thermal denaturation for gelation, surface rheology for whipping) there is also a need to understand the properties of the various individual components in a commercial isolate, how they influence each other's behaviour and, in the case of food systems, how their properties are affected by other food components and manufacturing processes. It is also worth noting that isolates prepared in a research laboratory often have significantly different properties to those prepared commercially, due

**Table 7.4** Target areas for improving legume seed globulins

| Nutrition[1] | Food functionality[2] |
| --- | --- |
| Increased digestibility | Increased foaming/emulsifying power |
| Removal of intolerance factors | Increased solubility at pH 4–7 |
| Greater levels of methionine and trytophan | More varied gelling and texturizing properties |
| Increased palatability | Less heat stable |
| Decrease nutrient binding to improve 'bioavailability' | Increased flavour/water/fat binding |

[1] For general references covering these nutritional properties see Waggle *et al.* 1989; [2] for general references covering the improvement of these functional properties see Wright and Bumstead 1984; Jiminez-Flores and Richardson 1987.

simply to scale-up factors. Given the desire to enhance qualities such as those listed in Table 7.4, there are two general strategies one can adopt. First, one can modify the material *postharvest*. For example, limited proteolysis of legume globulins and isolates enhances foaming/emulsifying and solubility properties and thermal denaturation increases digestibility (Jimenez-Flores and Richardson 1987; Nielsen *et al.* 1988). However, there are two drawbacks to this approach. First, the introduction of another processing step increases costs, and secondly, current legislation restricts the types of modification one can perform. For example, various chemical modifications, such as acetylation and succinylation, are known to have beneficial effects on many of the functional qualities listed in Table 7.4 (Wright and Bumstead 1984) but are precluded from use. Similarly, the chemical grafting of polymethionine to improve amino acid balance is also prohibited.

A second, more satisfactory approach would be to 'engineer' the proteins at the *crop level*. By achieving the ideal combination and composition of seed globulins, improvements in quality can be achieved. Such engineering is possible at several levels. Breeding/selecting cultivars with high or low 7S/11S ratios will influence quality parameters (Murphy and Resurrection 1984). For example, soya 7S and 11S have distinctive gelling, emulsification and thermal properties Saio and Watanabe 1978; Wright 1987), and pea 'isolates' with a high 7S/11S ratio have improved emulsification properties (Dagorn-Scaviner *et al* 1986). Furthermore, 11S globulins typically contain higher levels of sulphur amino acids. It has also been shown that the 7S/11S ratio can, to some degree, be influenced by nutrient availability during growth (Gayler and

Sykes 1985). At the next level, specific gene products (i.e. globulin subunits) will possess good/bad attributes, for example isolated soya 11S subunits possess different gelling properties (Mori *et al.* 1982; Nakamura *et al.* 1985). The ability to select cultivars enhanced in 'quality' polypeptides would be very desirable, such an approach parallels that of determining which polypeptides in gluten govern good/bad breadmaking quality or are associated with coeliac disease. As discussed on pp. 167–8, the genetic diversity of seed globulins is quite substantial, and is not yet fully understood. A major obstacle to this approach is our current inability to isolate truly homogeneous globulin polypeptides to determine correlations with specific properties. FPLC and electrophoretic techniques have been used to assess subunit heterogeneity of seed globulins with a view to correlating profiles with quality (Lambert *et al.* 1987; Buehler *et al.* 1989).

Advances in recombinant DNA technology during the 1980s offer a possible solution to the above problem. One could facilitate the production of homogeneous seed globulins by preparing cDNAs encoding specific subunits and expressing them in a suitable microbial host, subunit assembly being achieved either *in vitro* or *in vivo*. Characterization of such recombinant material should enable the identification of 'quality gene products' for which specific probes (DNA or monoclonal antibodies) could be made for use in breeding trials, etc. Recombinant DNA technology also enables us to modify existing genes and design the 'ideal' seed globulin. Protein engineering as such is currently limited with respect to seed globulins due to the absence of detailed crystal structures, and an understanding at the sequence level of many of the properties targeted in Table 7.4. This technology is currently being used to unravel the molecular basis of foaming and emulsification of other food proteins (Bacon *et al.* 1988) and for a model protein (Kato and Yutani 1988); the *de novo* design and synthesis of peptides with outstanding interfacial properties has recently been patented (Anon 1989b). Detailed structural information of this kind will undoubtedly be forthcoming for seed proteins in the future. In the meantime, educated guesses and/or shotgun engineering are the only options available. Such 'designer seed globulins' could be employed for producing probes to screen the existing genetic diversity of legumes.

It is very unlikely that the production of recombinant seed globulins, engineered or otherwise, in fermenters/bioreactors, would ever compete economically with the existing low cost (approximately £4000/tonne) bulk production of isolates, etc. The ultimate goal must surely be to generate transgenic legumes with improved seed globulins. This could be achieved by introducing either totally foreign genes (e.g. globulins from other species) or modified endogenous globulin genes. However, there are several considerations that must be taken into account:

(1) *Tissue specificity*. Not only must the new genetic material be introduced into the legume, but its expression must be correctly controlled and targeted towards the seed.

(2) *Background expression*. With multigene families such as seed globulins, for an introduced 'designer gene' to have a significant effect it must either be overexpressed relative to the endogenous genes, or the native genes must be repressed. This may be less of a problem for legume seed storage proteins (approximately 10 genes/family) compared with those of cereals, where the gene copy number if much higher.

(3) *Physiological constraints*. Altering existing globulins or introducing foreign globulins could affect the efficiency of packing and processing (glycosylation, proteolysis) of these storage proteins, functions necessary for seed viability. Similarly, changing the amino acid composition may adversely affect the digestibility and supply of nitrogen to the seedling.

(4) *Legislation*. The concept of farmers growing transformed crops raises ethical questions and is not currently favoured by many sectors of the UK population. It would undoubtedly lead to tight regulatory control from the Government of the day.

## Characterization of the genes encoding the major legume seed storage proteins

As described on pages 169–72, legume storage proteins are highly heterogeneous, resulting from the presence of multigene families and variable degrees of post-translational modification. This section summarizes briefly the classical and molecular genetics of the 11S and 7S globulins. Such information is obviously central to the task of producing transgenic legumes possessing stylized storage globulin contents. (For more detailed accounts see Croy and Gatehouse 1985; Casey *et al.* 1986.)

## 11S genes

Five soyabean subunit types have been identified and the genes sequenced (Moreira *et al.* 1979, 1981; Staswick *et al.* 1981; Nielsen *et al.* 1989). Genetic analysis has located these five genes at four different loci within the genome (Cho *et al.* 1989). *Gy1* and *Gy2* occur in tandem repeat at one locus, while single glycinin genes are present at each of three other loci. *Gy4* is characterized by two alleles, one of which causes absence of the subunit. Pea 11S globulins fall into two broad classes, the more abundant 'major' legumins and the less abundant 'minor' legumins. There are four to five major *LegA* genes mapping close to the *r* locus on chromosome 7, while a similar number of genes encoding minor legumins map to chromosome 1 (Domoney *et al.* 1986). The transcriptional units of these 11S subunits all conform to a general pattern. Important regulatory TATA, AGGA, and CAAT box sequences are located near the 5′ end of the sequence. Adjacent to the AGGA/CAAT boxes is a highly conserved region of about 30 bp denoted the 'legumin box' (Gatehouse *et al.* 1986; Baumlein *et al.* 1986), while a number of poly(A) addition sites are present at the 3′ end of the sequence. The *LegJ* family is somewhat exceptional in that, although possessing a legumin box, it does not have the AGGA/CAAT sequences. *LegA* and glycinin-type genes have three small introns located within the coding sequence; the *LegJ* types have only two introns.

## 7S genes

In soya the three subunit types of β-conglycinin are encoded by a multigene family of at least 15 genes located in six regions of the DNA. Ten genes are clustered in genomic regions designated A, B and C; D, E and F contain one gene each (Harada *et al.* 1989). Region A encompasses a 44 kb domain containing five genes separated by 1.5–4.0 kb; B and C contain three genes. The gene clusters are thought to be present on at least two chromosomes. Of the genes identified, three are α/α′-subunit genes, six are β-subunit genes and four show homology to both α/α′- and β-subunits. Seven genes, encoding three polypeptide families, have been identified in French bean (Talbot *et al.* 1984). These can be divided into two classes, α and β, the α being derived from the β class by the insertion of direct repeat sequences encoding 14 amino acids. Six different vicilin loci have been identified in pea, three of which are located close to the *r* locus on chromosome 7 (Ellis *et al.* 1986). Typically, 7S subunit transcription units are 2.0–2.5 kb in length and have five introns, which are significantly richer in A/T than the exons. 5′ to two adjacent ATG codons are three TATA sequences and a single CCAT sequence; the 3′ flanking sequence has three copies of AATAA sequence.

Although little is known about the regulation of either 7S or 11S subunit genes, studies with β-conglycinin genes, amongst others, suggest that each gene within a cluster is programmed by its own developmental control element (Harada *et al.* 1989). Many of the above genes have been cloned, sequenced and transformed into plant and microbial hosts, (see pp. 177–82). Since the latter fail to splice plant introns with any degree of fidelity, an intact cDNA is required in these systems. In addition, expression in transgenic plants requires both 5′ and 3′ flanking regulatory sequences derived from either the donor species or from another species, for example the cauliflower mosaic virus 35S CaMV viral sequence promoter and terminator.

# Expression of legume storage protein genes

As highlighted on pages 174–6, there are two distinct goals for pursuing such research. First, the production of reasonable amounts of homogeneous material via heterologous expression for correlating structure with 'quality' parameters. Secondly, the production of transgenic legumes with idealized seed globulin composition. These two systems will now be considered in turn.

## Expression in non-plants

Yeast (*Saccharomyces cerevisiae*) and *E. coli* have been the most widely adopted hosts for heterologous expression of seed globulins and recombinant proteins in general (Marston and King 1987). The genetics of these organisms are well advanced and the conditions necessary to effect transformation and maximize expression have been clearly

established (Buell and Panayotatos 1986; Kingsman *et al.* 1988). Introduction of foreign genetic material is accomplished via self-replicating host plasmids with appropriate control elements; promoters, terminators, etc.

## *S. cerevisiae*

Yeast offers several advantages for expressing seed globulin genes. Being eukaryotic, the mechanisms of transcription, translation, post-translational processing, targeting and secretion are similar to the host tissue. Disadvantages are that large proteins (>30 000 mol. wt) are not normally secreted, and that expression levels are often low. Using the shuttle vector pMA 91 (Mellor *et al.* 1983), which is derived from the 2 μm yeast plasmid, cDNAs encoding pea seed globulins have been introduced into yeast (Yarwood *et al.* 1987). pMA 91 vectors use promoter and terminator sequences of the highly expressed yeast glycolytic enzyme phosphoglycerate kinase (PGK). The presence or absence of signal sequences was found to have a profound effect upon seed globulin expression. (Fig. 7.4). Expression of two 11S cDNAs (*LegA* and *LegJ*) and one 7S cDNA, all possessing their own signal sequence resulted in the recombinant proteins being sequestered within the endoplasmic reticu-

**Fig 7.4**   Expression of cDNAs encoding *P. sativum* storage proteins in *S. cerevisiae*. Western blot of proteins expressed *in vitro* in *S. cerevisiae*. Lane (1), legumin standard; (2), vicilin standard; (3), supernatant proteins from untransformed yeast; (4), PGK-vicilin fusion protein; (5), vicilin lacking N-terminal signal sequence; (6), 60-amino acid deletion mutant of vicilin; (7), logumin; (8), legumin lacking N terminal signal sequence. Sample preparation, PAGE and western blotting were as detailed in Yarwood *et al.* 1987. The substrate was 3,3′-diaminobenzidine tetrahydrochloride.

lum (e.r.) and Golgi apparatus as insoluble aggregates. The levels of expression in all cases were relatively high, > 1% of total yeast protein. With *LegA* the signal sequence appeared to have been cleaved, but with *LegJ* a doublet was consistently found on Western blotting, indicative of incomplete processing. Both legumin cDNAs were expressed as 60 000 mol. wt polypeptides with no visible processing into A- and B-chains (see pp. 169–70). The vicilin cDNA did not have its signal peptide removed, nor was it glycosylated despite the presence of a single glycosylation site (Stewart 1989). Although informative, the above systems are not particularly suited to the objective of correlating structure to function, a process requiring reasonable amounts of soluble native-like protein. By expressing pea 7S and 11S cDNAs devoid of functional signal sequences the recombinant proteins were targeted to the cytosol where they were recoverable in a soluble form (Watson *et al.* 1988; Chambers *et al.* 1990b) accounting for up to 5% of soluble protein. Using the expression vector pMA 257 (Tuite *et al.* 1982) a fusion protein with a 22 amino acid extension was prepared, three residues being derived from the vicilin signal sequence and the rest from the PGK structural gene. The recombinant protein was recovered from the cytosol in a homogeneous form by a rapid purification procedure using hydroxyapatite chromatography with yields greater than 95% (Fig. 7.5). Little or no product was observed in the particulate fraction. Typically, 10 mg of pure protein was recovered from 1 litre of minimal medium.

CD analysis of the intact fusion protein revealed that the polypeptide was not correctly folded compared to native vicilin (Fig. 7.6) and gel filtration studies showed that it was not assembled into trimers but existed as a mixture of very large aggregates, with much of the material eluting in the excluded fraction. The possibility that the N-terminal extension was responsible for the non-native structure was eliminated when identical behaviour was observed for vicilin lacking this modification. Differential scanning calorimetry (DSC) analysis of this recombinant protein demonstrated that ordered structure was present, but considerably less compared to native vicilin and it also had a markedly lower denaturation temperature (G.W. Plumb and N. Lambert unpublished observations). Attempts were made to renature and reassemble the soluble aggregate from solutions of 6 M urea or 5 M guanidinium hydrochloride by dialysis but large soluble aggregates always re-

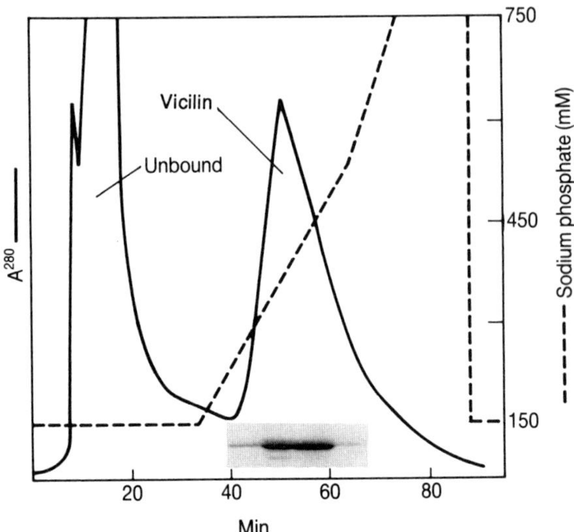


**Fig. 7.5**   Purification of recombinant vicilin from yeast. Yeast extract is passed down a hydroxyapatite-Ultrogel column under conditions that bind the recombinant protein only. Insert is SDS-PAGE profile of the eluted vicilin taken across the peak (taken from Watson *et al.* 1988).

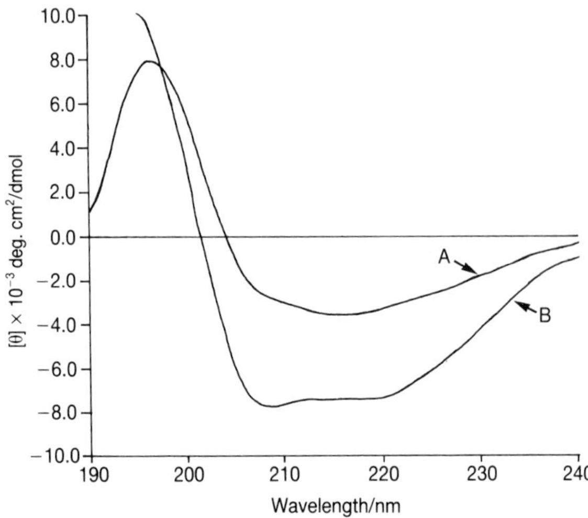


**Fig. 7.6**   CD spectra of native pea vicilin and recombinant yeast vicilin. A, Native pea vicilin. B, Recombinant yeast vicilin.

sulted, even when using methods that were successful in reassembling denatured analogous subunits from French bean (Chambers *et al.* 1990b).

The inability to convert the recombinant vicilins to a native form severely restricts their value for structure–function correlations. Despite these limitations, various mutant cDNAs were constructed by site-directed mutagenesis using the cDNA encoding the PGK-vicilin fusion protein as a template. Surface hydrophobicity/hydrophilicity is a significant structural factor in determining food functionality, thus a deletion mutant was constructed from which a 60 residue α-helical hydrophilic motif at the natural surface α/β processing site was removed (see p. 171). Expression of this cDNA in yeast resulted in a smaller subunit, again recoverable as large soluble aggregates from the cytosol at a level of expression slightly lower than the wild type. Interestingly, a second, less abundant species was observed in the particulate membrane fraction, with a slightly lower (2 kDa) molecular weight (see Fig. 7.4). These recombinant products were much less antigenic than the wild type subunits described above, indicating perhaps that the surface hydrophilic element contains a major epitope (G.W. Plumb, J.N. Yarwood and N. Lambert unpublished data).

Thiol–disulphide interactions have long been associated with food functionality, particularly gelation (Schmidt 1981); furthermore, sulphur amino acids are nutritionally limiting in legume proteins (see pp. 174–5). Native vicilin does not possess any such amino acids, hence the introduction of sulphur-containing residues by protein engineering provides a unique opportunity to assess their role in food protein functionality. Three mutant cDNAs were constructed, in which two serine residues believed to be on the surface were replaced by cysteines, as single and double mutants. The single mutants were expressed in a similar manner to the wild type cDNA, being characterized by large soluble aggregates and expression levels of > 5%. No evidence of interchain disulphide bonds was found, and turbidity and gelation measurements are currently in progress. In the case of the double mutant, 'Western blots' of crude yeast extracts maintained for up to 7 days at 4°C in the absence of protease inhibitors or reducing agents showed no sign of proteolytic cleavage or irreversible aggregation. However, the characteristic formation of aggregates occurred upon purification and there was also some evidence of disulphide bond formation (see Fig. 7.7). This material is also undergoing further analysis. Table 7.5 lists all the vicilin constructs that have been prepared with this yeast expression system (G.W. Plumb, J.N. Yarwood and N. Lambert unpublished data). This study clearly demonstrates the feasibility of producing designer seed globulins, although the current inability to convert the recom-

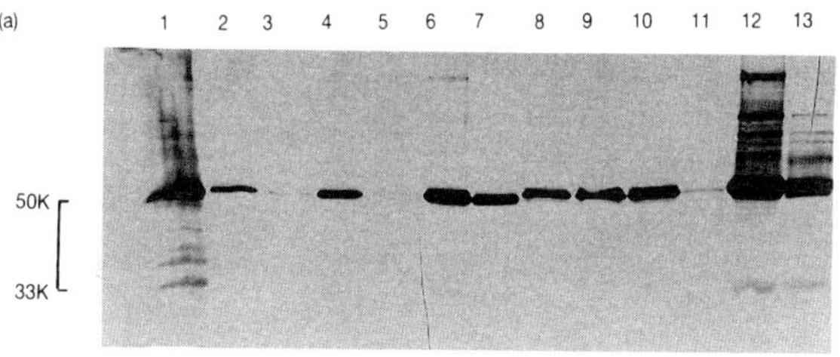

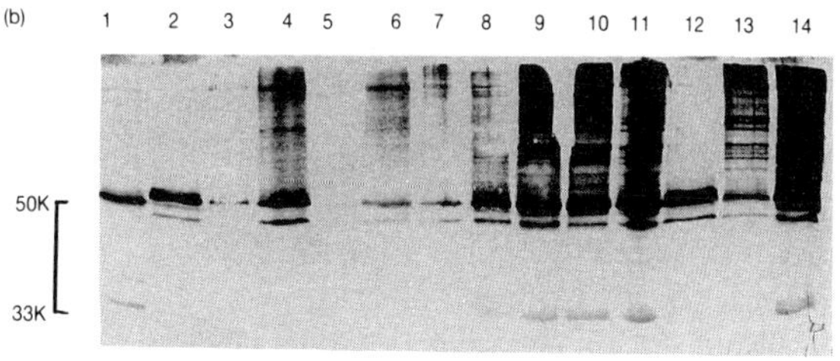

**Fig. 7.7**  Western blots showing comparative chromatographic behaviour of vicilin lacking N-terminal signal sequence and serine→cysteine vicilin mutants during purification from *S. cerevisiae*. (a) lane 1, vicilin standard; lanes 2–7, vicilin lacking N-terminal signal sequence: 2, yeast-soluble supernatant; 3, 45–95% saturation (NH₄)₂SO₄ pellet; 4, peak eluting from Sephadex G25; 5, unbound fraction from HA-Ultrogel column; 6, bound fraction from HA-Ultrogel column; 7, lyophilized HA-Ultrogel bound fraction; lanes 8–13, single serine→cysteine mutant: 8, yeast-soluble supernatant; 9, 45–95% saturation (NH₄)₂SO₄ pellet; 10, peak eluting from Sephadex G25; 11, HA-Ultrogel unbound fraction; 12, bound fraction from HA-Ultrogel; 13, material precipitating after HA-Ultrogel column. (b) Double serine→cysteine mutant vicilin. Lane 1, vicilin standard; 2, yeast-soluble supernatant; 3, 45–95% saturation (NH₄)₂SO₄ pellet; 4, peak eluting from G25; 5, HA-Ultrogel unbound fraction; 6, HA-Ultrogel bound fraction; 7, HA-Ultrogel peak after dialysis against 50 mM borate pH 8; 8, dialysed and concentrated HA-Ultrogel bound fraction; 9, dialysed concentrated, lyophilized HA-Ultrogel bound fraction; 10, insoluble precipitate after storage of bound, concentrated HA-Ultrogel material; 11, supernatant after storage of bound, concentrated, HA-Ultrogel eluant; 12, yeast soluble supernatant after prolonged storage at −80°C; 13 and 14, material eluting from FPLC superose 6 column. All samples were prepared as described in Yarwood *et al.* 1987 and PAGE and Western blotting was carried out in the absence of β-mercaptoethanol.

binant polypeptides into native conformations limits their use.

Other groups have used different systems for expressing seed globulins in yeast. For example, Cramer *et al.* (1985, 1987) successfully expressed a phaseolin gene in yeast using a repressible acid phosphatase (PH05) promoter in a multicopy expression plasmid pYE (Schaber *et al.* 1986). Not only was the signal sequence cleaved but the polypeptide was also glycosylated. The recom-

**Table 7.5**    Engineered vicilin subunits produced in yeast

| Recombinant vicilin subunit | Level of expression[a] (%) | Molecular weight[b] (kDa) | Immunocross-reactivity[c] (%) |
|---|---|---|---|
| 1. Wild type | 5.5 | 51 | 4 |
| 2. 1. with 22 residue N-terminal extension | 6.5 | 53 | 4 |
| 3. 2. with Ser4→Cys | 7.2 | 53 | 4 |
| 4. 2. with Ser84→Cys | 7.5 | 53 | 4 |
| 5. 2. with Ser4→Cys *and* Ser84→Cys | 6.0 | 52 | 4 |
| 6. 2. with 60 residue hydrophilic deletion | 4.0 | 47 ⎤[d]<br>45 ⎦ | 0.1 |

[a] Values expressed as percent of yeast soluble protein; [b] values determined from SDS-PAGE /Western blots; [c] values are relative to purified pea vicilin; [d] the 45 kDa species was much less abundant and resided in the membrane fraction, all the other species were located in the cytosol.

binant protein was recovered from the soluble fraction where it accounted for up to 3% of the total protein. No further characterization of the phaseolin (e.g. the aggregration state) was reported. Replacement of the phaseolin signal sequence with the PH05 signal or hybrid PH05/phaseolin signal sequences resulted in incomplete processing. Utsumi *et al.* (1988b) expressed preproglycinin in an analogous yeast system and also obtained efficient processing of the signal sequence, but the polypeptide was not glycosylated, despite the presence of a single glycosylation site. Expression levels were only 0.3% of total soluble protein.

## *E. coli*

Attempts to express pea seed globulin cDNAs in *E. coli* resulted in large insoluble aggregates (Delauney 1984) visible as inclusion bodies (Marston 1986). Recently, however, soya 11S subunits have been expressed successfully in *E. coli* using the Pharmacia expression vector pKK 233–2 (Fukazawa *et al.* 1987; Utsumi *et al.* 1987, 1988a). In the most detailed study, Utsumi *et al.* (1988a) expressed a cDNA encoding the glycinin A1aB1b-subunit precursor. It was shown that the level of expression was markedly dependent upon the amount of signal sequence or mature N-terminal that was retained. Full length cDNAs failed to accumulate any recombinant product, whereas a construct minus the entire signal peptide and the first three residues of the mature protein resulted in highest level of expression; generating 1.4–2.1% of total *E. coli* protein. It was further shown that the material had assembled into trimers of about 8S, although the fidelity of disulphide bond formation was not investigated. This work clearly demonstrates the potential of *E. coli* for producing recombinant seed

globulins, although it would not be feasible to generate glycosylated material in this host.

## Others

Attempts were made to utilize *Aspergillus nidulans* for expressing a genomic clone of a *LegB* type gene. (Schiemann *et al.* 1988). Although RNA transcripts were demonstrated in both mycelium and spores, no expressed legumin was detected, suggesting that, as in *S. cerevisiae*, the fungal system failed to splice plant introns. The potential of *Aspergillus* as an expression host for cDNAs is well recognized, being capable of secreting several g/litre of various proteins (Cullen *et al.* 1987), and is worthy of further attention with regard to seed globulin production, as is the yeast *Pichia pastoris*, which has also been used successfully to produce large amounts of recombinant material (Digan *et al.* 1989).

A novel approach was adopted by Bustos *et al.* (1988), who expressed a phaseolin cDNA in insect cells using a baculovirus expression vector. This system uses a strong viral polyhedrin promoter to drive expression and yielded high levels of protein (1–500 mg/l), which was efficiently glycosylated, processed and secreted. Such a system should obviously be subject to further investigation.

## Expression in plants

Most transgenic plant studies have used members of the Solanaceae family (Horsch *et al.* 1984)—tobacco (*Nicotiana* sp.) and petunia (*Petunia* sp.)—as hosts because of their relative ease of transformation and regeneration. The vectors most commonly used for introducing foreign DNA into these plants have been the disarmed Ti plasmids of *Agrobacter-*

*ium tumefaciens* and related binary vectors, e.g. Bin 19 (Bevan 1984). Early experiments demonstrated that individual 7S and 11S genes could be effectively introduced into tobacco and petunia. Such studies have, for example, been reported for genes encoding the β-subunit of phaseolin (Greenwood and Chrispeels 1985; Sengupta-Gopalan *et al.* 1985), the α′-subunit of β-conglycinin (Beachy *et al.* 1985) and the *LegA* subunit of pea 11S (Ellis *et al.* 1988). Approximately 1–2 kb of 5′ and 3′ non-coding flanking sequence is also required for precise temporal and spatial expression. The resultant proteins were correctly glycosylated in the case of the 7S species, the 11S proteins correctly processed into the A- and B-chains, and all were assembled into the correct multimeric species. Although the proteins were deposited in protein bodies the levels of expression were very low, in the range 0.1–1.0% of the total seed protein.

Hoffman *et al.* (1988) recently increased the methionine content of a β-phaseolin gene by *in vitro* mutagenesis and introduced this into tobacco. In common with the above studies correct translation and processing occurred but the protein was sequestered within the e.r. and Golgi vesicles. The level of expression was exceptionally low—0.2% of that observed with the unmodified gene—a fact attributed to selective degradation within the Golgi.

Transformed plants are morphologically normal and set viable seed in which the introduced genes segregate as a single Mendelian dominant. However, when transformation involves the simultaneous introduction of a single selectable marker (usually kanamycin resistance) accompanied by a number of non-selectable genes, such as storage protein genes, the resultant transgenic plants show varying combinations of the latter. This failure to express non-selected genes may be due to a number of factors (Ellis *et al.* 1988) but highlights the type of problem that may arise in trying to modify subunit combinations when several genes must be introduced into a host species.

The disappointing levels of expression have focused much effort on identifying and understanding the regulatory and control elements involved in plant gene expression in general, not only for seed globulins, although these provide a suitable model system. Integration of a gene and its associated 5′ and 3′ flanking sequences into a plant genome as described above is a random process and different transgenic lines will differ in both gene copy number and insertional location. These vari-

ables cannot currently be controlled because targeting genetic information to a precise genomic location by homologous recombination has not yet been reproducibly achieved for non-selectable markers in plants. The variability of gene expression due to 'position effect' has been variously reported as between 25–50-fold with the wheat gene Cαβ-1 (Nagy *et al.* 1986). However, Chen *et al.* (1989), studying 20 different transgenic plants with a single copy of the α′-subunit of β-conglycinin, found only a two-fold plant to plant variation in levels of expression. Linking α′ and β-subunit genes on a single fragment again resulted in only small differences in expression between plants. Obviously the level of variability will depend upon such factors as the nature of the genes, the promoter used and the level of selection applied to the transformants.

A further confusion with respect to position effect comes from Naito *et al.* (1988), who transformed petunia with a disarmed Ti plasmid carrying both α and β conglycinin genes. They demonstrated that when the two genes and their promoters were placed in divergent orientation (← →) the expression of the β-gene was enhanced several fold, whilst that of the α-gene was decreased. It was suggested that sequences of the α′-gene enhanced the level of expression of a nearby β-gene, but that when the genes were in direct orientation (→ →) the distance between α′- and β-promoters excluded such an enhancement. It is factors such as this that will eventually need to be considered when producing transgenic plants expressing acceptable levels of recombinant proteins.

The relative contribution of copy number to variation in levels of expression remains controversial and must to some extent be considered in conjunction with position effects. However, using two independent *LegA* constructs Shirsat *et al.* (1989a) concluded that variability of expression was a function of position effect and independent of gene copy number.

Considerable progress has been made in identifying 5′ regulatory sequences using sequence comparison of conserved elements, promoter deletions and nuclear binding assays. In addition to the TATA and CAAT sequences around the transcription start it is now known that both temporal and spatial regulation reside in specific conserved sequences lying within a few hundred base pairs of the transcription start. However, it remains a naïve assumption that such elements act independently to regulate transcription and a combined and cooperative interaction of both 'cis' and 'trans' acting factors is almost certainly involved in bringing about the structural chromatin changes involved in gene regulation. Several development-specific seed protein gene transcriptional control elements have been identified by promoter-deletion experiments.

In the 11S type genes from pea, soya, broad bean and sunflower, an approximately 28 bp consensus sequence, the 'legumin box' has been located within 250 bp of the transcription start (Gatehouse *et al.* 1986; Baumlein *et al.* 1986). Analysing a series of promoter deletions of the *LegA* gene expressed in tobacco, Shirsat *et al.* (1989b) found that a construct containing only 97 bp of 5' regulatory sequence, which included the TATA, CAAT boxes and the terminal 12 bp of the legumin box, failed to express, while a deletion with −549 bp of 5' flanking sequence, which included the entire legumin box, yielded a low level of expression. The quantitative increase in expression obtained when −1203 bp of sequence were included suggests that an enhancer-like element positioned upstream of the legumin box is required to upregulate levels of expression (Shirsat *et al.* 1989b). Using a DNA binding assay Meakin and Shirsat (1990 unpublished) failed to demonstrate the binding of nuclear proteins to the legumin box. It appears, therefore, that although such a sequence may be involved in the regulation of the legumin system, it does not activate transcription in isolation from the remaining 5' regulatory elements. Similar sequences have been identified in 7S genes, e.g. the 'vicilin box' in 5' region of pea, soya and French bean (Gatehouse *et al.* 1986).

Other putative control elements have been identified in the 5' region of β-conglycinin genes (Okamuro and Goldberg 1989). It seems likely that such elements act in a combinatorial way to generate the diverse array of developmentally and spatially specific patterns of genes expressed. Indeed Okamuro and Goldberg (1989) calculate that a set of 20 control elements taken three at a time can generate more than 1000 unique combinations of control elements, thereby generating unlimited developmental, temporal, cell and tissue specific patterns of gene expression.

To date no plant DNA binding proteins have been isolated but a number of workers have demonstrated specific binding of embryo nuclear protein extracts to precise 5' regions of storage globulin genes (Jofuku *et al.* 1987; Bustos *et al.* 1989; Jordano *et al.* 1989). What does seem clear is that the DNA binding and regulatory function of these proteins may represent separable domains

and that the regulation of gene activity by these proteins involves not just DNA binding but cooperative interaction with other transcription factors.

The major hurdle to engineering legume seed globulins remains our inability to reproducibly transform and regenerate legume species. Forage legumes, e.g. *Lotus corniculatus*, have proved more amenable to *Agrobacterium* transformation and subsequent regeneration (Forde *et al.* 1989) than grain/oilseed legumes. The limiting factor is the ease with which fully fertile plants can be regenerated from legumes (Mroginski and Kartha 1984). Reports of transgenic soyabean plants (Widholm 1988) produced by both *Agrobacterium* and 'particle gun' methods suggest that most legume species will be amenable to transformation and regeneration in the next few years.

## Future perspectives

It is clear from the previous sections that much progress has been made, especially during the 1980s, towards the goal of designer legume seeds with idealized globulin composition. What are the challenges immediately facing scientists in this area?

(1) Detailed three-dimensional structures of seed globulins are essential for any cogent understanding of how their functionality/quality attributes are manifested. Such molecular information is vital to any programme aimed at producing 'designer seed globulins'. The close sequence homologies between these proteins probably means that only one or two structures would have to be determined empirically, others could then be modelled using sophisticated molecular graphics techniques.

(2) A system for expressing large quantities (> 100 mg) of native and mutagenized seed globulins is essential for testing structure–function predictions; this work is allied to that described above. Furthermore, the ability to produce these levels of totally homogeneous material may in fact provide a means of generating protein crystals of a quality suitable for X-ray diffraction studies. The existing systems based on *E. coli* and baculovirus show much potential.

(3) The availability of a characterized panel of monoclonal antibodies to seed globulins would facilitate epitope mapping studies in support of

structural evaluations, and provide a basis for the rapid screening of genetic variability in breeding programmes. Furthermore, they could be employed as functional probes for transgenic plants.

(4) Legume transformation is still very much in its infancy and the current inability to regenerate transformed plants prevents the production of transgenic legumes with improved globulin composition. Associated with this is our present inability to target genes to their correct location via homologous recombination. The selective deletion of undesirable characteristics/subunits similarly awaits the development of transformation, regeneration and efficient targeting protocols but could prove possible either by selective mutagenesis to silence genes or by the insertion of genes encoding antisense RNA. This has already proved effective in partially inactivating polygalacturonidase activity in tomato development (Morris *et al.* 1989).

(5) Finally the possibility of using legumes as 'green factories' to synthesize pharmaceutical and commercially valuable proteins is now a tangible goal, as was recently demonstrated by the production of Leu-enkephalin as a removable insert from the 2S albumin of *Arabidopsis thaliana* (Vandekerchkhove *et al.* 1989).

Couple these scientific opportunities with a growing political awareness of the importance of land ecology, 'wholesome' foodstuffs and the significance of developing countries and one perceives that legumes in general, and their seed globulins in particular, will be a major focus of interest in the coming decade.

# References


Alexenko, A.Y., Nikolaev, I.V. and Vinetski, Y.P. (1988). Soyabean 11S globulin polypeptides have an antigenic homology with 11S globulins from various plants. *Theoretical and Applied Genetics.* **76:** 143–7.

Anon (1988). Food ingredients on show. *Chemistry and Industry.* **October:** 737.

Anon (1989a). Three ingredients in a pod. *Food F.I.P.P.* **11:** 43–5.

Anon (1989b). Designer ingredients for the foods of the future. *Food Manufacture,* **April:** 25–30.

Bacon, J.R., Lambert, N., Phalp, M., Plumb, G.W. and Wright, D.J. (1987). Resolution of pea legumin subunits by high-performance liquid chromatography. *Analytical Biochemistry.* **160:** 202–10.

Bacon, J.R., Hemmant, J.W., Lambert, N., Moore, R. and Wright, D.J. (1988). Characterisation of the foaming properties of lysozymes and α-lactalbumins: a structural evaluation. *Food Hydrocolloids.* **2:** 225–45.

Bailey, C.J. and Boulter, D. (1971). Urease, a typical seed protein of the Leguminosae. In *Chemotaxonomy of the Leguminosae,* (ed. J.B. Harborne, D. Boulter and B.L. Turner), pp. 485–502. Academic Press, London.

Baumlein, H., Wobus, V., Pustell, J. and Kafatos, F.C. (1986). The legume gene family: structure of a β-type gene of Vicia Faba and a possible legumin gene specific regulatory element. *Nucleic Acids Research.* **14:** 2707–20.

Beachy, R.N., Chen, Z.L., Horsch, R.B., Rogers, S.G., Hoffman, N.J. and Fraley, R.T. (1985). Accumulation and assembly of soyabean β-conglycinin in seeds of transformed petunia plants. *EMBO Journal.* **4:** 3047–53.

Bevan, M. (1984). Binary agrobacterium vectors for plant transformation. *Nucleic Acids Research.* **12:** 8711–21.

Bhatty, R.S. (1988). Composition and quality of lentil, a review. *Canadian Institute of Food Science and Technology Journal.* **21:** 144–60.

Bilidt, H. (1990). Functional ingredients from the yellow field pea. In *Food Technology International, Europe,* (ed. A. Furner), pp. 219–22.

Blagrove, R.J., Lilley, G.G., Van Donkelaar, A., Sun, S.M. and Hall, T.C. (1984). Structural studies of a French bean storage protein: phaseolin. *International Journal of Biological Macromolecules.* **6:** 137–41.

Blanshard, J.M.V. and Mitchell, J.R. (1988). *Food Structure—Its Creation and Evaluation.* Butterworths, London.

Borroto, K. and Dure, L. (1987). The globulin seed storage proteins of flowering plants are derived from two ancestral genes. *Plant Molecular Biology.* **8:** 113–31.

Bown, D., Ellis, T.H.N. and Gatehouse, J.A. (1988). The sequence of a gene encoding convicilin from pea (*Pisum sativum* L.) shows that convicilin differs from vicilin by an insertion near the N-terminus. *Biochemical Journal.* **251:** 717–26.

Buehler, R.E., McDonald, M.B., Van Toai, T.T. and St. Martin, S.K. (1989). Soybean cultivar identification using high performance liquid chromatography of seed proteins. *Crop Sciences.* **29:** 32–7.

Buell, G. and Panayotatos, N. (1986). Mechanism and practice. In *Maximizing Gene Expression,* (ed. W. Reznikoff and L. Gold), pp. 345–65.

Bustos, M.M., Luckow, V.A., Griffing, L.R., Summers, M.P. and Hall, T.C. (1988). Expression, glycosylation and secretion of phaseolin in a baculovirus system. *Plant Molecular Biology.* **10:** 475–88.

Bustos, M.A., Guiltinan, M.J., Jordano, J., Begun, D., Kalkan, F.A. and Hall, T.C. (1989). Regulation of β-glucuronidase expression in transgenic tobacco plants by an A/T rich cis-acting sequence found upstream of a french bean β-phaseolin gene. *The Plant Cell*. **1**: 839–53.

Casey, R., Domoney, C. and Ellis, N. (1986). Legume storage proteins and their genes. *Oxford Surveys of Plant Molecular and Cell Biology*. **3**: 1–95.

Chakraborty, P., Sosuliski, F. and Bose, A. (1979). Ultra centrifugation of salt-soluble proteins in ten legume species. *Journal of the Science of Food and Agriculture*. **30**: 766–71.

Chambers, S.J., Carr, H.J. and Lambert, N. (1990a). An investigation of the dissociation and denaturation of legumin by salts using laser light scattering and CD spectroscopy. *Biochimica et Biophysica Acta*. **1037**: 66–72.

Chambers, S.J., Carr, H.J., Plumb, G.W. and Lambert, N. (1990b). The problem of regolding and reassembling recombinant vicilin subunits. *Journal of Science Food and Agriculture*. **52**: 107–18.

Chen, Z.-L., Naito, S., Nakamura, I. and Beachy, R.N. (1989). Regulated expression of genes encoding soybean β-conglycinin in transgenic plants. *Developmental Genetics*. **10**: 112–22.

Cherry, J.P. (1981). *Protein Functionality in Food*. American Chemical Society, Washington.

Cho, T.-J., Davies, C.S., Fischer, R.L., Turner, N.E., Golberg, R.B. and Nielsen, N.C. (1989). Molecular characterisation of an aberrant allele for the Gyl, glycinin gene: a chromosomal rearrangement. *The Plant Cell*. **1**: 339–50.

Coates, J.B., Medeiros, J.S., Thanh, V.H. and Nielsen, N.C. (1985). Characterisation of the subunits of β-conglycinin. *Archives of Biochemistry and Biophysics*. **243**: 184–94.

Cramer, J.H., Lea, K. and Slightom, J.L. (1985). Expression of phaseolin cDNA genes in yeast under control of natural plant DNA sequences. *Proceedings of the National Academy of Science, USA*. **82**: 334-8.

Cramer, J.H., Lea, K., Schaber, M.D. and Kramer, R.A. (1987). Signal peptide specificity in post-translational processing of the plant protein phaseolin in *S. cerevisiae*. *Molecular and Cellular Biology*. **7**: 121–8.

Croy, R.R.D. and Gatehouse, J.A. (1985). Genetic engineering of seed proteins: current and potential applications. In *Plant Genetic Engineering* (ed. J.H. Dodds), pp. 143–268. Cambridge University Press, Cambridge.

Croy, R.R.D., Gatehouse, J.A., Tyler, M. and Boulter, D. (1980). A purification and characterisation of a third storage (convicilin) from the seeds of pea. *Biochemical Journal*. **191**: 509–16.

Cullen, D., Gray, G.L., Wilson, L.J., Hayenga, K.J., Lamsa, M.H., Rey, M.W., Norton, S. and Berka R.M. (1987). Controlled expression and secretion of bovine chymosin in Aspergillus nidulans. *Biotechnology*. **5**: 369–76.

Dagorn-Scaviner, C., Gueguen, J. and Lefebvre, J. (1986). A comparison of interfacial behaviours of pea legumin and vicilin at air/water interfaces. *Die Nahrung*. **30**: 337–47.

Delauney, A.J. (1984). Cloning and characterisation of cDNAs encoding the major pea storage proteins, and expression of vicilin in *E. coli*. *PhD.Thesis*, University of Durham, UK.

Derbyshire, E., Wright, D.J. and Boulter, D. (1976). Legumin and vicilin, storage proteins of legume seeds. *Phytochemistry*. **15**: 3–24.

Dieckert, J.W. and Dieckert, M.C. (1985). The chemistry and biology of seed storage proteins. In *New Protein Foods*, Vol. 5 (ed. A.M. Altschut), pp. 1–25. Academic Press, London.

Digan, M.E., Lair, S.V., Brierley, R.A., Siegel, R.S., Williams, M.E., Ellis, S.B., Kellaris, P.A., Provow, S.A., Craig, W.S., Velirelebi, G., Harpold, M.M. and Thill, G.P. (1989). Continuous production of a novel lysozyme via secretion from the yeast, pichia pastoris *Biotechnology*. **7**: 160–4.

Domoney, C., Ellis, T.H.N. and Davies, D.R. (1986). Organisation and mapping of legumin genes in Pisum. *Molecular and General Genetics*. **202**: 280–5.

Earle, F.R. and Jones, Q. (1962). Analyses of seed samples from 113 plant families. *Economic Botany*. **16**: 221–50.

Ellis, T.H.N., Domoney, C., Castleton, J., Cleary, W. and Davies, D.R. (1986). Vicilin genes of Pisum. *Molecular and General Genetics*. **205**: 164–9.

Ellis, J.R., Shirsat, A.H., Hepher, A., Yarwood, J.N., Gatehouse, J.A., Croy, R.R.D. and Boulter, D. (1988). Tissue specific expression of a pea legumin gene in seeds of *Nicotiana plubaginifolia*. *Plant Molecular Biology*. **10**: 203–14.

Forde, B.G., Day, H.M., Furton, J.F., Wen-Jun, S., Cullimore, J.V. and Oliver, J.E. (1989). The glutamine synthetase genes from *Phaseolus vulgaris* display contrasting developmental and spatial patterns of expression in transgenic Lotus conniculatus plants. *The Plant Cell*. **1**: 391–401.

Fukazawa, C., Udaka, K., Murayama, A., Higuchi, W. and Tosuka, A. (1987). Expression of soyabean glucinin subunit precursor cDNA in *E. coli*. *FEBS Letters* **224**: 125–7.

Ganesh-Kumar, K. and Venkataraman, L.V. (1980). Chick pea seed proteins: isolation and characterisation of 10.3S protein. *Journal of Agriculture and Food Chemistry*. **28**: 524–9.

Gatehouse, J.A., Lycett, G.W., Croy, R.R.D. and Boulter, D. (1982). The post-translational proteolysis of the subunits of vicilin from Pea. *Biochemical Journal*. **207**: 629–32.

Gatehouse, J.A., Croy, R.R.D. and Boulter, D. (1984). The synthesis and structure of pea storage proteins. *C.R.C. Critical Reviews in Plant Sciences*. **1**: 287–314.

Gatehouse, J.A., Evans, I.M., Croy, R.R.D. and Boulter, D. (1986). Differential expression of genes during legume seed development. *Philosophical Transactions of the Royal Society (London)*. **B 304**: 367–84.

Gayler, K.R. and Sykes, G.E. (1985). Effects of nutritional stress on the storage proteins of soybeans. *Plant Physiology.* **78**: 582–5.

Greenwood, J.S. and Crispeels, M.J. (1985). Correct targeting of bean storage protein phaseolin in the seeds of transformed tobacco. *Plant Physiology.* **79**: 65–71.

Gueguen, J. (1983). Legume seed protein extraction, processing and end-product characteristics. *Qualitas Plantarum, Plant Foods for Human Nutrition.* **32**: 267–303.

Gueguen, J., Chevalier, M., Barbot, J. and Schaeffer, F. (1988). Dissociation and aggregation of pea legumin induced by pH and ionic strength. *Journal Science Food and Agriculture.* **44**: 167–82.

Harada, J.J., Barker, S.J., Goldberg, R.B. (1989). Soyabean β-conglycinin genes are clustered in several DNA regions and are regulated by transcriptional and post-transcriptional process. *The Plant Cell.* **1**: 415–25.

Hara-Nishimura, I., Nishimura, M. and Akazawa, T. (1985). Biosynthesis and intracellular transport of 11S globulin in developing pumpkin cotyledons. *Plant Physiology.* **77**: 747–52.

Hasegawa, K., Tanaka, T. and Tamai, S. (1981). Preparation of hybrid proteins from subunits of sesame 13S and soya 11S globulins. *Agricultural and Biological Chemistry.* **45**: 809–15.

Haumann, B.F. (1988). Fats and oils industry changes. *Journal of the American Oil Chemists Society.* **65**: 702–15.

Haytowitz, D.B. and Matthews, R.H. (1989). Nutrient content of other legume products. In *Legumes: chemistry, technology and human nutrition* (ed. R.H. Matthews), pp. 219–44, Marcel Dekker Inc., New York.

Hirano, M., Kagawa, H., Kamata, Y. and Yamauchi, F. (1985). Structural homology among the major 7S globulin proteins. *Phytochemistry.* **26**: 41–5.

Hoffman, L.M., Donaldson, D.D. and Herman, E.M. (1988). A modified storage protein is synthesized, processed and degraded in the seeds of transgenic plants. *Plant Molecular Biology.* **17**: 717–29.

Horsch, R.B., Fraley, R.T., Rogers, S.G., Sanders, P.R., Lloyd, A. and Hoffman, N. (1984). Inheritance of functional foreign genes in plants. *Science.* **223**: 496–8.

I'Anson, K.J., Bacon, J.R., Lambert, N., Miles, M.J., Morris, V.J. and Wright, D.J. (1987). Synchrotron radiation wide-angle X-ray scattering studies of glycinin solutions. *International Journal of Biological Macromolecules.* **9**: 368–70.

I'Anson, K.J., Miles, M.J., Bacon, J.R., Carr, H.J., Lambert, N., Morris, V.J. and Wright, D.J. (1988). Structure of the 7S globulin from pea. *International Journal of Biological Macromolecules.* **10**: 311–17.

Jayarama-Setty, K. and Narasinga Rau, M.S. (1973). Studies on groundnut proteins: effect of SDS on the physico-chemical properties. *Indian Journal of Biochemistry and Biophysics.* **10**: 149–54.

Jimenez-Flores, R. and Richardson, T. (1987). Effects of chemical, genetic and enzymatic modifications on protein functionality. In *Food Biotechnology*, Vol. 1 (ed. R.D. King and P.S.J. Cheetham), pp. 87–137. Elsevier Publishers, London.

Jofuku, K.O., Okamuro, J.K. and Goldberg, R.B. (1987). Interaction of an embryo DNA binding protein with a soybean lectin gene upstream region. *Nature.* **328**: 734–7.

Joradno, J., Almoguera, C. and Thomas, T.L. (1989). A sunflower Helianthinin gene upstream sequence ensemble contains an enhancer and site of nuclear protein interaction. *The Plant Cell.* **1**: 855–66.

Kagawa, H. and Hirano, H. (1988). Identification and structural characterisation of the glycinin seed storage protein $A_7$ subunit of soyabean. *Plant Sciences (Limerick, Ireland).* **56**: 189–95.

Kato, A. and Yutani, K. (1988). Correlation of surface properties with conformational stabilities of wild-type and six mutant tryptophan synthase α-subunits substituted at the same position *Protein Engineering.* **2**: 153–6.

Kay, D.E. (1979). *Crop and Product Digest No. 3—Food Legumes.* Tropical Products Institute, London.

Kingsman, S.M., Kingsman, A.J., Dobson, M.J., Mellor, J. and Roberts, N.A. (1988). Heterologous gene expression in *Saccharomyces cerevisiae.* In *Yeast Biotechnology* (ed. G.E. Russel), pp. 113–51, Intercept Ltd., UK.

Lambert, N., Plumb, G.W. and Wright, D.J. (1987). Application of h.p.l.c. to the assessment of subunit heterogeneity in plant 11S storage globulins. *Journal of Chromatography.* **402**: 159–72.

Liener, I.E. (1983). Toxic constituents in legumes. In *Chemistry and biochemistry of legumes* (ed. S.K. Arora), pp. 217–57. Edward Arnold, London.

Liener, I.E. (1989). Antinutritional factors. In *Legumes: chemistry, technology and human nutrition* (ed. R.H. Matthews), pp. 339–82. Marcel Dekker Inc., New York.

Mackie, P. (1988). Exports are vital. *Journal of the American Oil Chemists' Society.* **65**: 1076–8.

Marston, F.A.O. (1986). The purification of eukaryotic polypeptides synthesized in *E. coli. Biochemical Journal.* **240**: 1–12.

Marston, F.A.O. and King, D.J. (1987). Production of heterologous proteins in *E. coli* and yeast. In *Chemical Aspects of Food Enzymes* (ed. A.T. Andrews), *Royal Society of Chemistry Proceedings.* 208 29.

McPherson, A. (1980). The three-dimensional structure of canavalin at 3.0 Å resolution by X-ray diffraction analysis. *Journal of Biological Chemistry.* **255**. 10472–80.

Mellor, J., Dobson, M.J., Roberts, N.A., Tuite, M.T., Emtage, J.S., White, S., Lowe, P.A., Patel, T., Kingsman, A.J. and Kingsman, S.M. (1983). Efficient synthesis of enzymatically active calf chymosin in *S. cerevisiae. Gene.* **24**: 1–14.

Momma, T., Negoro, T., Hirano, H., Matsumoto, A.,

Uduka, K. and Fukuzana, C. (1985). Glycinin $A_5$ $A_4$ $A_3$ mRNA: CDNA cloning and nucleotide sequencing of a splitting storage protein subunit of soyabean. *European Journal of Biochemistry.* **149:** 491–6.

Moreira, M.A., Hermodson, M.A., Larkins, B.A. and Nielsen, N.C. (1979). Partial characterisation of the acidic and basic polypeptides of glycinin. *Journal of Biological Chemistry.* **254:** 9921–6.

Moreira, M.A., Hermodson, M.A., Larkins, B.A. and Nielsen, N.C. (1981). Comparison of the primary structure of the acidic polypeptides of glycinin. *Archives of Biochemica Biophysica.* **210:** 633–42.

Mori, T., Nakamura, T. and Utsumi, S. (1982). Formation of pseudoglycinins and their gel hardness. *Journal of Agricultural and Food Chemistry.* **30:** 828–31.

Morris, P.C., Smith, C.J.S., Watson, C.W., Knapp, J.E., Davies, K., Picton, S., Grierson, D. (1989). Controlling the expression of the tomato polygalacturonase gene: promotion by regulatory as acting regions and inhibition by antisense RNA. *Journal of Cellular Biochemistry Supplement.* **13D M532:** 339.

Mossé, J. and Baudet, J. (1983). Crude protein content and amino acid composition of seeds: variability and correlations. *Qualitas Plantarium, Plant Foods for Human Nutrition.* **32:** 225–45.

Mossé, J. and Pernollet, J.C. (1983). Storage proteins of legume seeds. In *Chemistry and Biochemistry of Legumes* (ed. S.K. Arora), pp. 111–93. Edward Arnold, London.

Mroginski, L.A. and Kartha, K.K. (1984). Tissue culture of legumes for crop improvement. *Plant Breeding Reviews.* **2:** 215–64.

Murphy, P.A. and Resurreccion, A.P. (1984). Varietal and environmental differences in soybean glycinin and β-conglycinin content. *Journal of Agricultural and Food Chemistry.* **32:** 911–15.

Murray, D.R. (1979). A storage role for albumins in pea cotyledons. *Plant, Cell and Environment.* **2:** 221–6.

Nagy, F., Kay, S.A., Boutry, M., Hsu, M.-Y. and Chua, N.-H. (1986). Phytochrome-controlled expression of a wheat cab gene in transgenic tobacco seedlings. *EMBO Journal.* **5:** 119–24.

Naito, S., Dubé, P.H. and Beachy, R.N. (1988). Differential expression of Conglycinin α′ and β subunit genes in transgenic plants. *Plant Molecular Biology.* **11:** 109–23.

Nakamura, T., Utsumi, S. and Mori, T. (1985). Formation of pseudoglycinins from intermediary subunits of glycinin and their gel properties and network structure. *Agricultural and Biological Chemistry.* **49:** 2733–40.

Nielsen, N.C. (1984). The chemistry of legume storage proteins. *Philosophical Transactions of the Royal Society in London.* **B 304:** 287–96.

Nielsen, S.S., Desphpande, S.S., Hermodson, M.A. and Scott, M.P. (1988). Comparative digestibility of legume storage proteins. *Journal of Agricultural and Food Chemistry.* **36:** 896–902.

Nielson, N.C., Dickinson, C.D., Cho, T.-J., Thanh, V.H. and Scallon, B.J. (1989). Characterisation of the glycinin gene family. *The Plant Cell.* **1:** 313–28.

Okamuro, J.K. and Goldberg, R.B. (1989). Regulation of plant gene expression: general principles. In *The Biochemistry of Plants*, Vol. 15 (ed. A. Marcus), pp. 2–82, Academic Press, California.

Osborne, T.B. (1924). *The Vegetable Proteins.* Longmans, London.

Pernollet, J.-C. (1985). Biosynthesis and accumulation of storage proteins in seeds. *Physiologie Vègètable.* **23:** 45–59.

Paaren, H.E., Slighton, J.L., Hall, T.C., Inglis, A.S. and Blagrove, R.J. (1987). Purification of a seed glycoprotein: N-terminal and deglycosylation analysis of phaseolin, *Phytochemistry.* **26:** 335–43.

Plietz, P. and Damaschun, G. (1986). The structure of the 11S seed globulins from various plant species: comparative investigations by physical methods. *Studia Biophysica.* **116:** 153–73.

Plietz, P., Drescher, B. and Damaschun, G. (1987). Relationship between the amino acid sequence and the domain structure of the subunits of the 11S seed globulins. *International Journal of Biological Macromolecules.* **1:** 161–5.

Plumb, G.W., Carr, H.J., Newby, V.K. and Lambert, N. (1989). A study of the trypsinolysis of pea 11S globulin. *Biochimica et Biophysica Acta.* **999:** 281–8.

Prakash, V. and Nandi, P.K. (1976). Dissociation and denaturation behaviour of sesame α-globulin in SDS solution. *International Journal of Peptide and Protein Research.* **8:** 385–92.

Saio, K. and Watanabe, T. (1978). Differences in functional properties of 7S and 11S soyabean proteins. *Journal of Texture Studies.* **9:** 135–57.

Schaber, M.D., DeChiara, T.M. and Kramer, R.A. (1986). Yeast vectors for production of interferon. *Methods in Enzymology.* **119:** 416–23.

Schiemann, J., Baumlein, H. and Weigelt, A. (1988). Co-transformation of non-selectable higher plant genes into *Aspergillus nidulans*: transfer of a Vicia Faba legumin B gene. *Biochemie and Physicologie der Pflanzen.* **183:** 243–50.

Schwenke, K.D., Rauschal, E., Zirwer, D. and Linow, K.H. (1985). Structural changes of the 11S globulin from sunflower seed after succinylation. *International Journal of Peptide and Protein Research.* **25:** 347–54.

Sengupta-Gopalan, C., Reichert, N.A., Barker, R.F., Hall, T.C. and Kemp, J.D. (1985). Developmentally regulated expression of the bean β-phaseolin gene in tobacco seed. *Proceedings of the National Academy of Sciences, USA.* **82:** 3320–4.

Shirsat, A.H., Wilford, N. and Croy, R.R.D. (1989a). Gene copy number and levels of expression in transgenic plants of a seed specific gene. *Plant Science.* **61:** 75–80.

Shirsat, A.H., Wilford, N., Croy, R.R.D. and Boulter, D. (1989b). Sequences responsible for the tissue specific promoter activity of a pea legumin gene in tobacco. *Molecular and General Genetics.* **215**: 326–31.

Spencer, D. (1984). The physiological role of storage proteins in seeds. *Philosophical Transactions of the Royal Society in London.* **B 304**: 275–85.

Staswick, P.E., Hermodson, M.A. and Nielsen, N.C. (1981). Identification of the acidic and basic subunit complexes of glycinin. *Journal of Biological Chemistry.* **256**: 8752–5.

Stewart, G. (1989). The expression of pea vicilin in yeast. *PhD Thesis*, University of Durham, UK.

Suresh Chandra, B.R., Prakash, V. and Narasinga Rao, M.S. (1985). Association-dissociation of glycinin in urea, guanidine hydrochloride and SDS solutions. *Journal of Biosciences.* **9**: 177–84.

Talbot, D.R., Adang, M.J., Slightom, J.L. and Hall, T. (1984). Size and organisation of a multigene family encoding phaseolin, the major seed storage protein of phaseolin vulgaris. *Molecular and General Genetics.* **198**: 42–9.

Tuite, M.F., Dobson, M.J., Roberts, N.A., King, R.M., Burke, D.C., Kingsman, S.M. and Kingsman, A.J. (1982). Regulated high efficiency expression of human interferon-alpha in *S. cerevisiae. EMBO Journal.* **1**: 603–8.

Tulloch, P.A. and Blagrove, R.J. (1985). Electron microscopy of seed storage globulins. *Archives of Biochemistry and Biophysics.* **241**: 521–32.

Utsumi, S., Kim, C.-S., Kohno, M. and Kito, M. (1987). Polymorphism and expression of cDNAs encoding glycinin subunits. *Agricultural Biological Chemistry* **51**: 3267–73.

Utsumi, S., Kim, C.-S., Sato, T. and Kito, M. (1988a). Signal sequence of preproglycinin affects production of the expressed protein in *E. coli FEBS Letters* **71**: 349–58.

Utsumi, S., Sato, T., Kim, C.-S. and Kito, M. (1988b). Processing of preproglycinin expressed from cDNA encoding Ala B1b subunit in *S. cerevisiae. FEBS Letters.* **233**: 273–6.

Uzzan, A. (1988). Vegetable protein products from seeds. In *Developments in Food Proteins*, Vol. 6 (ed. B.J.F. Hudson), pp. 73–118. Elsevier Applied Science, London.

Vandekerckhove, J., Van Damme, J., Van Lijsebettens, M., Botterman, J., DeBlock, M., Vandewiele, M., De Clercq, A., Leemans, J., Van Montagu, M. and Krebbers, E. (1989). Enkephalins produced in transgenic plants using modified 2S seed storage proteins. *Biotechnology.* **7**: 929–38.

Waggle, D.H., Steinke, F.H. and Shen, J.L. (1989). Isolated soya proteins. In *Legumes: chemistry, technology and human nutrition* (ed. R.H. Matthews), pp. 99–138. Marcel Dekker, Inc., New York.

Watson, M.D., Lambert, N., Delauney, A., Yarwood, J.N., Croy, R.R.D., Gatehouse, J.A., Wright, D.J. and Boulter, D. (1988). Isolation and expression of a pea vicilin cDNA in the yeast *S. cerevisiae. Biochemical Journal.* **251**: 857–64.

Welland, M.E., Miles, M.J., Lambert, N., Morris, I., Coombs, J.H. and Pethica, J.B. (1989). Structure of the globular protein vicilin revealed by scanning tunnelling microscopy. *International Journal of Biological Macromolecules.* **11**: 29–32.

Widholm, J.M. (1988). Is soyabean transformation finally here? *Trends in Biotechnology.* **6**: 265–6.

World Fats and Oils Report. (1988). *Journal of the American Oil Chemists Society.* **65**: 1232–58.

Wright, D.J. (1987). The seed globulins. In *Developments in Food Proteins*, Vol. 5 (ed. B.J.F. Hudson), pp. 81–157. Elsevier Applied Science, London.

Wright, D.J. (1988). The seed globulins—Part II. In *Developments in Food Proteins*, Vol. 6 (ed. B.J.F. Hudson), pp. 119–78. Elsevier Applied Science, London.

Wright, D.J. and Bumstead, M.R. (1984). Legume proteins in food technology. *Philosophical Transactions of the Royal Society in London.* **B 304**: 381–93.

Yarwood, J.N., Harris, N., Delauney, A., Croy, R.R.D., Gatehouse, J.A., Watson, M.D. and Boulter, D. (1987). Construction of a hybrid cDNA encoding a major legumin precursor polypeptide and its expression and localisation in *S. cerevisiae. FEBS Letters.* **222**: 175–80.

Zirwer, D., Gast, R., Welgle, H., Schtesier, B. and Schwenke, K.D. (1985). Secondary structure of globulins from plant seeds: a re-evaluation from circular dichroism measurements. *International Journal of Biological Macromolecules.* **7**: 105–8.

# 8

# Expression of Wheat Gluten Proteins in Heterologous Systems

Pippa J. Madgwick, Kathryn A. Pratt and Peter R. Shewry

## Introduction

Wheat is the most widely grown crop in the world, with a total area of $220\times10^6$ hectares and yields exceeding $500\times10^6$ tonnes per annum (FAO 1988). Much is used for human consumption, particularly in the form of breads, noodles and pastas. The suitability of wheat for these products is determined, to a large extent, by the properties of the grain storage proteins. When wheat flour is wetted and kneaded to form dough these proteins form a network, which can be isolated in a form that is substantially pure by washing to remove starch and soluble components. This is called gluten, and is a cohesive mass with a unique combination of two physical properties: extensibility (i.e. viscous flow) and elasticity. It is this combination of properties that allows wheat doughs to be expanded by fermentation to give leavened bread, and also contributes to its functionality in other food systems. Gluten is not formed by the seed storage proteins of barley and rye, although they are closely related to those of wheat.

Wheat gluten can also be isolated from flour by an industrial process, and an increasing amount is used for food products in Europe and the USA. Most of this is used for breadmaking, to improve (fortify) poor quality flours and to increase the gluten content for speciality breads such as high fibre and wholemeal. Other food uses also exploit its unusual physical properties, for example as a binder in pet foods and processed meats.

Studies of wheat gluten proteins have traditionally been aimed at understanding the molecular basis for breadmaking and pastamaking quality, in order to improve the qualities of low grade flours. Although these considerations remain of great im-portance, it is now possible to look towards making more extreme changes to use gluten to replace animal proteins in other food systems (for example, as an emulsifier and foaming agent). The advent of molecular biology not only provides tools for exploring the structure and functionality of wheat gluten proteins but also, through genetic engineering, for manipulating the structure of the gluten produced by the wheat plant. The ultimate aim is to design wheats to produce glutens for a range of food and non-food uses, including traditional uses such as breadmaking.

## Wheat gluten proteins

Wheat gluten consists predominantly of proteins ($\simeq 70$–80% dry weight), other components including residual starch and lipids. The proteins are classically divided into two groups on their solubility in alcohol–water mixtures (usually 60–70% v/v aqueous ethanol or 50% v/v aqueous propan-1-ol). The soluble proteins, called gliadins, are present as monomers, some containing intrachain disulphide bonds. They are classified on the basis of their electrophoretic mobility at low pH into four groups: $\alpha$, $\beta$, $\gamma$ and $\omega$-gliadins, in order of decreasing mobility (Woychick et al. 1961) (Fig. 8.1a). They exhibit strong non-covalent protein–protein interactions (mainly hydrogen bonds and hydrophobic interactions), and are considered to be mainly responsible for the viscosity and extensibility of gluten. In contrast, the glutenins consist of polymers stabilized by interchain disulphide bonds, and are responsible for the elasticity. Separation of the reduced glutenin proteins by SDS-PAGE shows two major groups of subunits:

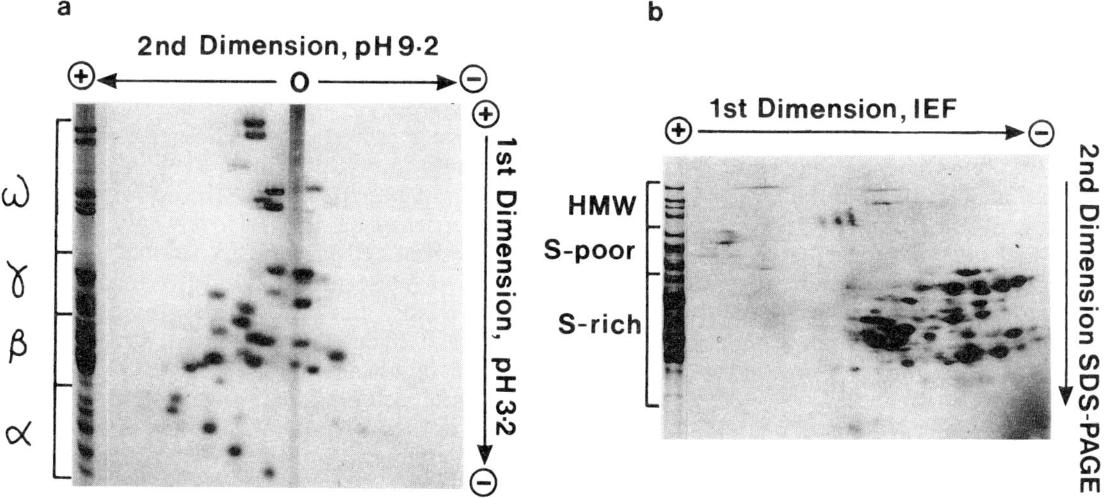



**Fig. 8.1**  Polymorphism of gluten proteins from wheat Chinese Spring. (a) Gliadins separated by electrophoresis at pH 3.2 followed by electrophoresis at pH 9.2. (b) Total gluten proteins separated by isoelectric focusing (pH range 3.5–10) followed by SDS-PAGE. Taken from Shewry and Tatham (1990).

the high molecular weight (HMW) and low molecular weight (LMW) groups.

Altogether over 50 individual components can be resolved by two-dimensional electrophoresis of total gluten proteins (Fig. 8.1b) and molecular and chemical analyses indicate that most, if not all, of these are the products of single genes. These genes are mostly clustered into complex loci, which have been mapped using classical Mendelian analysis as well as more modern approaches such as restriction fragment length polymorphisms and *in situ* hybridization (see papers in Miller and Koebner 1988).

Although the classification of gluten proteins into gliadins and glutenins is valid in terms of their physical and technological properties, it does not accurately reflect their evolutionary relationships (as shown by comparisons of amino acid sequences). Amino acid sequence comparisons have led to an alternative classification, into three groups (Shewry *et al.* 1986) (Fig. 8.2). Two of these, the HMW and S-poor prolamins, comprise the polymeric HMW subunits of glutenin and the monomeric ω-gliadins respectively. The third group, the S-rich prolamins, comprises both monomeric and polymeric components, which fall into three families: the α type gliadins (α- and β-gliadins), the γ-type gliadins and the LMW subunits of glutenin. All these components are defined as prolamins because they are soluble, either in the native state or after reduction of interchain disulphide bonds, in alcohol–water mixtures.

Wheat gluten proteins, like other prolamins, are rich in glutamine and proline, the combined proportions ranging from about 30–70 mol%. They also vary in the proportions of other residues, for example glycine (about 1–20 mol%), cysteine (about 0–2.5 mol%) and phenylalanine (about 0.5–9 mol%) (Shewry and Miflin 1985). Nevertheless, all have unusual compositions when compared with most globular proteins.

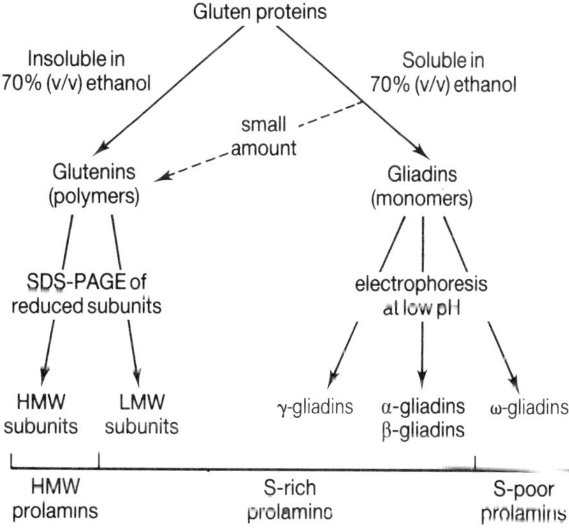



**Fig. 8.2**  Summary of the classification and nomenclature of wheat gluten proteins.

These characteristic amino acid compositions result from one particular structural feature: the presence of repeated amino acid sequences. These are based on one or more short peptide motifs (ranging in length from three to nine residues), and account for between 30% (in some S-rich prolamins) and almost 100% (in the S-poor ω-gliadins) of the proteins. They appear to be responsible for many of the unusual properties of prolamins, including solubility (Kreis and Shewry 1989).

The presence of repeated sequences results in a clear domain structure, with a single repetitive domain flanked by one or two non-repetitive domains (or short non-repetitive sequences). The repetitive and non-repetitive domains adopt different conformations, and appear to fold almost independently (Tatham *et al.* 1990b). Whereas the former are generally rich in β-turns, the latter are more like globular proteins, with elements of α-helix and β-sheet (Tatham *et al.* 1985, 1990a, b).

## Why express gluten proteins in heterologous systems?

It is clear from the brief summary presented above that wheat gluten is a highly complex system, with over 50 individual components interacting by covalent disulphide bonds and by strong non-covalent forces. The structures of the individual gluten proteins and their interactions are responsible for the physicochemical properties of gluten, and for its functionality in breadmaking and other food systems. It is essential to understand the structure of gluten at the molecular level if we are to manipulate its properties, whether by chemical modification or genetic engineering, for new food and industrial uses.

One limitation to the detailed analysis of individual gluten proteins is the difficulty of preparing large amounts of single homogeneous components. It is usually relatively easy to prepare groups of related components (for example total γ-gliadins), and these are suitable experimental material for some studies (see, for example, Tatham and Shewry 1985), but not for detailed characterization (for example, by X-ray crystallography). The individual components of these groups are highly homologous, and often extremely difficult to separate. An extreme example of this is the LMW subunits of glutenin. Crude fractions containing these proteins are relatively easy to prepare (Bietz and Wall 1980; Shewry *et al.* 1983; Tatham *et al.*

1987), but individual components have only recently been prepared in picomole quantities by transfer from two-dimensional gels (Tao and Kasarda 1989).

Expression in heterologous hosts provides a clear opportunity to produce single components in large quantities, especially because their unusual solubility properties make them relatively easy to separate from the host cell proteins. A second and related application is the preparation of individual structural domains. Because the repetitive and non-repetitive domains have different structures and may fold independently (see above) they can be most readily studied in isolation, but are often difficult to prepare from purified proteins owing to lack of suitable sites for proteolytic or chemical cleavage.

A more problematic application is the use of protein engineering to explore the structures, interactions and hence functionality of gluten proteins. It should be relatively straightforward to explore the structures of individual monomers, for example to determine the role of specific features in thermostability. Attempts to express monomeric α- and γ-gliadins in order to carry out studies of this type are described below.

The analysis of protein interactions and functionality is more difficult because the interactions are established during grain development (although they may be modified during processing) and cannot be precisely reproduced *in vitro*. Ideally the engineered protein should be produced in the developing wheat seed, but this is not currently possible due to lack of a stable transformation system for wheat. A logical alternative is to treat individual gluten proteins as a model system, in order to explore specific aspects of gluten functionality. We are currently attempting to carry out studies of this type on the HMW subunits of glutenin in order to determine the molecular basis for gluten elasticity (see pp. 196–7).

We will now discuss work carried out at Rothamsted and in other laboratories on the expression of wheat gluten proteins using three different host systems: *E. coli*, yeast and cultured insect cells. In addition we will review relevant work on the expression of a second group of cereal prolamins, the zeins of maize. These proteins share many of the properties of wheat gluten proteins (Tatham *et al.* 1990a) and their expression is therefore relevant to the main theme of this article.

# Expression in *Escherichia coli*

*E. coli* is, of course, widely used by molecular biologists and has many advantages as a host for expression. It grows rapidly, is easy to handle and manipulate, and foreign DNA can be readily introduced in plasmid vectors. In addition the expressed protein can be readily extracted. However, it has not always been successful for expression of active higher plant proteins. This may relate to failure to carry out post-translational processing (e.g. disulphide bond formation, glycosylation) or incorrect folding.

## Expression of maize prolamins

Results obtained using prolamin (zein) genes from maize indicated that the unusual properties of the protein might result in low expression, due either to toxicity to the bacterium or to rapid degradation of the protein. Wang and Esen (1985) constructed a cDNA expression library of developing maize endosperm using the plasmid pUC8 as a vector, and screened colonies with antibodies raised to two zein groups. These were the major zein group, called α-zeins, which are insoluble in water, and a quantitatively minor water-soluble group (γ-zeins). Screening with the anti-γ-zein antibody gave 23 positive colonies. SDS-PAGE analyses of total cell extracts showed that two clones had distinguishable stained bands corresponding to fusion proteins (γ-zeins fused with the first five residues of β-galactosidase) and densitometric scanning indicated that these accounted for about 5% of the protein applied to the gel. The remaining clones did not have bands that were visible after staining, but all had bands that reacted with the antibody after Western blotting. Almost all the clones also had bands with a higher mol. wt than would be expected for the fusion proteins. This may have arisen by association of the fusion protein with itself and/or with host proteins.

Very few colonies reacted when the same library was screened with antibody to α-zeins, although hybridization with cDNA probes showed that about twice as many colonies contained inserts related to α-zeins than to γ-zeins. The calculated percentage expression (i.e. number of colonies reacting with antibody compared with those hybridizing with cDNA) was about 6% for α-zeins compared to 30% for γ-zeins, suggesting that some

unique property of the α-zeins resulted in lower expression.

Norrander *et al.* (1985) also tried to express an α-zein cDNA in *E. coli* using M13mp and pUC-derived expression vectors. In all cases the levels of expression were low, around 1% of the total *E. coli* proteins or less.

## Expression of gluten proteins

The first attempt to express a wheat gluten protein in a heterologous host also used *E. coli* (Bartels *et al.* 1985). They used a partial cDNA, encoding part of an HMW subunit of glutenin. Less than half of the insert had been sequenced (Thompson *et al.* 1983) and it was thought to encode only the repetitive domain, lacking the N- and C-terminal domains. This clone was initially expressed as a fusion with β-galactosidase. It was estimated that *E. coli* produced native β-galactosidase at a level of 10–15% of the total protein labelled with [$^{35}$S]-methionine, but produced the fusion protein to a level of only 1–1.5%. Separation of the total cell proteins by SDS-PAGE showed three bands that reacted with antibodies raised against β-galactosidase or against purified HMW subunits. These were of the expected mol. wt—170 000—but were also of lower mol. wt (145 000 and 140 000). Pulse chase experiments indicated that the smaller bands were not formed by degradation of the larger one. It was also possible to produce a protein with only a small amount of β-galactosidase sequence by inserting a stop codon in frame with the C-terminal end of the HMW subunit sequence. This produced a protein with a mol. wt of 77 000–78 000, which was soluble in 50% propan-2-ol and precipitated by antibodies to HMW subunits. Although the mol. wt (77 000–78 000) was higher than that predicted on the basis of the length of the cloned cDNA (57 000–60 000), this is consistent with the well established anomalous mobility of the HMW subunits on SDS-PAGE (Bunce *et al.* 1985). The level of expression of the HMW subunit was not reported, but it was clearly not above that of the fusion protein.

These initially disappointing expression levels of maize and wheat prolamins in *E. coli* led workers to seek alternative expression systems. However, a recent report by Galili (1989) demonstrates that high level expression in *E. coli* is possible, using a different vector system.

Galili (1989) cloned genomic fragments encoding

HMW subunits of glutenin into a vector based on bacteriophage T7 RNA polymerase. The resulting plasmids directed the synthesis of large amounts of the mature forms of HMW subunits (Fig. 8.3); densitomer tracings indicating a level of about 7% of the total extracted protein. The expressed proteins reacted with antibodies raised against purified HMW subunit and were soluble at 60°C in aqueous alcohol solutions containing reducing agents such as 2-mercaptoethanol. The latter property facilitated their purification (Fig. 8.3). The sequence encoding the subunit signal peptide was removed before insertion into the vector and the author speculated that this may have contributed to the high level of expression. Galili (1989) also showed that protein fractions extracted from the *E. coli* in the absence of reducing agent contained some HMW subunit oligomers stabilized by interchain disulphide bonds, but their relationship to the behaviour of HMW subunits in wheat gluten is not known.

## Expression in yeast

Bakers yeast (*Saccharomyces cerevisiae*) is often used for the production of eukaryotic proteins because, unlike *E. coli*, it will often correctly cleave eukaryotic signal sequences (for example in prepro-glycinin, Utsumi *et al.* 1988) and will also sometimes correctly glycosylate the protein (e.g. the human granulocyte macrophage colony stimulating factor, Ernst 1988). Yeast is less easy to handle than *E. coli*, having a tough cell wall that makes it more difficult to transform and to extract DNA and protein, although the latter problem can be overcome by engineering vectors to secrete the protein into the medium. Yeast also grows more slowly than *E. coli*, with a generation time of over 2 h compared with about 20 min. Yeast has been used to produce many plant proteins including thaumatin (Edens *et al.* 1984), β-glucanase (Jackson *et al.* 1986), α-amylase (Rothstein *et al.* 1984) and ricin (Richardson *et al.* 1987).

## Expression of maize prolamins

Viotti and co-workers have reported high level expression of α-zeins in maize, using galactose-inducible expression vectors (Coraggio *et al.* 1986; Compagno *et al.* 1987). The level of zein production was very low (about 0.2% of the total produc-

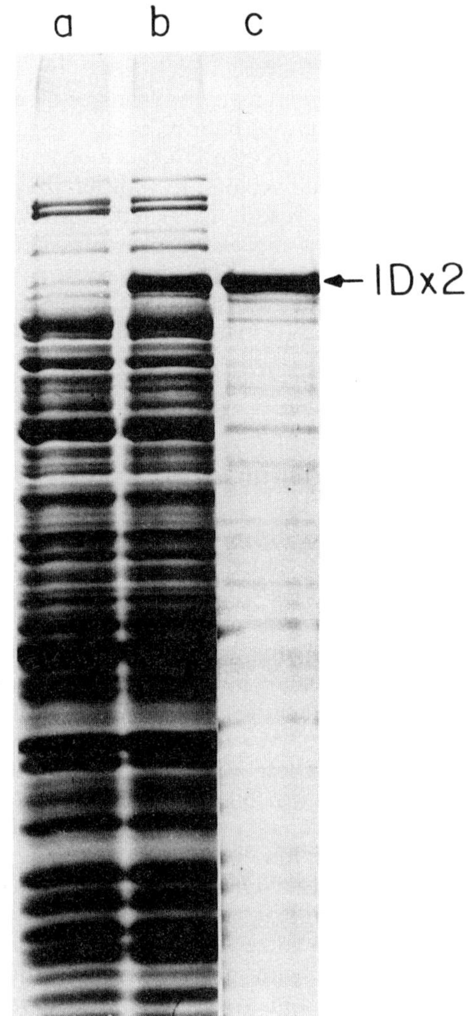



**Fig. 8.3**  Expression of a wheat HMW subunit of glutenin in *E. coli*. (a) Total proteins from induced cells harbouring the control expression vector pEt-3a. (b) Total proteins from induced cells harbouring the expression vector pEt-3a with the genomic fragment encoding subunit 1Dx2. (c) Soluble supernatant proteins isolated from cells expressing subunit 1Dx2 by sonication in 70% (v/v) ethanol/1% (v/v) 2-mercaptoethanol and incubation at 60°C for 30 min. Reproduced with permission from Galili (1989).

tion) during the exponential phase of growth in galactose minimal medium, but increased during the early stationary phase to reach 4–5% of the total yeast protein. Immunocytochemical analysis showed that in one case the zein was accumulated within the mitochondria of the transformed yeast cell (Faoro *et al.* 1987). These levels of expression were the highest that had been reported at that time for any cereal prolamins and influenced our decision to use a yeast expression system for wheat gluten proteins.

## Expression of gluten proteins

Attempts in other laboratories to express wheat gluten proteins in yeast have yielded disappointing results. Neill *et al.* (1987) cloned the entire coding sequence of an α-gliadin gene into an inducible vector under the control of the CYC-1 regulatory regions. A single mRNA terminating in the 3′ untranslated region of the α-gliadin was isolated from transformed yeast and α-gliadin protein of the expected mol. wt (about 31 000) could be detected immunochemically in total protein extracts. The mol. wt was the same as that of the mature α-gliadin protein, indicating that signal peptide cleavage had occurred. However, the expressed protein only accounted for about 0.1% of the total cell protein and the authors speculated that this may have been due to the lack of sufficient glutamine tRNAs to sustain synthesis of the glutamine-rich regions.

Low level expression was also reported by Scheets and Hedgcoth (1989), who cloned the complete coding region for a γ-type gliadin into an expression vector containing a yeast ADH1 promoter and CYC-1 transcription terminator. Northern blotting indicated that gliadin-coding mRNA was produced using the ADH1 transcription start site and the CYC-1 termination site, and that no termination occurred near the wheat polyadenylation signal. The presence of a low level of γ-gliadin was demonstrated by the immunoblotting of total protein extracts, but it was not possible to determine whether signal peptide cleavage had occurred. Analysis of the growth medium indicated the presence of 0.5–1.5 ng of gliadin/ml. Although the authors assumed that this resulted from secretion, it is not possible to rule out an origin from cell breakage and death.

## Expression of a wheat γ-gliadin at Rothamsted

Several yeast expression vectors that utilize the yeast phosphoglycerate kinase (PGK) promoter and terminator regions have been constructed. We have used a constitutive expression vector, pMA 91 (Mellor *et al.* 1983; Kingsman *et al.* 1990), and an

inducible derivative of it, pKV 49 (Kingsman *et al.* 1990). These vectors were generously provided by Dr S. Kingsman (University of Oxford), and have been used to produce eukaryotic proteins, including human interferon (Tuite *et al.* 1982) and plant proteins such as the B-chain of ricin (Richardson *et al.* 1987) and a legumin precursor polypeptide (Yarwood *et al.* 1987). However, we were not able to detect expression of the N-terminal repetitive domain of a wheat HMW glutenin subunit using either vector, although mRNA was present (P. Madgwick and P.R. Shewry, unpublished results).

More success was achieved with expression of a γ-type gliadin. The cDNA clone used for this study, pTAG 1436, encodes a $\gamma_3$-type gliadin of 295 residues (Bartels *et al.* 1986). This includes a signal peptide of 19 residues, and the mature protein has a mol. wt of 31 629. The protein has a clear domain structure (see Fig. 8.4), with a short N-terminal sequence of 12 residues and a C-terminal non-repetitive domain of 151 residues flanking a 113 residue proline-rich repetitive domain. The latter consists of repeated peptides that vary in precise sequence, but appear to be based on the consensus sequence ProGlnGlnProPheProGln(Gln). The γ-gliadin insert was unstable, even when maintained in an *E. coli rec A*⁻ host strain, a ladder of bands being visible on electrophoresis of restriction fragments from pTAG 1436. Care was therefore taken during the purification of the fragment to ensure that the full sequence was present.

The complete γ-gliadin coding sequence, including the signal peptide, was initially inserted into pMA 91. This is a shuttle vector that contains the *E. coli* origin of replication and the β-lactamase gene from the plasmid pBR 322. It has the LEU2 gene for selection in yeast and sequences from the yeast 2 μ plasmid to allow for replication and maintenance of plasmid copy number in yeast. The coding sequence for the complete γ-gliadin (including the signal peptide) was inserted at the *Bgl* II cloning site at position −2 with respect to the translation initiation codon of the PGK gene, and the recombinant plasmid (called pKAP 2, see Fig. 8.4) used to transform the LEU2⁻ host strain MD40-4C.

Total cell proteins were extracted from cells grown with shaking at 30°C, and analysed by SDS-PAGE. Although it was not possible to see an extra band corresponding to the γ-gliadin on the stained gel (Fig. 8.5a, tracks 1–6), this was clearly shown by Western blotting using an antiserum raised against purified γ-gliadin (Festenstein *et al.*

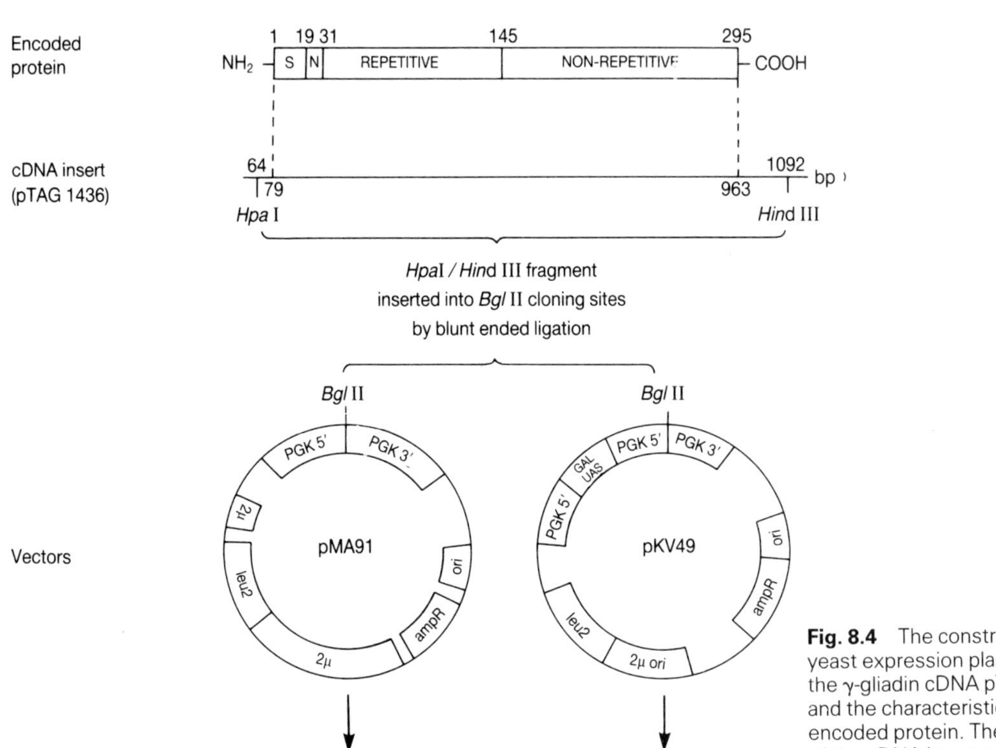



**Fig. 8.4** The construction of yeast expression plasmids using the γ-gliadin cDNA pTAG1436, and the characteristics of the encoded protein. The numbering of the cDNA insert corresponds to that used by Bartels *et al.* 1986.

1987) (Fig. 8.5b, tracks l–o). The maximum level of γ-gliadin was present in the log phase. The absolute amount of gliadin produced was difficult to determine, but it probably accounted for less than 1% of the total extracted proteins.

The same constitutive expression vector was also transformed into a protease deficient (Pep⁻) host strain, MT302/1C, to determine whether the low levels of gliadin present in MD40-4C resulted from proteolysis of the expressed protein. The level of gliadin extracted from cells of this transformant was approximately the same as that extracted from the transformed cells of MD40-4C, indicating that protease activity was not a significant problem.

The low yield of the γ-gliadin in pMA 91 led us to explore the use of other vector systems. In particular, cells transformed with pKAP 2 had a generation time of about 4 h compared with about 2 h for cells transformed with control plasmid, indicating that the production of gliadin had adverse effects on cell growth. We decided, therefore, to use an inducible vector, pKV 49. This possesses all the sequences present in pMA 91 for maintenance and replication of the plasmid, and the expression of the gene inserted into the *Bgl* II

cloning site is again under the control of the yeast PGK promoter and transcription termination sequences. However, a 50 bp region of the PGK UAS sequence is replaced by 144 bp of UAS from the intergenic region between the *GAL 1* and *GAL 10* genes of yeast, which confers inducibility by galactose.

The γ-gliadin fragment was inserted into the *Bgl* II cloning site of pKV 49 (Fig. 8.4), and the resulting plasmid (pKAP 3) transformed into the host strain DBY 745. Cells were initially grown for about 24 h (corresponding to late log/early stationary phase) on glucose medium, and then induced by transfer to galactose medium.

Analysis of total extracted proteins by SDS-PAGE showed an additional band of mol. wt about 37 000 to 38 000 on the gel stained with Coomassie BBR 250 (Fig. 8.5a, tracks c–f, see arrow in track f), which, on Western blotting reacted with antibodies raised against γ-gliadins (Fig. 8.5b, tracks c–f). Production was greatest after about 28 h induction (Fig. 8.5a, track f), but some degradation appeared to have occurred at 43 h (not shown). Induction of the cultures in early or mid-log phase resulted in lower levels of production, similar to

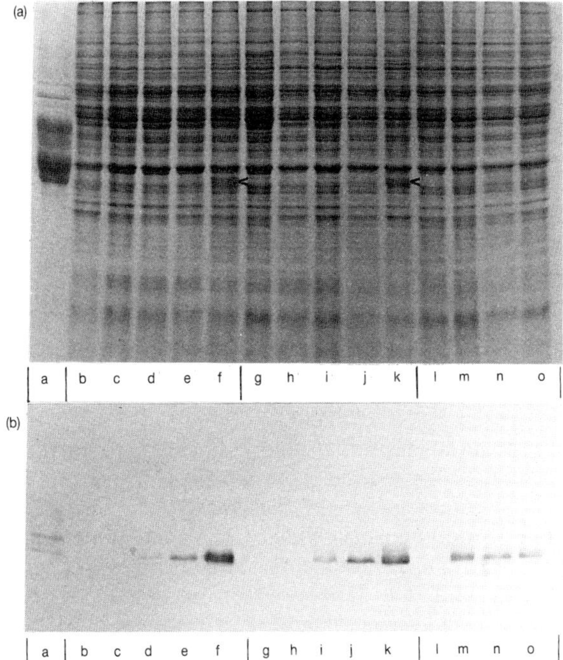



**Fig. 8.5**   SDS-PAGE (a) and Western blotting using a
γ-gliadin antiserum (b) of wheat γ-gliadin expressed in yeast.
a, Standard total gliadin fraction; b, total proteins from cells of
DBY745 with the control plasmid pKV 49; c–f, total proteins
from cells DBY745 with the inducible vector pKAP3 at 0, 4, 10
and 28 h after induction, respectively; g, total proteins from
cells DJ2168 with the control plasmid pKV 49; h–k, total
proteins from the Pep-host strain BJ2168 with the inducible
vector pKAP3, at 0, 4, 10 and 28 h after induction,
respectively; l, total proteins from cells of MD40-4C with the
control plasmid PMA 91, sampled 48 h after inoculation;
m–o, total proteins from cells of MD40-40 with the
constitutive vector pKAP 2, at 20, 30 and 48 h after
inoculation, respectively. Approximately equal amounts of
total cell proteins (100 μg) are loaded in tracks b–o. The
arrows in part (a), tracks f and k, indicate the stained bands
corresponding to the γ-gliadin.

those observed with the constitutive expression
system. The best levels of expression (with cells
induced in early stationary phase) corresponded to
about 3% of the total extracted proteins, which
represents a great advance on the yields of gliadins
obtained with other yeast expression systems. A
slightly lower level of expression was obtained
when pKAP 3 was grown in the Pep⁻ host strain
BJ 2168 (Fig. 8.5a, b, tracks h–k).

Although the mol. wt calculated for the expressed γ-gliadin by SDS-PAGE (37 000–38 000)
was higher than that expected from the absolute
molecular weight of the mature γ-gliadin encoded
by the cDNA clone (31 629) this is not a cause for

great concern. It is well known from other studies
that many prolamins migrate anomalously on SDS-
PAGE, giving incorrect molecular weights when
compared to standard proteins (Bunce *et al.* 1985).

The expressed γ-gliadin was readily purified
from lysed yeast cells by solution in 50% (v/v)
propan-1-ol at 60°C followed by precipitation by
the addition of aqueous NaCl (Fig. 8.6a). N-
terminal amino acid sequencing showed a single
sequence corresponding to that of the mature protein. This indicates that the wheat signal sequence
was recognized and cleaved by the yeast cells. The
protein differed in its properties from those of
γ-type gliadins purified from wheat, being only
sparingly soluble in 0.1 M acetic acid in absence of
a reducing agent. This indicates that at least some
of the protein had incorrectly formed disulphide
bonds, and this was confirmed by SDS-PAGE of
the unreduced fraction. This showed the presence

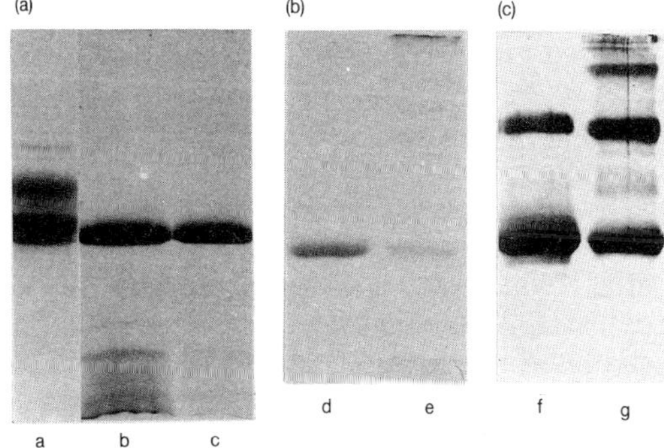



**Fig. 8.6**   SDS-PAGE of a γ-gliadin expressed in yeast. (a)
Purification. a, Total gliadin fraction prepared from wheat cv.
Chinese Spring; b, fraction extracted from yeast cells with
50% (v/v) aqueous propan-1-ol at 60°C; c, fraction
precipitated from the alcoholic extract by addition of 1.5 M
NaCl and standing at 4°C, and redissolved in 50% (v/v)
aqueous propan-1-ol. All samples were run under reducing
conditions. (b) Comparison of the SDS-PAGE patterns of the
expressed γ-gliadin under reducing and non-reducing
conditions, after staining with Coomassie BBR250. d, Protein
reduced and carboxamidomethylated; e, protein not reduced
but carboxamidomethylated to prevent disulphide bond
rearrangement. (c) Western blot analysis of the gel shown in
part (b) using antiserum raised against γ-gliadins. Tracks f and
g correspond to d and e respectively in part (b). Although the
reduced sample gives a single major band corresponding to
the reduced monomer on the stained gel (track d), 'western
blotting' reveals that some unreduced dimer is also present
(track f). The stained separation of the non reduced sample
(track e) clearly shows the presence of dimer, oligomers and
polymers, including some protein which only just enters the
gel. The dimers and oligomers are also shown by 'western
blotting' (track g), but the higher mol. wt polymers are less
clearly shown, probably due to inefficient transfer.

of several HMW bands corresponding to protein oligomers stabilized by interchain disulphide bonds, which were converted into monomer in the presence of 2-mercaptoethanol (Fig. 8.6b).

It is clearly necessary to develop an effective *in vitro* folding system if the yeast expression system is to be used for protein engineering studies, and we are currently attempting to do this based on the system of Creighton (1986).

## Expression in eukaryotic cell cultures

Mammalian cell cultures have not, to the best of our knowledge, been used to express cereal prolamins, and are beyond the scope of this article. However, over the past few years a new cell culture method has been developed for the production of a wide variety of foreign proteins, using a recombinant baculovirus to introduce the foreign gene into cultured insect cells (see, for example, Miyamoto *et al.* 1985; Emery and Bishop 1987; Matsuura *et al.* 1987).

The *Autographa californica* nuclear polyhedrosis virus is a double-stranded DNA virus. It is released into the environment encased in a crystalline protein matrix, which provides protection until ingested by caterpillars of the host species. Digestion within the caterpillar releases the virus, resulting in infection. The crystalline matrix is composed predominantly of a protein called polyhedrin, which is encoded by the viral gene and produced during the later stages of infection. The polyhedrin gene has a very efficient promoter and polyhedrin can account for about 50% of all cellular proteins during the later stages of infection. By replacing the polyhedrin gene with a foreign gene it is possible to produce infected insect cells in which the foreign protein accounts for up to 50% of the total, for example the lymphocytic choriomeningitis virus nucleocapsid protein (Emery and Bishop 1987). This system has also been used to produce the bean storage protein phaseolin (Bustos *et al.* 1988).

The attraction of the baculovirus system lies partly in the efficiency of the polyhedrin promoter, which results in exceptionally high levels of expression. However, there are also other advantages resulting from the eukaryotic nature of the insect cells. The cells should recognize the processing signals of the expressed eukaryotic proteins, and carry out post-translational modifications such as disulphide bond formation, glycosylation and phosphorylation. This should dispense with the need for costly downstream processing when used on an industrial scale (Wright 1986). A further attraction for the expression of wheat gluten proteins is that polyhedrin forms insoluble masses (inclusions) within the cell, indicating that the unusual solubility properties of the gluten proteins might pose less of a problem than with other expression systems.

## Expression of a wheat HMW subunit of glutenin

We have attempted to express wheat gluten proteins in cultured cells of *Spodoptera frugiperda*, using vectors based on the *Autographa californica* nuclear polyhedrosis virus (AcNPV). This work has been in collaboration with Professor David Bishop and Dr Hilary Overton at the NERC Institute of Virology and Environmental Microbiology (Oxford).

The gene initially selected for expression in our laboratory encodes a high molecular weight (HMW) subunit called 1By9 (Halford *et al.* 1987). It encodes a protein of 705 residues, which includes a signal peptide of 21 residues. The molecular weight of the mature protein is 73 578. It has a clear domain structure (Fig. 8.7) with non-repetitive N- and C-terminal domains of 104 and 42 residues respectively flanking a repetitive domain of 538 residues. The latter consists of repeats based on two motifs, a hexapeptide (consensus ProGlyGlnGlyGlnGln) and a nonapeptide (consensus (GlyTyrTyrProThrSerLeuGlnGln).

The HMW subunits are of particular interest because they appear to be largely responsible for the elastic properties of gluten. In particular, allelic variation in the HMW subunit composition of different cultivars of wheat is closely correlated with differences in dough elasticity and breadmaking quality (Payne 1987). This correlation is almost certainly related to the formation of HMW (above $1 \times 10^6$) polymers that are stabilized by interchain disulphide bonds (Field *et al.* 1983). Our studies are, therefore, directed towards determining the relationships between the structures and interactions of the HMW subunits and gluten elasticity.

The entire coding sequence (Fig. 8.7) of the 1By9 gene was initially cloned into the transfer vector pAcYM1 (Matsuura *et al.* 1987). This construct, and the wild type baculovirus, were used to

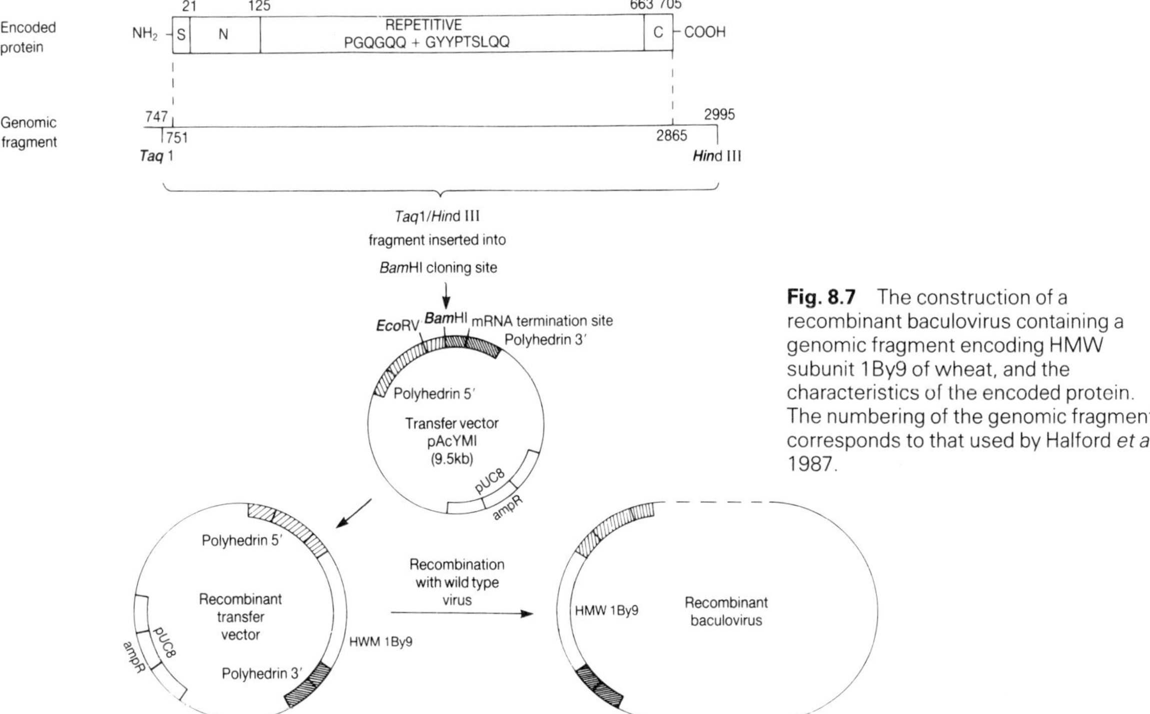



**Fig. 8.7** The construction of a recombinant baculovirus containing a genomic fragment encoding HMW subunit 1By9 of wheat, and the characteristics of the encoded protein. The numbering of the genomic fragment corresponds to that used by Halford *et al.* 1987.

coinfect cultured insect cells, which were screened for cells lacking polyhedra. Pure virus was obtained after several rounds of plaque purification and was used to reinfect cells. Labelling the proteins with [³⁵S]-methionine at 24 and 48 h after infection showed a major novel protein (Fig. 8.8), which bound to antibodies raised against the HMW subunits. However, the mol. wt of the novel protein (about 36 000) was much smaller than that of subunit 1By9, which has a mol. wt, by SDS-PAGE, of about 95 700 (Ng and Bushuk 1989). Analysis of virus isolated from the infected cells demonstrated that only the 5′ half of the HMW subunit gene was present, due to spontaneous truncation. A second recombinant virus, which was identified using a hybridization procedure, also produced a low mol. wt (40 000) polypeptide, but the nature of the rearrangement or mutation within the HMW subunit gene was not established. Similarly, more recent attempts to transfer all or specific parts of the 1By9 gene into the transfer vector have been unsuccessful because of rearrangement and plasmid instability (unpublished results of S. Thompson, P. Madgwick, P.R. Shewry and D.H.L. Bishop).

The disappointing results with the HMW subunit gene do not necessarily mean that baculovirus vectors cannot be used for other gluten proteins. In fact, more recent studies using a gene encoding a LMW subunit of glutenin (Colot *et al.* 1989) have given high level expression of an unrearranged protein (unpublished results of S. Thompson, P. Madgwick, P.R. Shewry and D.H.L. Bishop). This may relate to the fact that the repetitive domain of the LMW subunit is less well conserved than that of the HMW subunit, reducing the frequency of recombination.

The conclusion to be drawn from these studies is that gene instability may be a problem when expressing highly repetitive cereal prolamins using baculovirus vectors. Such instability is, as we have seen, also a problem in other expression systems, even in *E. coli* hosts with a *rec A⁻* system.

## Conclusions

It is clear that the development of high level expression systems for cereal prolamins is still in its infancy, and that systems that give good expression of other eukaryotic proteins are often unsuccessful. This may relate to instability of the prolamin genes, toxicity of the expressed protein, or other unidentified factors such as codon usage and availability of

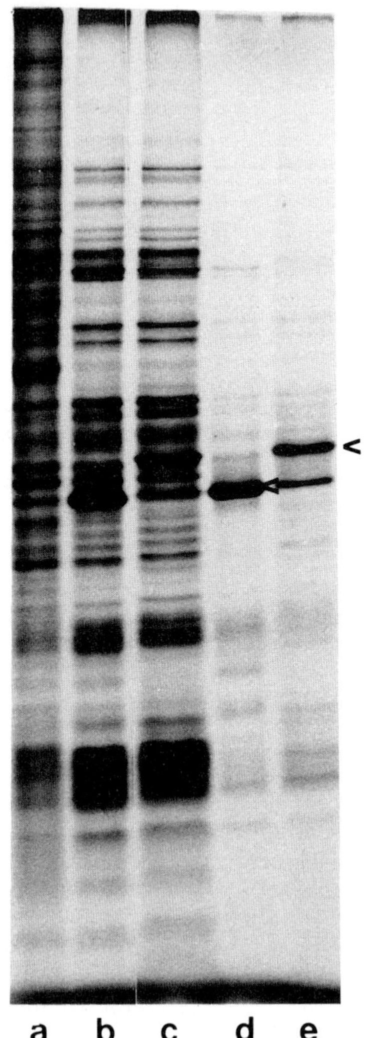



**Fig. 8.8** Analysis of protein synthesis in cultured cells of *Spodoptera frugiperda* by incubation with [$^{35}$S]-methionine followed by SDS-PAGE and autoradiography. a, Proteins from uninfected cells; b and d, proteins from cells infected with wild type baculovirus, and labelled at 24 and 48 h after infection respectively; c and e, proteins from cells coinfected with wild type and recombinant baculovirus, and labelled at 24 and 48 h after infection, respectively. The arrows in tracks d and e indicate the polyhedrin and truncated HMW subunit protein respectively.

tRNAs for some amino acids. Nevertheless, adequate levels of expression of some proteins (HMW subunits in *E. coli*, γ-gliadin in yeast, LMW subunit in baculovirus) have been achieved, and it is now possible to initiate protein engineering studies. These will enable us to explore the relationship between the structural and functional properties of these economically important proteins and, together with improved methods for plant transformation and regeneration, lead to the production of 'designer' wheat varieties for specific end uses.

## Acknowledgements

We are indebted to Dr S.M. Kingsman and her colleagues (notably Dr C. Stanway and Dr S. Fulton) at the University of Oxford for supplying yeast expression vectors and for advice on yeast expression, and to Professor D.H.L. Bishop and Dr H. Overton (NERC, Oxford) for advice and assistance with the baculovirus studies, and for supplying vectors and cell lines. We are also grateful to Professor R.B. Freedman and Dr M. Tuite (University of Kent, Canterbury) for advice on the expression of the γ-gliadin in yeast. The γ-gliadin cDNA was supplied by Dr R. Thompson (IPSR, Cambridge Laboratory) and the HMW subunit gene by Dr N.G. Halford (Rothamsted). We are also grateful to Dr G. Galili (Weizmann Institute) for supplying Figure 8.3.

## References

Bartels, D., Thompson, R.D. and Rothstein, S. (1985). Synthesis of a wheat storage protein subunit in *Escherichia coli* using novel expression vectors. *Gene*. **35**: 159–67.

Bartels, D., Altosaar, I., Harberd, N.P., Barker, R.F. and Thompson, R.D. (1986). Molecular analysis of γ-gliadin gene families at the complex *Gli-1* locus of bread wheat (*T. aestivum* L.). *Theoretical and Applied Genetics*. **72**: 845–53.

Bunce, N., White, R.P. and Shewry, P.R. (1985). Variation in estimates of molecular weights of cereal prolamins by SDS-PAGE. *Journal of Cereal Science*. **3**: 131–42.

Bustos, M.M., Lucknow, V.A., Griffing, L.R., Summers, M.D. and Hall, T.C. (1988). Expression, glycosylation and secretion of phaseolin in a baculovirus system. *Plant Molecular Biology*. **10**: 475–88.

Bietz, J.A. and Wall, J.S. (1980). Identity of higher molecular weight gliadin and ethanol-soluble glutenin subunits of wheat: relation to gluten structure. *Cereal Chemistry*. **57**: 415–21.

Colot, V., Bartels, D., Thompson, R. and Flavell, R. (1989). Molecular characterization of an active wheat LMW glutenin gene and its relation to other wheat and barley prolamin genes. *Molecular and General Genetics.* **216**: 81–90.

Compagno, C., Coraggio, I., Ranzi, B.M., Alberghina, A.L., Viotti, A. and Martegani, E. (1987). Translational regulation of the expression of zein cloned in yeast under an inducible GAL promoter. *Biochemical and Biophysical Research Communications.* **146**: 809–14.

Coraggio, I., Compagno, C., Martegani, E., Ranzi, B.M., Sala, E., Alberghina, L. and Viotti, A. (1986). Transcription and expression of zein sequences in yeast under natural plant or yeast promoters. *EMBO Journal.* **5**: 459–65.

Creighton, T.E. (1986). Folding of proteins absorbed reversibly to ion-exchange resins. In *Protein Structure, Folding and Design* (ed. D.L. Oxender), pp. 249–57. Alan R. Liss Inc., New York.

Edens, L., Bom, I., Ledeboer, A.M., Maat, J., Toonen, M.Y., Visser, C. and Verrips, C.T. (1984). Synthesis and processing of plant protein thaumatin in yeast. *Cell.* **37**: 629–33.

Emery, V.C. and Bishop, D.H.L. (1987). The development of multiple expression vectors for high level synthesis of eukaryotic proteins: expression of LCMV-N and AcNPV polyhedrin protein in a recombinant baculovirus system. *Protein Engineering.* **1**: 359–66.

Ernst, J.F. (1988). Efficient secretion and processing of heterologous proteins in *Saccharomyces cerevisiae* is mediated solely by the pre-segment of a α-factor precursor. *DNA.* **7**: 355–60.

FAO. (1988). *Production Yearbook*, Vol. 41, 1987. FAO, Rome.

Faoro, F., Tornaghi, R., Coraggio, I., Martegani, E. and Viotti, A. (1987). Immunocytochemical localization of zein in transformed yeast cells *Atti del XVI Congresso di M.E. Bologna 1987*, pp. 73–4.

Festenstein, G.W., Hay, F.C. and Shewry, P.R. (1987). Immunochemical relationships of the prolamin storage proteins of barley, wheat, rye and oats. *Biochemica et Biophysica Acta.* **912**: 371–83.

Field, J.M., Shewry, P.R. and Miflin, B.J. (1983). Solubilization and characterization of wheat gluten proteins; correlations between the amount of aggregated proteins and baking quality. *Journal of the Science of Food and Agriculture.* **34**: 370–7.

Galili, G. (1989). Heterologous expression of a wheat high molecular weight glutenin gene in *Escherichia coli. Proceedings of the National Academy of Science, USA.* **86**: 7756–60.

Halford, N.G., Ford, J., Anderson, D.D., Greene, F.C. and Shewry, P.R. (1987). The nucleotide and deduced amino acid sequences of an HMW glutenin gene from chromosome 1B of bread wheat (*Triticum aestivum* L.) and comparison with those of genes from chromosomes 1A and 1D. *Theoretical and Applied Genetics.* **75**: 117–26.

Jackson, E.A., Ballance, G.M. and Thomsen, K.K. (1986). Construction of a yeast vector directing the synthesis and release of barley (1-3, 1-4)-β-glucanase. *Carlsberg Research Communications.* **51**: 445–58.

Kingsman, S.M., Cousens, D., Stanway, C.A., Chambers, A., Wilson, M. and Kingsman, A.J. (1990). High efficiency yeast expression vectors based on the promoter of the phosphoglycerate kinase gene. *Methods in Enzymology.* **185**: 329–44.

Kreis, M. and Shewry, P.R. (1989). Unusual features of seed protein structure and evolution. *Bio-Essays.* **10**: 201–7.

Matsuura, Y., Posse, R.D., Overton, H.A. and Bishop, D.H.L. (1987). Baculovirus expression vectors: the requirements for high level expression of proteins, including glycoproteins. *Journal of General Virology.* **68**: 1233–50.

Mellor, J., Dobson, M.J., Roberts, N.A., Kingsman, A.J. and Kingsman, S.M. (1983). Efficient synthesis of enzymatically active calf chymosin in *Saccharomyces cerevisiae. Gene.* **24**: 1–14.

Miller, T.E. and Koebner, R.M.D. (ed. 1988). *Proceedings of the Seventh International Wheat Genetics Symposium*, Vols 1 and 2. IPSR, Cambridge, UK.

Miyamoto, C., Smithe, G.E., Farrel-Towt, J., Chizzonile, R., Summers, M.D. and Ju, G. (1985). Production of human *c-myc* protein in insect cells infected with a baculovirus expression vector. *Molecular and Cell Biology.* **5**: 2860–5.

Neill, J.D., Litts, J.C., Anderson, O.D., Greene, F.C. and Stiles, J.I. (1987). Expression of a wheat α-gliadin gene in *Saccharomyces cerevisiae. Gene.* **55**: 303–17.

Ng, P.K.W. and Bushuk, W. (1989). Concerning the nomenclature of the high molecular weight glutenin subunits. *Journal of Cereal Science.* **9**: 53–60.

Norrander, J.M., Vieira, J., Rubenstein, I. and Messing, J. (1985). Manipulation and expression of the maize zein storage proteins in *Escherichia coli. Journal of Biotechnology.* **2**: 157–75.

Payne, P.I. (1987). Genetics of wheat storage proteins and the effect of allelic variation on breadmaking quality. *Annual Review of Plant Physiology.* **38**: 141–53.

Richardson, P.T., Roberts, L.M., Gould, J.H., Smith, A.B. and Lord, M. (1987). The expression of ricin β-chain in *Saccharomyces cerevisiae. Biochemistry Society Transactions.* **15**: 903–4.

Rothstein, S.J., Lazarus, C.M., Smith, W.R., Baulcombe, D.C. and Gatenby, A.A. (1984). Secretion of wheat α-amylase expressed in yeast. *Nature, London.* **308**: 662–5.

Scheets, K. and Hedgcoth, C. (1989). Expression of a wheat γ-gliadin in *Saccharomyces cerevisiae* from a yeast ADH1 promoter. *Journal of Agricultural and Food Chemistry.* **37**: 829–33.

Shewry, P.R. and Miflin, B.J. (1985). Seed storage proteins of economically important cereals. In *Advances in Cereal Science and Technology*, Vol. VII. (ed. Y. Pomeranz), pp. 1–83. AACC, St. Paul, Minnesota.

Shewry, P.R. and Tatham, A.S. (1990). The structures of wheat gluten proteins and of related proteins from barley, rye and oats: relationship to coeliac disease. In *Proceedings of the 4th International Coeliac Symposium*. In press.

Shewry, P.R., Miflin, B.J., Lew, E.J.-L. and Kasarda, D.D. (1983). The preparation and characterization of an aggregated gliadin fraction from wheat. *Journal of Experimental Botany*. **34**: 1403–10.

Shewry, P.R., Tatham, A.S., Forde, J., Kreis, M. and Miflin, B.J. (1986). The classification and nomenclature of wheat gluten proteins: a reassessment. *Journal of Cereal Science*. **4**: 97–106.

Tatham, A.S. and Shewry, P.R. (1985). The conformation of wheat gluten proteins. The secondary structures and thermal stabilities of $\alpha,\beta,\gamma$, and $\omega$-gliadins. *Journal of Cereal Science*. **3**: 103–13.

Tatham, A.S., Miflin, B.J. and Shewry, P.R. (1985). The $\beta$-turn conformation in wheat gluten proteins: relationship to gluten elasticity. *Cereal Chemistry*. **62**: 405–12.

Tatham, A.S., Field, J.M., Smith, S.J. and Shewry, P.R. (1987). The conformation of wheat gluten proteins 2. Aggregated gliadins and low molecular weight subunits of glutenin. *Journal of Cereal Science*. **5**: 203–14.

Tatham, A.S., Shewry, P.R. and Belton, P.S. (1990a). Structural studies of cereal prolamins, including wheat gluten. *Advances in Cereal Science and Technology*, Vol. **10**: 1–78. (ed. Y. Pomeranz) AACC, St. Paul, Minnesota. In press.

Tatham, A.S., Pineau, F. and Popineau, Y. (1990b). Conformational studies of peptides derived by the enzymic hydrolysis of a $\gamma$-type gliadin. *Journal of Cereal Science*. **11**: 1–13.

Tao, H.P. and Kasarda, D.D. (1989). Two-dimensional gel mapping and N-terminal sequencing of LMW glutenin subunits. *Journal of Experimental Botany*. **40**: 1015–20.

Thompson, R.D., Bartels, D., Harberd, N.P. and Flavell, R.B. (1983). Characterization of the multigene family coding for the HMW glutenin subunits in wheat using cDNA clones. *Theoretical and Applied Genetics*. **67**: 87–96.

Tuite, M.F., Dobson, M.J., Roberts, N.A., King, R.M., Burke, O.C., Kingsman, S.M. and Kingsman, A.J. (1982). Regulated high efficacy expression of human interferon-alpha in *Saccharomyces cerevisiae*. *EMBO Journal*. **1**: 603–8.

Utsumi, S., Sato, T., Kim, C.S. and Kito, M. (1988). Processing of preproglycinin expressed from cDNA-encoding $A_{1a}B_{1b}$ subunit in *Saccharomyces cerevisiae*. *FEBS Letters*. **233**: 273–6.

Wang, S.-Z. and Esen, A. (1985). Expression of maize prolamins in *Escherichia coli*. *Plant Science*. **42**: 49–54.

Woychick, J.H., Boundy, J.A. and Dimler, R.J. (1961). Starch gel electrophoresis of wheat gluten proteins with concentrated urea. *Archives of Biochemistry and Biophysics*. **94**: 477–82.

Wright, K. (1986). Insect virus as super-vector? *Nature*. **321**: 718.

Yarwood, J.N., Harris, N., Delauney, A., Croy, R.R.D., Gatehouse, J.A., Watson, M.D. and Boulter, D. (1987). Construction of a hybrid cDNA encoding a major legumin precursor polypeptide and its expression and localization in *Saccharomyces cerevisiae*. *FEBS Letters*. **222**: 175–180.

# 9

# Exploring the Structure and Assembly of Wheat Storage Proteins Using an *in vitro* Transcription/Translation System

Neil J. Bulleid, Peter R. Shewry and Robert B. Freedman

## Introduction

Early studies of the details of the biosynthesis of wheat cereal storage proteins were limited by the complex nature of this group of proteins and the difficulty in isolating individual mRNA species (Forde and Miflin 1983; Shewry *et al.* 1988). However, these early studies did show that wheat cereal storage proteins are synthesized on membrane-bound polysomes; when the mRNA was translated *in vitro* in absence of endoplasmic reticulum (e.r.) membranes, it gave rise to polypeptides with a slightly higher molecular weight than the native proteins, due to uncleaved signal sequence (Donovan *et al.* 1982; Greene 1981).

Immunocytochemical studies have indicated that these proteins are synthesized on the e.r. and routed via the Golgi apparatus before being deposited in the cell as high molecular weight (HMW) aggregates called protein bodies (Graham *et al.* 1962; Kim *et al.* 1988). Analysis of these protein bodies reveals that they are a heterogeneous mixture of gliadins and glutenins (Field *et al.* 1983) that are stabilized by both inter- and intramolecular cross-links, including disulphide bonds. Information on the formation of disulphide bonds in these proteins is limited by the insoluble nature of this complex and the difficulty in purification of individual components.

A number of genes coding for HMW- and LMW-glutenins and for gliadins have now been isolated and sequenced, and their deduced amino acid sequences reveal the positions and numbers of cysteine residues (Fig. 9.1). The gliadins contain even numbers of cysteine residues in their C-

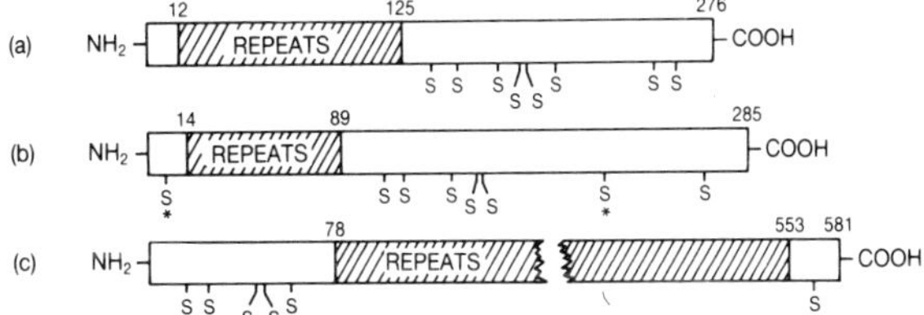



**Fig. 9.1** Schematic structures of wheat prolamins. (a) γ-gliadin. (b) LMW-glutenin. (c) HMW-glutenin subunit 12. Based on Bartels *et al.* (1986), Okita *et al.* (1985) and Thompson *et al.* (1985). The cysteine residues in the LMW-subunit, indicated by asterisks are predicted to be unpaired.

terminal domain and are thought to form only intramolecular disulphide bonds, giving rise to soluble monomeric proteins. The LMW-glutenins have a similar overall structure to the gliadins, but some components, including the one studied here, have a cysteine residue within the N-terminal domain (Okita *et al.* 1985; Colot *et al.* 1989; Tao and Kasarda 1989). This is unlikely to form intramolecular disulphide bonds because of the rigidity imposed by the repeat region, and is more likely to form intermolecular disulphide bonds. In addition, the LMW subunits have seven cysteine residues in their C-terminal domains and comparison with the γ-gliadins allows us to predict which of these is unpaired and therefore available to form intermolecular bonds. The formation of intermolecular disulphide bonds by these two cysteine residues will give rise to extended oligomeric structures. The HMW-glutenins contain cysteine residues, mainly at the ends of the molecule, suggesting that these proteins could form linear or branched polymers joined by intermolecular disulphide bonds (Ewart 1972, 1977; Payne 1986).

In order to understand more about the formation of disulphide bonds during protein synthesis, we have used a cell-free system to study their formation during the *in vitro* translation of mRNA coding for individual cereal storage proteins (Bulleid and Freedman 1988a). Genes coding for these proteins have been subcloned into vectors containing bacteriophage promoters, which allow the *in vitro* transcription of cloned DNA (Kreig and Melton 1984). The resulting RNA transcript can be used to direct the synthesis of the polypeptide coded by the gene, in a message-dependent cell-free translation system such as rabbit reticulocyte lysate. Using this system we have been able to study the formation of disulphide bonds in a γ-gliadin, LMW-glutenin and several HMW subunits. We have investigated the role that the enzyme protein disulphide isomerase has in the catalysis of disulphide bond formation. A study has also been made of the role of sequence in determining structural differences between two HMW-glutenin subunits, using truncated and chimeric coding sequences.

## The formation of disulphide bonds in cereal storage proteins

Wheat storage proteins are synthesized on the endoplasmic reticulum, and are thus cotranslationally translocated into the lumen where signal peptide cleavage occurs. We can reconstruct these early events in protein synthesis by employing a cell-free system. The advantage of this approach is that individual components of the system can be manipulated, enabling their roles in protein synthesis, translocation and post-translational modification to be evaluated. Conventionally a nuclease-treated, mRNA-dependent rabbit reticulocyte lysate (Pelham and Jackson 1976) is used in the presence of dog pancreas microsomes (Walter and Blobel 1983); this system will translate exogenously added mRNA, and cotranslationally transfer the protein product to the interior of the microsomal vesicles where the signal sequence is removed by signal peptidase to yield the mature polypeptides. Although the most stable and active preparation is derived from animal sources (rabbit and dog) the evolutionary conservation of the system is such that these preparations are entirely effective towards plant mRNA species (Dobberstein and Blobel 1977; Bassuner *et al.* 1984). Folding and disulphide bond formation are not normally monitored for two reasons:

(1) Thiol reductants (e.g. DTT) are normally added to the reticulocyte lysate because several components are susceptible to oxidative inactivation.
(2) The analysis of radiolabelled translation products, to determine whether signal peptide cleavage has occurred, is normally carried out by SDS-PAGE (in the presence of reducing agents) to obtain good resolution and accurate determination of protein molecular weight.

Hence, translation is carried out in conditions in which disulphide bonds are unlikely to form, and analysis of translation products is performed in conditions where any disulphide bonds would be broken.

The formation of disulphide bonds in proteins at synthesis can be followed *in vitro* by using specialized cell-free systems. Scheele and Jacoby (1982) titrated oxidized glutathione (GSSG) into a gel-filtered reticulocyte lysate that had been supplemented with DTT. They demonstrated that in these conditions, the products of translation of pancreatic zymogen mRNAs were correctly processed, translocated and disulphide bonded. The methods exploited differences in solubility and electrophoretic properties of the reduced, unfolded and oxidized, and folded forms of the zymogens. Kaderbhai and Austen (1985) made similar observations on the cotranslational folding and oxidation of prolactin using a non-gel-filtered lysate in the presence of DTT and GSSG. We have subsequently prepared a reticulocyte lysate using standard procedures, but without adding any exogenous reducing agents. The lysate prepared in this fashion was characterized for disulphide bond formation by using pituitary poly($A^+$) RNA which codes predominantly for prolactin (Bulleid and Freedman 1988a).

In both our study and the work of Kaderbhai and Austen (1985), the formation of disulphide bonds by prolactin was monitored by performing SDS-PAGE under non-reducing conditions. The disulphide bonded product has a greater mobility than the reduced protein, owing to the formation of a more compact structure, which decreases its hydrodynamic volume (Goldenberg and Creighton 1984). Hence the reduced and disulphide bonded forms of the protein can be distinguished, and the formation of disulphide bonds can be monitored.

We found that prolactin, synthesized in our lysate in the presence of microsomes to cleave the signal sequence, could form intramolecular disulphide bonds without the addition of GSSG, presumably using dissolved $O_2$ as ultimate oxidant (Fig. 9.2). We have used this lysate to study the formation of disulphide bonds in cereal storage proteins.

Because most of the analyses of the translation products involved changes of mobility on SDS-PAGE, it was necessary to study the synthesis of individual proteins. This was achieved by transcribing cDNA and genomic clones *in vitro* to yield RNA transcripts coding for single proteins. Genes of interest were subcloned into one of the commercially available vectors (pGEM, Bluescribe or Bluescript) that contain bacteriophage promoters flanking the multiple cloning site. pGEM is available from Promega Biotech and contains the SP6 and T7 promoters; Bluescribe and Bluescript are available from Stratagene and contain the T3 and T7 promoters. Full details of these plasmids can be found in the manufacturers' catalogues. The resulting constructs were linearized downstream from the coding region and the cloned DNA transcribed in the presence of ribonucleotides and the appropriate bacteriophage RNA polymerase. A variety of genes were subcloned and transcribed in this fashion, as shown in Table 9.1.

The RNA transcribed from these genes was translated in our cell-free system under conditions that allow the formation of disulphide bonds. As reported previously (Bulleid and Freedman 1988a), the γ-gliadin can form intramolecular disulphide bonds *in vitro*. This is consistent with the physical properties of the gliadins in that they are monomeric globular proteins. The LMW-glutenin could also form intramolecular disulphide bonds as

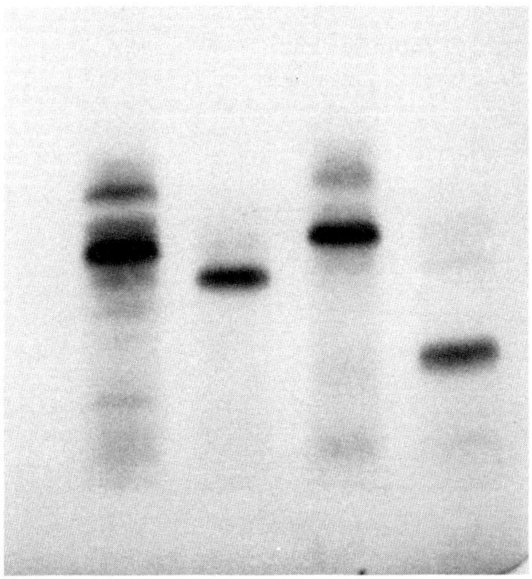

**Fig. 9.2**   Cell-free translation of pituitary poly(A)$^+$ RNA under conditions favouring the formation of disulphide bonds. Cell-free translation was carried out in the absence (lanes 1 and 3) or presence (lanes 2 and 4) of dog pancreas microsomes, in a rabbit reticulocyte lysate prepared without the addition of DTT. Translation products were separated by SDS-PAGE on a 11%-(w/v)-polyacrylamide gel after being reduced and carboxyamidomethylated (lanes 1 and 2) or after being carboxyamidomethylated (lanes 3 and 4).

shown in Fig. 9.3 (compare the mobility difference between lanes 3 and 4). No well resolved oligomeric structures were formed. The LMW-glutenins have a very similar structure to the γ gliadins and it was not unexpected that they could form intramolecular disulphide bonds within the C-terminal domain of the protein (see Fig. 9.1). However, within their N-terminal domain there is an additional Cys residue which is thought to be involved in the formation of intermolecular disulphide bonds. That we did not see any oligomeric

**Table 9.1**   Subcloned and transcribed genes

| Original clone name | Source/ reference | pUKC vector no | Restriction enzyme used to linearize | Protein encoded |
|---|---|---|---|---|
| pSBB17 | Chinese Spring[1] | 1000 | Sal1 | HMW-glutenin subunit 12 full length[2] |
|  |  |  | Fok1 | HMW-glutenin subunit 12-truncation N-terminal domain and some repeat region[3] |
| pTag1436 | Chinese Spring[4] | 1001 | HindIII | γ-gliadin with repeat deletion[5] |
| λCH3501 | Chinese Spring[6] | 1003 | EcoRI | LMW-glutenin subunit |
| PTag1436 | Chinese Spring[4] | 1005 | HindIII | γ-gliadin full-length |
| pWS10 | HOPE[7] | — | Sal1 | HMW glutenin subunit 10 |
| pWSH10/12 |  |  | Sal1 | HMW-glutenin subunit 10/12 hybrid |
| pWSH12/10 |  |  | Sal1 | HMW-glutenin subunit 12/10 hybrid |
| λHMW47 | Cheyonne[8] | 1008 | Sal1 | HMW-glutenin subunit 9 |

References [1]Thompson *et al.* 1985; [2]Bulleid and Freedman 1988a; [3]Goldsbrough *et al.* 1988; [4]Bartels *et al.* 1986; [5]Bulleid and Freedman 1988b; [6]Colot *et al.* 1989; [7]Goldsbrough *et al.* 1989; [8]Halford *et al.* 1987.

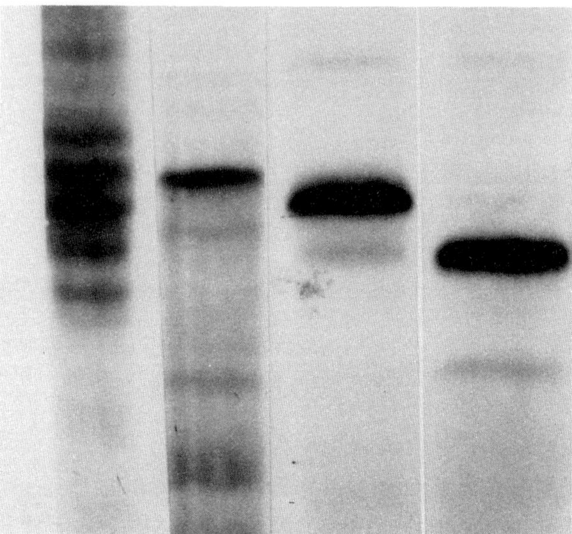

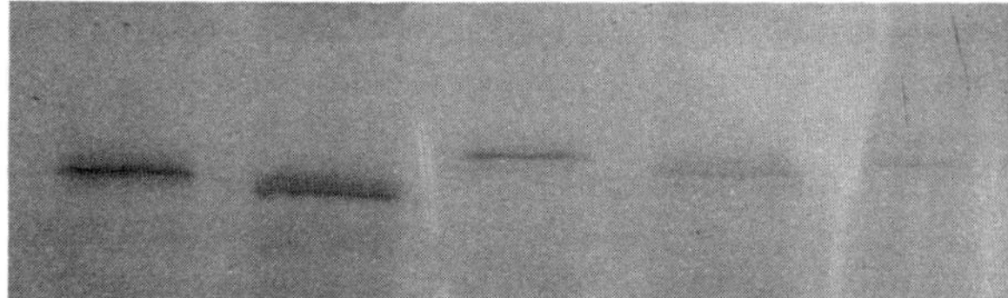

**Fig. 9.4**    Cell-free translation of RNA coding for HMW-glutenin subunits 10 and 12. Translations were carried out in the presence of microsomal vesicles and products of translation separated by SDS-PAGE either in the presence (lanes 1 and 3) or absence (lanes 2 and 4) of reducing agent DTT. Lanes 1 and 2, products of translation of subunit 10 mRNA; lanes 3 and 4, products of translation from subunit 12 mRNA.

**Fig. 9.3**    Cell-free translation of RNA coding for LMW-glutenin. Translation was carried out as described previously (Bulleid and Freedman 1988a). Lane 1, [$^{14}$C]-labelled wheat cereal storage proteins; lane 2, products of translation in the absence of added microsomes; lane 3, products of translation in the presence of microsomes; lane 4, as lane 3 but samples were separated in the absence of added DTT. The LMW-subunit gene was isolated by Colot *et al.* (1989).

structures may well reflect the low concentrations of translation products within the microsomal vesicles or the short time course over which translation *in vitro* is performed. Alternatively, it may be a consequence of the use of microsomal vesicles from dog pancreas; development of a fully wheat-derived *in vitro* translation system would resolve this question. A further alternative is that intermolecular disulphides in this protein may not be formed in the lumen of the e.r. *in vivo*, but in a later subcellular compartment.

Translation of the HMW-glutenin subunits 10 and 12 resulted in the synthesis of proteins that had slightly faster mobilities on SDS-PAGE carried out under non-reducing conditions than the fully-reduced protein (Fig. 9.4). This may be due to the formation of the intramolecular disulphide bonds within the N-terminal domain of these proteins. Again, no oligomeric structures were observed, possibly for the reasons outlined above. Interestingly, Galili has recently shown that he could identify oligomeric structures when HMW-glutenin subunit 2 was expressed in *Escherichia coli* (Galili 1989), indicating that this subunit can self-assemble. He has also expressed HMW-glutenin

subunit 12 in *E. coli*, but has not been able to identify oligomeric structures formed by this protein (personal communication). It remains to be seen whether HMW-glutenin subunit 2 can form oligomeric structures when translated *in vitro*.

## The role of protein disulphide isomerase in the formation of disulphide bonds

As mentioned earlier, many cereal storage proteins are present in the developing seed as disulphide-linked aggregates. Very little is known about the mechanism of formation of disulphide bonds during the biosynthesis of cereal storage proteins. However, it now seems quite clear that the rapid formation of disulphide bonds in animal secretory proteins at synthesis requires catalysis and that the enzyme protein disulphide isomerase (PDI) (EC5.3.4.1.) performs this role (Freedman 1984; Freedman *et al.* 1989). This enzyme has been found in a wide variety of mammalian tissues active in the biosynthesis and secretion of disulphide-bonded proteins. Its activity has also been detected in the mature seeds of wheat, both in the germ and endosperm fractions (Grynberg 1978). The enzyme is located in the endoplasmic reticulum of developing wheat endosperm (Roden *et al.* 1982), a location consistent with a role in the formation of disulphide bonds during the synthesis of cereal storage proteins. We have recently purified this protein from wheat endosperm and shown that it has very similar properties to those of the mammalian enzyme. Thus it has a similar isoelectric point, it exists as a dimer with a mol. wt of 105 000 by gel filtration and has a monomeric mol. wt of approximately 60 000.

The purification of wheat PDI follows a similar procedure to that devised for mammalian PDI. Wheat endosperm is first extracted in a buffer containing protease inhibitors and centrifuged to separate insoluble material. The supernatant is subjected to an ammonium sulphate fractionation with the precipitate between 50 and 85% saturation being collected. This is resuspended and dialysed against citrate buffer at pH 5.3 and applied to a cation exchange column equilibrated in this buffer. The void fraction is collected and dialysed against phosphate buffer, pH 6.3, and applied to an anion exchange column equilibrated with this buffer. The PDI is eluted with a salt gradient and finally purified by gel filtration. Analysis of the purified protein by SDS-PAGE (Fig. 9.5) shows that it is slightly larger than the mammalian protein.

We have used the biosynthesis and disulphide bond formation of γ-gliadin as a model system to study the role that PDI has in this process. The enzyme is abundant in the dog pancreas microsomes that are used in our translation system. We can selectively deplete these microsomes of PDI by a mild alkaline wash (Paver *et al.* 1989). The resulting washed microsomes, which retain < 5% of the original level of PDI, are still active in cotranslational translocation and processing. When γ-gliadin mRNA is translated either in the presence of untreated dog pancreas microsomes or of PDI-depleted microsomes a clear difference is observed in the percentage of translation products that are disulphide bonded (Fig. 9.6a, b). In the presence of control microsomes, 90% of the translocated translation products are disulphide bonded, compared to less than 30% in the presence of PDI-depleted microsomes. Furthermore, reincorpora-

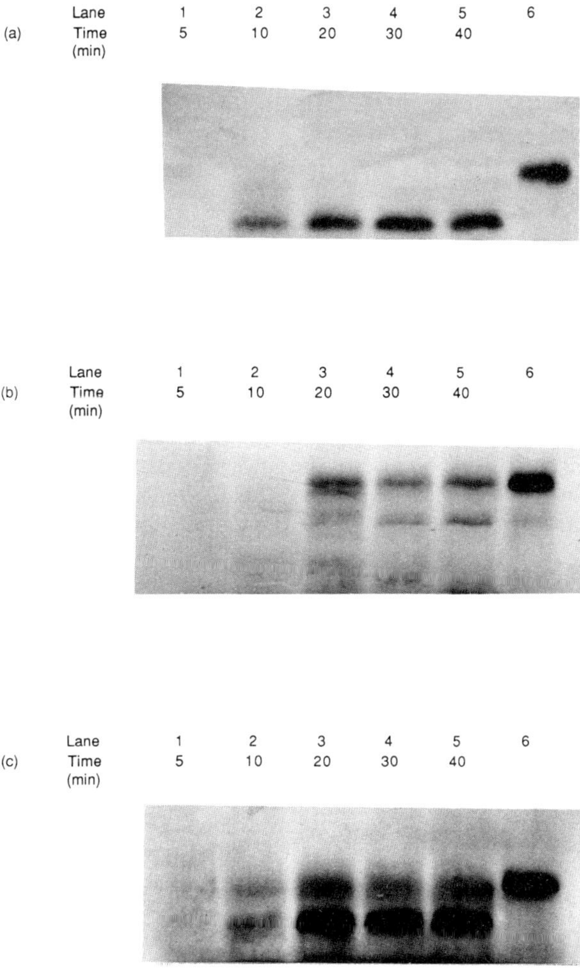

**Fig. 9.6** Time course of disulphide bond formation in newly synthesized γ-gliadin. mRNA coding for γ-gliadin, obtained by transcription *in vitro*, was translated in the presence of: (a) native microsomes; (b) pH 9-washed microsomes; (c) pH 9-washed microsomes reconstituted with purified PDI. Samples were taken at various time points after initiation of translation and separated through an 11% polyacrylamide gel under non-reducing conditions (lanes 1–5) or reducing conditions (lane 6).

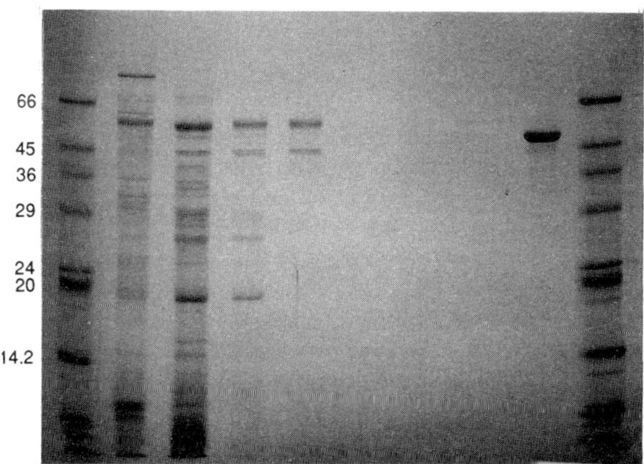

**Fig. 9.5** SDS-PAGE analysis of the stages of purification of wheat PDI. Samples were separated by SDS-PAGE and stained with Coomassie Blue. Samples were loaded as follows: lanes 1 and 10, molecular weight markers (Sigma), mol. wt as indicated; lane 2, crude endosperm extract; lane 3, ammonium sulphate precipitate (50–85%); lane 4, cation exchange column void fraction; lane 5, anion exchange column, salt eluate; lanes 6, 7 and 8, purified wheat PDI after gel-filtration with 1, 2 and 3 μg, respectively; lane 9, purified bovine PDI.

tion of purified bovine PDI into these microsomes reversed the defect in disulphide bond formation (Fig. 9.6c) (Bulleid and Freedman 1988b). This indicates that PDI has a role in the formation of disulphide bonds in this cereal storage protein when it is expressed in a mammalian cell-free system. The presence of PDI in the endoplasmic reticulum of wheat endosperm suggests that the enzyme also has a role in this process *in vivo*. Further work is required to characterize wheat PDI and to establish the level of homology between the mammalian and wheat proteins.

## A comparison of HMW-glutenin subunits 10 and 12 by analysis of translation products

It has been known for some time that the HMW-glutenin subunit 10 has a lower mobility relative to subunit 12 in SDS-PAGE after extraction from wheat seeds (Payne *et al.* 1981). When subunits 10 and 12 genes were sequenced it was realized that subunit 10 was twelve amino acids shorter than subunit 12 (Thompson *et al.* 1985; Goldsbrough *et al.* 1988). This anomalous mobility can be demonstrated by transcription and translation of the isolated genes *in vitro*. Thus, the products of translation *in vitro* have the same physical properties in terms of electrophoretic mobility, as the authentic proteins (Goldsbrough *et al.* 1989). This anomalous mobility can be destroyed if the chaotropic agent urea is added to the sample and the SDS-containing gels. In this case, subunit 10 has the higher mobility, consistent with its smaller size. This demonstrates that the anomalous mobility in SDS-PAGE is due to the incomplete denaturation of the proteins in the presence of SDS.

We have investigated the role that sequence has in maintaining this SDS-resistant structure. By the construction of chimeric genes we have been able to pinpoint a region of 171 amino acids containing just six amino acid differences between the two proteins. This is responsible for the anomalous electrophoretic mobility (Goldsbrough *et al.* 1989). These two HMW-glutenin subunits are closely related in sequence but are associated with different baking quality (Pogna *et al.* 1987). It remains to be seen whether the differences in electrophoretic mobilities are a reflection of those differences in the physical properties of these proteins that determine their breadmaking quality or viscoelasticity.

## Conclusions

We have used the *in vitro* synthesis of individual wheat cereal storage proteins to study both the formation of disulphide bonds and the proteins involved in the catalysis of this process. We have also shown that the cell-free expression of two HMW-glutenin subunits can be used to analyse conformational differences between these proteins. These techniques have obvious limitations but provide an alternative to expression of proteins in heterologous systems and can provide some information on the type of disulphide bonds formed and on the physical properties of the proteins as determined by electrophoretic techniques.

## Acknowledgements

We are grateful to Dr R.D. Thompson, Dr V. Colot and Dr N.G. Halford for supplying wheat cDNAs and genes used in this study. The work is supported by a Linked Research Group grant (LRG 174) from the Agricultural and Food Research Council.

## References

Bartels, D., Altesaar, I., Harberd, N.P., Barker, R.F. and Thompson, R.D. (1986). Molecular analysis of γ-gliadin gene families at the complex *Gli-1* locus of bread wheat (*T. aestivum* L.) *Theoretical and Applied Genetics*. **72**: 845–53.

Bassuner, R., Wobus, U. and Rapoport, T.A. (1984). Signal recognition particle triggers the translocation of storage globulin polypeptides from field beans (*Vicia faba* L.) across mammalian endoplasmic reticulum membrane. *FEBS*. **166**: 2: 1188.

Bulleid, N.J. and Freedman, R.B. (1988a). The transcription and translation *in vitro* of individual cereal storage-proteins from wheat (*Triticum aestivum*, cv-Chinese Spring). *Biochemical Journal*. **254**: 805–10.

Bulleid, N.J. and Freedman, R.B. (1988b). Defective co-translational formation of disulphide bonds in protein disulphide-isomerase-deficient microsomes. *Nature*. **335**: 649–51.

Colot, V., Bartels, D., Thompson, R.D. and Flavell, R. (1989). Molecular characterization of an active LMW glutenin gene and its relation to other wheat and barley prolamin genes. *Molecular and General Genetics*. **216**: 81–90.

Dobberstein, B. and Blobel, G. (1977). Functional interaction of plant ribosomes with animal microsmal membranes. *Biochemical and Biophysical Research Communications*. **74**: 1675–82.

Donavan, G.R., Lee, J.W., Longhurst, T.J. (1982). Cell-free synthesis of wheat prolamins. *Australian Journal of Plant Physiology*. 9: 59–68.

Ewart, J.A.D. (1972). A modified hypothesis for the structure and rheology of glutenins. *Journal of the Science of Food Agriculture*. 23: 687–99.

Ewart, J.A.D. (1977). Re-examination of the linear glutenin hypothesis. *Journal of the Science of Food Agriculture*. 281: 191–9.

Field, J.M., Shewry, P.R., Burgess, S.R., Forde, J., Parmar, S. and Miflin, B.J. (1983). The presence of high molecular weight aggregates in the protein bodies of developing endosperms of wheat and other cereals. *Journal of Cereal Science*. 1: 33–41.

Forde, J. and Miflin, B.J. (1983). Isolation and identification of mRNA for the high-molecular-weight storage proteins of wheat endosperm. *Planta*. 157: 567–76.

Freedman, R.B. (1984). Native disulphide bond formation in protein biosynthesis: evidence for the role of protein disulphide isomerase. *Trends in Biological Sciences*. 9: 438–41.

Freedman, R.B., Bulleid, N.J., Hawkins, H.C. and Paver, J.L. (1989). Role of protein disulphide-isomerase in the expression of native proteins. *Biochemical Society Symposium*. 55: 167–92.

Galili, G. (1989). Heterologous expression of a wheat high molecular weight glutenin gene in *Escherichia coli. Proceedings of the National Academy of Sciences, USA*. 86: 7756–60.

Goldenberg, D.P. and Creighton, T.E. (1984). Gel electropheresis in studies of protein conformation and folding. *Analytical Biochemistry*. 138: 1–18.

Goldsbrough, A.P., Robert, L., Schmick, D. and Flavell, R.B. (1988). Molecular comparisons between breadmaking quality determining high molecular weight glutenin subunits of wheat gluten. *Proceedings of the 7th International Wheat Genetics Symposium, Cambridge*. 727–33.

Goldsbrough, A.P., Bulleid, N.J., Freedman, R.B. and Flavell, R.B. (1989). Conformational differences between two wheat (*Triticum aestivum*) 'high-molecular-weight' glutenin subunits are due to a short region containing six amino acid differences. *Biochemical Journal*. 263: 837–42.

Graham, J.S.D., Jennings, A.C., Morten, R.K., Palk, B.A. and Raison, J.K. (1962). Protein bodies and protein synthesis in developing wheat endosperm. *Nature (London)*. 196: 967–9.

Greene, F.C. (1981). *In vitro* synthesis of wheat (*Triticum aestivum* L.) storage proteins. *Plant Physiology*. 68: 778–83.

Grynberg, A., Nicolas, J. and Drapron, R. (1978). Some characteristics of protein disulphide isomerase (EC.5.3.4.1.) from wheat (*Triticum vulgare*) embryo. *Biochimie*. 60: 547–51

Halford, N.G., Forde, J., Anderson, D.D., Greene, F.C., Shewry, P.R. (1987). The nucleotide and deduced amino acid sequences of an HMW-glutenin gene from chromosome IB of bread wheat (*Triticum aestivum* L.) and comparison with those of genes from chromosomes 1A and 1D. *Theoretical and Applied Genetics*. 75: 117–26.

Kaderbhai, M.A. and Austen, B.M. (1985). Studies on the formation of intrachain disulphide bonds in newly biosynthesised bovine prolactin. *European Journal of Biochemistry*. 153: 167–78.

Kim, W.T., Franceschi, V.R., Krishnan, H. and Okita, J.W. (1988). Formation of wheat protein bodies: involvement of the Golgi apparatus in gliadin transport. *Planta*. 176: 173–82.

Krieg, P.A. and Melton, D.A. (1984). Functional mRNAs are produced by SP6 *in vitro* transcription of cloned cDNAs. *Nucleid Acids Research*. 12: 7057–70.

Okita, T.W., Cheesebrough, V. and Reeves, C.D. (1985). Evolution and heterogeneity of the α/β and γ-type DNA sequences. *Journal of Biological Chemistry*. 26: 8203–13.

Paver, J.L., Hawkins, H.C. and Freedman, R.B. (1989). Preparation and characterization of dog pancreas microsomal membranes specifically depleted of protein disulphide-isomerase. *Biochemical Journal*. 257: 657–63.

Payne, P.I. (1986). Varietal improvement in the breadmaking quality of wheat: contributions from biochemistry and genetics, and future prospects from molecular biology. *British Crop Protection Council Monogram No. 34*. 69–81.

Payne, P.I., Corfield, K.G., Holt, L.M. and Blackman, J.A. (1981). Correlations between the inheritance of certain high-molecular weight subunits of glutenin and bread-making quality in progenies of six crosses of bread wheat. *Journal of the Science of Food Agriculture*. 32: 51–60.

Pelham, H.R.B. and Jackson, R.J. (1976). An efficient mRNA-dependent translation system from reticulocyte lysates. *European Journal of Biochemistry*. 67: 247–56.

Pogna, N.E., Mellini, F. and Dal Belin Peruffo, A. (1987). Glutenin subunits of Italian common wheats of good breadmaking quality and comparative effects of high molecular weight glutenin subunits 2 and 5, 10 and 12 on flour quality. In *Agriculture: hard wheat: agronomic, technological, biochemical and genetic aspects* (ed. B. Barghi), pp. 53–69. Commission of the European Communities, Milano, Italy.

Roden, L.T., Miflin, B.J. and Freedman, R.B. (1982). Protein disulphide-isomerase is located in the endoplasmic reticulum of developing wheat endosperm. *FEBS Letters*. 138: 121–4.

Scheele, G. and Jacoby, R. (1982). Conformational changes associated with proteolytic processing of pre-secretory proteins allow glutathione-catalysed formation of native disulphide bonds. *Journal of Biological Chemistry*. 257. 12277–82.

Shewry, P.R., Parmar, S. and Field, J.M. (1988). Two

dimensional electrophoresis of cereal prolamins: applications to biochemical and genetic analysis. *Electrophoresis*. **9**: 727–37.

Tao, H.P. and Kasarda, D.D. (1989) Two-dimensional gel mapping and N-terminal sequencing of LMW-glutenin subunits. *Journal of Experimental Botany*. **40**: 1015–20.

Thompson, R.D., Bartels, D. and Harberd, N.P. (1985). Nucleotide sequence of a gene from chromosome ID of wheat encoding a HMW-glutenin subunit. *Nucleic Acids Research*. **13**: 6833–46.

Walter, P. and Blobel, G. (1983). Preparation of microsomal membranes for co-translational protein translocation. *Methods in Enzymology*. **96**: 84–93.

# 10

# Synthesis of Zeins and their Potential for Amino Acid Modification

Craig R. Lending, John C. Wallace and Brian A. Larkins

## Introduction

The storage proteins of maize, like many other cereals, including sorghum, wheat and barley, are a group of alcohol-soluble proteins classified as prolamines (Shotwell and Larkins 1989). Maize prolamines, also called zeins, account for 50 – 60% of the total seed protein, and hence determine much of the nutritional value of the seed. Like most prolamines, zeins are devoid of lysine and low in tryptophan, both essential amino acids for monogastric animals. As a consequence, maize protein is of poor nutritional value for humans and certain livestock (Nelson 1969). However, certain of the zein proteins contain high concentrations of sulphur amino acids (Shotwell and Larkins 1989), so that maize protein is a good complement to the sulphur-poor proteins of legumes.

Because maize is an important grain product worldwide, much research and plant breeding effort has been devoted to increasing its lysine content. The discovery of the mutant *opaque-2* (Mertz *et al.* 1964), which decreases the synthesis of zein proteins and results in an increased percentage of lysine in the seed, raised expectations for its use in the nutritional improvement of maize. However, *opaque-2* mutants were found to have several inferior agronomic qualities, including low yields, a soft, chalky endosperm and decreased disease resistance (Glover and Mertz 1987). As a consequence, they have not been widely used, nor have other mutations that have similar effects on storage protein synthesis, such as *floury-2* (Glover and Mertz 1987).

Because zein genes do not encode lysine, it is not possible by conventional plant breeding to increase the lysine content of these storage proteins. However, the development of recombinant DNA procedures, site-specific mutagenesis and genetic engineering provides an alternative approach for increasing the lysine content of zeins. While this biotechnological approach appears feasible, we do not know whether zein proteins that contain lysine will be processed and accumulate in the seed. Therefore, it is important to understand the structural relationships of zein proteins, their organization within protein bodies and the mechanisms regulating their synthesis and deposition into protein bodies.

## Synthesis and structural relationships of zein proteins

Synthesis of zein proteins is initiated in developing maize endosperm shortly after the period of free nuclear division, and it continues until the seed reaches maturity. Zeins are synthesized by polysomes bound to the rough endoplasmic reticulum (r.e.r.), and they accumulate in the lumen of the r.e.r. as insoluble aggregates called protein bodies (Larkins and Hurkman 1978).

Zein proteins are traditionally extracted from mature endosperm meal by differential solubility in alcohol solutions containing reducing agents such as 2-mercaptoethanol. With procedures similar to that of Landry and Moureaux (1970), the proteins are recovered in a number of solubility fractions. This has made it difficult to compare the relative amounts of the various proteins and has also led to a complex nomenclature for zeins.

A simpler and more quantitative method to separate the total zein fraction from the non-zein protein was recently described (Wallace *et al.*

1990). This procedure involves dissolving the total endosperm protein in an alkaline buffer containing 2-mercaptoethanol and SDS. With the subsequent addition of ethanol, the non-zein proteins precipitate while the zeins remain in solution. With this approach, the relative concentrations of the zein proteins can be estimated by immunological assays.

Examination of the alcohol-soluble proteins separated by SDS-PAGE, after extraction of endosperm with 70% ethanol and 2% 2-mercaptoethanol, reveals polypeptides with mol. wt of 27 000, 22 000, 19 000, 16 000, 14 000, 12 000, and 10 000 (Fig. 10.1). Sequence analysis of the genes encoding these proteins has shown that zeins comprise four structurally distinct types (Larkins *et al.* 1989). The proteins with mol. wt 22 000 and 19 000 are designated as α-zeins, the protein of mol. wt 14 000 as β-zein, and the proteins of mol. wt 27 000 and 16 000 as γ-zeins. Preliminary characterization of the 12 000 mol. wt polypeptide indicates that its amino acid composition is similar to that of the 27 000 and 16 000 mol. wt. γ-zeins (M. Lopes and B.A. Larkins, unpublished observations). The 10 000 mol. wt protein is designated the δ-zein. The β-, γ- and δ-zeins are all sulphur-rich proteins, while the α-zeins contain only low amounts of cysteine and methionine (Larkins *et al.* 1989). The sulphur-rich zeins are encoded by genes present in one or two copies, but the α-zeins are encoded by a large multigene family that consists of between 75 and 150 genes, although many of these genes may not be transcriptionally active (Shotwell and Larkins 1989).

## Distribution of zeins in endosperm

Immunocytochemical techniques have been used to study the distribution of the various zeins within the endosperms of developing kernels. The various zeins are distributed heterogeneously within isolated protein bodies (Ludevid *et al.* 1984; Lending *et al.* 1988), and isolated protein bodies of different sizes contain different amounts of the zein types (Lending *et al.* 1988). Figure 10.2 shows an ultrathin section through endosperm tissue 18 days

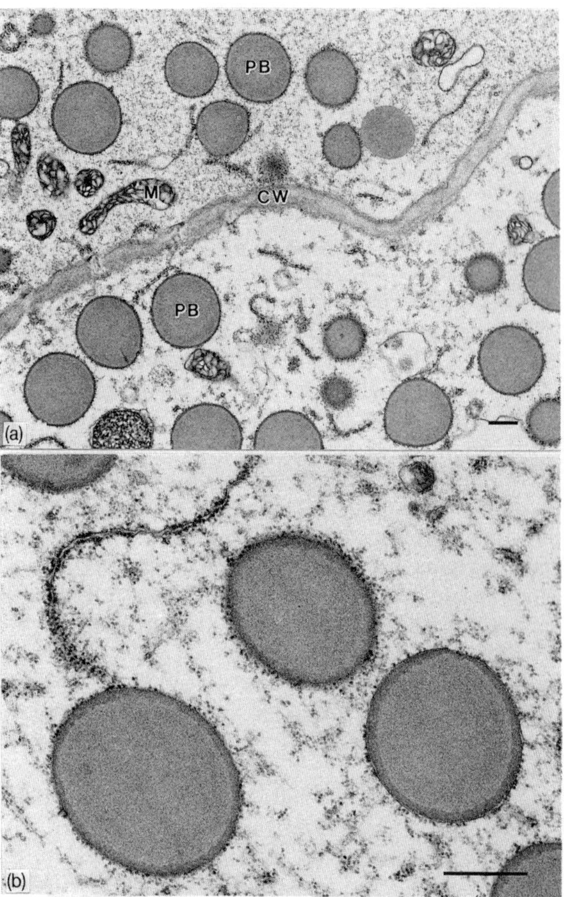

**Fig. 10.2**  Electron micrograph showing protein bodies in developing endosperm. Protein bodies within the fifth subaleurone layer of developing maize endosperm. Ultrathin section of the maize inbred W64A, 18 DAP, fixed in 4% formaldehyde and 1% glutaraldehyde, embedded in LR white resin, and post-stained with uranyl acetate and lead citrate (see Lending and Larkins 1989 for details). The abbreviations used are: PB, protein body; M, mitochondrion; CW, cell wall. Bar = 0.5 μm. (a) Low magnification micrograph, showing numerous protein bodies within the cells. (b) High magnification micrograph, showing protein bodies within r.e.r. The light-staining central region is surrounded by a peripheral band that stains darkly.

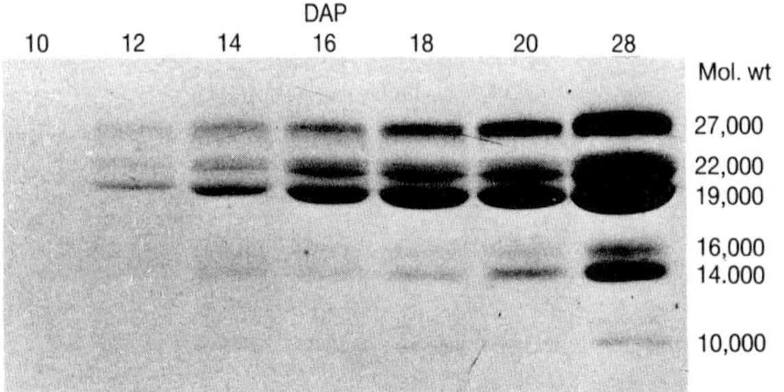

**Fig. 10.1**  Developmental regulation and tissue specificity of zein gene expression in endosperm. SDS-PAGE of zeins extracted with 70% EtOH and 2% 2-mercaptoethanol from endosperm of the inbred W64A. The developmental stage is listed across the top and the apparent molecular weights are given along the side. Each sample represents the amount of zein extracted from 1 mg flour from kernels. (Source: Kodrzycki *et al.* 1989.)

after pollination (DAP), stained with uranyl acetate and lead citrate; protein bodies are found within the r.e.r. They consist of both light- and dark-staining regions, with the darker regions typically forming a peripheral band, as shown in the figure. These patterns reflect the distribution of the various zeins, as discussed below.

## Light microscopy

When semithin sections of developing maize endosperm at 18 DAP were immunostained for either $\alpha$-, $\beta$- or $\gamma$-zein and examined by light microscopy, the $\alpha$-zein had a distinctly different tissue distribution from the $\beta$- and $\gamma$-zeins (Fig. 10.3). Sparse immunolabelling with the $\alpha$-zein antibody was observed within the subaleurone layer and the first starchy endosperm layer, although it is not readily apparent at the magnification shown (Fig. 10.3b) (see legend for Fig. 10.3 for endosperm terminology). The staining intensity increased in cells progressively deeper in the endosperm, and became more or less constant by the third starchy endosperm layer. The immunostaining formed solid circular deposits, and the diameter of these deposits increased from the subaleurone layer until it became constant in the third starchy endosperm layer (Fig. 10.3b, inset).

Unlike the pattern observed for $\alpha$-zein at 18 DAP, both $\beta$- and $\gamma$-zein antisera heavily immunostained protein bodies within the subaleurone and first starchy endosperm layers (Figs 10.3c and 10.3d, respectively). Patterns for both $\beta$- and $\gamma$-zein were similar, although the intensity of the $\gamma$-zein staining was considerably less than that

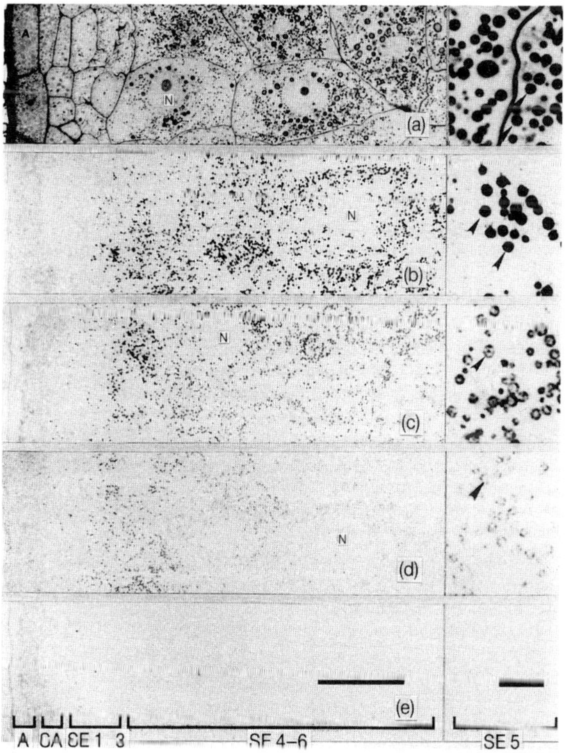

**Fig. 10.3**  Light micrographs of endosperm 18 DAP, illustrating the immunostaining patterns with antibodies directed against $\alpha$, $\beta$, and $\gamma$-zeins; for all micrographs, bar = 50 µm. All insets, bar = 5 µm. The abbreviations used are: A, aleurone; N, nucleus. Brackets at bottom indicate approximate cell position: A, aleurone; SA, subaleurone; SE, starchy endosperm. (a) Sections stained with basic fuchsin illustrating a region similar to those immunostained with the various antibodies. (a, inset) High magnification view of the fifth starchy endosperm layer. Most protein bodies from this region of the endosperm are from 0.8 to 1.4 µm in diameter (arrowheads). (b) Sections incubated with $\alpha$-zein antibody followed by goat anti-rabbit/colloidal gold and enhanced with silver intensification reagent. Staining deposits in the subaleurone layer (not visible at this magnification) gradually increase in diameter and intensity between SE-1 and SE 3. At SE-4, and deeper cell layers, the staining becomes more or less constant. (b, inset) High magnification view of the fifth starchy endosperm layer. Solid circular deposits of silver occur over protein bodies (arrowheads). (c) Sections incubated with $\beta$-zein antibody followed by goat anti-rabbit/colloidal gold and enhanced with silver intensification reagent. Staining patterns in the subaleurone layer were solid circles (not visible at this magnification); protein bodies further from this layer generally immunostained with open circular patterns. The labelling intensity was greatest in cell layers SE-2–4, and decreased in deeper layers of the endosperm. (c, inset) High magnification view of the fifth starchy endosperm layer; $\beta$-zein is peripherally distributed in the protein bodies; some internal inclusions also immunostain (arrowhead). (d) Sections incubated with $\gamma$-zein antibody followed by goat anti-rabbit/colloidal gold and enhanced with silver intensification reagent. The immunostaining patterns are as described for $\beta$-zein (plate c), but the staining intensity was lower. (d, inset) High magnification view of the fifth starchy endosperm layer, demonstrating the peripheral distribution of $\gamma$-zein; some internal inclusions also immunostain (arrowhead). The staining intensity is generally lower than for the $\beta$-zein antibody (see plate c). (e) Section immunostained with preimmune serum followed by goat anti-rabbit/colloidal gold and enhanced with silver intensification reagent. Little specific staining is observed in these sections; a slight darkening of the aleurone occurs in all sections, and is caused by the reaction of the silver-enhancement reagent with the lipid bodies. (e, inset) High magnification view of the fifth starchy endosperm layer. Slight background labelling is seen, but no specific immunostaining is observed over protein bodies. (Source: Lending and Larkins 1989.)

observed with β-zein. This difference in intensity may indicate quantitative differences in the accumulation of the two proteins, or may merely reflect a difference in antibody avidity. The labelling intensities for both of these proteins were greatest in the second through to the fourth subaleurone layers, and became progressively less in more interior cell layers. The immunostaining appeared as solid circular patterns within the first few cells beneath the aleurone layer (not apparent at the magnification shown). The labelling pattern in cells further from the aleurone layer was that of open circles of varying intensity (Figs 10.3c and 10.3d, insets). Inclusions were often observed within the central regions of these circles (arrowhead, Fig. 10.3c, inset).

## Electron microscopy

Because of the different distribution patterns observed with the immunocytochemical staining at the level of the light microscope, a high-resolution analysis of the several cell layers beneath the aleurone was done to provide a clearer understanding of the initial stages of protein body growth. Medial sections of protein bodies found in the various cell layers were examined after ultrathin sections were immunostained with either α-, β- or γ-zein antibodies and detected with a goat anti-rabbit/colloidal gold conjugate. These immunostainings showed that the protein bodies increased in size from the first subaleurone layer inwards; additionally, the zein composition of the protein bodies changed dramatically within the first several cell layers.

The protein bodies within the subaleurone layer were predominantly dark-staining (Fig. 10.4a), with little or no α-zein (Fig. 10.4b), and consisted mostly of β- and γ-zein (Figs 10.4c and d, respectively). In cells progressively further from the subaleurone layer, protein bodies increased in size (compare Figs 10.4a, e and m) and light-staining material was observed within the central region of these protein bodies. This light-staining material correlated with the appearance of α-zein (Figs 10.4f, j and n).

As protein bodies increased in size with distance from the aleurone layer, the α-zein filled the interior region of the protein bodies, and the dark-staining β- and γ-zeins became peripherally located around the protein bodies (Figs 10.4g, k, o and Figs 10.4h, l and p, respectively). The immuno-labelling for the β- and γ-zeins was preferentially

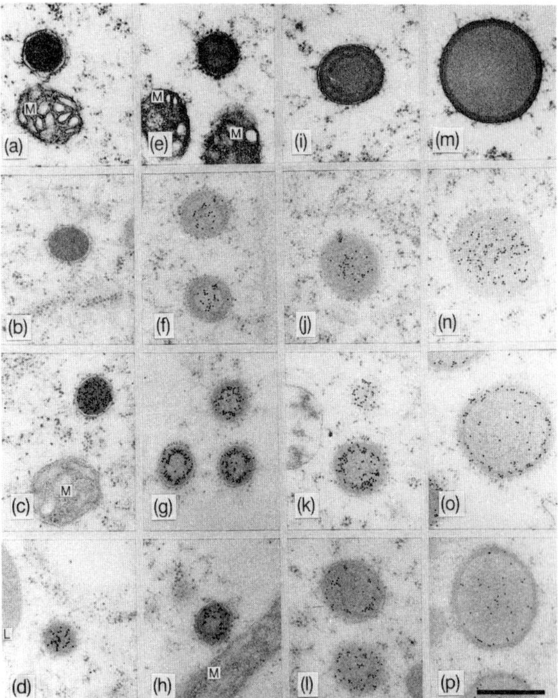

**Fig. 10.4** Electron micrographs showing representative protein bodies from different cell layers of the endosperm at 14 DAP. The protein bodies are representative of the morphology and size of protein bodies within the cell layer indicated, and represent a developmental series (from left to right). For immunostaining, sections were incubated in the indicated primary antibody and then in goat anti-rabbit/colloidal gold (10 nm); Bar = 0.5 μm. M, mitochondrion. (a–d) Protein bodies from the subaleurone layer. (a) Section stained with uranyl acetate and lead citrate. (b) Section immunostained for α-zein. (c) Section immunostained for β-zein. (d) Section immunostained for γ-zein. (e–h) Protein bodies from the first starchy endosperm layer. (e) Section stained with uranyl acetate and lead citrate. (f) Section immunostained for α-zein. (g) Section immunostained for β-zein. The labelling is preferentially distributed over the dark-staining region near the interface with the light-staining material. (h) Section immunostained for γ-zein. (i–l) Protein bodies from the second starchy endosperm layer. (i) Section stained with uranyl acetate and lead citrate. (j) Section immunostained for α-zein. (k) Section immunostained for β-zein. (l) Section immunostained for γ-zein. (m–p) Protein bodies from the fifth starchy endosperm layer. (m) Section stained with uranyl acetate and lead citrate. (n) Section immunostained for α-zein. (o) Section immunostained for β-zein. (p) Section immunostained for γ-zein. (Source: Lending and Larkins 1989.)

observed at the interface between the light- and dark-staining regions (Fig. 10.4g) and was often associated with small, dark-staining deposits within the light-staining regions of the protein bodies (Figs 10.4g and h).

The pattern of cell development within corn endosperm has been previously described; the subaleurone cells are meristematic and function as a cambial layer (Fisk 1925; Randolph 1936). Mitosis within the endosperm becomes limited to the region directly beneath the aleurone layer by 10 DAP, with few mitotic cells in the interior regions of the endosperm (Kiesselbach 1949). A gradient in protein body size and zein content throughout development was consistently observed in the immunocytochemical studies described above (Lending and Larkins 1989). As the cells matured, the size of protein bodies increased, concomitant with an increase in the amount of α-zein. Also, as the seed matured, larger protein bodies were observed progressively closer to the aleurone layer. Thus, even though static images were observed, the gradient of protein body size from the subaleurone region to the interior of the endosperm reflects the pattern of cell maturation.

These studies suggest that protein body formation occurs by a sequential deposition of zeins during endosperm development, as illustrated in Figure 10.5. Initial accretions within the r.e.r. are composed of dark-staining deposits of both β- and γ-zeins; little or no α-zein is contained within these small protein bodies. Subsequently, α-zein begins to accumulate and is observed as discrete, light-staining locules within the β- and γ-zeins. The locules of α-zein fuse and form the central region of the protein body, while most of the β- and γ-zein forms a continuous layer on the periphery.

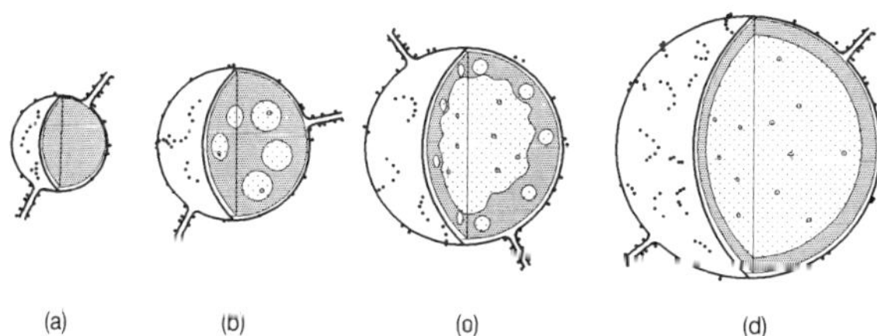


**Fig. 10.5** Diagram illustrating the development of protein bodies in maize endosperm (not to scale). The heavily stippled regions correspond to regions that are rich in β- and γ- zeins and the lightly stippled regions correspond to regions rich in α-zein. The protein body is surrounded by rough endoplasmic reticulum (dark dots are ribosomes). Some β and γ-zeins are found within the regions that consist primarily of α-zein, and may be strands of these proteins that run through the protein body, or small, individual aggregates. (Source: Lending and Larkins 1989.)

# Zein modification and expression

One goal of the above investigations has been to determine the nature of the interactions within and between the different zein molecules contained in protein bodies. Specifically, if the primary amino acid sequences of the proteins are modified to increase the lysine content, will the integrity and function of protein bodies be maintained? Globulins, the major storage proteins of dicots, have been modified to increase their content of sulphur amino acids (Hoffman *et al.* 1988; Saalbach *et al.* 1988). Unfortunately, reliable methods for the transformation of maize are only now becoming available (Fromm *et al.* 1990; Gordon-Kamm *et al.* 1990), so modified zeins have not yet been tested in developing maize seeds. Therefore, heterologous systems have been used to study protein body formation; these systems include the *Xenopus* oöcyte and transgenic petunia and tobacco plants.

## Expression of zeins in *Xenopus* oöcytes

To create modified zeins, a cloned cDNA for an α-zein protein was altered using techniques of *in vitro* mutagenesis (Wallace *et al.* 1988a). The resulting clones encode the modified zeins shown in Figure 10.6, which shows the primary amino acid sequence and the amino acid substitutions. There are single and double substitutions of lysines for other (neutral) amino acids, insertions of lysine- and tryptophan-rich oligopeptides and one large peptide insertion. These changes were made at various places within the molecule, bearing in mind the three-dimensional structure for α-zein proposed by Argos *et al.* (1982). The large peptide insertion construct was made as a negative control because it seemed likely that the insertion of a 17 kDa hydrophilic peptide into a 19 kDa hydrophobic protein (the α-zein) would profoundly disrupt its normal structure.

For the experiments using the *Xenopus* oöcytes, these modified zein cDNA clones were inserted into a plasmid vector between a promoter for RNA polymerase SP6 and a polyadenylate sequence. Conditions were then established for the *in vitro* production of capped, polyadenylated mRNAs that encoded the normal and modified zeins. These artificial mRNAs were injected into stage 6 oöcytes (Kawata *et al.* 1988); [³H]-leucine was injected as a radioactive tracer to label the synthesized proteins.

Hurkman *et al.* (1981) showed that the injection

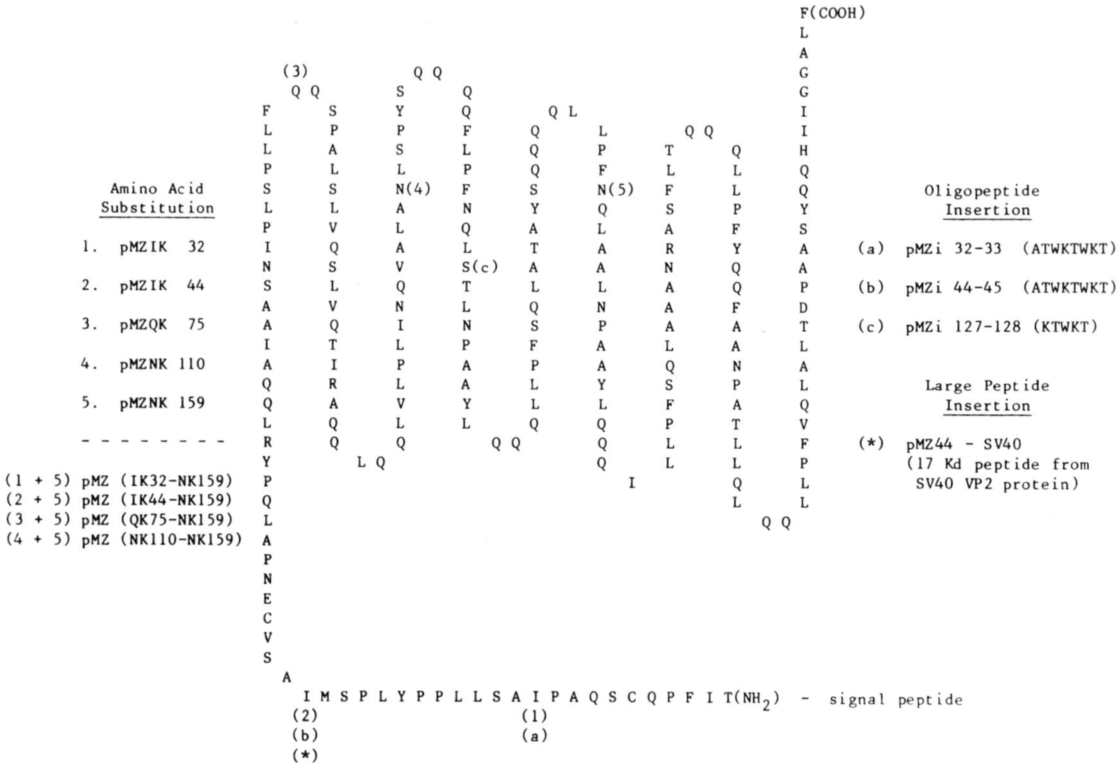

**Fig. 10.6**   Modified zein constructs showing the positions of amino acid substitutions, oligopeptide insertions and a large peptide insertion made in a mature zein amino acid sequence. The sequence corresponds to the 19-kDa zein cDNA clone cZ19c1. It is arranged according to the proposed secondary structure of Argos *et al.* 1982, where the parallel regions represent α-helix and the non-parallel section the NH₂-terminal turn region. The signal peptide of 21 amino acids is not shown. The names of the constructs (at the side of the figure) indicate the sites of the substitution or insertion (amino acid number is that of the preprotein) and the amino acids substituted or inserted. Amino acid substitution mutants were synthesized by oligonucleotide-mediated mutagenesis. (Source: Wallace *et al.* 1988.)

of authentic zein mRNA into oöcytes resulted in the synthesis and processing of zein proteins and the formation of structures that resembled maize protein bodies. Injection of synthetic α-zein mRNA also produced protein bodies according to two criteria: the zein protein produced was found in membrane-bound vesicles, and these vesicles were very dense, indicating that the proteins were very tightly packed together (Wallace *et al.* 1988a).

The density of the protein bodies made in frog oöcytes was determined as follows: the oöcyte components were first resolved on a metrizamide density gradient, the alcohol-soluble protein fraction was separated by SDS-PAGE to assay for the presence of zeins, and the labelled proteins were detected by fluorography. Figure 10.7a shows the distribution of proteins after injection of authentic zein mRNAs; much of the zein was found at a density corresponding to that of maize protein bodies. When the synthetic mRNA for a 19 kDa α-zein was injected, the protein also aggregated into dense vesicles (Fig. 10.7b). That these protein bodies are not as dense as authentic ones may be due to the lack of β- and γ-zeins; as described above, these proteins are apparently necessary for the proper formation of protein bodies.

Single and double lysine substitutions and oligopeptide insertions had no effect on the ability of the zein protein to aggregate into dense particles (Figs 10.7d–g). Only the large peptide insertion (Fig. 10.7h) interfered with the ability of the zein to form protein bodies in oöcytes. Based on observations in this heterologous system, lysine substitutions and additions in α-zein proteins can be tolerated without detrimental effects on protein body formation.

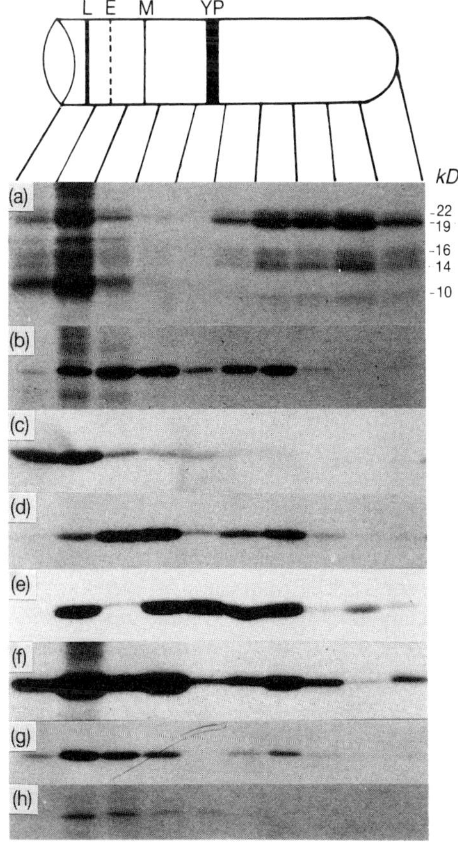

**Fig. 10.7**   Densities of protein bodies from *Xenopus* oöcytes. Eight to 12 oöcytes were injected and the organelles were separated by density gradient centrifugation. The banding positions of oöcyte components were clearly visible, and are shown in the drawing (E, endomembranes; L, lipids; M, mitochondria; YP, yolk platelets). Fractions were collected and zeins were extracted; the zeins were separated by SDS-PAGE, and analysed by fluorography. Injection of (a) native zein mRNA. (b) SP6 mRNA for wild type 19-kDa zein. (c) Exogenous [³H]-labelled zein added to oöcyte homogenate. (d) pMZIK 32 mRNA. (e) pMZ (NK110-NK159) mRNA. (f) pMZi 32-33 mRNA. (g) pMZi 127-128 mRNA. (h) pMZ44-SV40 mRNA. The sizes of the zeins in (a) are shown; the migration of all others was equivalent at 19 kDa except for that of the pMZ44-SV40 zein (h), which was 35 kDa. (Source: Wallace *et al.* 1988.)

## Expression of zeins in transgenic petunia and tobacco plants

In an effort to test the function of normal and modified zein peptides in an organism more closely related to maize, zein genes have also been introduced into transgenic tobacco and petunia plants (Hoffman *et al.* 1987; Schernthaner *et al.* 1988; Ueng *et al.* 1988; Williamson *et al.* 1988; Wallace *et*

*al.* 1989). Zein promoters do not function efficiently (Schernthaner *et al.* 1988) or with normal tissue specificity (Ueng *et al.* 1988) in dicots. Thus, in order to effect high-level expression of the maize protein in the seeds of the transgenic plants, gene constructs were made that placed zein coding regions under the control of a dicot promoter: either the seed-specific promoter for the bean storage protein β-phaseolin (Sengupta-Gopalan *et al.* 1985) or the non-specific 35S promoter from cauliflower mosaic virus (Schernthaner *et al.* 1988). When the β-zein coding region was placed under the regulation of the French bean β-phaseolin promoter, the β-zein peptide accumulated to high levels in transgenic tobacco plants (Hoffman *et al.* 1987). The protein was found in the vacuolar protein bodies (Fig. 10.8).

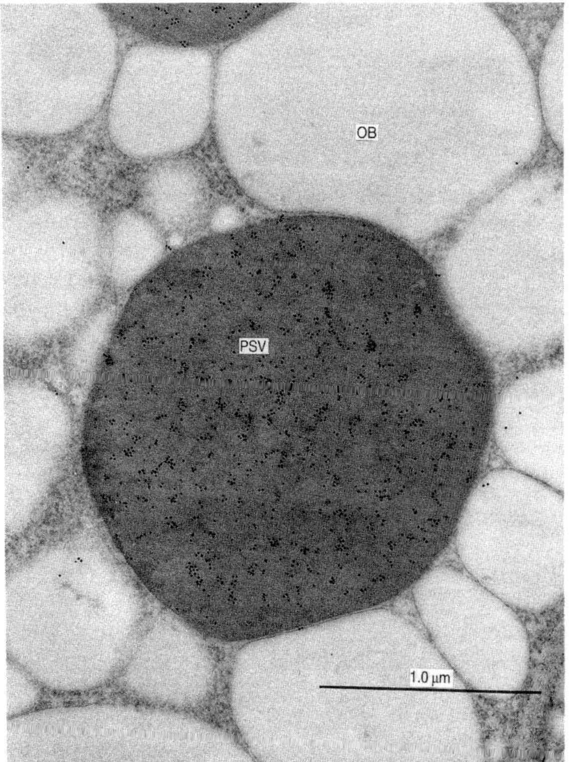


**Fig. 10.8**   Immunogold labelling of β-zein in transgenic tobacco plants. The β-zein gene was placed under the regulation of the French bean β-phaseolin gene flanking region and tobacco plants were transformed by *Agrobacterium tumefaciens*. Transgenic plants synthesized the β-zein in a tissue-specific manner during the latter half of seed development. Electron microscopic immunogold labelling shows that the protein accumulates in the crystalloid component of vacuolar protein bodies. (Source: Hoffman *et al.* 1987.)

When an α-zein coding sequence was placed under the control of the phaseolin promoter, its mRNA was produced in abundant amounts in seeds of transgenic petunia plants ≏ 1–3% of the total poly A(+) RNA. However, the zein protein did not accumulate and was barely detected by the most sensitive western blotting techniques (Williamson *et al.* 1988). Either the α-zein mRNA was very inefficiently translated in the heterologous system or the protein was rapidly degraded. The small amount of zein protein that was made did not appear to form protein bodies; immunolocalization studies showed that the protein appeared in inappropriate places within the cells of petunia seeds, mainly in inclusions that were appressed against the cell walls (Wallace *et al.* 1989).

Until normal α-zein proteins can be synthesized at sufficiently high levels to form protein bodies in dicots, the question of how the lysine-containing zeins will behave in this system cannot be addressed. The microscopic examination of authentic maize protein bodies described above, and the biochemical investigation in the *Xenopus* oöcyte system (Wallace *et al.* 1988b) point to the importance of the β- and γ-zeins in protein body formation and structure. Perhaps coexpression of these proteins in a single transgenic plant will result in 'normal' protein body formation and provide an experimental system for testing the effects of the modifications. Alternatively, the ability to synthesize genetically engineered proteins in maize itself may soon be a possibility (Fromm *et al.* 1990; Gordon-Kamm *et al.* 1990).

## Transcriptional considerations

Should one of the modified zeins prove to function normally in maize, the problem remains of how to express it at a level high enough to significantly improve the amino acid balance in the kernel. Since α-zeins are encoded by a large multigene family (Shotwell and Larkins 1989), expression of any one gene under its own promoter may not have much effect. For example, if α-zeins represent 50% of the total protein in the seed, and one of 50 α-zein genes were to direct the synthesis of a protein containing two lysines

$$50\% \text{ protein} \times \frac{1}{50} \text{ genes} \times \frac{2 \text{ lysines}}{220 \text{ amino acids}}$$

$$= {\sim}0.01\%$$

the amount of lysine would only increase by about 0.01%.

Three approaches may help overcome this problem:

(1) Increase the strength of the promoter controlling the transcription of the modified zein.
(2) Insert multiple copies of the modified gene into the host genome.
(3) Utilize a mutant strain of maize in which endogenous α-zein transcription is decreased but that of the genetically engineered one is not.

All three of these approaches are potentially useful, as described below.

### Increase promoter strength

Kodrzycki *et al.* (1989) measured the transcription rates of different classes of zeins and other endosperm-specific genes. They found that, of the genes tested, the γ-zein was transcribed at a rate four to five times higher than any other, and more than 50 times that of the average α-zein gene. Since transcription of γ-zein displays essentially the same tissue-specificity and temporal regulation as α-zeins, its promoter would appear to be a good candidate for enhancing the expression of a modified α-zein. Of course, as more is learned about the cis elements controlling expression of zein genes, it may become possible to design an even better promoter.

### Introducing multiple gene copies

Introduction of multiple copies of genes encoding modified zein proteins would be expected to increase the level of synthesis.

### Utilization of zein regulatory mutants

Several maize mutants that result in reduced zein synthesis are known. The best characterized are *opaque-2* and *floury-2*. Due to their decreased zein content, these mutants show increased percentages of lysine and tryptophan and are sometimes referred to as 'high-lysine maize'. Unfortunately, poor agronomic characteristics resulting from the lack of zeins have prevented their widespread commercial development. However, if the zein content of these strains could be increased specifically with a lysine-containing zein, both agricultural and nutritional properties would improve. The γ-zein promoter is unaffected by the *opaque-2* mutation (Pérez-Grau *et al.* 1986; Kodrzycki *et al.* 1989), so it might be particularly useful in this context.

Of course, expression of the mRNA for a lysine-containing zein that can form normal protein bodies does not necessarily mean that the protein will accumulate to high levels. Sufficient amounts of the appropriate amino acids and tRNAs would be needed for the mRNA to be efficiently translated, and the mRNA must remain stable in the developing endosperm throughout maturation. Finally, the modifications must be shown to produce no detrimental effects in the germinating seedling. After all, the true function of storage proteins lies not in the nutrition of animals, but rather in the nutrition of developing plants.

# References


Argos, P., Pedersen, K., Marks, M.D. and Larkins, B.A. (1982). A structural model for maize zein proteins. *Journal of Biological Chemistry*. **257**: 9984–90.

Fisk, E.L. (1925). The chromosomes of *Zea mays*. *American Journal of Botany*. **14**: 53–75.

Fromm, M., Klein, T., Goff, S.A., Roth, B., Morrish, F. and Armstrong, C. (1990). Transient expression and stable transformation of maize using microprojectiles. In *Plant Molecular Biology; Proceedings of The 6th NATO Advanced Studies Institute on Plant Molecular Biology*, (ed. R.C. Herrmann and B.A. Larkins), Plenum Press, London.

Glover, D.V. and Mertz, E.T. (1987). Corn. In *Nutritional Quality of Cereal Grains: genetic and agronomic improvement. Agronomy Monograph No. 28*, (ed. R.A. Olson and K.J. Frey), pp. 183–336. American Society of Agronomy, Madison, Wisconsin.

Gordon-Kamm, W.J., Spencer, T.M., Mangano, M.L., Adams, T.R., Daines, R.J., Start, W.G., O'Brien, J.V., Chambers, S.A., Adams, W.R., Willetts, N.G., Rice, T.B., Mackey, C.J., Krueger, R.W., Kausch, A.P. and Lemaux, P.G. (1990). Transformation of maize cells and regeneration of fertile transgenic plants. *The Plant Cell*. **2**: 603–18.

Hoffman, L.M., Donaldson, D.D., Bookland, R., Rashka, K. and Herman, E.M. (1987). Synthesis and protein body deposition of maize 15-kd zein in transgenic tobacco seeds. *EMBO Journal*. **6(11)**: 3213–21.

Hoffman, L.M., Donaldson, D.D. and Herman, E.M. (1988). A modified storage protein is synthesized, processed, and degraded in the seeds of transgenic plants. *Plant Molecular Biology*. **11**: 717–29.

Hurkman, W.J., Smith, L.D., Richter, J. and Larkins, B.A. (1981). Subcellular compartmentalization of maize storage proteins in *Xenopus* oocytes injected with zein messenger RNAs. *Journal of Cell Biology*. **89**: 292–9.

Kawata, E.E., Galili, G., Smith, L.D. and Larkins, B.A. (1988). Translation in *Xenopus* oocytes of mRNAs transcribed in vitro. In *Plant Molecular Biology Manual* (ed. S.B. Gelvin, R.A. Schilperoort and D.P.S. Verma), pp. 1–22. Kluwer Academic Publishers, Dordrecht, The Netherlands.

Kiesselbach, T.A. (1949). *The Structure and Reproduction of Corn*. University of Nebraska Agricultural Experiment Station, Research Bulletin. **161**: 1–96.

Kodrzycki, B., Boston, R.S. and Larkins, B.A. (1989). The *opaque-2* mutation of maize differently reduces zein gene transcription. *The Plant Cell*. **1**: 105–14.

Landry, J. and Moureaux, T. (1970). Hétérogénéité des glutélines du grain de maïs: extraction sélective et compositon en acides aminés des trois fractions isolées. *Bulletin Societe de Chimie Biologique*. **52**: 1021–37.

Larkins, B.A. and Hurkman, W.J. (1978). Synthesis and deposition of zein in protein bodies of maize endosperm. *Plant Physiology*. **62**: 256–63.

Larkins, B.A., Lending, C.R., Wallace, J.C., Galili, G., Kawata, E.E., Geetha, K.B., Martin, D.N. and Bracker, C.E. (1989). Zein gene expression during maize endosperm development. In *The Molecular Basis of Plant Development* (ed. R.B. Goldberg), pp. 109–20. Alan R. Liss, Inc., New York.

Lending, C.R. and Larkins, B.A. (1989). Changes in the zein composition of protein bodies during maize endosperm development. *The Plant Cell*. **1**: 1011–23.

Lending, C.R., Kriz, A.L., Larkins, B.A. and Bracker, C.E. (1988). Structure of maize protein bodies and immunocytochemical localization of zeins. *Protoplasma*. **143**: 51–62.

Ludevid, M.D., Torrent, M., Martinez-Izquierdo, J.A., Puigdomènech, P. and Palau, J. (1984). Subcellular localization of glutelin-2 in maize (*Zea mays* L.) endosperm. *Plant Molecular Biology*. **3**: 227–34.

Mertz, E.T., Bates, L.S. and Nelson, O.E. (1964). Mutant gene that changes protein composition and increases lysine content of maize endosperm. *Science*. **145**: 279–80.

Nelson, O.E. (1969). Genetic modification of protein quality in plants. *Advances in Agronomy*. **21**: 171–94.

Pérez-Grau, L., Cortadas, J., Puigdomènech, P. and Palau, J. (1986). Accumulation and subcellular localization of glutelin-2 transcripts during maturation of maize endosperm. *FEBS Letters*. **202**: 145–8.

Randolph, L.F. (1936). Developmental morphology of the caryopsis in maize. *Journal of Agricultural Research*. **53**: 881–916.

Saalbach, G., Jung, R. Saalbach, I. and Muntz, K. (1988). Construction of storage protein genes with increased number of methionine codons and their use in transformation experiments. *Biochemie und Physiologie der Pflanzen*. **183**: 211–18.

Schernthaner, J.P., Matzke, M.A. and Matzke, A.J.M.

(1988). Endosperm-specific activity of a zein gene promoter in transgenic tobacco plants. *EMBO Journal.* **7:** 1249–55.

Sengupta-Gopalan, Reichert, N.A., Barker, R.F., Hall, T.C. and Kemp, J.D. (1985). Developmentally regulated expression of the bean β-phaseolin gene in tobacco seed. *Proceedings of the National Academy of Sciences, USA.* **82:** 3320–4.

Shotwell, M.A. and Larkins, B.A. (1989). The biochemistry and molecular biology of seed storage proteins. In *The Biochemistry of Plants: a comprehensive treatise,* Vol. 15 (ed. A. Marcus), pp. 296–345. Academic Press, New York.

Ueng, P., Galili, G., Sapanara, V., Goldsbrough, P.B., Dube, P., Beachy, R.N. and Larkins, B.A. (1988). Expression of a maize storage protein gene in petunia plants is not restricted to seeds. *Plant Physiology.* **86:** 1281–5.

Wallace, J.C., Galili, G., Kawata, E.E., Cuellar, R.E., Shotwell, M.A. and Larkins, B.A. (1988a). Aggregation of lysine-containing zeins into protein bodies in *Xenopus* oocytes. *Science.* **240:** 662–4.

Wallace, J.C., Galili, G., Kawata, E.E., Lending, C.R., Kriz, A.L., Bracker, C.E. and Larkins, B.A. (1988b). Location and interaction of the different types of zeins in protein bodies. *Biochemie und Physiologie der Pflanzen.* **183:** 107–15.

Wallace, J.C., Ohtani, T., Lending, C.R., Lopes, M., Williamson, J.D., Shaw, K.L., Gelvin, S.B. and Larkins, B.A. (1989). Factors affecting physical and structural properties of maize protein bodies. In *Plant Gene Transfer, UCLA Symposium on Molecular and Cellular Biology, New Series,* Vol. **129** (ed. C. Lamb and R.N. Beachy), pp. 205–16. Alan R. Liss Inc., New York.

Wallace, J.C., Lopes, M.A., Paiva, E. and Larkins, B.A. (1990). New methods for extraction and quantitation of zeins reveal a high content of gamma-zein in modified *opaque-2* maize. *Plant Physiology.* **92:** 191–6.

Williamson, J.D., Galili, G., Larkins, B.A. and Gelvin, S.B. (1988). The synthesis of a 19 kilodalton zein protein in transgenic *Petunia* plants. *Plant Physiology.* **88:** 1002–7.

# 11

# Structure and Protein Engineering of Thaumatin and other Sweet Proteins

C. Theo Verrips

## Introduction

### Sweeteners and sweet taste reception

Four basic taste sensations can be experienced by man in chemoreception: sweet, bitter, sour and salt. Specialized cells are concentrated in structures called taste buds present in the oral cavity—on the tongue, on the palate and in the pharynx. These taste buds serve as the organ for the sense of taste in mammals.

A taste bud consists of 40–60 taste cells and in adults there are 4000–5000 taste buds present on the tongue. The cells of the taste buds are in contact with the oral cavity by taste pores, about 2 μm long and the taste buds are grouped in circumvallate, foliate and fungiform papillae (Figs 11.1a and b).

The neural afferent fibres from the different taste papillae are the nervus vagus, nervus facialus and nervus glossopharyngeus. The taste compound probably adsorbs to a receptor protein inducing a conformational change, rather than penetrating the receptor cell. This change in conformation results, via a complex series of reactions, in the polarization of the receptor cell, which triggers the nerve system. Koshland (1974) pointed out that the interaction of a taste compound and a receptor and the subsequent depolarization of the membrane are quite similar to the reactions of a micro-organism towards a chemotactic attractant.

As shown in a recent review on sweeteners (Van der Wel et al. 1987) the number of classes of compounds that are perceived as sweet is very large. This diversity of sweet compounds, together with the results of a number of behaviour and competition studies, suggests that different recep-

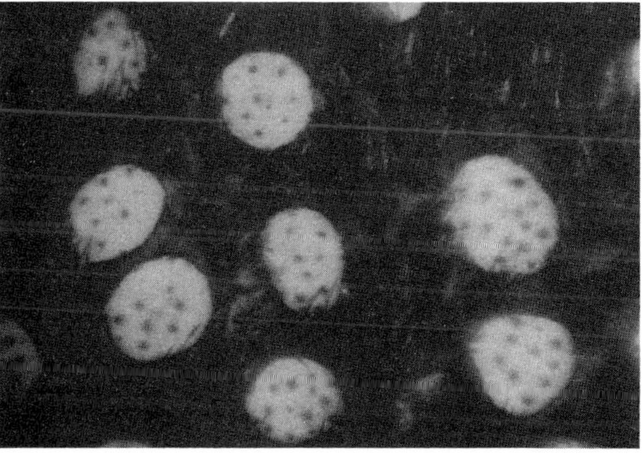

**Fig. 11.1(a)** Fungiform papillae of the cynomologus monkey (dorsal view). Taste buds are visible as coloured spots on the dorsal surface of the papillae.

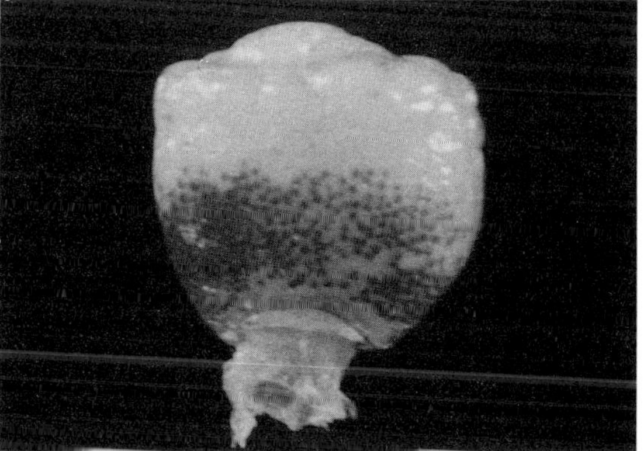

**Fig. 11.1(b)** Circumvallate papillae of the cynomologus monkey (lateral view). The lower half of the papillae is covered with taste buds.

tor proteins are involved for various classes of sweet compounds, or that the same receptor protein interacts at different sites or in different ways with various sweeteners. Hough and Edwardson (1978) found that antibodies raised against thaumatin recognize non-proteineous sweeteners like di-hydrochalcone, tetrachlorosucrose and the dipeptide L-Asp-L-Phe-O-Me, better known under its trade name Aspartam. They concluded that antibodies against thaumatin have binding sites quite similar to those of taste cells and that these sites can bind quite different sweeteners. This hypothesis is supported by the finding that thaumatin or monellin, covalently coupled to a support, bind the same two proteins (156 and 47 kDa) from lingual tissue extract (gustatory papillae) and that this binding can be broken by the well known sweet taste modifier gymnemic acid or by Aspartam (Persuad *et al.* 1988). Although not proved, these proteins may be receptor proteins for sweeteners. The 47 kDa protein was also isolated from circumvallate papillae. These tests were carried out with cow, pig and rat tissue extracts and indicate that, if various classes of receptor proteins are involved, it is not unlikely that they have evolved from one, or a few, common ancestor molecule(s).

The large number of sweeteners and the possibility that various receptor proteins are involved in sweet taste perception account for the fact that, in spite of an enormous amount of research, we still have no clear picture of taste perception.

## Various sweet taste compounds present in the plant kingdom

Beside the common sweet components like glucose, fructose and saccharose, plants contain a large variety of components that are perceived by mammals as sweeteners. These range from small molecules like stevioside, a (1′→2) linked disaccharide diterpene present in the leaves of *Stevio rebaudiana*, which is marked in Japan, to large proteins like thaumatin, marketed under the trade name Talin. The role of these compounds in the metabolism or reproduction of the plants is not known, although many hypotheses have been proposed.

Among the sweet compounds present in plants three proteins have attracted much attention. These are:

(1) *Miraculin.* Isolated from the pulp of the red berries of *Richadella dulcicia* (in older literature *Synsepalum dulcificum,* first described by Daniell in 1852).

(2) *Monellin.* Present in fruit of *Dioscoreophyllum cumminsii* Diels.

(3) *Thaumatin.* Found in the arils of the fruit of *Thaumatococcus danielli* Benth (Fig. 11.2), also described first by Daniell in 1855.

All three plants grow in the rain forests of Africa and were rediscovered by Inglett and May (1968).

A sweet protein has recently been isolated from the Malaysian tree *Curculigo latifolia* by Yamashita *et al.* (1990). They called this protein *curculin.*

Two of these proteins, monellin and thaumatin, are extremely sweet, about 100 000 times as sweet as sucrose on a molar basis. On a weight basis thaumatin is about 3000 times sweeter than sucrose (a 10% solution) and monellin 2000 times sweeter (Hough and Edwardson, 1978). Miraculin has the intriguing property of changing a sour to a sweet perception, and curculin is sweet on its own (about 10 000 times sweeter than sucrose on molar basis or 285 times on weight basis), but is also able to change sour to sweet.

Sweeteners like sucrose are required in high concentrations (0.1 to 0.001 mol/litre) to be perceived as sweet and this makes them less suitable for studying the interaction between sweetener and taste receptor. The intense sweetness of monellin and thaumatin, registered at a level of 0.01 μmol/ litre, makes these components very suitable for the analysis of taste perception, and techniques similar to those used in the study of the interactions

**Fig. 11.2**  Opened fruit of *Thaumatococcus danielli.* The arils, containing the sweet protein are the lighter regions to the right of the dark fruit.

between hormones and receptor proteins can be used.

# Sweet proteins

## Miraculin

*Richadella dulcifica* grows in West Africa and carries red, olive-shaped berries of 1–2 cm. The fruit consists of a large seed surrounded by a thin layer of pulp that contains miraculin.

Its isolation and purification was described first by Brouwer *et al.* (1968) and Theerasilp and Kurihara (1988) showed that miraculin is a glycoprotein with a molecular mass of 24 kDa, 13.9% of which is the glyco-moiety.

Miraculin as such is not sweet, but after exposing the tongue to miraculin all sour-tasting substances taste sweet for 1–2 h. The amino acid sequence of miraculin has recently been determined (Fig. 11.3). It consists of 191 amino acids and contains two glycosylated asparagines (42 and 186) and seven cysteines (Theerasilp *et al.* 1989). The established disulphide bridges are 47–92, 148–159 and 152–155, leaving Cys (138) as an unpaired residue. In contrast with thaumatin, miraculin contains two histidine residues and modification of these residues destroys the sweetness-inducing property (Y. Kurihara, personal communication). Another remarkable feature of miraculin is that it contains two amino acid stretches with a striking homology to soybean trypsin inhibitor.

## Curculin

*Curculigo latifolia* grows in the tropical forests of Malaysia. Yamashita et al. (1990) recently found that the fruits of this bamboo-like tree have protein bodies containing a sweet protein, now called curculin.

This protein is relatively small (114 amino acids) and contains no glyco-groups. In contrast to the other sweet proteins, which all have high pI values, curuculin has a pI value of 7.1. It contains seven arginines, and three lysines and is rich in aspartic acid, leucine and glycine. Curculin is normally present as a dimer with a molecular mass of 27 800. The most remarkable feature of curculin is that it is sweet as such (like thaumatin and monellin), but it is also able to change sour to sweet (like miraculin). The sweetness of curculin can be suppressed completely by the addition of 1 mmol/litre $Ca^{2+}$ or $Mg^{2+}$.

Curculin does not react with polyclonal antibodies raised against miraculin (Y. Kurihara, personal communication).

## Monellin

*Dioscoreophyllum cumminsii* Diels grows in Central African countries. The red berries are approximately 1 cm long and grow in grape-like clusters on the stem of the plant. The tough outer skin encloses the white mucilaginous pulp and the seed. Monellin is found in the pulp (Inglett and May

```
1                          10                          20                          30
D   S   A   P   N   P   V   L   D   I   D   G   E   K   L   R   T   G   T   N   Y   Y   I   V   P   V   L   R   D   H

31                         40                          50                          60
G   G   G   L   T   V   S   A   T   T   P   N   G   T   F   V   C   P   P   R   V   V   Q   T   R   K   E   V   D   H

61                         70                          80                          90
D   R   P   L   A   F   F   P   E   N   P   K   E   D   V   V   R   V   S   T   D   L   N   I   N   F   S   A   F   M

91                         100                         110                         120
P   C   R   W   T   S   S   T   V   S   R   L   D   K   Y   D   E   S   T   G   Q   Y   F   V   T   I   G   G   V   K

121                        130                         140                         150
G   N   P   G   P   E   T   I   S   S   W   F   K   I   E   E   F   C   G   S   G   F   Y   K   L   V   F   L   P   T

151                        160                         170                         180
V   C   G   S   C   K   V   K   C   G   D   V   G   I   Y   I   D   Q   K   G   R   R   R   L   A   L   S   D   K   P

181                        191
F   A   F   F   F   N   K   T   V   Y   F
```

**Fig. 11.3**   The amino acid sequence of mature miraculin as determined by Theeraslip *et al.* (1989).

1969). The protein was first purified by Morris and Cagan (1972) and consists of two chains, with 44 and 50 amino acids respectively (Frank and Zuber 1976). The amino acid sequences of both chains are depicted in Figure 11.4.

The three dimensional structure has recently been reported at the 3 Å resolution (Ogata *et al.* 1987) and a schematic drawing of the backbone is given in Figure 11.5.

*A chain:*

1

FREIKGYEYQLYVYASDKLFRADISEDYK

30

TRGAKLLRFNGPVPPP

*B chain:*

1

GEWEIIDIGPFTQNLGKFAVDEENKIGQY

30

GRLTFNKVIRPCMKKTITENE

**Fig. 11.4**   The amino acid sequence of the two peptide chains of monellin as determined by Frank and Zuber (1976).

**Fig. 11.5**   The C-α backbone of monellin derived from the X-ray diffraction pattern of this protein as determined by Ogata *et al.* (1987).

## Thaumatin

*Thaumatococcus danielli* Benth grows in West African rainforests from Sierra Leone to Angola. The plant grows best in well drained soil and prefers shade from tall trees. Flowering normally occurs twice a year (March–June and August–October) and 20–40% of the trees bear clusters of mature fruits, normally protected against light by leaves or mud. The fruit contains one to three hard black seeds and arils containing the sweet protein (see Fig. 11.2).

The sweet protein can be isolated easily from the arils (van der Wel and Loeve 1972) and the amino acid sequence was determined by Iyengar *et al.* (1979). Of the five different thaumatin molecules found (I, II, III, a and b (van der Wel and Loeve 1972; Higgenbotham and Hough 1977)); thaumatin I and II are the most abundant proteins and consist of 207 amino acids, differing from each other at five positions (Fig. 11.6). The various forms of thaumatin have extremely high pI values, ranging from 11.5 to 12.5 and all contain eight disulphide bridges.

The three-dimensional structure of thaumatin I has been determined with a resolution of 3.1 Å (de Vos *et al.* 1985) and is depicted in Figure 11.7. The nucleotide sequence of the cDNA of thaumatin II has been determined by my group (Edens *et al.* 1982) and some features of the structure of genes have also been described (Ledeboer *et al.* 1984). The presence of introns in the genes leads us to speculate whether thaumatin is a storage protein in which introns can be present (Fischer and Goldberg 1982). However, as will be shown later, thaumatin is more closely related to a class of stress proteins.

Comparison of the amino acid sequence of the isolated protein and the sequence deduced from the cDNA shows that thaumatin II is synthesized in a precursor form that is processed at both the N- and C-termini (Fig. 11.6). The N-terminal extension is similar to other sequences found in plant protein (Lazaro *et al.* 1988) and is most probably involved in the translocation of the protein into the endoplasmic reticulum.

An interesting property of thaumatin is that only old-world monkeys and humans perceive the molecule as sweet. Although thaumatin does not induce a clear electrophysiological response in rats, it must have some attractiveness to mammals other than monkeys and men, as rats preferred water containing only 0.5 μmol/litre thaumatin over pure water

-22                                                                                                                                              -1

H₂N-Met  Ala  Ala  Thr  Thr  Cys  Phe  Phe  Phe  Leu  Phe  Pro  Phe  Leu  Leu  Leu  Leu  Thr  Leu  Ser  Arg  Ala

|  | 1 | 46 | 6? | 67 | 76 | 113 | 207 |
|---|---|---|---|---|---|---|---|
| Thaumatin II Ala |  | _y | Arg | Arg | Glu | Asp | Ala |
| Thaumatin I Ala |  | Asn | Ser | _, | Ary | Asn | Ala |

(208)
Leu  Glu  Leu  Glu  Asp  Glu-COOH

**Fig. 11.6**  The N- and C-terminal extensions of thaumatin II as derived from the nucleotide sequence (Edens *et al.* 1982). The N-terminal is quite similar to signal sequences in general and homologous to the signal sequence of α-amylase inhibitor (Lazaro *et al.* 1988). The N-terminal sequence is recognized by signal sequence peptidases in yeast (Edens *et al.* 1984). The role of the C-terminus is unknown and it is not cleaved off in yeasts (Verrips, unpublished results). The differences between thaumatin I (Iyengar *et al.* 1979) and thaumatin II (compare with Fig. 11.9) are shown.

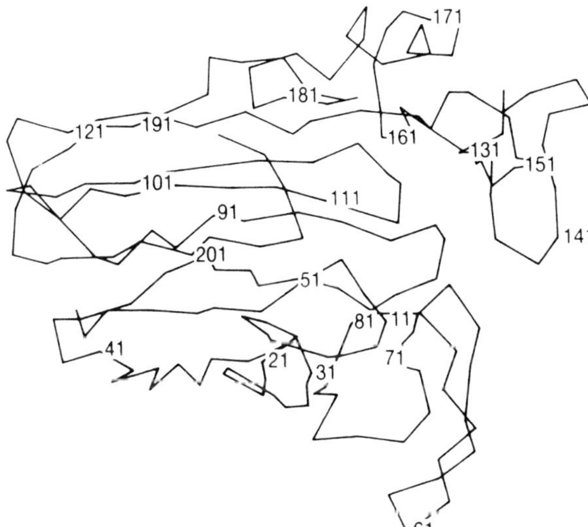

**Fig. 11.7**  The C-α backbone of thaumatin I derived from the X-ray diffraction pattern of this protein as determined by De Vos *et al.* (1985).

(Dugas du Villard *et al.* 1980). Thaumatin was also able to enhance the sweetness of sucrose for rats and the results of these behaviour studies were supported by electrophysiological measurements (Brouwer *et al.* 1973). The enhancement of the sweetness of sucrose by thaumatin suggests that thaumatin interacts with the receptor protein of the rat, but not at the recognition site for sucrose. By binding it could induce a conformational change in the recognition site, thereby facilitating the binding of sucrose. Alternatively, the basic protein thaumatin may interact with the membrane of the taste cell introducing a change in the surface charge, which results in a modification of the membrane and makes the sucrose receptor more sensitive.

Later, van der Wel (personal communication) proved that thaumatin blocks the sucrose receptor on the tongue of monkeys and showed that this block can be neutralized by adding monoclonal antibodies against thaumatin. These results suggest that the binding sites for sucrose and thaumatin are either identical or so closely related that the sucrose receptor site is blocked when thaumatin is bound. The fact that monoclonal antibodies can neutralize this block means they recognize either the same or nearly the same site on thaumatin as the receptor. Another possibility is that the monoclonal antibodies interact with a different site on thaumatin, thereby introducing a conformational change resulting in the destruction of the binding site on thaumatin for the receptor cell. The latter hypothesis is less likely because the eight disulphide bridges give thaumatin a rather rigid structure.

In relation to the competition between thaumatin and low molecular weight compounds for the binding site of receptor molecules and Koshland's hypothesis on the similarity between taste perception and chemotaxis of micro-organisms (Koshland 1974), it is important to refer to the work of Barreau and van der Wel (1983) on the effect of thaumatin on the chemotactic behaviour of *Escherichia coli*. Thaumatin does not induce any chemotactic response, but a level of 0.2 μmol/litre completely blocks the chemotactic response of *E. coli* towards L-serine and galactose. The effect of acetylation of thaumatin was dependent upon the number of modified ε-amino groups of lysine. One or two modifications of the thaumatin amino groups induced a chemotactic response in *E. coli* towards thaumatin but with modification of four or more ε-amino groups, thaumatin showed neither a stimulatory nor an inhibitory effect on the chemotactic effect of L-serine.

A remarkable observation was that the addition of 0.025 mol/litre phosphate was sufficient to eliminate the effect of thaumatin on the mobility of *E. coli* towards attractants. Phosphate also has a remarkable effect on chemoreception of thaumatin by humans (van der Wel 1981). Thaumatin was unable to induce a response when present in a 0.025 mol/litre phosphate buffer. However, the sweet taste was immediately perceived on rinsing the tongue with water.

From these experiments it may be concluded that, in the presence of phosphate, thaumatin binds to the receptor protein or to the membrane surrounding the receptor protein, but that phosphate blocks the interaction of thaumatin with the side of the receptor protein that is involved in the initiation of the polarization of the taste cell. The indication that thaumatin may interact, via its lysine or arginine residues, with the membrane of the taste cell suggests a mechanism similar to that of signal sequences of proteins that are important for translocation in micro-organisms and plants. The positively charged amino acids of the signal sequences interact, via their amino group, with the phosphate group of the phosholipids of membranes (Saier *et al.* 1989). Low molecular mass sweeteners, like Aspartam (L-Asp-L-Phe-O-Me, Fig. 11.8), often have a polar region, consisting of a proton donor and acceptor group, and a hydrophobic region; the proton donor and acceptor groups of many sweet compounds are approximately 3 Å apart (Schallenberger and Acree 1967).

Kier (1972) subsequently pointed out that, in addition to the donor and acceptor, a hydrophobic group is essential. Studies of a large number of small molecules showed that the hydrophobic group must be approximately 5.5 Å from the proton acceptor group and approximately 3.5 Å from the protein donating group. A very nice demonstration of the importance of the hydrophobic group is the sweetness of the Aspartam analogue L-aspartyl-aminomalonic fenchylester (Fig. 11.9). Whereas the $NH_2^+$ and $COO^-$ groups of the aspartyl residue function as proton donor and acceptor, respectively, the structure of the terpene–alcohol group seems to fit well into the hydrophobic part of the receptor, resulting in a sweetness which is 33 000 times that of sucrose (van der Wel *et al.* 1987.

## TL or PR proteins

With the advent of computerized data bases of protein and nucleotide sequences, it has become routine to compare newly determined sequences with existing sequences. This has resulted in the identification of some proteins with a remarkable homology to thaumatin, called thaumatin-like proteins (TL proteins) or pathogenesis-related proteins (PR proteins). The first TL protein was discovered by Cornelissen *et al.* (1986). This protein could be induced in tobacco by infection with

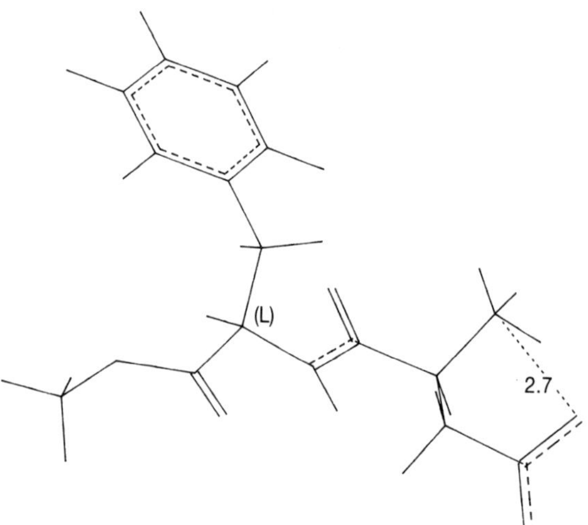


**Fig. 11.8** A possible structure of Aspartam (L-Asp-L-Phe-O-Me) as predicted with molecular mechanics (Discover programme). The distance between the proton donor and acceptor is in agreement with the model for sweet molecules as proposed by Schallenberger and Acree (1967).

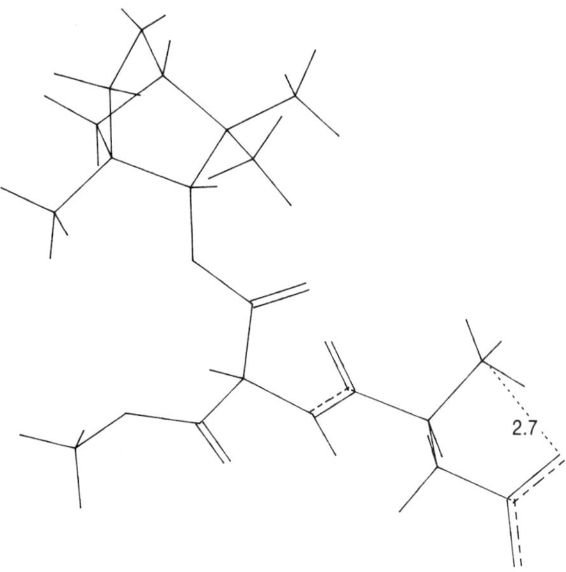


**Fig. 11.9** A possible structure for the sweetest component on weight basis, L-aspartyl-aminomalonic fenchylester (Van der Wael *et al.* 1987) predicted with molecular dynamics (Discover programme).

tobacco mosaic virus (TMV). Although it showed extensive sequence homology (Fig. 11.10) it did not cross-react with antibodies raised against thaumatin. TMV also induced a TL protein in maize, which proved to have inhibitory action against bovine trypsin and α-amylase of *Tribolium castaneum* beetles (Richardson *et al.* 1987). This protein also contains 16 cysteine residues out of 206 amino acids.

A number of other proteins with a high homology to thaumatin II have recently been described (Singh *et al.* 1987; King *et al.* 1988; Payne *et al.* 1988). Their amino acid sequences and their similarities with thaumatin II are shown in Figure 11.10. One of the most remarkable facts is that all these proteins contain 16 cysteines and that the distance (defined here as number of amino acids) between these cysteines is quite invariant, which may mean that the structures between neighbouring cysteine residues are also quite conserved. Miraculin also has sequences rich in cysteine in its C-terminal part. Curculin contains 7 Cys residues and shares with thaumatin a sequence of two subsequent Cys residues.

Such conserved structures between cysteine residues have also been described by Drenth *et al.* (1980) for snake venom postsynaptic neurotoxins, and certain domains of wheat germ agglutinin and bacterial spore coat proteins may have conserved loops between cysteine residues (Donovan *et al.* 1980).

## Chemical modification of monellin and thaumatin

Before protein engineering by changing the nucleotide sequence of the gene became technically possible, chemical modification was the only way to determine whether certain amino acids were involved in the catalytic or binding properties of a protein. Thaumatin and monellin have been modified in this way by several groups.

Methylation of the side chain NH₂ groups of lysine did not greatly affect the sweetness of thaumatin (van der Wel and Bel 1976) or monellin (Cagan and Morris 1979), but acetylation of four or more of the ten lysine residues destroyed the sweetness of both proteins completely (van der Wel and Bel 1976). The lysine residues at positions 97 and 106 of thaumatin seem to be of particular importance for the sweet taste. Modification of the guanidino groups of six of the eleven arginines of thaumatin with 1,2-cyclohexanedione did not re-

sult in a significant decrease of the sweetness (van der Wel and Bel 1978).

Modification of tyrosine residues by iodination showed that two tyrosines could be modified without affecting the sweetness of thaumatin but that further modification destroyed the sweetness. Tyrosine 95 is considered to be especially important for the sweetness (van der Wel 1980). Iodination also destroyed the sweetness of monellin (van der Wel and Bel 1976).

Both monellin and thaumatin contain only one methionine and modification of this amino acid did not affect the intensity of the sweetness of thaumatin, but changed the character of its sweetness in the direction of sucrose (van der Wel 1980). Modification of the carboxylic groups increased the sweetness by a factor of six, but the introduction of the sweet taste sensation became extremely slow, although it lasted much longer (van der Wel personal communication).

Selective reduction of thaumatin, either by dithiothreitol or by cysteine, resulted in an unexpected hydrolytic activity of thaumatin (van der Wel and Bel 1980). The selective reduction was done in such a way that only two (of the four) cysteine residues located between amino acids 138 and 163 were reduced, which means that either only disulphide bridge 149–158 had been broken or that at the most, two of the bridges 134–145, 149–158 or 159–164 were broken. All these bridges are located in domain III (Fig. 11.11). The hydrolytic activity (measured as protease, esterase and amidase activity) could be enhanced by acetylation of the ε-amino groups of lysine. The hydrolytic activity of reduced thaumatin is even more remarkable because the protein does not contain any histidine and has no homology with cysteine proteases, such as papain.

Monellin, containing only one cysteine, does not show any hydrolytic activity in the presence of a reducing agent. However, after acetylation of about half the lysine residues, the addition of a reducing agent changed the acetylated monellin into an amidase (van der Wel and Bel 1980).

Hardly any information is available on the chemical modification of miraculin or curculin. However, modification of the histidine residues of miraculin completely destroys its capability to change a sour taste into a sweet one.

A recent publication described the production of a fragment of thaumatin by proteolysis with an enteropeptidase (Tanaka and Yamamoto 1988). This fragment had a molecular mass of about

| | 1 | | | | | | | | | 10 | | | | | | | | | | 20 |
|---|---|---|---|---|---|---|---|---|---|---|---|---|---|---|---|---|---|---|---|---|
| Thaumatin II | A | T | F | E | I | V | N | R | C | S | Y | T | V | W | A | A | A | S | K | G |
| Mai (30)* | A | V | F | T | V | V | N | Q | C | P | F | T | V | W | A | A | A | S | V | P | V |
| PROB-12 (29) | A | T | F | D | I | V | N | Q | C | T | Y | T | V | W | A | A | A | S | P | - |
| NP24 (31) | A | T | I | E | V | R | N | N | C | P | Y | T | V | W | - | A | A | S | - | - | - |
| PR-R (32) | A | T | F | D | I | V | N | K | C | T | Y | T | V | W | A | A | A | S | P | - |
| Osmotin (33) | A | T | I | E | V | R | N | N | C | P | Y | T | V | W | - | A | A | S | P | I |

| | 21 | | | | | | | | | 30 | | | | | | | | | | 40 |
|---|---|---|---|---|---|---|---|---|---|---|---|---|---|---|---|---|---|---|---|---|
| Thaumatin | D | A | A | L | D | A | G | G | R | Q | L | N | S | G | E | S | W | T | I | N |
| Mai | - | - | - | - | - | - | G | G | G | R | Q | L | N | R | G | E | S | W | R | I | T |
| PROB-12 | - | - | - | - | - | - | G | G | G | R | Q | L | N | S | G | Q | S | W | S | I | N |
| NP24 | - | - | T | P | I | G | G | G | R | R | L | N | R | G | Q | T | W | V | I | N |
| PR-R | - | - | - | - | - | - | G | G | G | R | R | L | D | S | G | Q | S | W | S | I | N |
| Osmotin | - | - | - | - | - | - | G | G | G | R | R | L | D | R | G | Q | T | W | V | I | N |

| | 41 | | | | | | | | | 50 | | | | | | | | | | 60 |
|---|---|---|---|---|---|---|---|---|---|---|---|---|---|---|---|---|---|---|---|---|
| Thaumatin | V | E | P | G | T | K | G | G | K | I | W | A | R | T | D | C | Y | F | D | D |
| Mai | A | P | A | G | T | T | A | A | R | I | W | A | R | T | G | C | Q | F | D | A |
| PROB-12 | V | N | P | G | T | V | Q | A | R | I | W | G | R | T | N | C | N | F | D | G |
| NP24 | A | P | R | G | T | K | M | A | R | I | W | G | R | T | G | C | N | F | N | A |
| PR-R | V | N | P | G | T | V | Q | A | R | I | W | G | R | T | N | C | N | F | D | G |
| Osmotin | A | P | R | G | T | K | M | A | R | V | W | G | R | T | N | C | N | F | N | A |

| | 61 | | | | | | | | | 70 | | | | | | | | | | 80 |
|---|---|---|---|---|---|---|---|---|---|---|---|---|---|---|---|---|---|---|---|---|
| Thaumatin | S | G | R | G | I | C | R | T | G | D | C | G | G | L | L | Q | C | K | R | F |
| Mai | S | G | R | G | S | C | R | T | G | D | C | G | V | V | V | Q | C | T | G | Y |
| PROB-12 | S | G | R | G | N | C | E | T | G | D | C | N | G | M | L | E | C | Q | G | Y |
| NP24 | A | G | R | G | T | C | Q | T | G | D | C | G | G | V | L | Q | C | T | G | W |
| PR-R | S | G | R | G | N | C | E | T | G | D | C | N | G | M | L | E | C | Q | G | Y |
| Osmotin | A | G | R | G | T | C | Q | T | G | D | C | G | G | V | L | Q | C | T | G | W |

| | 81 | | | | | | | | | 90 | | | | | | | | | | 100 |
|---|---|---|---|---|---|---|---|---|---|---|---|---|---|---|---|---|---|---|---|---|
| Thaumatin | G | R | P | P | T | T | L | A | E | F | S | L | Y | Q | Y | G | K | D | Y | I |
| Mai* | G | R | A | P | N | T | L | A | E | Y | A | L | K | Q | F | N | $N_L$ | D | F | F |
| PROB-12 | G | K | P | P | N | T | L | A | E | F | A | L | N | Q | - | P | $N_Q$ | D | F | V |
| NP24 | G | K | P | P | N | T | L | A | E | Y | A | L | D | $Q_F$ | S | N | L | D | F | W |
| PR-R | G | K | A | P | N | T | L | A | E | F | A | L | N | Q | P | N | Q | D | F | V |
| Osmotin | G | K | P | P | N | T | L | A | E | Y | A | L | D | Q | F | S | $G_L$ | D | F | W |

| | 101 | | | | | | | | | 110 | | | | | | | | | | 120 |
|---|---|---|---|---|---|---|---|---|---|---|---|---|---|---|---|---|---|---|---|---|
| Thaumatin | D | I | S | N | I | K | G | F | N | V | P | N | D | F | S | P | T | T | R | G |
| Mai | D | I | S | I | L | D | G | F | N | V | P | Y | S | F | L | P | D | G | $G_S$ | G |
| PROB-12 | D | I | S | I | V | D | G | F | N | I | P | M | E | F | S | P | T | N | G | G |
| NP24 | D | I | S | L | V | D | G | F | N | I | P | M | T | F | A | P | $T_K$ | $P_S$ | S | $G_K$ |
| PR-R | D | I | S | L | V | D | G | F | N | I | P | M | E | F | S | P | T | N | G | G |
| Osmotin | D | I | S | L | L | D | G | F | N | I | P | M | T | F | | P | T | $N_S$ | G | $G_K$ |

```
           121                         130                            140
Thaumatin   C    R  G  V  R  C  A  A  D  I  V  G  Q  C  P  A  K  L  K  A
Mai*        Cs   R  G  P  R  C  A  V  D  V  N  A  R  C  P  A  E  L  R  -
PROB-12     C    R  N  L  R  C  T  A  D  I  N  E  Q  C  P  A  Q  L  K  -
NP24        C    H  A  I  H  C  T  A  N  I  N  G  E  C  P  R  A  L  K  V
PR-R        C    R  N  L  R  C  T  A  P  I  N  E  Q  C  P  A  Q  L  K  T
Osmotin     C    H  A  L     C  T  A     I  N  G  E  C  P  A  E  L  R  V

           141                         150                            160
Thaumatin   P    G  G  G  C  N  D  A  C  T  V  F  Q  T  S  E  Y  C  C  T
Mai         Q    D  G  V  C  N  N  A  C  P  V  I  K  K  D  E  Y  C  Cvgs A
PROB-12     I    Q  G  G  C  N  N  P  C  T  V  F  K  T  N  E  F  C  Ctn  G
NP24        P    -  G  G  C  N  N  P  C  T  T  F  G  G  Q  Q  Y  C  C  T
PR-R        Q    -  G  G  C  N  N  P  C  T  V  I  K  T  N  E  Y  C  G  T
Osmotin     P    -  G  G  C  N  N  P  C  T  T  F  G  G  Q  Q  Y  C  C  T

           161                         170                            180
Thaumatin   T    G  K  C  G  P  T  E  Y  S  R  F  F  K  R  L  C  P  D  A
Mai*        A    N  N  C  H  P  T  N  Y  S  R  Y  F  K  G  Q  C  P  D  A
PROB-12     P    G  S  C  G  P  T  D  L  S  R  F  F  K  A  R  C  P  D  A
NP24        Q    G  P  C  G  P  T  E  L  S  K  F  F  K  K  R  C  P  D  A
PR-R        N    G  Pgs C  G  P  T  D  L  S  R  F  F  K  E  R  C  P  D  A
Osmotin     Q    R  P  C  G  P  T  F  F  S  K  F  F  K  Q  R  C  P  D  A

           181                         190                            200
Thaumatin   F    S  Y  V  L  D  K  P  T  T  V  T  C  P  G  S  S  N  Y  R
Mai         Y    S  Y  P  K  D  D  A  Ts I  F  T  C  P  A  G  T  N  Y  K
PROB-12     Y    S  Y  P  Q  D  D  P  Ps T  F  T  C  P  P  G  I  N  Y  R
NP24        Y    S  Y  P  Q  D  D  P  Ts T  F  T  C  P  Gg S  T  N  Y  R
PR-R        Y    S  Y  P  Q  D  D  P  Ts L  F  T  C  P  S  G  T  N  Y  R
Osmotin     Y    S  Y  P  Q  D  D  P  Ts T  F  T  C  Pg G  S  T  N  Y  R

           201                  207
Thaumatin   V  T  F  C  P  T  A
Mai         V  V  F  C  P
PROB-12     V  V  F  C  P
NP24        V  V  F  C  P  N  G  V  A  D  P  N  F  P  L  E  M  P  A  S  T  D  E  V  A  K
PR-R        V  V  F  C  P
Osmotin     V  I  F  C  P  N  G  Q  A  H  P  N  F  P  L  E  M  P  G  S     D  E  V  A  K
```

*References

**Fig. 11.10**   Comparison of the amino acid sequences of a number of PR- or TL-like proteins and thaumatin II. Mai (a maize protein that inhibits bovine trypsin and α-amylase of beetles (Richardson *et al.* 1987)); PROB 12 (a virus-induced TL protein in tobacco (Cornelissen *et al.* 1986)); NP24 (PR protein induced by salt in tomato (King *et al.* 1988)); PR-R (the major PR protein in tobacco leaves (Payne *et al.* 1988)); Osmotin (a PR-protein induced by salt in tobacco cells (Sing *et al.* 1987)).

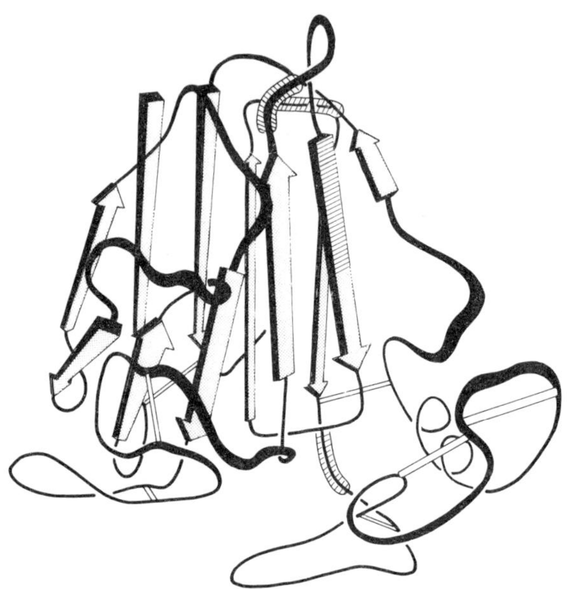

**Fig. 11.11**  Schematic drawing of the backbone structure of thaumatin I. The open bars represent the disulphide bonds. The three domains as described by De Vos *et al.* (1985) are given.

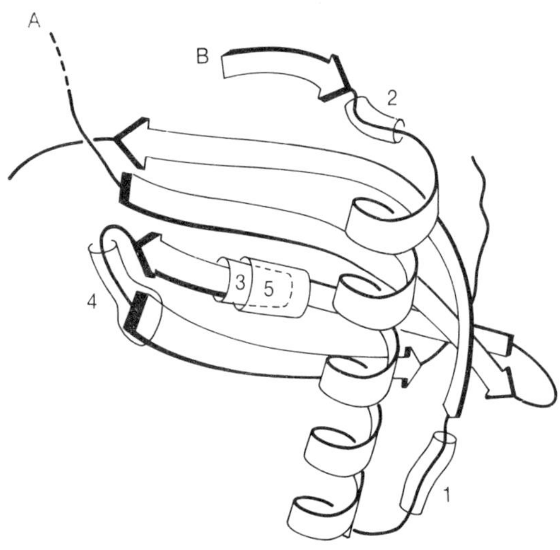

**Fig. 11.12**  Schematic drawing of the backbone structure of monellin. Tubed sections are regions of homology with thaumatin.

12 kDa and proved to be sweet. However, no data were presented to show which part of the protein was responsible for the sweet character.

Although miraculin, curculin, monellin and thaumatin all have intensive sweet characters, strong homologies cannot be detected by comparisons of the amino acid sequences or three-dimensional structures. However, in spite of this lack of homology, antibodies raised against thaumatin recognize monellin and antibodies raised against monellin recognize thaumatin (Hough and Edwardson 1978). This shows that epitopes of extreme high similarity must be present on the two molecules. The longest identical sequences in the two proteins are tripeptides (Fig. 11.12).

Although tripeptides are quite small to serve as epitopes, they may be sufficient if they are exposed by the surrounding amino acids in a similar way on the surfaces of proteins. In this respect it should be noticed that three amino acids (SerLysLeu) at the C-terminus of luciferase seem to be sufficient for the translocation of the protein into peroxisomes (Gould *et al.* 1988) and it is generally accepted that protein–protein recognition is necessary for translocation.

Thaumatin and miraculin have a single identical tetrapeptide in common but, allowing for some

conserved substitutions, it is possible to recognize a stretch of nine amino acids of which five are identical (sequences 198–206 of thaumatin and 143–151 of miraculin). However, whether this stretch has a role in determining sweetness is uncertain. It is notable that the stretch 143–151 of miraculin is homologous with a trypsin inhibitor, although the homologous sequence is not the sequence involved in the interaction with trypsin (Creighton and Charles 1987). It is also remarkable that some homology exists between the signal peptides of thaumatin and unrelated cereal α-amylase/trypsin inhibitors (Lazaro *et al.* 1988).

No clear sequence homology exists between monellin and miraculin, nor do antibodies raised against thaumatin or monellin recognize miraculin or curculin.

Chemical modification showed that the basic character of thaumatin, as such, is not important but that particular basic residues, notably Lys (97) and (106) are. Tyr (95) also seems to be essential for the sweet character of thaumatin.

Partial reduction changed thaumatin into a hydrolytic enzyme despite the absence of a histidine in the molecule.

Miraculin, monellin and thaumatin are basic proteins with pI values of 9.3, 8.7 and 11.5–12.5, respectively but curculin is a neutral protein (pI = 7.1).

# Protein engineering of sweet proteins

## Cloning and expression of thaumatin genes in micro-organisms

The isolation of polyadenylated RNA from the arils of *Thaumatococcus daniellii* Benth, the construction of a cDNA library, the selection of clones and the determination of the nucleotide sequence of thaumatin II, and the expression of thaumatin in micro-organisms (Edens *et al.* 1982) were the first steps towards the modification of thaumatin by protein engineering and were patented in 1981 (Verrips *et al.* 1980). As shown in Figure 11.6, thaumatin II proved to be a protein with N- and C-terminal extensions. The 22-amino acid N-terminal extension showed high similarity with signal sequences of lower eukaryotes and prokaryotes and we subsequently showed that, indeed, this plant N-terminal sequence was cleaved exactly in yeasts (Edens *et al.* 1984). The 6-amino acid C-terminal extension was not removed in the yeast and was therefore removed by truncating the gene (unpublished data). The expression in *Saccharomyces cerevisiae* under control of the glyceraldehyde-3-phosphate dehydrogenase promoter was quite low (about 100 molecules per cell).

To increase the amount of thaumatin produced, the thaumatin gene was integrated into the rDNA locus of *S. cerevisiae*, resulting in about 50–100 copies of the gene and a more than ten-fold increase in production (Lopes *et al.* 1989). The thaumatin II gene was also cloned in the yeast *Kluyveromyces lactis* (Edens *et al.* 1983), but none of these systems gave a correctly folded protein.

The nucleotide sequence showed that the codon usage in *Thaumatococcus daniellii* Benth was quite different from the preferred codon usage in yeast, which prompted Huang *et al.* (1987) to synthesize variants of the published thaumatin genes using the preferred yeast codons. Together with the powerful phosphoglycerokinase (PGK) promoter this considerably enhanced the production of thaumatin (to 20% of the cellular protein), but although the protein showed cross-reactivity with antibodies raised to thaumatin, it was not sweet. Refolding resulted in a sweet protein (Lee *et al.* 1988).

Whether the lack of correct folding was due to the fact that the mature form was placed directly after the PGK promoter and therefore did not follow the pathway via the e.r. and Golgi before secretion (as is the case in the plant) was not clear.

The expression in *S. cerevisiae* of genetically engineered thaumatin I has been described in a patent application (Weickman and Ghoshdasti 1987). Modification of amino acid residues Asp (113) and Lys (46) + Asp (113) did not influence the sweetness of the protein. Huang *et al.* (1987) also carried out N-terminal modifications of thaumatin to study the cleavage of Met (1) and the N-terminal acetylation. The modified thaumatins were recognized by antibodies but their sweetness was not described.

Illingworth *et al.* (1988) described the cloning of the mature thaumatin II gene behind the α-amylase signal sequence and under control of the α-amylase promoter in *Bacillus subtilis*. They obtained an excreted protein (about 1 mg/litre) that was recognized by antibodies raised against plant thaumatin. The construction of the mature thaumatin behind the α-amylase signal sequence introduced a change in the N-terminal sequence of rDNA thaumatin (AlaThrPheGlu to LeuThrAlaProSer).

Treatment of thaumatin with aminopeptidase resulted in loss of sweetness and this change in the N-terminus may explain why the protein excreted by the *Bacillus* and recognized by antibodies was not sweet (U.R.L. Vlaardingen, unpublished results).

Illingworth *et al.* (1989) also reported the expression of thaumatin in *Streptomyces lividans*.

## Expression of the thaumatin gene in other plants

Cloning of thaumatin in micro-organisms did not always result in a correctly folded protein and hence the usefulness of microbiological productions systems is questionable.

M. Witty has cloned prepro-thaumatin behind the CaMV promoter in *Solanum tuberosum* and, although data on the precise processing have not been provided, the product was perceived as sweet—a strong indication that the processing and folding were correct (personal communication).

M. Tanaka, T. Endo and A. Kaji have recently cloned prepro-thaumatin in *Nicotiana tabacum* and have shown that at least the N-terminal processing took place. However, no indication of the sweetness of the thaumatin produced was given (personal communication).

Witty's results demonstrated that *Solanum tuberosum* is a suitable host for producing genetically engineered and also potentially modified thaumatin. As long as the yield of correctly folded thaumatin in micro-organisms is low, these host systems cannot be used for the production of modified thaumatins.

## Cloning, expression and modification of monellin genes in micro-organisms

Monellin consists of two peptide chains, but this may not be an obstacle for its synthesis in micro-organisms (as shown in the case of insulin: Goeddel *et al.* 1979). However, such a production route requires additional processing. Therefore Kim *et al.* (1989) used knowledge of the crystal structure of monellin to construct a single chain monellin gene, using different linkers to connect the two peptide chains without significant interference to the tertiary structure. These genes were cloned in *E. coli* and all the chimeric constructs proved to be sweet. One of the proteins lacking Phe(1) of the A-chain proved to be as sweet as normal monellin, but was more pH stable and more easily renatured after heat denaturation. The linkers used by Kim *et al.* (1989) were identical to loop structures already present in monellin. One of these linkers, GluAspTyrLysThrArgGlyArg, contains the sequence ThrArgGly, which is also found in thaumatin. This homochimeric monellin has only 10% of the sweetness of natural monellin, but deletion of GlyArg in the loop restored the sweetness to 40% of natural monellin. From this Kim *et al.* (1989) concluded that this sequence is not involved in the sweetness of monellin, although it may indicate that the new loop interacts with the receptor via ThrArgGly, and that the drastic decrease in sweetness is caused by competition between the original ThrArgGly and the new ThrArgGly sequence, the latter not resulting in sweetness.

This approach, of making chimeric constructs of monellin, has also been described in patent applications (Cho 1988; Kim 1988). Kim's group also reported the expression of the chimeric monellin in a number of other micro-organisms, including *Bacillus*, *Saccharomyces*, *Neurospora*, *Streptomyces* spp., and in plant cell culture (Cho and Kim 1988).

## Possible targets for protein engineering of thaumatin

Although at present protein engineering of monellin seems to be more successful than that of

thaumatin, the finding of a number of TL proteins that are probably homologous in their tertiary structures but are neither sweet nor recognized by antibodies against thaumatin opens the way to detailed analysis of thaumatin. In particular, the differences between thaumatin and the TL and PR proteins can be explored to determine the site(s) giving thaumatin its sweetness. For example, in spite of this high degree of homology, four differences in charged amino acids between these TL proteins and thaumatin are located in the short region of thaumatin depicted in Figure 11.13. Moreover, the fact that monellin and thaumatin are recognized by their reciprocal antibodies and bind to the same two proteins of the lingual tissue despite large difference in amino acid sequence, offers the possibility of defining the sequence(s) that are similar between thaumatin and monellin with reasonable confidence. This is probably the sequence located between amino acids 90 and 120 of thaumatin; perhaps even the sequence 94 to 106, although it cannot be ruled out that the sequence 118–120, together with the sequence 94–100, forms a structural entity quite similar to that of the monellin sequence 19–31. Figure 11.13 shows the C-α backbones of the sequences monellin 24–35 and thaumatin 90–120, which may interact with the antibodies and/or the receptor proteins. The likelihood of the involvement of this sequence in the sweetness of thaumatin is further supported by the fact that the differences between thaumatins I and II (Fig. 11.6) and the amino acid substitutions Asp

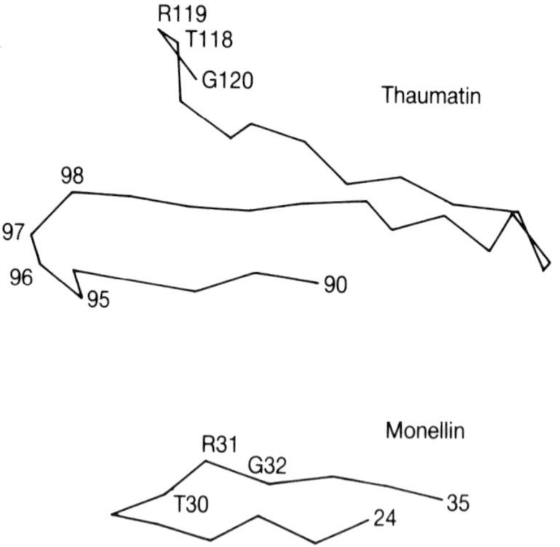


**Fig. 11.13**   Enlarged drawings of the C-α backbones of the monellin sequence 24–35 and the thaumatin I sequence 90–120.

(113) and Lys (46) + Asp (113) (that do not affect the sweetness) are located outside this region. Furthermore, the comparable monellin structure is present in the chimeric monellin, and at least one of these constructs proved to be sweet.

It became clear from chemical modification studies that modification of certain Lys and Tyr residues destroys the sweetness of thaumatin and as long ago as 1976 van der Wel and Bel (1976) pointed out that Lys residues 97 and 106 were probably involved in the sweetness of thaumatin. Kim *et al.* (1989) showed that the Lys and Tyr residues are involved in the interactions between adjacent β-sheets in monellin. If this is also true for thaumatin, it is unlikely that these β-sheets are really the site of interaction with the taste receptor or with the antibodies. The most likely remaining candidates are the structure TyrGlyLysAsp (95–98) and/or ThrArgGly (118–120). The computer-modelled conformations of the former peptide, using the information of the coordinates of the C-α atoms, is given in Figure 11.14.

The ThrArgGly sequence is also present in monellin (30–32) and is connected to the sequence SerGluAspTyrLys (25–29). Although no obvious sequence homology exists between miraculin and monellin, miraculin has the reverse structure Lys-TyrAspGluSer (104–108). However, whether this is related to the sweet-including character of miraculin is unknown. It should also be noted that neither this sequence nor the sequence homology to thaumatin (198–206) is in the vicinity of one of the

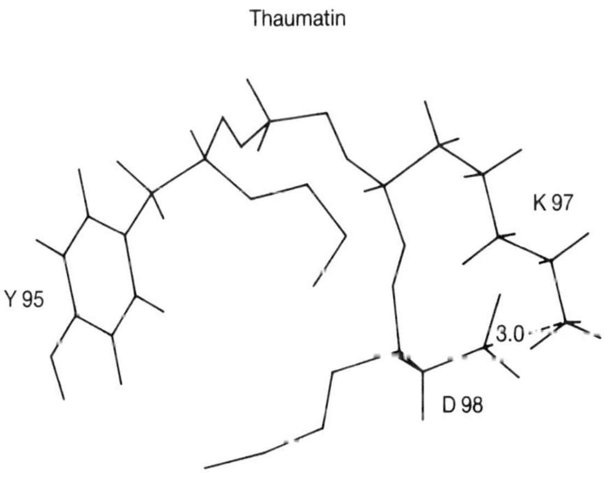

Thaumatin

Y 95

K 97

3.0

D 98

**Fig. 11.14**  A computer aided design of the thaumatin I sequence 95–98, based on the known C-α backbone, showing that the distance between a potential proton donor and acceptor group complies with the model of Schallenberger and Acree (1967).

His residues, which Kurihara considers as essential for the sweet inducing character of miraculin (Y. Kurihara personal communication).

The TL and PR proteins have very high similarity to the thaumatin region 90–120, 13 amino acids being identical between all TL and PR proteins and thaumatin and seven either identical or structurally related amino acids (see Fig. 11.10). However, the amino acids Tyr (95), Lys (97), Tyr (99), Lys (106) and Arg (119) are only present in thaumatin. The difference in the small loop (95–100) between two β-sheets is especially remarkable as the essential structure for sweet compounds, proton donor and acceptor structures at a distance of 3 Å, may be present in this loop in thaumatin (Fig. 11.14) but absent in the TL and PR proteins. Therefore, mutations in this region, and especially of these three amino acids, are worthwhile.

Another fascinating aspect of thaumatin is its enzymic (i.e. its proteolytic) activity when it is partially reduced. Although Beynon and Cusack (1990) recently published data that questioned the proteolytic activity of thaumatin and proposed that the measured effect was due to contamination of Talin with a protease, now called thaumatopain, protein engineering can solve this problem unambiguously. Evidence exists that this reduction occurs between Cys (149) and Cys (158) and/or Cys (134) and Cys (145) and/or Cys (159) and Cys (164). Systematic modification of these Cys residues should demonstrate which reduced disulphide bridge(s) are essential for the acquired catalytic property. Such a mutant will be extremely useful for protein engineers involved in proteases and thaumatin. In this respect it is important to refer to the work of Skern *et al.* (1990) on picornaviral proteases. These proteases contain a stretch of amino acids—GlyAspCysGlyGly—perfectly homologous to thaumatin (69–74) but also to the maize inhibitor (Richardson *et al.* 1987). De Vos *et al.* (1985) proposed that small loops stabilized by disulphide bridges are often present in proteins that bind to membrane-bound receptors and suggested that domains II and III of thaumatin (Fig. 11.11) may be involved in the sweetness. However, domains II and III of thaumatin are very similar to the non-sweet TL and PR proteins. The only consistent difference is Lys (137), which is replaced by Glu or Gln in the TL and PR proteins. Replacement of this Lys residue can also be of interest. Another question that can be solved by protein engineering is whether the β-sheets just below the 'sweet region' are important for the correct exposure of this region on the surface of the thaumatin molecule. If this is not the case, and the mutation studies with the Cys residues prove that the domains II and III are also unimportant for the sweet taste, it may be possible to develop a short peptide that has the same shape as the sweet domain. Takada and Yamamoto (1988) claimed that at least half the thaumatin molecule can be deleted without affecting the sweetness.

The targets for designing such peptides are clear, they should have an intensity of sweetness comparable with that of thaumatin, but should lack the long lasting aftertaste. They should have better heat stability and have all these properties over a pH range from 3 to 7, which covers nearly all food products. That such an approach is realistic has been demonstrated for monellin by Kim *et al.* (1989). This is, of course, quite demanding, but by understanding the principles underlying the interactions between receptor proteins, thaumatin and other sweeteners it may be possible to design various peptides, each displaying their optimal sweetness at certain pH (which will also depend on the type of acid used). Various blends of these sweet peptides can then be used for various food products to get the optimal effect in the product.

## Acknowledgements

The author wishes to thank Professor Yoshie Kurihara (Yokohama National University, Japan) for providing a number of unpublished data on miraculin and curculin and my colleagues Aart van Beusekom, Hans Peters, Maarten Egmond for preparing the molecular graphics and Jac Klerx and Jan Maat for their valuable comments.

## References

Barreau, C. and van der Wel, H. (1983). The effect of thaumatins on the chemotactic behaviour of *Escherichia coli*. *Chemical Senses*. **8**: 71–9.

Beynon, R. and Cusack, M. (1990). Thaumatin not proteolytic. *Nature*. **344**: 498.

Brouwer, J.N., van der Wel, H., Francke, A. and Henning, G.J. (1968). Miraculin, the sweetness-inducing protein from miracle fruit. *Nature*. **220**: 373–4.

Brouwer, J.N., Hellekant, G., Kasahara, H., van der Wel, H. (1973). Electrophysiological study of the gustatory effects of the sweet proteins monellin and thaumatin in monkey, guinea pig and rat. *Acta Physiology Scandinavica*. **89**: 550–7.

Cagan, R.H. and Morris, R.W. (1979). Biochemical studies of taste sensation: binding to taste tissue of $^{3}$H-labeled monellin, a sweet-tasting protein. *Proceedings National Academy of Sciences, USA*. **76**: 1692–6.

Cho, J.M. (1988). DNA sequence encoding monellin type peptide sweetener—in which the two sub-units are joined covently; and new expression vectors and transformed cells. Patent WO 8810303.

Cho, J.M. and Kim, S.H. (1988). Sweetener composition containing monellin analog—protein engineering and transformation of *Escherichia, Bacillus, Neurospora, Saccharomyces and Streptomyces* spp. and plant cell culture. Patent WO 8810271.

Cornelissen, B.J.C., Hooft van Huijsduijnen R.A.M. and Bol, J.F. (1986). A tobacco mosaic virus-induced tobacco protein is homologous to the sweet-tasting protein thaumatin. *Nature*. **321**: 531–2.

Creighton, T.E. and Charles, I.G. (1987). Biosynthesis, processing, and evolution of bovine pancreatic trypsin inhibitor. *Cold Spring Harbor Symposium on Quantitative Biology*. **LII**: 511–19.

De Vos, A.M. Hatada, M., van der Wel, H., Krabbendam, H., Peerdeman, A.F. and Sung-Hou, K. (1985). Three-dimensional structure of thaumatin I, an intensely sweet protein. *Proceedings National Academy of Sciences, USA*. **82**: 1406–9.

Donovan, W., Zheng, L., Sandman K. and Losick, R. (1980). Genes encoding spore coat polypeptides from *Bacillus stubtilis*. *Journal of Molecular Biology*. **196**: 1–10.

Drenth, J., Low, B.W., Richardson, J.S. and Wright, C.S. (1980). The toxin-agglutinin fold. *Journal of Biological Chemistry*. **255**: 2652–5.

Dugas du Villard, X., van der Wel, H. and Brouwer, J.N. (1980). Enhancement of the perceived sucrose sweetness in the rat by thaumatin. *Chemical Senses*. **5**: 93–8.

Edens, L., Heslinga, L., Klok, R., Ledeboer, A.M., Maat, J., Toonen, M.Y., Visser, C. and Verrips, C.T. (1982). Cloning of cDNA encoding the sweet-tasting plant protein thaumatin and its expression in *Escherichia coli*. *Gene*. **18**: 1–12.

Edens, L., Ledeboer, A.M., Verrips, C.T. and van de Berg, J.A. (1983). *Kluyveromyces* yeast containing rDNA plasmid for cultivation in production of pre-prothaumatin or its modified forms. Patent application WO 83 04051.

Edens, L., Bom, I., Ledeboer, A.M., Maat, J., Toonen, Y., Visser, C. and Verrips, C.T. (1984). Synthesis and processing of the plant protein thaumatin in yeast. *Cell*. **37**: 629–33.

Fischer, R.L. and Goldberg, R.B. (1982). Structure and flanking regions of soybean seed protein genes. *Cell*. **29**: 651–60.

Frank, G. and Zuber, H. (1976). The complete amino acid sequences of both subunits of the sweet protein monellin. *Hoppe-Seyler's Zeitschrifft fur Physiologische Chemie.* **357**: 585–92.

Goeddel, D.V., Kleid, D.G., Bolivar, F., Heyncher, H.L., Yansura, D.G., Crea, R., Hirose, T., Kraszewski, A., Itakura, K. and Riggs, A.D. (1979). Expression in *E. coli* of chemically synthesized genes for human insulin. *Proceedings National Academy of Sciences, USA.* **76**: 106–10.

Gould, S.J., Keller, G.A. and Subramani S. (1988). Identification of peroxisomal targeting signals located at the carboxy terminus of four peroxisomal proteins. *Journal of Cell Biology.* **107**: 897–905.

Higgenbotham, J.D. and Hough, C.A.M. (1977). Useful taste properties of amino acids and proteins. In *Sensory properties of foods* (ed. G.G. Birch, J.G. Brennan and K.J. Parker), pp. 129–49. Applied Sciences, London.

Hough, C.A.M. and J.A. Edwardson (1978). Antibodies to thaumatin as a model of the sweet taste receptor. *Nature.* **271**: 381–3.

Huang, S., Elliott, R.C., Liu, P.S., Koduri, R.K., Weickmann, J.L., Lee, J.H., Blair, L.C., Ghosh-Dastidar, P., Bradshaw, R.A., Bryan, K.M. Einarson, B., Kendall, R.L., Kolacz, K.H. and Saito, H. (1987). Specificity of cotranslational amino-terminal processing of proteins in yeast. *Biochemistry.* **26**: 8242–6.

Illingworth, C., Larson, G. and Hellekant, G. (1988). Secretion of the sweet-tasting plant protein thaumatin by *Bacillus subtillis. Biotechnology Letters.* **10**: 587–92.

Illingworth, C., Larson, G. and Hellekant, G. (1989). Secretion of the sweet-tasting plant protein thaumatin by *Streptomyces lividans. Journal of Industrial Microbiology.* **4**: 37–42.

Inglett, G.E. and May, J.F. (1968). Tropical plants with unusual taste properties. *Economic Botany.* **22**: 326–31.

Inglett, G.E. and May, J.F. (1969). Serendipity berries—source of a new intense sweetener. *Journal of Food Science.* **34**: 408–11.

Iyengar, R.B., Smits, P., van der Ouderaa, F. van der Wel, H., van Brouwershaven, J., Ravestein, P., Richeters, G. and van Wassenaar, P.D. (1979). The complete amino-acid sequence of the sweet protein thaumatin I. *European Journal of Biochemistry.* **96**: 193–204.

Kier, L.B. (1972). A molecular theory of sweet taste. *Journal of Pharmaceutical Sciences.* **61**: 1394.

Kim, S.H. (1988). Single chain analogues of monellin—comprising A and B chain and covalent linker retain sweetness 50 times that of sucrose under denaturing conditions. Patent application WO 8810265.

Kim, S.H., Kang, C.H., Kim, R., Cho, J.M., Lee, Y.B., Lee, T.K. (1989). Redesigning a sweet protein: increased stability and renaturability. *Protein Engineering.* **2**: 571–5.

King, G.J., Turner, V.A., Hussey, C.E., Wurtele, E.S., Lee, S.M. (1988). Isolation and characterization of a tomato cDNA clone which codes for a salt-induced protein. *Plant Molecular Biology.* **10**: 401–12.

Koshland, D.E. (1974). Chemotaxis as a model for sensory systems. *FEBS Letters.* **40**: S3–S9.

Lázaro, A., Rodriguez-Palenzuela, P., Maraña, Carbonero, P. and Garcia-Olmedo, F. (1988).Signal peptide homology between the sweet thaumatin II and related cereal α-amylase/trypsin inhibitors. *FEBS Letters.* **239**: 147–50.

Ledeboer, A.M., Verrips, C.T. and Dekker B.M. (1984). Cloning of the natural gene for the sweet-tasting plant protein thaumatin. *Gene.* **30**: 23–32.

Lee, J.H., Weickmann, J.L., Koduri, R.K., Gosh-Dastidar, P., Saito, K., Blair, L.C., Date, T., Lai, J.S., Hollenberg, S.M. and Kendall, R.L. (1988). Expression of synthetic thaumatin genes in yeast. *Biochemistry.* **27**: 5101–7.

Lopes, T.S., Klootwijk, J., Veenstra, A.E., van der Aar, P.C., van Heerikhuizen, H., Raué, A. and Planta, R.J. (1989). *Gene.* **79**: 199–206.

Morris, J.A. and Cagan, R.H. (1972). Purification of monellin the sweet principle of *Dioscoreophyllum cumminsii. Biochimica Biophysica Acta.* **261**: 114–22.

Ogata, C., Hatada, M., Tomlinson, G., Shin, W.C. and Kim, S.H. (1987). Crystal structure of the intensely sweet protein monellin. *Nature.* **328**: 739–42.

Payne, G., Middlesteadt, W., Williams, S., Desai, N., Dawn Parks, T., Dincher,S., Carnes, M. and Ryals, J. (1988). Isolation and nucleotide sequence of a novel cDNA clone encoding the major form of pathogenesis-related protein R. *Plant Molecular Biology.* **11**: 223–4.

Persuad, K.C., Chiavacci, L. and Pelosi, P. (1988). Binding proteins for sweet compounds from gustatory papillae of the cow, pig, and rat. *Biochimica Biophysica Acta.* **967**: 65–75.

Richardson, M., Valdes-Rodriguez, S. and Blanco-Labra, A. (1987). A possible function for thaumatin and a TMV-induced protein suggested by homology to a maize inhibitor. *Nature.* **327**: 432–4.

Saier, M.H., Werner, P.K. and Muller, M. (1989). Insertion of proteins into bacterial membranes: mechanism, characteristics, and comparisons with the eucaryotic process. *Microbiological Reviews.* **53**: 333–66.

Schallenberger, R.S. and Acree, T.E. (1967). Molecular theory of sweet taste. *Nature.* **216**: 480–3.

Singh, N.K., Nelson, D.E., Kuhn, D., Hasegawa, P.M. and Bressan, R.A. (1987). Molecular cloning of osmotin and regulation of its expression by ABA and adaptation to low water potential. *Plant Physiology.* **85**: 529–36.

Skern, T., Zorn, M., Blaas, D. and Kuechler, E. (1990). Protease or protease inhibitor? *Nature.* **344**: 26.

Takada, C. and Yamamoto, A. (1988). Changes of residual sweetness and heat stability of thaumatin proteolysates hydrolyzed by various proteases. *Teikoku Gukuen Kiyo.* **14**: 9–19.

Theerasilp, S. and Kurihara, Y. (1988). Complete purification and characterization of the taste-modifying protein, miraculin, from miracle fruit. *Journal of*

*Biological Chemistry*. **263**: 11536–9.

Theerasilp, S., Hitotsuyu, H., Nakaku, S., Nakaya, K., Nakamura, Y. and Kurihara, Y. (1989). Complete amino acid sequence and structure characterization of the taste-modifying protein, Miraculin. *Journal of Biological Chemistry*. **264**: 6655–9.

van der Wel, H. (1980). The role of organic chemistry in taste perception. In *Olfaction and Taste VII*. 13–21 (ed. van der Starre, H.). Information Retrieval Ltd., London.

van der Wel, H. (1981). Psychophysical studies of thaumatin and monellin. In *Criteria of Food Acceptance*. 292–5 (ed. Solms, J. and Hall, R.L.). Forster Verlag AG, Switzerland.

van der Wel, H. and Bel, W.J. (1976). Effect of acetylation and methylation on the sweetness intensity of thaumatin I. *Chemical Senses and Flavors*. **2**: 211–18.

van der Wel, H. and Bel, W.J. (1978). Structural investigation of the sweet-tasting proteins thaumatin and monellin by immunological studies. *Chemical Senses and Flavors*. **3**: 99–104.

van der Wel, H. and Bel, W.J. (1980). Enzymatic properties of the sweet-tasting proteins thaumatin and monellin after partial reduction. *European Journal of Biochemistry*. **104**: 413–18.

van der Wel, H. and Loeve, K. (1972). Isolation and characterization of thaumatin I and II, the sweet-tasting proteins from *Thaumatococcus daniellii* Benth. *European Journal of Biochemistry*. **31**: 221–5.

van der Wel, H., van der Heijden, A. and Peer, H.G. (1987). Sweeteners. *Food Reviews International*. **3**: 193–268.

Verrips, C.T., Ledeboer, A.M., Edens, L., Klok, R. and Maat, J. (1990) DNA sequence, encoding the various allelic forms of mature thaumatin, and cloning vectors, etc. US Patent 4.891316.

Weickmann, J.L. and Ghoshdasti, P. (1987). Recombinant thaumatin I analogues which can be re-folded to give native conformations eliciting a sweet taste sensation. Patent application WO 87 03007.

Yamashita, H., Theeraslip, S., Aiuchi, T., Nakaya, K., Nakamura, Y. and Kurihara, Y. (1990). Purification and complete amino acid sequence of a new type of sweet protein with taste modifying activity, curculin. *Journal of Biological Chemistry*. **265**: 15770–5.

# 12

# Plant Enzymes for the Food Industry

Richard W. Pickersgill

## The role of plant enzymes in the food industry

### Enzymes and the food industry

Biological catalysts have long been exploited in making cheese, bread and alcoholic drinks. Enzymes have also found a number of other roles in the processing and protection of foods. The major advantages of biological catalysts over chemical catalysts are the specificity and stereospecificity of enzymes, the relatively mild conditions necessary for efficient catalysis and the ability to engineer the enzymes for improved performance.

Enzymes are used in foods to improve the chemical composition and quality of a food or food ingredient and to reduce the manufacturing cost. The food ingredients may be stabilizers, flavours, colourings or flavour enhancers. In addition, enzymes themselves can be used as antimicrobials or antioxidants to improve the quality and storage life of a packaged food. Enzymes can also be used to remove components from foods and drinks (e.g.

oxidoreductases for the removal of alcohol from beer and wine). The current extent of enzyme use in foods is indicated in Table 12.1. Three types of enzymes are used, hydrolases, oxidoreductases and isomerases. About 50% of all commercially produced enzymes are used in foods and the estimated sales of important food enzymes are given in Table 12.2.

## Industrial potential of engineered enzymes

Enzymes have evolved for efficient catalysis in biological systems, not evolved for industrial use, and consequently improvements in operating performance may be possible. The most successful industrial processes exploit the natural advantages of the enzyme to the full, but often the enzyme will not operate efficiently under the conditions that prevail in the process. The industrially important parameters are activity, specificity, stereospecificity, stability and half life. These will vary primarily

**Table 12.1**  The extent of current enzyme use in foods

|  | Protease | Lipase | Phospho-lipase | Amylase | Cellulase | Pectinase | Oxido-reductase |
|---|---|---|---|---|---|---|---|
| Brewing | + |  |  | + | + |  |  |
| Baking | + |  | + | + |  |  | + |
| Dairy | + | + |  | + |  |  | + |
| Meat | + | + |  |  |  |  |  |
| Fats and oils |  | + | + |  |  |  |  |
| Juice | + |  |  |  |  | + |  |
| Starch modification |  |  |  | + |  |  |  |
| Protein modification | + |  |  |  |  |  |  |
| Sweeteners |  |  |  | + | + |  |  |
| Alcohol removal |  |  |  |  |  |  | + |
| Botanicals extraction | + | + | + | + | + | + |  |

In addition isomerases are used in starch modification.

**Table 12.2**   Estimated annual sales of important food enzymes

| Enzyme | Estimated sales ($ million) | |
| --- | --- | --- |
| | 1985 | 1990 |
| Glucose isoamylase | 75.5 | 92.4 |
| Bacterial amylase | 54.9 | 67.8 |
| Glucoamylase | 54.9 | 65.6 |
| Rennets | 31.8 | 42.5 |
| Papain | 16.5 | 18.2 |
| Other | 23.5 | 30.8 |

From Wasseman (1984).

as a function of pH, ionic strength, temperature, substrate concentration, product concentration and the presence of other molecules (e.g. specific inhibitors or denaturants). In principal the activity of any enzyme can be optimized for a given set of processing conditions.

## Plant enzymes and food processing

Many enzymes used in food processing come from bacteria and fungi, fewer from plants and animals. Notable exceptions are the cysteine proteases from plants, which find a wide variety of uses in foods, and the amylases from malted wheat and barley, which are of great importance in brewing and baking. Other plant enzymes are used in foods but have not yet been characterized in sufficient detail to allow knowledge-based protein engineering. Correct folding of recombinant plant cysteine proteases is actively sought in a number of laboratories (including that of the author) and engineered proteases will follow directly. Knowledge-based protein engineering of fungal and plant amylases is in progress, although the results of these experiments are preliminary.

## Cysteine proteases in foods

Industrial 'papain' is a partially purified preparation of the latex from *Carica papaya*. This preparation can contain significant amounts of enzymes, other than cysteine proteases, which may be important in some uses. 'Papain' is used in beer making, for meat tenderization, in biscuit production, for making fish protein concentrates and in a number of other applications. Unlike other proteases, 'papain' does not produce large quantities of bitter peptides and this feature, combined with reason-

able stability and activity, is often the reason for choosing it for food purposes. In beer making, enzymes present in the malt act on the barley to release soluble carbohydrates and nitrogen sources required for good yeast growth. Added 'papain' and amylases are used so that the amount of malt, and hence production costs, can be reduced. When chilled, beer develops an intense chill-haze. This results from the precipitation of protein–polyphenol and protein–tannin complexes that are formed during brewing. 'Papain' prevents chill-haze by partial hydrolysis of the proteins that form the precipitate and higher activity against these complexes would be useful.

In the United States 'papain' is used to tenderize meat. Many different methods of applying the enzyme are possible and improving the distribution of the enzyme in meat would be useful. These might include swelling the meat at low pH, which would require the enzyme to be active at lower pH values.

## Amylases in foods

Amylases are important in bread making and brewing because they increase the available fermentable sugar. Traditionally, flour contained sufficient $\alpha$- and $\beta$-amylase to produce baked products with good colour, volume and texture. However, mechanical harvesting and milling techniques are so rapid that germination does not occur and the amylase levels are too low to hydrolyse the starch granules damaged during milling. Hence there is insufficient sugar available in the fermentation period that follows the milling process. The amylase level is therefore supplemented by additional amylases from malted wheat and barley or from bacteria (*Bacillus subtilis*) or fungi (*Aspergillus oryzae*). These amylases have good activity at pH 5.0–5.5 but have different thermostabilities; the fungal amylase is the least stable and the bacterial enzyme the most stable. The thermostability of the amylase is crucial during baking because the starch is gelatinized at the high temperatures used and becomes available for hydrolysis. If the activity of the $\alpha$-amylase is not controlled (e.g. using the thermostable bacterial amylase), over-liquefaction will occur and the bread will be sticky. This problem is overcome by using a heat sensitive $\alpha$-amylase, which provides a good amount of fermentable sugar and produces a bread with soft crumb, deep crust colour, good volume and improved grain and texture.

As a counter example, an α-amylase with high thermostability is needed for the liquefaction of starch where temperatures well in excess of the starch gelling point (about 80°C) are used. Thermostable amylases are also used in brewing to supplement low intrinsic enzyme levels. Modifying the pH- and temperature-activity profiles, stability and half life will aid existing uses of amylases. Engineering the specificity of amylases is expected to open up new uses for these enzymes in foods because the size and composition of the carbohydrate determines its properties.

## Rational protein engineering

Rational protein engineering demands knowledge of protein structure and of the interactions involved in enzyme stability and action. This knowledge base is discussed on pages 237–41, pages 241–7 discuss the cysteine proteases and pages 247–50 the amylases.

# Knowledge-based protein engineering

## Protein structure and modelling

### Protein structure

About 100 non-homologous protein structures have been determined by X-ray analysis. The covalent geometry of these proteins is similar to that of amino acids and other small molecules and the angles, lengths and torsion angles are close to the most favourable values. There is considerable interest in the nature of the non-covalent interactions, including van der Waals and electrostatic interactions, in proteins because it is these interactions that determine their compact folded structures and consequently their function. Some general observations on the nature of water soluble, globular protein structures are:

(1) The presence of an amino acid side chain (i.e. all amino acids except glycine) greatly reduces the conformational possibilities of the neighbouring main chain torsion angles (Ramachandran and Sasisekharan 1968).
(2) Close packing results in exclusion of all but a

few water molecules from the hydrophobic interior of the protein and results in globular structure (Richards 1977).
(3) Buried hydrogen bonding groups normally occur as hydrogen donor–acceptor pairs (Baker and Hubbard 1984).
(4) Ionized groups tend to be on the surface of the protein, but when buried they are solvated by hydrogen bonding or by oppositely charged groups (Janin 1979; Rashin and Honig 1984).
(5) Side chain torsion angles in proteins have preferred values (Janin *et al.* 1978; Ponder and Richards 1987).

In addition to these general observations, recurring themes are found at all levels of protein structure.

### Structure prediction

The protein folding problem poses difficult challenges for protein design and engineering. A full understanding of protein folding would enable protein sequences to be folded into three-dimensional structures without the need for experimental elucidation of structure by X-ray or NMR analysis. There are three broad approaches to the prediction of the tertiary structure of a polypeptide from sequence:

(1) By homology with a protein of known structure (Browne *et al.* 1969; Blundell *et al.* 1987).
(2) By prediction of secondary structures and assembling these structures into a compact globular structure (Cohen *et al.* 1983).
(3) Using empirical energy functions *ab initio* to find the tertiary structure with the lowest potential energy (Levitt and Warshel 1975).

The most successful of these methods for modelling homologous protein structures occurring in nature is by homology. Such modelling is based on the observation that structures with homologous sequences have similar protein structures (e.g. the cysteine proteases papain and actinidin, Fig. 12.1). For sequences with greater than 50% identity, modelling by homology can yield a structure that is probably correct to within 1 Å, although individual side chains may be more in error. The hierarchical approach (2), relies upon accurate secondary structure prediction, which is not possible. Current methods of secondary structure prediction are accurate to only about 60% and this is unlikely to improve to the required accuracy because the terti-

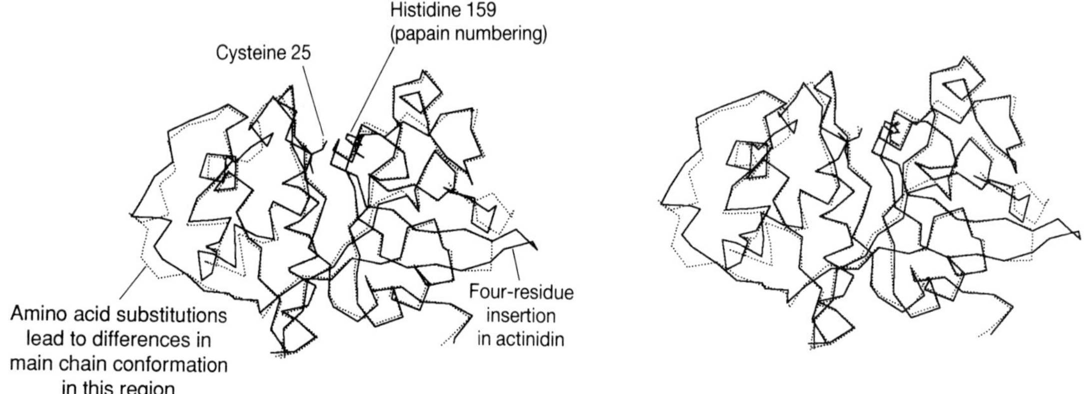



**Fig. 12.1**   A comparison of the three-dimensional structures of the cysteine proteases papain (broken lines) and actinidin (full line) indicates that the structural cores are very similar despite only about 50% amino acid sequence identity. The figure shows the α-carbon backbone and active site cysteine and histidine residues of these enzymes. Differences in the solvent accessible side chains confer the different isoelectric points, activities and reactivities of these enzymes.

ary environment has an effect on the secondary structure formed. This, and the formidable problem of packing many units of secondary structure on the atomic scale, makes this approach difficult. The most fundamental approach to protein folding is to minimize the potential energy of the polypeptide–solvent system and obtain the observed equilibrium structure, (3). The major problems here are the accuracy of the potential functions used and the amount of conformational space that can be searched in a realistic amount of computing time (see the discussion of molecular dynamics simulation below).

A simpler task is to predict how an amino acid substitution (or a deletion/insertion) affects the structure. However, this is not quite the same problem as predicting the structure of a homologous protein because an engineered protein may not fold correctly, whereas a homologous protein is known to fold correctly. This demands the development of methods to address the problems posed by protein engineering. Comparison of homologous sequences and structures can give a good indication of what sequences are compatible with a given protein fold and therefore what engineered changes can be accommodated and which key residues with special properties should not be substituted.

## Modelling the structure of a modified protein

The design of modified proteins relies on knowledge of protein structure as revealed by X-ray analysis of proteins. Starting from the X-ray structure of the native protein, a model structure of the mutant is constructed by substituting amino acid side chains. The substitution is often made using a graphics system that allows the change to be visualized. The precise geometry of the side chain is not always crucial for protein design or for the outcome of a protein engineering experiment; for instance, if the amino acid is remote from the active site and involves a long range electrostatic interaction with an active site group. However, when the side chain is intimately involved in molecular recognition, or is implicated in catalysis, then precise conformation is likely to be absolutely crucial.

## Energy minimization and molecular dynamics

Energy minimization produces a model more consistent with the potential function used (it usually includes van der Waals and electrostatic terms) but it will not correct an incorrect model. The combination of conformational search, or different starting conformations, with energy minimization searches more conformational space and can therefore indicate the relative merits of different conformations. The search of configurational space can be usefully expanded by using the molecular dynamics simulation technique. Solvent, and consequently entropic effects, can be included. In molecular simulation the inertia of the atoms enables the system to cross energy barriers of the order kT; this is essential when the flexibility of the protein and rearrangement of solvent are important. The agreement between X-ray and average dynamics structures is about 1.0 Å for backbone atoms and 1.5–2.0 Å for all atoms. The limitations of the method are the accuracy of the potential functions and the simulation time necessary to search sufficient conformational space. These limitations are significant when modelling changes in protein structure but make *ab initio* protein folding impossible (see van Gunsteren 1988 for a review of the role of molecular simulation in protein engineering).

## Databases and fragments

Databases and database derived rules are invaluable guides for rational protein engineering and have also been extensively used for modelling by homology (Sutcliffe *et al.* 1987a, b). In the substitution of one residue for another, databases can provide precedents for such substitutions or indicate that certain substitutions are most unusual and therefore probably unfavourable. In addition, database derived rotamer angles are regularly used to model substituted side chains (Fig. 12.2). Fragments of protein structure can be extracted from databases and used to replace parts of the existing structure to engineer binding sites, stability and the physical properties of proteins (e.g. solubility and size). Graphics based databases (Jones and Thirup 1986) enable fragments to be examined for compatibility of the fragment with the rest of the protein molecule (Fig. 12.3).

## Modelling and X-ray analysis

Modelling the structure of mutant proteins is an important part of protein engineering and suggests which mutants to construct. Although models can satisfactorily explain certain features of protein structure and function, particularly when the precise geometry of the substituted residue is not crucial, modelling is not a substitute for high-resolution X-ray analysis of mutant protein that reveals the precise nature of the time and space averaged structure.

## Protein structure and function

### Stability and function

Protein stability and function are mainly determined by the same non-covalent forces that deter-

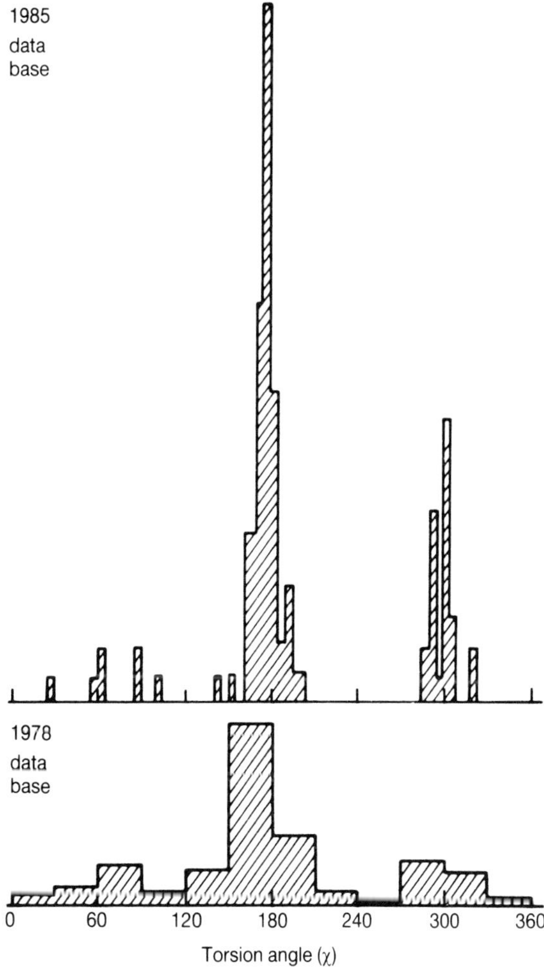



**Fig. 12.2**   The distribution of side chain torsion angles of amino acids in globular proteins is well defined. The distributions shown are for the chi 1 torsion angle of valine residues. The 1978 database is based on 19 protein structures and 238 valines. The 1985 database consists of 19 well refined proteins (determined to 1.8 Å or better and refined to a crystallographic R-factor of 18% or less) and 151 valines. Sample sizes are normalized so the total areas under each curve are equal. This figure shows the improvement in the accuracy of protein structures from 1978 to 1985, largely due to the widespread use of crystallographic refinement. The limited number of conformers occurring in proteins is obvious in the 1978 and even more clearly defined in the 1985 distribution (reproduced with permission from Ponder and Richards 1987).

mine protein structure. However, covalent linkages, e.g. disulphide bridges and covalently linked enzyme intermediates, are also important for some proteins and enzymes. X-ray and NMR studies can indicate which amino acids in the protein are

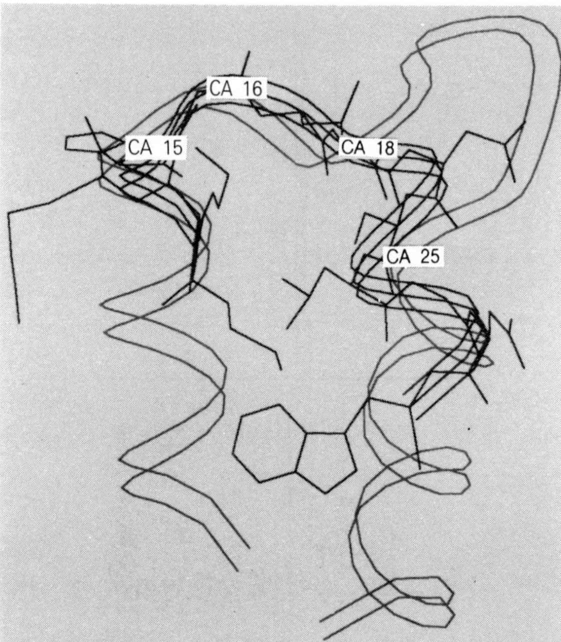



**Fig. 12.3** Comparing loops connecting α-helices using a graphics based database program. The backbone of the starting structure is represented as the two-stranded ribbon and has the longer loop. A shorter loop has been found in the database the backbone of which is shown as three-stranded ribbon. The single line shows all atoms of the replacement loop. New properties and functions can be conferred to a protein by loop replacements.

involved in binding a substrate or an inhibitor and the nature of the non-covalent interactions involved. Tight binding is characterized by complementary molecular shapes, matched hydrogen bonding patterns and hydrophobic patches, and complementary electrostatic fields.

Different modelling methods are suitable for different types of problems and the choice is often determined by the accuracy needed and the time available. However, the level of detail chosen for the modelling should be appropriate to the specific problem to be solved.

### Electrostatics

Engineering electrostatic complementarity is a key factor in effective enzyme engineering and developments in electrostatic methods will have important results for understanding proteins and engineering novel proteins. There are two approaches to the modelling of electrostatic interactions in proteins; macroscopic models that assume a dielectric constant (e.g. numerical solution of the Poisson or the Poisson–Boltzmann equation by the finite difference method, Warwicker and Watson 1982) and microscopic methods that include solvent explicitly (e.g. the Langevin dipole model of Warshel and Levitt 1976). The choice of approach depends upon the problem to be addressed, see below and, for reviews, Harvey (1989) and Matthew *et al.* (1985).

### Molecular dynamics and the energetics of binding and catalysis

Molecular dynamics (see p. 238) has been used to simulate the effect of protein engineering experiments on subtilisin in a semiquantitative way (Warshel and Sussman 1986). The effect of the Asn155 to Thr substitution on the free energy of hydrolysis of amides by subtilisin was 19.7 kJ/mol (Wells *et al.* 1986) and the simulation gave 21 kJ/mol, which is in good agreement (Hwang and Warshel 1987). Substitution of Asn155 by Ala has a free energy change of hydrolysis of 15.5 kJ/mol and simulations predicted values of 14.3 ± 4.6 kJ/mol (Rao *et al.* 1987) and 17.2 ± 2.5 kJ/mol (Hwang and Warshel 1987). These predicted values are in remarkably good agreement with the experiment but should still be viewed as semiquantitative because of the current limitations of the simulation method. The free energy perturbation method used in these calculations involves changing the molecular mechanical parameters during a simulation and determining the free energy change during this process. A physical process can be simulated in this way but it is easier to simulate a non-physical process. A thermodynamic cycle is then used to calculate the free energy change for the physical process.

## Protein engineering

### Protein stability

The free energy of folding is the difference in the free energies of the folded and unfolded states and is of the order of 20–90 kJ/mol for most biologically occurring proteins. A major problem is that the structure of the unfolded state is unknown and so design must be based only on the folded state as defined by X-ray analysis. This means that we assume the change in the free energy of stabilization of the unfolded state is negligible, an assumption that will not always be true. An alternative assumption is that an extended conformation adequately represents the unfolded state. Methods of designing more stable structures include:

(1) Replacing glycine by any other amino acid and replacing any other amino acid by proline to increase stability by decreasing the configurational entropy of unfolding (Matthews *et al.* 1987).
(2) Stabilizing $\alpha$-helical dipoles by interaction with a charged side chain (Nicholson *et al.* 1988; Sali *et al.* 1988).
(3) Increasing hydrophobic surface inaccessible to solvent (Matsumura *et al.* 1988; Kellis *et al.* 1988; 1989).
(4) Introducing disulphide bridges (Perry and Wetzel 1984).
(5) A combination of random mutagenesis, design of buried hydrophobics, design of electrostatic interactions at calcium binding sites, sequence homology consensus and serendipity (Pantoiano *et al.* 1988).
(6) Charge engineering according to calculation of electrostatic energy (Pickersgill *et al.* 1991).

A particularly impressive increase in the stability of subtilisin was achieved by Pantoiano *et al.* (1989), (5), in the above list. This modified enzyme has a 14°C increase in the midpoint of the thermal unfolding transition and a 300-fold decrease in the rate of thermal inactivation. The six mutations involved yielded approximately additive increases in stability with only small structural perturbations as a result of each mutation.

### Activity and specificity

A key enzyme engineering paper (Leatherbarrow *et al.* 1985) showed that catalytic rate enhancement is achieved by transition state stabilization. The manipulation of enzyme–substrate interactions by protein engineering can result in modified substrate specificity (Estell *et al.* 1986; Wells *et al.* 1987a, b; Fersht *et al.* 1985; Craik *et al.* 1985; Murali and Creaser 1986). Engineering of tissue type plasminogen activator (TPA) has resulted in a protein that is not inhibited by its serpin inhibitor (Madison *et al.* 1989) and is potentially a valuable therapeutic agent. This is an exciting result because the structure of the serpin was modelled on a homologous structure and engineering was based on this model.

### Modifying pH-activity profiles

Fersht and co-workers have made detailed studies of the effect of charge changes on the pK value of the active site histidine of subtilisin (Russell and Fersht 1987; Russell *et al.* 1987). Making the surface more negatively charged raises the $pK_a$ of acidic groups in proteins because they all lose a proton on ionization. Conversely, making the surface more positive lowers the $pK_a$ of acidic groups. The pK shifts are maximized at low ionic strength. In addition, significant pK shifts may be observed at moderate ionic strengths provided that the mutations do not concentrate counter-ions in the active site cavity and that ionic strengths below 0.1 M are used for multiply-charged counter-ions (Russell and Fersht 1987). Macroscopic electrostatic models have been reasonably successful in predicting the pK shifts at active site groups due to protein charge engineering (Gilson and Honig 1987; Sternberg *et al.* 1987).

## Plant cysteine proteases

### Introduction

#### Overview

Cysteine proteases are present in plants, animals, bacteria and fungi and many belong to a family of homologous enzymes of similar structure and function. Certain details of the mechanism and the role of specific amino acid residues in determining the detailed activities of these enzymes remain to be elucidated by biochemical and structural studies of engineered proteins. Novel cysteine proteases with improved pH-activity profiles, temperature-activity profiles and protease-protein specificities will both improve and extend the use of these proteases in foods.

#### Molecular properties

The best characterized cysteine proteases are from plant sources and, of these, papain is the most extensively studied. They are endopeptidases with molecular weights of about 24 000. The dried latex of the papaya fruit contains at least four cysteine proteases; papain (Cohen *et al.* 1986), papaya protease omega (Dubois *et al.* 1988), protease IV (Ritonja *et al.* 1989) and chymopapain (Jacquet *et al.* 1989). Other enzymes of this family include actinidin from kiwifruit, bromelain from pineapple, ficin from the latex of fig, asclepain from the roots or latex of milkweed, calotropin and caltropain from the latex of the Indian madar plant and aleurain from barley. Most cysteine protease sequences are sufficiently homologous with the sequence of papain to indicate that the enzymes share the same three-dimensional framework structure. The exception is clostripain, which consists of two chains (mol. wt 43 000 and 12 500) and has no sequence homology with papain (Gilles *et al.* 1983). The isoelectric points of the plant enzymes are highly variable, which indicates that there are considerable differences in their molecular surface charge despite the conservation of the framework structure. The papaya latex enzymes, the ficins and stem bromelain are very basic, whereas actinidin and fruit bromelain are very acidic. The reason for the different isoelectric points is not known but may reflect the pH, ionic strength and protein–substrate and protein–inhibitor interactions that have evolved in any one system. Unfortunately, the precise biological role of the plant cysteine proteases is unknown but these enzymes have fairly broad specificities and attack a variety of substrates. Their function may therefore be to protect ripening fruit or plant tissues from attack by insects or fungi. The multiple forms of these enzymes may therefore reflect an evolutionary battle between the plant producing the proteases and inhibitors produced by the pest.

## Molecular biology and protein folding

The nucleotide sequence of a cDNA derived from the aleurone cells of barley codes for a protein of 361 amino acids, corresponding to the native cysteine protease (aleurain) of 218 amino acids plus a pre-peptide of about 23 residues and a pro-sequence of about 120 residues (Rogers *et al.* 1985). A cDNA clone for papain (Cohen *et al.* 1986) also encodes a long pro-sequence corresponding to 133 amino acids. Attempts to unfold and refold papain have not proved successful (Hernandex-Aranda and Soriano-Garcia 1988) and this suggests that the pro-sequence may be required for correct folding of the molecule. Requirement of a pro-sequence for correct folding has been shown for subtilisin E (Ikemura *et al.* 1987) and for α-lytic protease (Silen and Agard 1989). A synthetic papain has been expressed in *E. coli* but correct folding has not yet been achieved (Vernet *et al.* 1989). Processing of

the correctly folded pro-enzyme expressed in and secreted by yeast is expected to yield active recombinant protein.

## Structure

### X-ray analysis of cysteine proteases

The structures of papain and actinidin were determined at 2.8 Å resolution by multiple isomorphous replacement methods (Drenth *et al.* 1968, 1971; Baker 1977). The structure of papain in orthorhombic crystals has since been refined at 1.65 Å resolution to a crystallographic R-factor of 16.1% (Kamphuis *et al.* 1985) and that of actinidin at 1.70 Å resolution to 16.5% (Baker 1980; Baker and Dodson 1980). In addition, the structure of papain has been determined in monoclinic crystals at a resolution of 1.60 Å and R-factor of 16.0% (Pickersgill *et al.* 1991). Caltropin has been studied at 3.2 Å resolution (Heinemann *et al.* 1982) but unfortunately there is no amino acid sequence data for this enzyme. Protease omega, from papaya latex, has been crystallized (Pickersgill *et al.* 1990) and the structure solved by molecular replacement and refined at 1.9 Å to an R-factor of 15.2%.

The structure of inactive papain in orthorhombic crystals grown at pH 9.3 is almost identical to that in monoclinic crystals that were grown from partially active papain at pH 6.0. The root mean square (rms) deviation of the main chain atoms after superposition is only 0.287 Å, indicating that the structures are almost identical (the error in both structures is about 0.1 Å for atoms with average mobility). The similar thermal motion of the protein main chain in the two different crystalline environments (Fig. 12.4) shows that this is a property of the molecule and not of the crystal. Small differences in side chain conformation (Fig. 12.5) can be attributed to highly mobile side chains or to the different packing of molecules in the two different crystals. The papain used to grow the orthorhombic crystals was inactive enzyme with oxidized active site cysteine, and three oxygen atoms attached to cysteine 25 are clearly seen in the electron density map. Monoclinic crystals were grown from activatable papain and a sulphur atom attached to cysteine 25 can be seen in the electron density map of this structure. These different covalent modifications of the active site cysteine affect the detailed hydrogen bonding pattern at the active site.

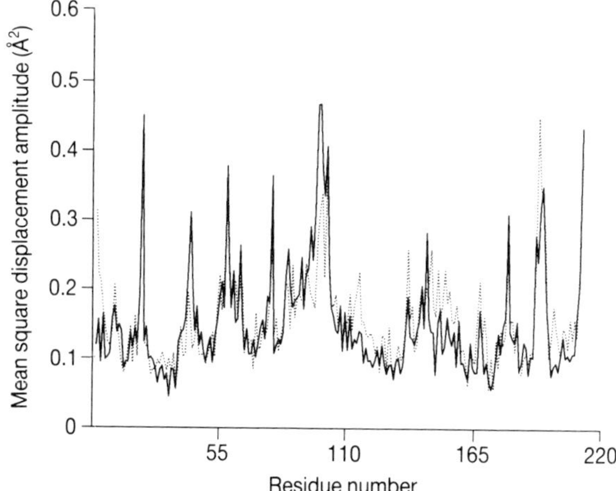

**Fig. 12.4**  Mean square displacement amplitudes (a measure of thermal motion) for the backbone atoms of papain in monoclinic (full line) and orthorhombic (broken line) crystals. The similarity of the graphs indicates that the motion is a property of the protein and not of the crystal.

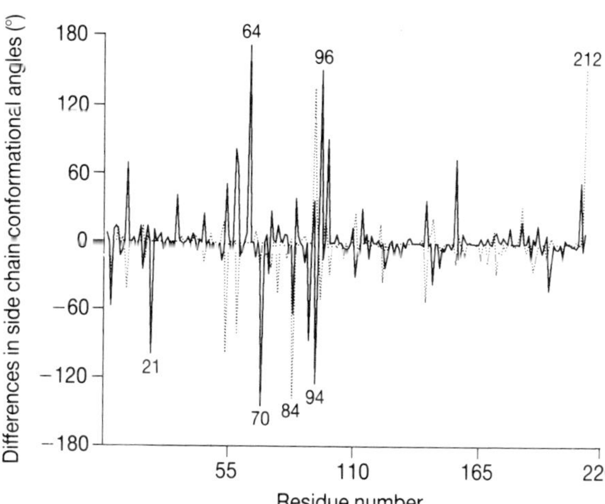

**Fig. 12.5**  Differences in the side chain torsion angles chi 1 (full line) and chi 2 (broken line) for papain in monoclinic compared with orthorhombic crystals. Most of the differences are due to thermal motion but a few differences can be traced to differences in the crystal packing.

## The active site of papain

The active site cysteine 25 of papain is at the N-terminal end of the long central α-helix and at the bottom of the cleft formed between the two domains (see Fig. 12.1). The cleft is relatively deep and narrow on one side of Cys25 and wider and more open on the other side (Fig. 12.6). The walls of the deep part of the cleft that make the S1 and S2 binding subsites are formed by residues 64–67 on one side and 156–159 on the other. Residues 64–67 are extended chain with the carbonyl groups of residues 64 and 66 and the amide of residue 66 pointing into the cleft. On the other side, the carbonyl of 158 is available for hydrogen bonding to the substrate. The floor of the wide, less deep part of the cleft is composed of the side chains of Gln19, His159, Trp177 and Trp181. The sides are formed by residues 20–23 and 136–142. His159 is adjacent to Cys25, and is also implicated in catalysis. His159 interacts with the cysteine on one side and with Asn175 on the other. Unfortunately, because the cysteine is always covalently modified in the crystals, the precise interaction of active site Cys(Sγ) and His(Nδ1) has not been observed. The hydrogen bond between His(Nε2) and Asn(Oδ1) is shielded from solvent by Trp177 and this interaction orientates the histidine and limits its range of motion to that needed for catalysis. The electron density map of monoclinic crystals reveals an additional atom bound to Cys25 (Fig. 12.7). This atom, presumably sulphur, makes no hydrogen bonds with adjacent groups in the protein. In the oxidized enzyme (Kamphuis *et al.* 1985) three oxygens are attached to Cys25. In this structure the histidine rotates slightly about the Cα–Cβ bond and moves to make a hydrogen bond with one of these oxygens. The histidine and cysteine in the active enzyme interact directly. Drenth *et al.* (1976) suggest that if the cysteine and histidine are present as a thiolate–imidazolium ion-pair then the cysteine and histidine would be coplanar, involving a rotation of the histidine by 30°. Gln19 lies where the active side cleft becomes shallower and wider. The NH₂ group of Gln19 points upwards into the active site cleft, one hydrogen interacts with the carbonyl of 22 but the other is free to bind to the substrate or intermediate. The amide hydrogen of residue 25 is also available for hydrogen bonding.

## Modelling by homology

The structures of papain and actinidin can be used to model other related cysteine proteases. Protease omega and papain have 69% sequence identity and the structural framework is therefore very similar. The only insertion in protease omega is between residues 168 and 169 of papain. Modelling this insertion is relatively simple, as a loop of the same length occurs at this position in actinidin. Protease omega is therefore structurally similar to papain

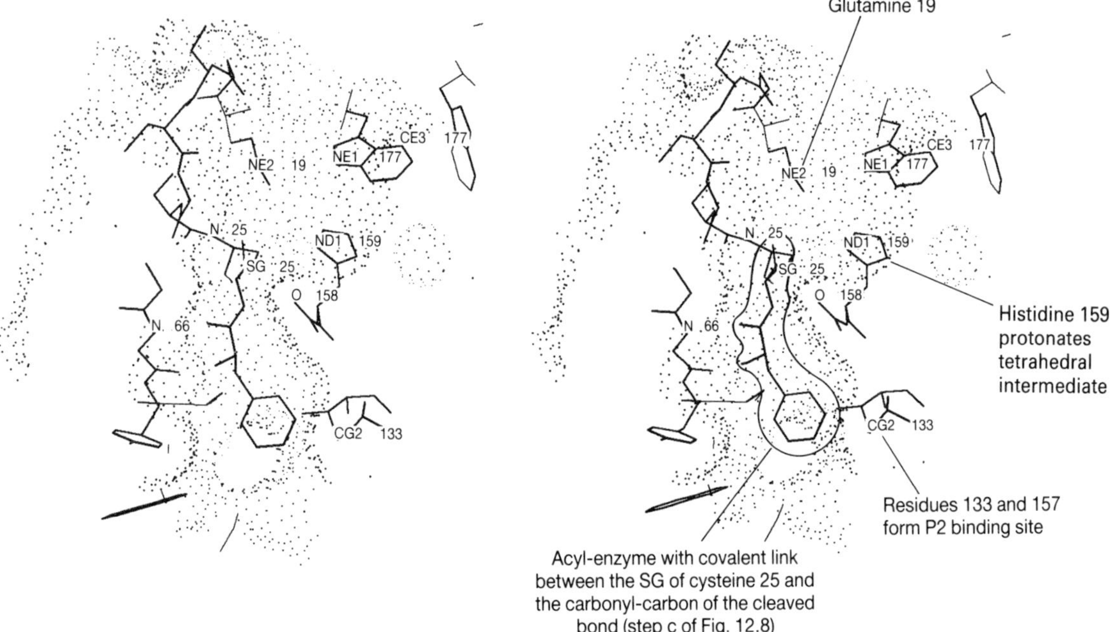



**Fig. 12.6**   The active site cleft of papain (in monoclinic crystals) with the solvent accessible surface shown as a dotted surface. The binding of the acyl-enzyme intermediate to the enzyme is modelled into the active site cleft.

but with an additional four-residue loop, as found in actinidin. Although the frameworks are similar, many side chains differ and it is these different side chains that affect the details of the protein's interaction with other molecules and that may affect the activity, stability and solubility of the enzyme.

## Catalysis, action and inhibition

### Crystallographic binding studies

A series of X-ray studies using chloromethyl ketone inhibitors of papain (Drenth *et al.* 1976) have given a clear picture of how these molecules bind subsites on the acyl side of the cleaved bond (subsites S1, S2 . . .) but there is no experimental information about the binding on the amino side of the cleaved bond (subsites S1′, S2′ . . .). The chloromethyl ketone inhibitors that were used all bind covalently to the sulphur atom of Cys25, not as true isosteres but with an additional methylene-group between the carbonyl carbon of the substrate and the sulphur of the cysteine. The acyl–enzyme complex can be modelled from these studies (see Fig. 12.6). The carbonyl of the cleaved bond points towards the

amide group of residue 25 and the $NH_2$ group of Gln19. The P1 site is up and out of the active site cleft. The P1–P2 peptides form a very short piece of antiparallel β-sheet with the carbonyls of residues 66 and 158 and with the amide of residue 66. The side chain of the P2 residue lies in a shallow hydrophobic pocket formed by the side chains of Val133 and Val157 together with Tyr67, Trp69 and Phe207. The binding of the P3 and P4 sites is not well defined and this may reflect an ability to bind a number of different residues in different ways. In addition, Drenth *et al.* (1976) report that there are distinct positive and negative features seen in difference electron density maps, which indicate a movement of the region 59–68 and less pronounced movement of 158–159.

The epoxide 1-{N-[ (L-3-trans-carboxyoxirane-2-carbonyl)-L-leucyl]-amino}-4-guanidinobutane (E-64) is a potent inhibitor of papain and several other cysteine proteases. The structure of the papain–E-64 complex has recently been determined at 2.4 Å resolution and an R-factor of 23.3% (Varughese *et al.* 1989) using crystals grown at pH 6.3, and at similar resolution using crystals grown at pH 9.2 (Matsumoto *et al.* 1989). The E-64 molecule interacts with the same subsites as the chloromethyl

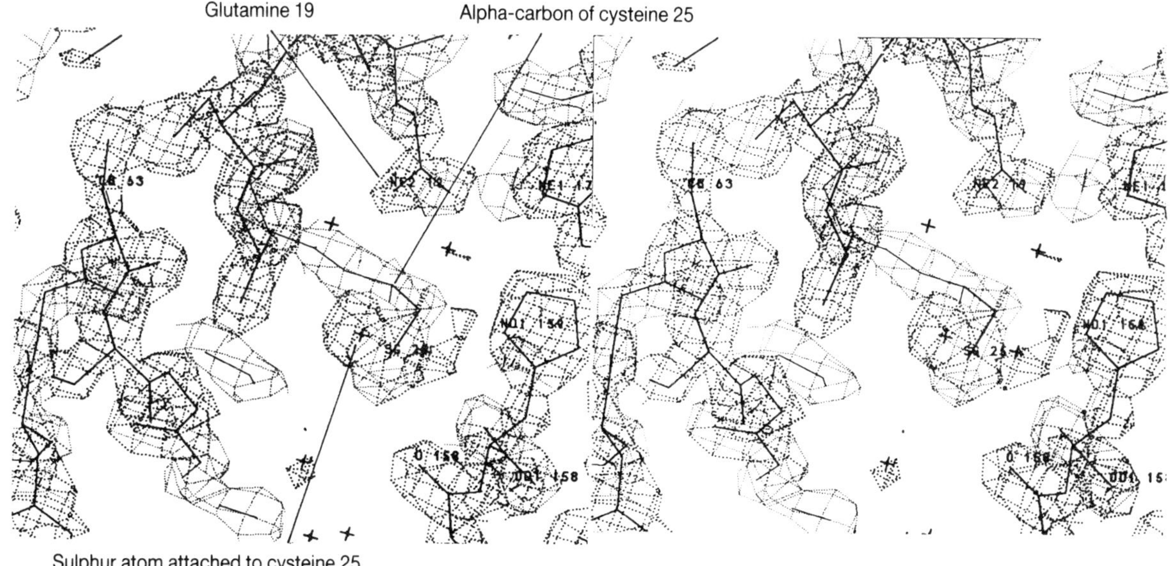



**Fig. 12.7**   Electron density map of papain (3Fobs–2Fcalc map) revealing the sulphur atom covalently bound to cysteine 25 in monoclinic crystals.

ketone derivatives. Comparison of the structure of the E-64 complex and the native papain molecule (Varughese *et al.* 1989) indicates a widening of the active site cleft in this complex as observed by Drenth *et al.* (1976) for the chloromethyl ketone inhibitors.

## Electrostatics, mechanism and activity

Hydrolysis of a peptide by a cysteine protease demands a potent nucleophile to attack the carbonyl carbon of the bond to be cleaved. The thiol group is a poor nucleophile, but the thiolate ion is a good one, and there is now considerable evidence to suggest that active cysteine proteases have an ion-pair, rather than a neutral pair, at their active sites (Lewis *et al.* 1981). Theoretical studies show that the ion-pair is stabilized by the electrostatic dipole of the active site helix and by other partially and fully charged groups (Hol *et al.* 1978; van Duijnen *et al.* 1979, 1980: Pickersgill *et al.* 1988; Pickersgill 1988; Pickersgill *et al.* 1989; Rullmann *et al.* 1989). The pH-activity profile is ultimately limited by the pH-range over which the ion-pair is formed. However, the work of Brocklehurst and co-workers clearly shows that the pH-activity of the proteases and reactivity of the cysteine is not only dependent on ion-pair formation, but also on other factors determined by the ionization of groups on

the protein and by specific binding interactions (Brocklehurst *et al.* 1989; Kowlessur *et al.* 1989). The proposed mechanism of hydrolysis of peptide substrates by the cysteine proteases based on biochemical, structural and theoretical data is shown in Figure 12.8.

## Interaction with macromolecular inhibitors

The cystatin molecules are tight, reversible inhibitors of the cysteine proteases (Barratt 1987). The structure of the 108-residue Gly9i form of chicken egg-white cystatin has been solved by X-ray analysis (Bode *et al.* 1988) and the molecule can be successfully docked with papain (Fig. 12.9). The inhibitor does not bind in the same way that substrate binds to the active site cleft, but with two conserved β-hairpin loops.

## Protein engineering

Biochemical studies on engineered protein will reveal the effect of Asp158 on the pK values of the active site groups and on the activity and reactivity of papain. Engineering other carefully selected charge groups will elucidate the effect of charge sign and position relative to the ion pair on ionization of the catalytic groups and the activity of the

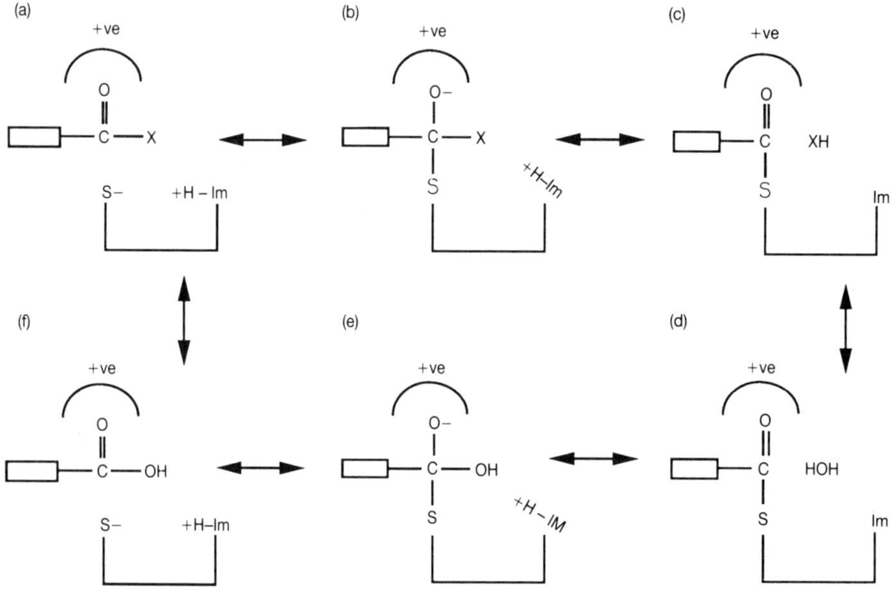

**Fig. 12.8**　A schematic representation of the proposed mechanism of hydrolysis by papain. X represents the part of the peptide on the amino side of the cleaved bond and the box represents the part of the peptide on the acyl side of the cleaved bond. S- and +H-Im are the thiolate and imidazolium ions respectively. Hydrolysis proceeds from (a) to (f); the proposed tetrahedral oxyanion intermediates (b) and (e) are stabilized by the positive potential in this region of the active site cleft (indicated as +ve) due to the electric dipole of the active site helix and the amide of residue 25.

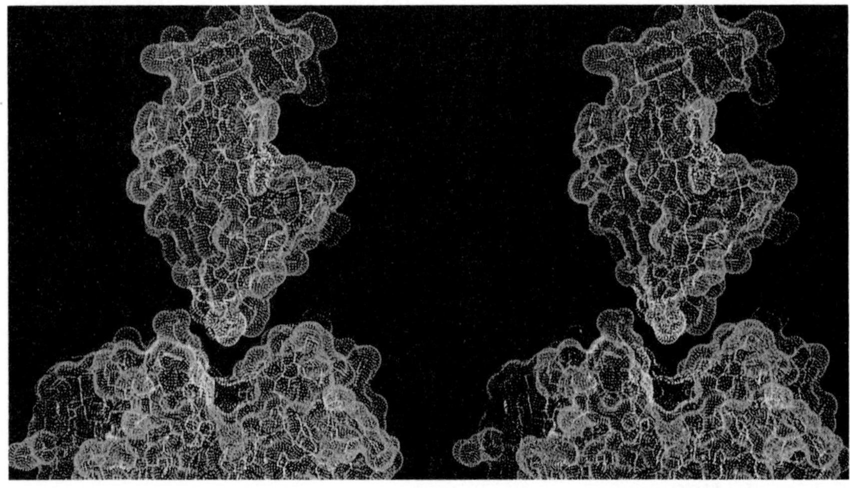

**Fig. 12.9**　Proposed mode of interaction of cystatin (top) and papain (bottom); the inhibitor is translated up and out of the complex (reproduced with permission from Bode *et al.* 1988).

enzyme. The importance of Gln19 in contributing to the 'oxyanion hole' of papain can also be tested by protein engineering.

There are also interesting structural problems to be addressed, such as the interaction of substrates with the active site cleft on the amino side of the cleaved bond and the structure of pro-papain, the large inactive precursor of the native enzyme.

Biochemical and crystallographic studies on engineered proteins will experimentally answer these key questions and calculations will help the design and understanding of the function of mutant enzymes.

The answers will provide the framework for modifying the properties of the cysteine proteases for function. There are demanding stability requirements for food processing and a need to make the enzymes more active at low pH, where bacterial growth is reduced and where certain foodstuffs take up water and allow the enzyme better access to substrate.

# The amylases

## Introduction

### Overview

α-amylase is widely distributed in plants, mammalian tissues and micro-organisms. The principal commercial sources are fungi, bacteria and cereals (malted wheat and barley). The best studied α-amylases are from porcine pancreas, *Aspergillus oryzae* (Taka-amylase), *Aspergillus niger* (acid-stable α-amylase) and barley. β-amylase is from higher plants, such as barley, wheat, soybean and sweet potato. The amylases are applied with great advantage in baking, brewing and in the liquefaction of starch to glucose syrups.

### Molecular properties

The substrates for amylases are starch (amylose and amylopectin), glycogen and various oligosaccharides. Amylose is a straight chain polysaccharide in which glucose units are linked by α-D-(1,4) glycosidic bonds. Amylopectin is branched with linear chains linked through α-D-(1,6) glycosidic bonds. α-amylase hydrolyses the (1,4) bonds of amylose and amylopectin in a non-ordered way and liberates smaller units with free, non-reducing end groups. This enzyme cannot cleave the (1,6) linkages in amylopectin and produces dextrins of low molecular weight. β-amylase can also cleave the (1,4) linkages of starch components, but it is an exo-hydrolase and releases maltose units from the non-reducing end of starch in an ordered way. Thus, β-amylase can completely hydrolyse amylose to maltose, but its activity on amylopectin is stopped short of complete hydrolysis by the (1,6) linkages at the branch points. These resistant residues, high molecular weight dextrins, can be rendered accessible to further hydrolysis by β-amylase by the action of α-amylase. Consequently, starch can be almost completely hydrolysed to maltose by use of both α- and β amylases. α-amylases require at least one calcium ion per enzyme molecule for both structural stability and catalytic activity (Vallee *et al.* 1959).

## Plant amylases; molecular biology and structural motifs

α-amylases from different sources, despite poor sequence similarity, are expected to have substantial structural similarity. The porcine, *Aspergillus oryzae* and *Aspergillus niger* enzymes possess an eight-fold α/β barrel domain and the plant α-amylases are also expected to have an α/β barrel domain. Indeed, it has been suggested that the α/β barrel may be a common structure in all α-1,4-D-glucan-cleaving enzymes (Macgregor and Svensson 1989).

The amino acid sequences of two barley α-amylase isoenzymes have been deduced from cDNA sequences (Rogers and Milliman 1983; Rogers 1985a). The homology between these isoenzymes is 78% and they show a distant homology to other α-amylases (Rogers 1985b; Nakajima *et al.* 1986). Three regions appear to be highly conserved in all α-amylases (Toda *et al.* 1982) and the barley, porcine and *Aspergillus oryzae* sequences are aligned in Figure 12.10a. α-amylase 2 is the most abundant of the two isoenzymes in barley malt and is distinguished from α-amylase 1 by its higher isoelectric point. α-amylase 1 is more active against starch granules than α-amylase 2, and this may reflect their different physiological roles (MacGregor and Balance 1980; Brown and Jacobsen 1982; Bertoft *et al.* 1984). In addition, only α-amylase 2 is inhibited by the endogenous barley α-amylase/subtilisin inhibitor (Weselake *et al.* 1985). Barley β-amylase has also been cloned and the amino acid sequence deduced (Kreis *et al.* 1987).

## Structure

### X-ray analysis of amylases

The structure of Taka-amylase was determined at 3.0 Å resolution by multiple isomorphous replacement methods (Matsuura *et al.* 1984). This structure has subsequently been determined in another crystal form at a resolution of 2.1 Å refined to a crystallographic R-factor of 24% (Dodson 1989). Porcine pancreatic α-amylase was solved and refined at 2.9 Å resolution (Buisson *et al.* 1987) and the structure of the acid-stable α-amylase from *Aspergillus niger* has been determined at 2.1 Å and 16% (Dodson 1988). A low resolution (4.5 Å) structure of soybean β-amylase has been determined (Aibara *et al.* 1984) and α-amylase isoenzymes 1 and 2 from germinated barley seeds have been crystallized (Svensson *et al.* 1987).

## Description of the structure

Taka-amylase consists of 478 amino acids folded into two domains. The first 380 residues make up the first domain, which is an eight-fold α/β barrel. There is an excursion involving three β-strands between the third β-strand and the third α-helix of the eight-fold α/β barrel (Fig. 12.10b). The second domain (381–478) is an eight-stranded antiparallel β-sandwich structure. The interface between the two domains is mainly composed of hydrophobic residues and a large cleft is located at the carboxyl end of the β-strands of the eight-fold α/β domain. A carbohydrate molecule, covalently bound to Asn197, and an isolated peak of electron density close to the cleft and assumed to be the calcium, were seen in the electron density map.

The porcine enzyme has the first 407 residues folded into an eight-fold α/β barrel, again with an excursion between the third β-strand and third α-helix. This excursion is longer than that found in Taka-amylase and consists of two antiparallel β-strands and a long irregular loop. The calcium ion binds at the interface between the α/β barrel and the excursion. The second domain (408–96) is an eight-stranded antiparallel β-barrel with Greek key topology. Residual electron density close to Asn412 suggests that a carbohydrate molecule may be covalently attached to this asparagine.

## The active site of α-amylase

The active site of Taka-amylase was identified by diffusing maltotriose, a poor substrate, into the crystals (Matsuura *et al.* 1984). Density corresponding to maltose was found in a difference Fourier map. Presumably the maltotriose is hydrolysed to maltose and glucose and the maltose binds to the active site. The maltose binds close to Glu230, which is located close to the bottom of the cleft at the carboxy terminal end of the α/β barrel. Other residues that might be directly involved in catalysis are His122, His296 and Asp297.

The active site of the porcine α-amylase was identified at 5 Å resolution by difference Fourier maps between substrate–analogue soaked and native crystals (Payan *et al.* 1980). These analogues bind at the carboxy terminal end of the α/β barrel. Asp197, Lys200 and Asp300 are close to the active site (Fig. 12.11). A second binding site for substrate analogues was also identified between the first and second domains. This binding site may anchor the hydrolase to the substrate. The calcium binding site has been identified by substituting differently scattering divalent metal ions for the calcium. The calcium ion binding site was unambiguously defined by these studies, showing binding by the Oδ1 of Asn100 and the main chain carbonyl of 201 from the α/β domain, and by Asp159 and Asp167 from the excursion. The calcium ion provides a novel ionic bridge between two β-structures. Thus the calcium ion may be essential

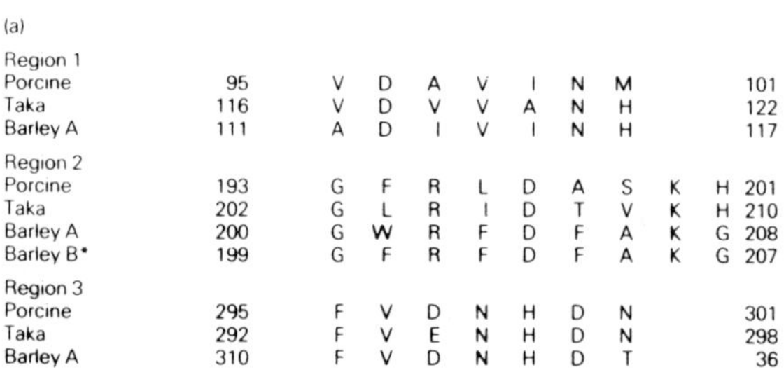

(a)

| Region 1 | | | | | | | | | | | |
|---|---|---|---|---|---|---|---|---|---|---|---|
| Porcine | 95 | V | D | A | V | I | N | M | | | 101 |
| Taka | 116 | V | D | V | V | A | N | H | | | 122 |
| Barley A | 111 | A | D | I | V | I | N | H | | | 117 |
| **Region 2** | | | | | | | | | | | |
| Porcine | 193 | G | F | R | L | D | A | S | K | H | 201 |
| Taka | 202 | G | L | R | I | D | T | V | K | H | 210 |
| Barley A | 200 | G | W | R | F | D | F | A | K | G | 208 |
| Barley B* | 199 | G | F | R | F | D | F | A | K | G | 207 |
| **Region 3** | | | | | | | | | | | |
| Porcine | 295 | F | V | D | N | H | D | N | | | 301 |
| Taka | 292 | F | V | E | N | H | D | N | | | 298 |
| Barley A | 310 | F | V | D | N | H | D | T | | | 36 |

*included because of W/F difference

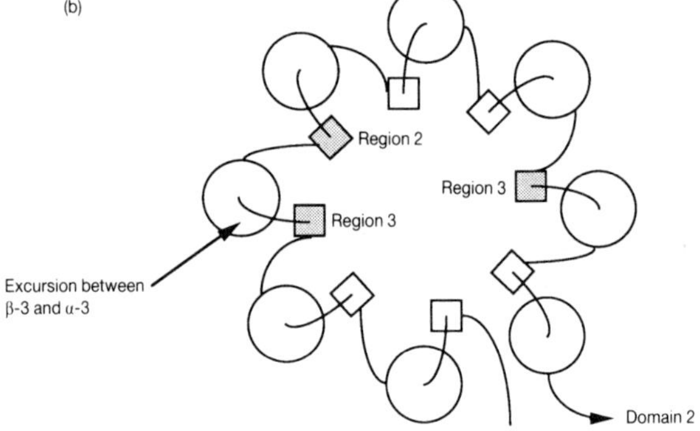


**Fig. 12.10** (a) Three regions of the sequences of Taka, porcine and barley α-amylases aligned; residues in these sequences have been implicated in catalysis. (b) Schematic representation of the eight-fold α/β barrel with the strands running up and out of the paper and helices into the paper (helices and strands represented as circles and squares respectively). The regions implicated in catalysis and calcium binding are at the C-terminus of the barrel.

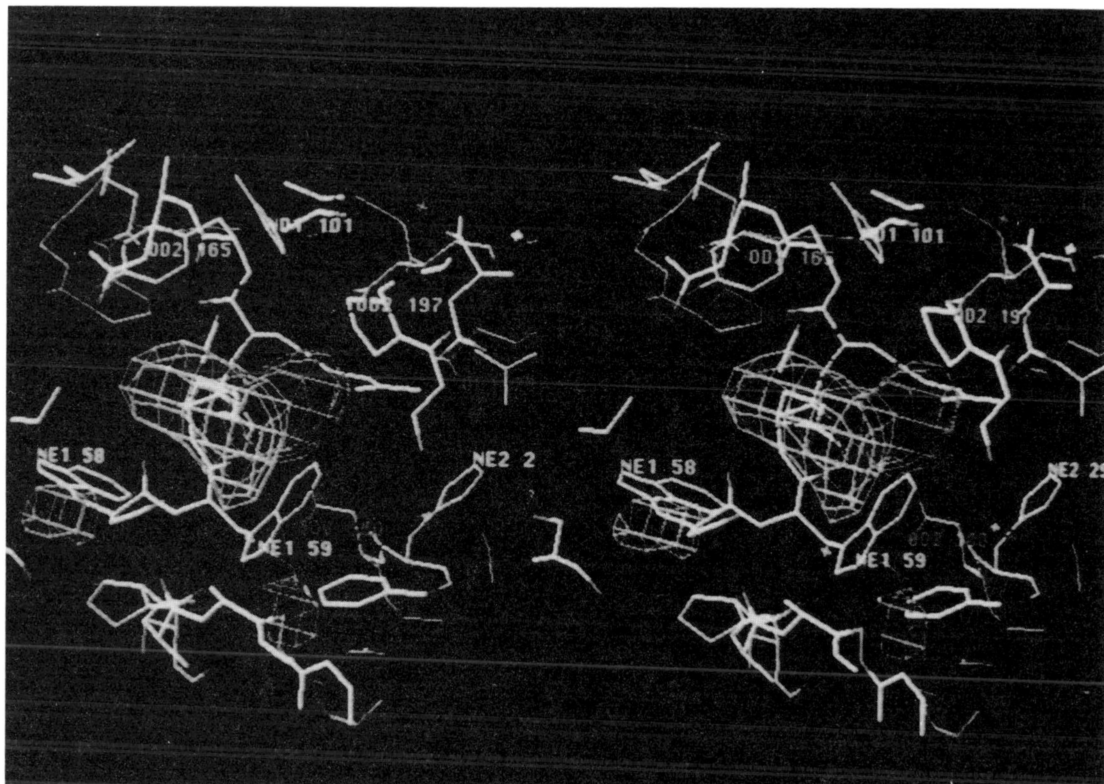

**Fig. 12.11**   The active site of porcine α-amylase with electron density corresponding to a modified maltotriose compound shown. These are close contacts with Asp197 and Asp300 (reproduced with permission from Buisson *et al.* 1988).

in stabilizing the active site cleft and the binding site is apparently similar in both the Taka-amylase and porcine α-amylase structures. Figure 12.10 identifies some of the active site residues of the barley α-amylases; Asp204 (equivalent to Asp197 of porcine), Asp315 (300), Lys207 (200) and Asn116 (100).

## Catalysis, action and inhibition

### Crystallographic binding studies

Matsuura *et al.* (1984) have shown that the reducing end of the maltose points towards Glu230 and that the hydroxyl group at C6 of the reducing end glucose apparently hydrogen bonds to His122. This hydrogen bond may be essential for catalysis because C6 substituted phenyl-α-maltosides are not hydrolysed by Taka-amylase. In this binding mode, the hydroxyl groups at C2 and C3 of the same glucose unit hydrogen bond to Oγ of Asp297

and Nε1 of His296, respectively. Thus, Glu230 has been proposed to be the general acid in the mechanism and Asp297 the group that stabilizes the carbonium ion intermediate in the reaction by analogy with the mechanism of lysozyme. Buisson *et al.* (1987) argue that Asp197 and Asp300 of the porcine enzyme are the catalytic residues. The two proposals agree on the role of Asp300 (Asp297 of Taka-amylase) but do not agree on the residue that is the general acid.

### Macromolecular α-amylase inhibitors

The structure of an α-amylase inhibitor, Hoe-476A from *Streptomyces tendae* 4158 has been determined (Pflugrath *et al.* 1986). It has been suggested that this inhibitor could be used to treat obesity and diabetes (Meyer *et al.* 1983). The structure of the enzyme–inhibitor complex will reveal the detailed intermolecular interactions involved in inhibition. An α-amylase inhibitor from wheat has also been crystallized (Maeda *et al.* 1987).

## Protein engineering

Protein engineering, together with biochemical and structural studies will identify the general acid in the mechanism of Taka, porcine and acid-stable α-amylase. Additionally, these experiments will characterize the extended active site that interacts with polysaccharides.

Amylases are primarily used in baking, brewing and for the liquefaction of starch. Different properties are needed for each application, for baking the thermostability of the amylase must not be too great because the enzyme must be destroyed during baking or over-liquefaction occurs. In contrast, high activity and survival of the enzyme at high temperatures is an important requirement for the use of amylases in brewing and in the starch industry (see pp. 236–7). A second requirement is ensuring optimum activity at the pH of the material to be converted. Enhanced thermostability by optimizing intramolecular interactions and modulated pH-optimum by manipulating electrostatic interactions are now achievable by protein engineering (see p. 241). Current studies will enable the specificity of the enzyme to be modified so as to yield a product of specific size. Such products will find a variety of uses in foods.

# Conclusion

## Summary

Knowledge-based protein engineering is an efficient method of tailoring enzymes for industrial function. Modelling the structure of proposed mutants, based on the structure of the native protein, can suggest the best substitution to alter function in the desired way. The purpose of modelling is to suggest useful mutants to construct, because modelling can be much quicker than the construction and testing of a mutant. However, modelling is not a substitute for high resolution X-ray or NMR analysis, which are used to experimentally determine the precise effect of the mutation on structure.

Structure-based calculations can indicate trends in the properties of a series of mutants. Of increasing importance is the use of these calculations to target residues for modification. Different models and methods of calculation are appropriate for different problems and the choice of method is often determined by the accuracy needed and the time available. The lack of knowledge of how proteins fold is the most severe problem in protein design. However, protein engineering is being applied to this problem and intermediates in the folding of barnase are being characterized (Matouschek *et al.* 1989). In addition to this, the careful work of Fersht and Matthews (see p. 241) has established a library of interaction energies in proteins that are invaluable in further protein design and engineering.

## Engineering enzymes for the food industry

The knowledge base described can be exploited in the engineering of enzymes for industrial use. In principle, the activity of any enzyme can be optimized for operation under a given set of industrial conditions. Trends in pH-activity profile, stability and specificity of engineered proteins have been experimentally observed and predicted with some success. These trends can be exploited in the engineering of plant cysteine proteases and amylases for use in foods, as described in this chapter.

## Plant enzymes, industry and the future

Plant enzymes sometimes have natural advantages that can be exploited in industrial uses. These advantages can be optimized and any disadvantages minimized by protein engineering. Engineered amylases or cysteine proteases are expected to be the first engineered plant enzymes used in foods. Modified enzymes, from both plants and other sources, are expected to have a great impact on food processing, quality and safety. Experimental measurements of engineered proteins will continue to provide information about the nature of the sequence–structure and structure–function relationship. Further insights will lead to computer-based methods of increased accuracy of prediction and design capability. The knowledge already available is sufficient to enable the design of simple proteins from first principles and enzymes designed *ab initio* are expected within the next few years. These enzymes will have no biological restraints, other than the requirement to fold correctly, and may therefore prove to be the most powerful catalysts of all.

# References

Aibara, S., Yamashita, H. and Morita, Y. (1984). Molecular shape and packing in crystals of soybean β-amylase. *Agriculture and Biological Chemistry*. **48**: 1575–9.

Baker, E.N. (1977). Structure of actinidin: details of the polypeptide chain conformation and active site from an electron density map at 2.8 Å resolution. *Journal of Molecular Biology*. **115**: 263–77.

Baker, E.N. (1980). Structure of actinidin after refinement at 1.7 Å resolution. *Journal of Molecular Biology*. **141**: 441–84.

Baker, E.N. and Dodson, E.J. (1980). Crystallographic refinement of the structure of actinidin at 1.7 Å resolution by fast Fourier least-squares methods. *Acta Crystallography*. **A36**: 559–72.

Baker, E.N. and Hubbard, R.E. (1984). Hydrogen bonding in globular proteins. *Progress in Biophysics and Molecular Biology*. **44**: 97–179.

Barrett, A. (1987). The cystatins: a new class of peptide inhibitors. *Trends in Biological Science*. **12**: 193–6.

Bertoft, E., Andtfolk, C. and Kulp, S. (1984). Effect of pH, temperature and calcium ions on barley malt α-amylase isoenzymes. *Journal of the Institute of Brewing*. **90**: 298–302.

Blundell, T.L., Sibunda, B.L., Sternberg, M.J.E. and Thornton, J.M. (1987). Knowledge-based prediction of protein structures and the design of novel molecules. *Nature*. **326**: 347–52.

Bode, W., Engh, R., Musil, D. *et al.* (1988). The 2.0 Å X-ray structure of chicken egg white cystatin and its possible mode of interaction with cysteine proteases. *EMBO Journal*. **7**: 2593–9.

Brocklehurst, K., O'Driscoll, M., Kowlessur, D. *et al.* (1989). The interplay of electrostatic and binding interactions determining active centre chemistry and catalytic activity in actinidin and papain. *Biochemical Journal*. **257**: 309–10.

Brown, A. and Jacobsen, J. (1982). Genetic basis and natural variation of α-amylase isozymes in barley. *Genetical Research*. **40**: 315–24.

Browne, W.J., North, A.C.T., Phillips, D.C. *et al.* (1969). A possible three dimensional structure of bovine α-lactalbumin based on that of hen's egg-white lysozyme. *Journal of Molecular Biology*. **42**: 65–86.

Buisson, G., Duee, E., Haser, R. and Payan, F. (1987). Three dimensional structure of porcine pancreatic α-amylase at 2.9 Å resolution. Role of calcium in structure and activity. *EMBO Journal*. **6**: 3903–16.

Cohen, F.E., Abarbanel, R.M., Kuntz, I.D. and Fletterick, R.J. (1983). Secondary structure assignment for α/β proteins: a combinatorial approach. *Biochemistry*. **22**: 4894–904.

Cohen, L.W., Coglan, V.M. and Dihel, L.C. (1986). Cloning and sequencing of papain-encoding cDNA. *Gene*. **48**: 219–27.

Craik, C.S., Largman, C., Fletcher, T. *et al.* (1985). Redesigning trypsin: alteration of substrate specificity. *Science*. **228**: 291–7.

Dodson, G. (1988). Structure function relationships in a peptide hormone and enzymes. The application of protein engineering. *Biotechnology Action Programme. Progress report (1988)*. **2**: 653–6.

Dodson, G. (1989). Structure function relationships in a peptide hormone and enzymes. Final meeting of contractors, protein design, Troia, Portugal, *Book of Abstracts, 14–17 November 1989*; 222–3.

Drenth, J., Jansomius, J.N., Koekoek, R. *et al.* (1968). Structure of papain. *Nature*. **218**: 929–32.

Drenth, J., Jansonius, J.N., Koekock, R. *et al.* (1971). The structure of papain. *Advances in Protein Chemistry*. **25**: 79–115.

Drenth, J., Kalk, K. and Swen, H. (1976). Binding of chloromethyl ketone substrate analogues to crystalline papain. *Biochemistry*. **15**: 3731–7.

Dubois, T., Kleinschmidtt, T., Schnek, A.G. *et al.* (1988). The thiol proteinases from the latex of *Carica papaya* L. II: The primary structure of proteinase omega. *Biological Chemistry Hoppe-Seyler*. **369**: 741–54.

Estell, D.A., Graycar, T.P., Millar, J.V. *et al.* (1986). Probing steric and hydrophobic effects on enzyme-substrate interactions by protein engineering. *Science*. **233**: 659–63.

Fersht, A.R., Shi, J.-P., Knill-Jones, J. *et al.* (1985). Hydrogen bonding and biological specificity analysed by protein engineering. *Nature*. **314**: 235–8.

Gilles, A.-M. DeWolf, A. and Keil, B. (1983). Amino acid sequences of the active site sulfhydryl peptide and other thiol peptides from the cysteine protease α-clostripain. *European Journal of Biochemistry*. **130**: 473–9.

Gilson, M.K. and Honig, B.H. (1987). Calculation of electrostatic potentials in an enzyme active site. *Nature*. **330**: 84–6.

Harvey, S.C. (1989). Treatment of electrostatic effects in macromolecular modelling. *Proteins: structure, functions and genetics*. **5**: 78–92.

Heinemann, U.P., Pal, G.P., Hilgenfeld, R. and Saenger, W. (1982). Crystal and molecular structure of the sulfhydryl protease calotropin D1 at 3.2 Å resolution. *Journal of Molecular Biology*. **161**: 591–606.

Hernandez Arana, A. and Soriano-Garcia, M. (1988). Detection and characterization by circular dichroism of a stable intermediate state formed in the thermal unfolding of papain. *Biochimica et Biophysica Acta*. **954**: 170–5.

Hol, W.G., van Duijnen, P.T. and Berendsen, H.J.C. (1978). The α-helix dipole and the properties of proteins. *Nature*. **273**: 443–6.

Hwang, J.K. and Warshel, A. (1987). Semiquantitative calculations of catalytic free energies in genetically modified enzymes. *Biochemistry*. **26**: 2669–73.

Ikemura, H., Takagi, H. and Inouye, M. (1987). Requirement of pro-sequence for the production of active subtilisin E in *Escherichia coli*. *The Journal of Biological Chemistry*. **262**: 7859–64.

Jacquet, A., Kleinschmidt, T., Schnek, A.G. *et al.* (1989). The thiol proteinases from the latex of *Carica papaya* L. III. The primary structure of chymopapain. *Biological Chemistry Hoppe-Seyler*. **370**: 425–34.

Janin, J. (1979). Surface and inside volumes in globular proteins. *Nature*. **277**: 491–2.

Janin, J., Wodak, S., Levitt, M. and Maigret, B. (1978). Conformation of amino acid side chains in proteins. *Journal of Molecular Biology*. **125**: 357–86.

Jones, T.A. and Thirup, S. (1986). Using known substructures in protein model building and crystallography. *The European Molecular Biology Organisation Journal*. **5**: 819–22.

Kamphuis, I.G., Kalk, K.H., Swarte, M.B.A. and Drenth, J. (1985). Structure of papain refined at 1.65 Å resolution. *Journal of Molecular Biology*. **179**: 233–56.

Kellis, Jnr., J.T., Nyberg, K., Sali, D. and Fersht, A.R. (1988). Contribution of hydrophobic interactions to protein stability. *Nature*. **333**: 784–6.

Kellis, J.T., Nyberg, K. and Fersht, A.R. (1989). Energetics of complementary side-chain packing in a protein hydrophobic core. *Biochemistry*. **28**: 4914–22.

Knowlessur, D., O'Driscoll, M., Topham, C.M. *et al.* (1989). The interplay of electrostatic fields and binding interactions determining catalytic-site reactivity in actinidin. A possible origin of differences in the behaviour of actinidin and papain. *Biochemical Journal*. **259**: 443–52.

Kreis, M., Williamson, M., Buxton, B. *et al.* (1987). Primary structure and differential expression of β-amylase in normal and mutant barleys. *European Journal of Biochemistry*. **169**: 517–25.

Leatherbarrow, R.J., Fersht, A.R. and Winter, G. (1985). Transition-state stabilisation in the mechanism of tyrosyl-tRNA synthetase revealed by protein engineering. *Proceedings of the National Academy of Sciences, USA*. **82**: 7840–4.

Levitt, M. and Warshel, A. (1975). Computer simulation of protein folding. *Nature*. **253**: 694–8.

Lewis, S.D., Johnson, F.A. and Shafer, J.A. (1981). Effect of cysteine-25 on the ionization of histidine-159 in papain as determined by proton nuclear magnetic resonance spectroscopy. Evidence for a his-159-cys-25 ion pair and its possible role in catalysis. *Biochemistry*. **20**: 48–51.

MacGregor, A. and Ballance, D. (1980). Hydrolysis of large and small starch granules from normal and waxy barley cultivars by α-amylases from barley malt. *Cereal Chemistry*. **57**: 397–402.

MacGregor, E.A. and Svensson, B. (1989). A supersecondary structure predicted to be common to several enzymes. *Biochemical Journal*. **259**: 145–52.

Madison, E.L., Goldsmith, E.J., Gerard, R.D. *et al.* (1989). Serpin resistant mutants of human tissue-type plasminogen activator. *Nature*. **339**: 721–3.

Maeda, K., Sato, M., Kato, Y. *et al.* (1987). Crystallographic study of endogenous α-amylase inhibitor from wheat. *Journal of Molecular Biology*. **193**: 825–6.

Matthew, J.B., Gurd, F.R.N., Garcia-Moreno, B. *et al.* (1985). pH dependent processes in proteins. *CRC Critical Reviews in Biochemistry*. **18**: 91–197.

Mathews, B.W., Nicholson, H. and Bektel, W.J. (1987). Enhanced protein thermostability from site directed mutations that decrease the entropy of unfolding. *Proceedings of the National Academy of Sciences, USA*. **84**: 6663–7.

Matouschek, A., Kellis, Jnr., J.T., Serrano, L. and Fersht, A.R. (1989). Mapping the transition state and pathway of protein folding by protein engineering. *Nature*. **340**: 122–6.

Matsumura, M., Bektel, W.J. and Mathews, B.W. (1988). Hydrophobic stabilization in T4 lysozyme determined directly by multiple substitutions of Ile 3. *Nature*. **334**: 406–20.

Matsumoto, K., Yamamoto, D., Ohishi, H. *et al.* (1989). Mode of binding of E-64-c, a potent thiol protease inhibitor, to papain as determined by X-ray crystal analysis of the complex. *FEBS Letters*. **245**: 177–80.

Matsuura, Y., Kusunoki, M., Harada, W. and Kakudo, M. (1984). Structure and possible catalytic residues of taka-amylase A. *Journal of Biochemistry*. **95**: 697–702.

Meyer, B., Muller, F., Kruger, J. and Grigoleit, H. (1983). Inhibition of starch absorption by α-amylase inactivator given with food. *Lancet*. **i**: 934.

Murali, C. and Creaser, E.H. (1986). Protein engineering of alcohol dehydrogenase–1. Effects of two amino acid changes in the active site of yeast ADH-1. *Protein Engineering*. **1**: 55–8.

Nakajimia, R., Iminaka, T. and Aiba, S. (1986). Comparison of amino acid sequences of eleven different α-amylases. *Applied Microbiology and Biotechnology*. **23**: 355–60.

Nicholson, H. Bektel, W.J. and Mathews, B.W. (1988). Enhanced protein thermostability from designed mutations that interact with α-helix dipoles. *Nature*. **336**: 651–6.

Pantoliano, M.W., Whitlow, M., Wood, J.F. *et al.* (1988). The engineering of binding affinity at metal ion binding sites for the stabilization of proteins: subtilisin as a test case. *Biochemistry*. **27**: 8311–17.

Payan, F., Haser, R., Pierrot, M. *et al.* (1980). Three dimensional structure of α-amylase from porcine pancreas at 5 Å resolution—active site location. *Acta Crystallography*. **B36**: 416–21.

Perry, L.J. and Wetzel, R. (1984). Disulfide bond engineering into T4 lysozyme: stabilization of the protein toward thermal inactivation. *Science*. **226**: 555–7.

Pflugraph, J., Wiegand, G. and Huber, R. (1986). Crystal structure determination, refinement and the molecular model of the α-amylase inhibitor hoe-467A. *Journal of Molecular Biology*. **189**: 383–6.

Pickersgill, R.W. (1988). A rapid method for calculating charge–charge interaction energies in proteins. *Protein Engineering*. 2: 247–8.

Pickersgill, R.W., Goodenough, P.W., Sumner, I.G. and Collins, M.E. (1988). The electrostatic fields in the active site clefts of actinidin and papain. *Biochemical Journal*. 254: 235–8.

Pickersgill, R.W., Sumner, I.G., Collins, M.E. *et al.* (1989). Structural and electrostatic differences between actinidin and papain account for differences in activity. *Biochemical Journal*. 257: 310–12.

Pickersgill, R.W., Sumner, I.G. and Goodenough, P.W. (1990). Preliminary crystallographic data for protease omega. *European Journal of Biochemistry*. 190: 443–4.

Pickersgill, R.W., Sumner, I.G., Collins, M.E. *et al.* (1991a). Modification of the stability of phospholipase A2 by charge engineering. *FEBS Letters*. 281: 219–22.

Pickersgill, R.W., Harris, G.W. and Garman, E. (1991b). Structure of monoclinic papain at 1.6 Å resolution. *Acta Crystallographica* (in press).

Ponder, J.W. and Richards, F.M. (1987). Tertiary templates for proteins: use of packing criteria in the enumeration of allowed sequences for different structural classes. *Journal of Molecular Biology*. 193: 775–91.

Ramachandran, G.N. and Sasisekharan, V. (1968). Conformation of polypeptides and proteins. *Advances in Protein Chemistry*. 23: 283–438.

Rao, S.N., Singh, U.C., Bash, P.A. and Kollman, P.A. (1987). Free energy perturbation calculations on binding and catalysis after mutating Asn 155 in subtilisin. *Nature*. 328: 551–4.

Rashin, A.A. and Honig, B. (1984). On the environment of ionizable groups in globular proteins. *Journal of Molecular Biology*. 173: 515–21.

Richards, F.M. (1977). Areas, volumes, packing and protein structure. *Annual Review of Biophysics and Bioengineering*. 6: 151–75.

Ritonja, A., Buttle, D.J., Rawlings, N.D. *et al.* (1989). Papaya proteinase IV amino acid sequence. *FEBS Letters*. 258: 109–12.

Rogers, J. (1985a). Two barley α-amylase gene families are regulated differently in aleurone cells. *Journal of Biological Chemistry*. 260: 3731–8.

Rogers, J. (1985b). Conserved amino acid sequence domains in α amylases from plants, mammals and bacteria. *Biochemistry and Biophysics Research Communications*. 128: 470–8.

Rogers, J. and Milliman, C. (1983). Isolation and sequence analysis of a barley α-amylase cDNA clone. *Journal of Biological Chemistry*. 258: 8169–74.

Rogers, J., Dean, D. and Heck, G (1985). Aleurain: a barley thiol protease closely related to mamalian cathepsin H. *Proceedings of the National Academy of Sciences, USA*. 82: 6512–16.

Rullmann, J.A.C., Bellido, M.N. and van Duijnen, P. (1989). The active site of papain. All-atom study of interactions with protein matrix and solvent. *Journal of Molecular Biology*. 206: 101–18.

Russel, A.J. and Fersht, A.R. (1987). Rational modification of enzyme catalysis by engineering surface charge. *Nature*. 328: 496–500.

Russel, A.J., Thomas, P.G. and Fersht, A.R. (1987). Electrostatic effects on modification of charged groups in the active site cleft of subtilisin by protein engineering. *Journal of Molecular Biology*. 193: 803–13.

Sali, D., Bycroft, M. and Fersht, A.R. (1988). Stabilization of protein structure by interaction of α-helix dipole with a charged side chain. *Nature*. 335: 740–3.

Silen, J. and Agard, D. (1989). The α-lytic protease pro region does not require a physical linkage to activate the protease domain *in vivo*. *Nature*. 341: 462–4.

Sternberg, M.J.E., Hayes, F.R.F. and Russel, A.J. (1987). Prediction of electrostatic effects of engineering of protein charges. *Nature*. 330: 86–8.

Sutcliffe, M.J., Haneef, I., Carney, D. and Blundell, T.L. (1987a). Knowledge based modelling of homologous proteins, part I: three dimensional frameworks derived from the simultaneous superposition of multiple structures. *Protein Engineering*. 1: 377–85.

Sutcliffe, M.J., Hayes, F.R.F. and Blundell, T.L. (1987b). Knowledge based modelling of homologous proteins. Part II: rules for the conformations of substituted sidechains. *Protein Engineering*. 1: 385–92.

Svensson, B., Gibson, R.M., Haser, R. and Astier, J.P. (1987). Crystallization of barley malt α-amylases and preliminary X-ray diffraction studies of the high pI isozyme, α-amylase 2. *The Journal of Biological Chemistry*. 262: 13682–4.

Toda, H., Kondo, K. and Narita, K. (1982). The complete amino acid sequence of taka-amylase A. *Proceedings of the Japanese Academy*. 58 Ser B. 208–12.

Vallee, B.L., Stein, E.A., Sumerwell, W.N. and Fisher, E.H. (1959). Metal content of α-amylases of various sources. *Journal of Biological Chemistry*. 234: 2901–29.

van Duijnen, P.T., Thole, B.T. and Hol, W.G.J. (1979). On the role of the active site helix in papain, an *ab initio* molecular orbital study. *Biophysical Chemistry*. 9: 273–80.

van Duijnen, P.Th., Thole, B.Th., Broer, R. and Nieuwpoort, W.C. (1980). Active site α-helix in papain and the stability of the ion pair. *International Journal of Quantum Chemistry*. 17: 651–71.

van Gunsteren, W.F. (1988). The role of computer simulation techniques in protein engineering. *Protein Engineering*. 2: 5–13.

Varughese, K., Ahmed, F.R., Carey, P.R. *et al.* (1989). Crystal structure of a papain-E-64 complex. *Biochemistry*. 28: 1330–2.

Vernet, T., Tessier, D.C., Laliberté, F. *et al.* (1989). The expression in *Escherichia coli* of a synthetic gene coding for the precursor of papain is prevented by its own putative signal sequence. *Gene*. 77: 229–36.

Warshell, A. and Levitt, M. (1976). Theoretical studies of enzyme reactions: dielectric, electrostatic and steric stabilization of the carbonium ion in the reaction of lysozyme. *Journal of Molecular Biology*. **103**: 227–49.

Warshell, A. and Sussman, F. (1986). Toward computer-aided site directed-mutagenesis of enzymes. *Biophysics*. **83**: 3806–10.

Warwicker, J. and Watson, H.C. (1982). Calculation of the electric potential in the active site cleft due to α-helix dipoles. *Journal of Molecular Biology*. **157**: 671–9.

Wasserman, B.P. (1984). Thermostable enzyme production. *Food Technology*. February 1984: 78–98.

Wells, J.A., Cunningham, B.C., Graycar, T.P. and Estell, D.A. (1986). Importance of hydrogen bond formation in stabilizing the transition state of subtilisin. *Philosophical Transactions of the Royal Society of London*. **317**: 415–23.

Wells, J.A., Powers, D.B., Bott, R.R. *et al.* (1987a). Designing substrate specificity by protein engineering of electrostatic interactions. *Proceedings of the National Academy of Sciences, USA*. **84**: 1219–23.

Wells, J.A., Cunningham, B.C., Graycar, T.P. and Estell, D.A. (1987b). Recruitment of subtrate-specificity properties from one enzyme into a releated one by protein engineering. *Proceedings of the National Academy of Sciences, USA*. **84**: 5167–71. **258**: 109–12.

Weselake, R.J., MacGregor, A.W. and Hill, R.D. (1985). Effect of endogenous barley alpha amylase inhibitor on hydrolysis of starch under various conditions. *Journal of Cereal Science*. **3**: 249–59.

# Part IV  EXPLORING PROTEIN INTERACTIONS

# 13

# Protein Engineering of the Barley Chymotrypsin Inhibitor 2

Alison F. Campbell

## Introduction

Chymotrypsin inhibitor 2 (CI-2) is a low molecular weight (9200 Da) protein inhibitor of serine proteases. As these enzymes play a significant role in the regulation of many biological functions, including digestion, fertilization and blood clotting, the ability to modulate proteolytic activity is desirable. Through study of inhibitors of these enzymes, and their mode of action, it will become possible to design therapeutically useful molecules.

Protein engineering can be used to probe structure/function relationships. The work described will illustrate the strategy developed to study the inhibitory specificity and mechanism of CI-2, starting from expression of a cDNA clone of the gene through to kinetic analysis of the gene product. Design of site-specific mutants of the inhibitor will be discussed, and the development of a suitable purification scheme outlined. The enzyme–inhibitor complexes are studied kinetically, and will be covered in some detail to indicate the type of data that can be obtained and how these can be interpreted to understand the molecular basis of inhibition.

## Chymotrypsin inhibitor 2

CI-2 is found in the endosperm of barley grains and is enriched in the mutant line *Hiproly*. The physiological role of this inhibitor is not known, but it is postulated to play a part in defence against microbial or pest attack (Ryan 1989). The protein has been isolated and fully sequenced (Svendsen *et al.* 1980) and is found to be a member of the potato inhibitor 1 family, sharing homology with a number of well characterized proteins involved in the regulation of proteolytic activity (Svendsen *et al.* 1982). These include the wound-induced tomato leaf inhibitor 2 (Graham *et al.* 1985) and eglin c, an inhibitor of elastase and cathepsin G produced in the medicinal leech (Seemuller *et al.* 1977). The alignment of reactive site sequences of members of this family is shown in Figure 13.1. Common features within this family include a leucine or methionine residue at the P1 position and an acidic residue (Asp, Glu) at the P1′ position (nomenclature according to Schecter and Berger 1967). CI-2 is a strong inhibitor of chymotrypsin and has also been shown to be a potent inhibitor of the microbial serine protease, subtilisin (Svendsen *et al.*

|  | P4 | P3 | P2 | P1 | P1′ | P2′ | P3′ | P4′ |
|---|---|---|---|---|---|---|---|---|
| Barley CI-2 | Ile | Val | Thr | Met | Glu | Tyr | Arg | Ile |
| Potato inhibitor 1 | Pro | Val | Thr | ML | Asp | TFL | Arg | Cys |
| Leech eglin c | Pro | Val | Thr | Leu | Asp | Leu | Arg | Tyr |
| Tomato leaf inhibitor 1 | Pro | Ile | Thr | Leu | Asp | Leu | Arg | Tyr |
| Barley CI-1 | Met | Val | Pro | Leu | Asn | Phe | Asn | Pro |

**Fig. 13.1** Alignment of the reactive site sequences for members of the potato inhibitor I family. The P1 residues are boxed. Sequences are taken from: CI-2, Svendsen *et al.* (1980); Potato Inhibitor 1 and leech eglin c, Seemuller *et al.* (1977); Tomato Leaf Inhibitor 1, Graham *et al.* (1975); CI-1, Jonassen and Svendsen (1982).

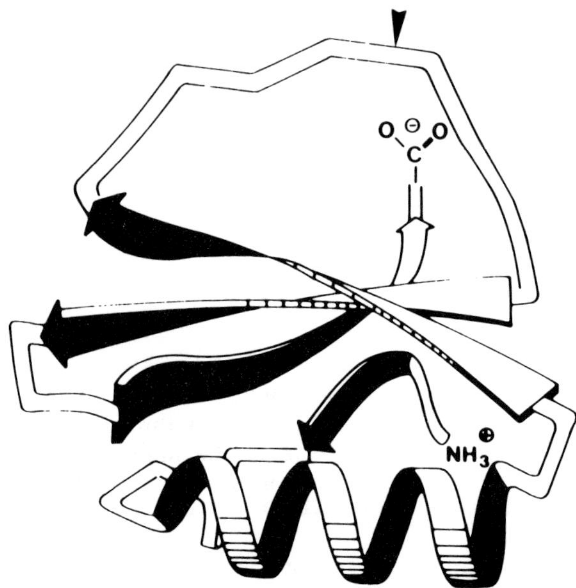

**Fig. 13.2** Basic structure of CI-2 showing secondary structural elements. Large arrows denote the β-strands and wide ribbons represent the α-helix. The position of the reactive site, in the extended loop, is indicated by a small arrow. The figure is taken from McPhalen *et al.* 1985.

1982). CI-2 is an attractive target for protein engineering because of the availability of detailed structural information, which permits a more informed choice of the site-specific mutations to be made. Computer graphics facilities can be used to model these changes in an attempt to predict and, sometimes, to explain the outcome of such experiments. High-resolution crystal structures have been solved for both the inhibitor alone (McPhalen and James 1987) and complexed with subtilisin BPN' (McPhalen *et al.* 1985). NMR solution structures have also been solved and are in close agreement with those determined in the solid state (Clore *et al.* 1987a, b). CI-2 is a wedge-shaped disc, and it is the narrow end of the wedge that interacts with the enzyme. This region of the inhibitor is an extended loop (Fig. 13.2), supported by two hydrogen bonds. These are formed from Glu60 and Thr58, both within the loop, which pair with two arginine side chains in the core protein, at positions 65 and 67, respectively.

## Serine protease inhibition

Binding within the active site of the serine proteases occurs through the reactive site region of the substrate or inhibitor, and subsequent hydrolysis occurs at a specific peptide bond within this region. The cleaved bond lies between the P1 and P1' residues, which lie to the N-terminus and C-terminus of the reactive site bond, respectively (Fig. 13.3). Following hydrolysis, the cleaved products may be released from the active site, in the case of inhibitors this occurs significantly more slowly than for hydrolysed substrates. The serine protease inhibitors have stable, rigid structures and exhibit tight binding to their cognate enzymes. The highly refined crystal structures of many of the complexes formed indicate that the inhibitor is not much distorted from its unbound state and can be said to resemble the Michaelis complex (Huber and Bode 1978; Fujinaya *et al.* 1982; Chen and Bode 1983; McPhalen and James 1987). This suggests that the inhibitory mechanism does not involve the formation of a tetrahedral intermediate or the creation of new covalent bonds. A mechanism has been proposed in which cleavage of the reactive site bond is an integral step (Laskowski and Kato 1980). It is suggested that, upon binding, the enzyme and inhibitor form a stable complex within which the inhibitor can be hydrolysed, at the reaction site, to form modified (cleaved) inhibitor. An equilibrium is then established between virgin and modified inhibitor, which favours the intact molecule. This may be accounted for by considering the rigidity of the molecule and the complex hydrogen-bonding network formed when the inhibitor is bound in the active site. Upon hydrolysis, the newly formed termini would be maintained in close proximity, facilitating resynthesis of the peptide bond. However, this theory does not explain why chemically inactivated enzymes can bind inhibitors (Haynes and Feeney 1968). Thus, to establish the full mechanism of serine protease inhibition it is necessary to understand why the reactive site bond hydrolysis takes place so slowly or, in some cases, not at all.

The structure of the CI-2–subtilisin complex (Fig. 13.4) shows the limited interaction between the enzyme and inhibitor. The side chain of the P1 residue protrudes from the surface of CI-2 into the active site of the enzyme, and the influence of this side chain upon inhibitory binding can be appreciated. Indeed, it has been shown that this side chain is a major determinant of specificity. For example, the mammalian α-1-antitrypsin is a serine protease inhibitor that controls the activity of granulocytic proteolytic enzymes, particularly elastase in the lungs (Carrell *et al.* 1982). A naturally occurring

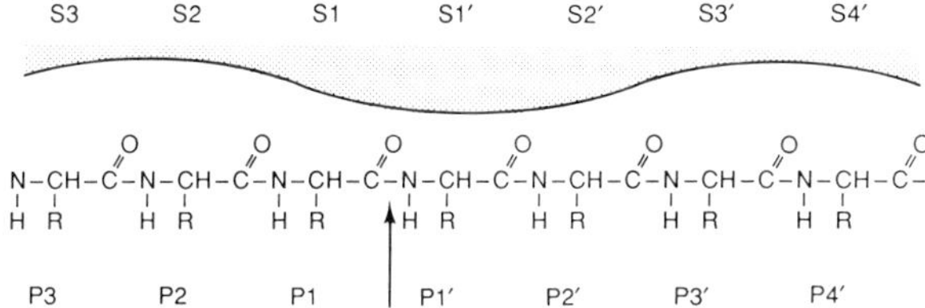



**Fig. 13.3**   Active site binding of substrate, or inhibitor, according to the nomenclature of Schecter and Berger (1967). Substrate (inhibitor) residues are identified P, whilst enzyme binding sites are denoted by S. The position of the scissile bond is indicated by an arrow.

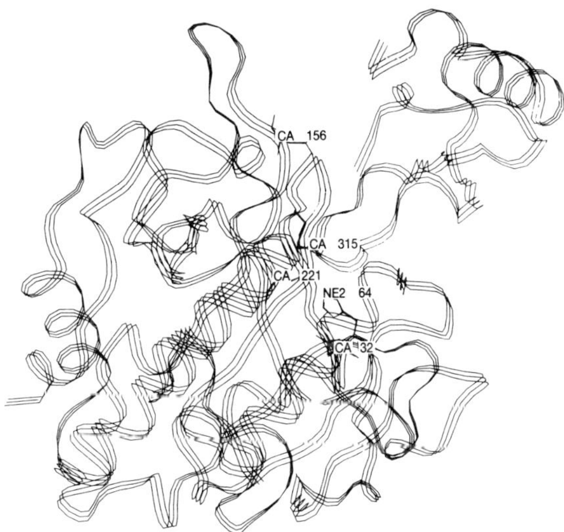



**Fig. 13.4**   Complex formed between CI-2 and subtilisin BPN'. There are limited interactions between enzyme and inhibitor, contact occurring through the reactive site loop of CI-2 within the active site of subtilisin. The side chain of Met59, the reactive site residue of CI-2, is displayed in bolder print (labelled at position 315 in the cocrystal). The positions of the catalytic triad residues in the active site of subtilisin are also indicated (His64, Glu156, Ser221). The figure is drawn from the structure determined by McPhalen *et al.* (1985).

lethal variant of this protein, α-1-antitrypsin Pittsburgh, has an arginine at the P1 position instead of the usual methionine residue. This single amino acid change alters specificity from elastase towards thrombin (Owen *et al.* 1983). Site-directed mutagenesis of the gene for α-1-antitrypsin has been used to confirm this finding, and extensive mutagenesis of the P1 residue has produced functional variants of the inhibitor (Courtney *et al.* 1985). For example, with a leucine at the P1 site the inhibitor binds more strongly to elastase whilst also inhibiting cathepsin G, but with phenylalanine in place the protein is specific for cathepsin G.

The studies presented here further the work on serine protease inhibitors by tailoring the specificity of a plant protease inhibitor. CI-2 is targeted towards a panel of serine proteases through changing only the P1 residue in the inhibitor. The altered binding properties of the mutant inhibitors form a basis for the investigation of enzyme–inhibitor interactions.

## Design and synthesis of mutants

CI-2 is known to strongly inhibit subtilisin and chymotrypsin, whilst having no inhibitory effect upon trypsin. The first objective was to increase the potency of the inhibitor towards chymotrypsin by the substitution of a tyrosine for the methionine at the P1 site. Tyrosine was chosen because chymotrypsin exhibits a preference for bulky hydrophobic side chains at this position in substrates (Stroud *et al.* 1974). Again, using known substrate preferences, a new inhibitor of trypsin was designed by replacing methionine with the positively charged lysine, whilst elastase was targeted by placing an alanine at the P1 site (Shotton and Hartley 1970). Enzymes and their preferred substrates are tabulated (Table 13.1).

Site-directed mutagenesis was performed on the protein coding region of the cDNA clone for the gene (Williamson *et al.* 1987) that had been transferred into the bacteriophage M13mp8. The method of double priming was used (Zoller and Smith 1984), with mutagenic oligonucleotides of 21 bases and using the M13 universal primer acting as the 5' protecting oligonucleotide. On synthesis of

**Table 13.1**   Enzyme specificity

| Enzyme | Preferred P1 residue |
| --- | --- |
| Subtilisin | Large hydrophobics (broad specificity) |
| Chymotrypsin | Hydrophobics |
| Trypsin | Basic (Lys, Arg) |
| Elastase | Small/medium sized hydrophobics |

the second strand, the phage DNA was propagated in the mismatch repair deficient strain of *Escherichia coli* BmH71.18mutL (Kramer *et al.* 1984). Colonies harbouring mutant copies of the gene were identified by hybridization screening using a radiolabelled form of the mutagenic oligonucleotide as probe (Maniatis *et al.* 1982). Mutations were confirmed by fully sequencing the gene, to ensure that no additional, unwanted, mutations had occurred.

The well characterized bacterium, *E. coli*, was chosen as the host for the high level expression of the gene product (the strain used was TG2). The gene encoding CI-2 was expressed from the vector pINIIIompA3 (Ghrayeb *et al.* 1984) (Fig. 13.5). This vector is based upon pBR 322, expression of the foreign gene is directed from a strong promoter, the lpp promoter. Expression is regulated through the lac promoter/operator system, and is induced upon addition of isopropyl-β-D-thiogalactoside (IPTG) to the growth medium (2TY). The CI-2 gene was placed immediately downstream of the ompA signal peptide sequence, and the protein produced was a fusion between ompA and CI-2. OmpA is a periplasmic protein and so directs secretion of the nascent protein into the periplasmic space, where the signal peptide is cleaved to release intact CI-2 (Lehnhardt *et al.* 1987). This fusion ensures a high level of CI-2 expression in a system that has already been optimized for the signal peptide. In addition, secretion

protects the protein from endogenous proteases that are present within the intracellular compartment and which might normally degrade the small inhibitor. Mutants of CI-2 were similarly expressed using this system.

Varying yields of the different CI-2s were found. Both wild type and the lysine mutant (CI-2MK59) were expressed well and initial yields were, routinely, greater than 20 mg/litre. However, the tyrosine and alanine mutants were expressed very poorly, to 0.1 mg/litre, in the same strain. Interestingly, yields of CI-2MA59 were increased to 15 mg/litre when transferred to the *E. coli* strain JM101, a phenomenon that cannot be accounted for solely by the faster growth of this strain.

## Purification and characterization

CI-2 (and its mutants) were purified from cellular proteins using a procedure that was developed to exploit its properties, such as its small size and extremely stable nature. After concentration of the cells, the whole cell contents were released by sonication. As CI-2 lies within the periplasm it was possible to rupture only the outer membrane by a process such as osmotic shock (Neu and Heppel 1965). However, yields were no higher than after sonication (some protein was found to remain membrane-bound) and practical problems were also encountered with the volumes to be handled in subsequent steps. Thus the pINIIIompA vector was used only as a high level expression vector.

The first step after sonication was to lower the pH of the solution to pH 4.4 with 1 M acetic acid. CI-2 had been found to be stable down to pH 1 and so the alteration of pH was used to precipitate many unwanted proteins whilst leaving CI-2 intact and in solution. The protein was further purified and concentrated by a series of ammonium sulphate precipitations, and, upon gel filtration, a fraction containing only low molecular weight protein was obtained. CI-2 was then isolated in a pure form by passing through an anion exchange, fast protein liquid chromatography (FPLC) column.

Before starting binding studies, the wild type protein was verified as authentic CI-2 in a series of comparisons against native CI-2, prepared from barley grains (a gift from F. Poulsen, Carlsberg Laboratories, Denmark). Both proteins had the same molecular weight as judged by SDS-PAGE, indicating that the recombinant protein had been correctly processed. Furthermore, N-terminal se-

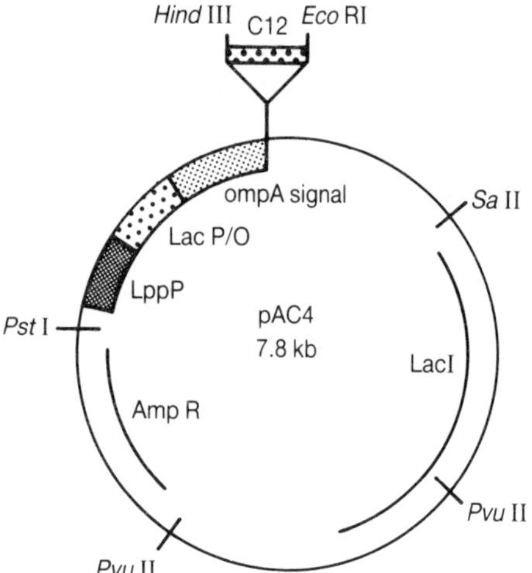



**Fig. 13.5**  Secretion vector used to express CI-2wt. Mutant CI-2 genes were also expressed from this vector, inserted as *Hind*III/*Eco*RI gene cassettes.

quencing revealed that the first eight residues of the recombinant protein corresponded to the published sequence of CI-2, SerSerValGluLysLysProGlu (Svendsen *et al.* 1980). Isoelectric focusing of the two proteins showed a single band for the recombinant CI-2 with a pI of 6.8, whilst there were six forms of the native protein. These have been noted previously and are thought to be due to N-terminal degradation of the protein during purification (Svendsen *et al.* 1980). Finally, both proteins reacted with an antiserum raised against CI-2 and produced precipitant arcs of antibody–antigen complex when subjected to immunodiffusion assay (Ouchterlony 1958).

# Kinetic analysis

Accurate kinetic methods were developed to study the enzyme–inhibitor interactions in detail. These permitted both the formation and the dissociation of complex to be followed and were applied to all the enzyme–inhibitor systems studied but were developed based on the interaction of wild type CI-2 with subtilisin BPN'. To illustrate how meaningful data could be obtained, the principles behind the analysis are explained below.

## Slow binding inhibition

CI-2 is a competitive inhibitor of subtilisin and binds tightly to the enzyme, with an equilibrium constant ($K_i$) for this reaction in the order of nM (Svendsen *et al.* 1980). The inhibition is represented as follows:

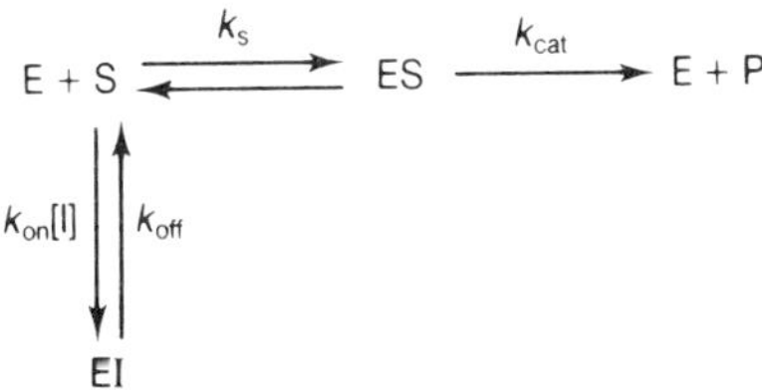

$K_i$ for the reaction is dependent on both the association rate constant ($k_{on}$) and the dissociation rate constant ($k_{off}$) for the equilibrium. That is, the strength of inhibition is a reflection of the formation and of the breakdown of the enzyme–inhibitor complex and

$$K_i = k_{off}/k_{on}$$

Therefore, it is not sufficient to measure $K_i$ alone to determine the events occurring at the molecular level.

In binding tightly to subtilisin, CI-2 can also be classed as a slow binding inhibitor, according to Williams and Morrison (1979) and Cha (1975). Inhibition can be described as slow binding if the approach to equilibrium for the inhibition can be observed. In the case of CI-2 and subtilisin, provided that concentrations of the enzyme and inhibitor are kept low, the steady state inhibited rate is established over a period of hours. Thus, the setting up of equilibrium can be followed spectrophotometrically as the decrease in the rate of substrate hydrolysis as more enzyme–inhibitor complex is formed. For the simplest treatment, the reaction is maintained as first order where the lowest inhibitor concentration is at least ten times the enzyme concentration. The enzyme concentration is kept low, in the order of nM, but at a sufficient level for substrate hydrolysis. The substrate releases a chromogenic moiety upon hydrolysis, and the corresponding change in absorbance can be followed. The concentration of substrate used is high, to ensure that it is not depleted over the long time course of the reaction. The high substrate concentration also prevents autolysis of the enzyme.

## Complex association

A mixture of inhibitor and substrate was allowed to equilibrate in the cuvette at the desired temperature for the reaction (25°C). Enzyme was added to initiate the reaction, which was immediately recorded. A high rate of substrate hydrolysis was observed initially, but the rate of product formation slowed over time as the equilibrium for inhibition was reached. For CI-2 and subtilisin BPN' the reaction was followed for over 16 h. A series of inhibitor concentrations was used, ranging from 1 to 13 nM, and the concentration of enzyme in the reaction was 0.25 nM. The substrate used was 1 mM succinyl-AlaAlaProPhe-*p*-nitrophenol (sAAPF*p*NA) and the reactions were performed in 0.1 M Tris–HCl; pH 8.6. A family of curves was produced, relating to the different inhibitor concentrations, and the approach to equilibrium can be clearly seen in Figure 13.6.

Slow binding can be described in terms of the initial and steady state rates in the following

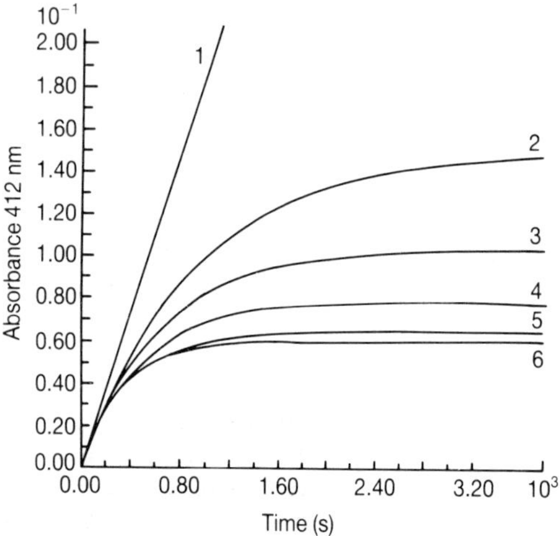

**Fig. 13.6**  Family of slow binding inhibition curves for subtilisin BPN' and CI-2. Uninhibited enzyme is represented by line 1 and each curve shows the time course of enzyme inhibition with increasing inhibitor concentrations (5, 7, 9, 11 and 13 nM; curves numbered 2–6, respectively). Reaction was initiated by the addition of enzyme solution (0.25 ml) to inhibitor and substrate (0.75 ml), conditions used were as described in the text. Data were collected for 16.6 h, only the early stages of the reaction are shown for clarity.

equation (Cha 1975):

$$v = v_{\mathrm{s}} + (v_{\mathrm{o}} - v_{\mathrm{s}})\, e^{-k't}$$

where $v$ represents the rate of reaction in time, $t$; $v_{\mathrm{o}}$ and $v_{\mathrm{s}}$ are the initial and final rates respectively; and $k'$ is the apparent first order rate constant for the transition to the steady state. Data for the progress of the inhibition reactions are fitted to the integrated form of this equation and values for $k_{\mathrm{on}}$ and $k_{\mathrm{off}}$ can be determined from a plot of $k'$ against [I].

This approach was applied to all enzyme–inhibitor interactions, although only the equilibrium established between CI-2wt and subtilisin BPN' can be described as true slow binding, in that initial rates of reaction are independent of the inhibitor concentration. However, although the approach to the steady state could not be followed in the other cases, the data could still be fitted to the same equations. Values for $k'$ varied within each experiment, as expected, as the transition from presteady-state to steady-state was not observed. Hence, a value for $k_{\mathrm{on}}$ could not be determined. This method ensured that the steady state was reached before the $K_{\mathrm{i}}$ was determined, which has not always been appreciated in determining

inhibition constants. Following the approach to equilibrium also permitted any deviations to be observed, such as sudden increase in proteolysis which might indicate cleavage of the inhibitor.

## Complex dissociation

The dissociation rate constant can also be directly determined by dilution of the preformed enzyme–inhibitor complex, below its apparent $K_{\mathrm{i}}$, into substrate. Dissociation needs to occur on a measurable time scale, as is the case for the complex formed between CI-2wt and chymotrypsin. A direct determination of $k_{\mathrm{off}}$ for CI-2wt and subtilisin BPN' was not possible because the dissociation occurred too slowly. Practically, the complex (1–2 nM) formed in the presence of excess inhibitor was diluted 500-fold into 1 mM substrate, and the release of free enzyme was followed as the substrate was hyrolysed. The rate of appearance of product increases as the complex slowly dissociates, as is seen in Figure 13.7. The substrate used was as for the subtilisin assays, and the buffer was 0.144 M Tris; pH 7.78.

The data were fitted to an integrated form of the rate equation for complex dissociation which states:

$$[E] = [E_{\mathrm{o}}](e^{-k_{\mathrm{off}}t})$$

where [E] is the enzyme concentration at time $t$;

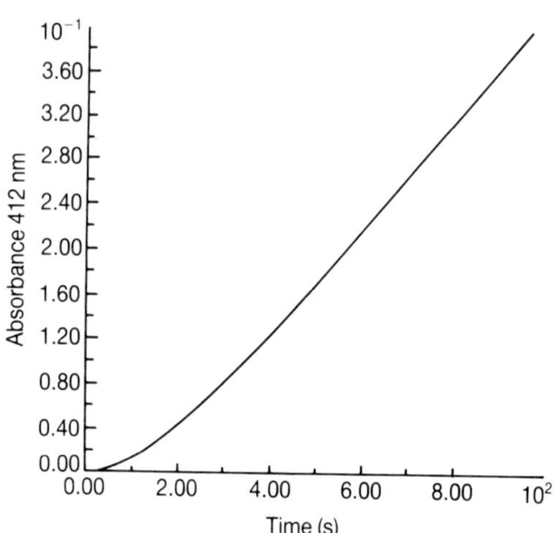

**Fig. 13.7**  Dissociation of CI-2wt–chymotrypsin complex. Preformed complex was diluted into substrate and complex dissociation was followed by monitoring release of free enzyme over 20 min.

[$E_o$] represents the initial enzyme concentration; and $k_{off}$ is the dissociation rate constant.

## Engineering specificity

Wild type CI-2 was found to strongly inhibit subtilisin BPN' with a $K_i$ of 2.9 pM. Although still a good inhibitor of chymotrypsin, the $K_i$ for the binding of CI-2wt to this enzyme is 1.6 nM. By dissecting the individual rate constants for the equilibrium, it was found that the weaker inhibition of chymotrypsin was a result of more rapid dissociation of the enzyme–inhibitor complex, $k_{off}$ was $5.2 \times 10^{-6}\,\mathrm{s}^{-1}$ for the inhibition of subtilisin BPN' and $6.1 \times 10^{-3}\,\mathrm{s}^{-1}$ for the CI-2–chymotrypsin complex. The association rate constants for these reactions were not significantly different. All results are presented in Table 13.2.

A tyrosine residue was substituted into the reactive site of the inhibitor, at the P1 position, to increase the strength of inhibition of chymotrypsin. As chymotrypsin exhibits a preference for substrates with this residue in this position, a more potent inhibitor of the enzyme was successfully generated. The $K_i$ for the equilibrium was reduced to 0.47 nM and this was primarily due to a reduction in the rate of dissociation for the complex, $k_{off}$ was decreased 6.5-fold in comparison to the CI-2wt–chymotrypsin interaction.

A new inhibitor of trypsin was designed by replacing Met59 with lysine. Lysine is a basic isostere of methionine and so the effect of a charged residue in this side chain could be appreciated. CI-2wt does not inhibit trypsin but appears to be non-specifically hydrolysed within the active site of the enzyme. However, CI-2MK59 is a strong inhibitor of trypsin, with a $K_i$ of 5.6 nM, although the reaction follows classical inhibition kinetics rather than slow binding kinetics. Upon prolonged in-

cubation of trypsin and CI-2MK59 (greater than 16 h) breakdown of the inhibitor into a small number of discrete fragments can be observed on SDS-PAGE analysis. Despite this, it can be said that a novel inhibitor of trypsin has been created and that non-specific proteolysis of the inhibitor has been reduced.

Kinetic analysis showed that CI-2wt was a reasonable inhibitor of elastase with a $K_i$ of approximately 30 nM. The $K_i$ was more difficult to determine for these reactions because the enzyme is inherently unstable and more than 60% of enzyme activity was lost over a 75 min experiment. However, the presence of CI-2 has a stabilizing effect on elastase and, over the time course of the experiments performed, the rate of reaction in the absence of inhibitor was eventually exceeded by the rates in the presence of CI-2. This finding may have clinical applications if proteolytic enzymes can be stabilized *in vivo* by exogenous inhibitors.

Elastase was believed to have specificity for small non-polar residues at the P1 site and hence alanine was substituted for Met59 (Shotton and Hartley 1970). Unexpectedly, the inhibition of elastase was found to be 20-fold weaker than in the presence of wild type CI-2, the $K_i$ for the reaction being 600 nM. The mutant was still tight binding and it was possible to construct stoichiometric binding curves for the interaction of both CI-2wt and CI-2MA59 with elastase. Again, the phenomenon of enzyme stabilization was observed in the presence of CI-2MA59. These findings suggest that the weaker inhibition is unlikely to be due to gross structural perturbation, and a more likely explanation is that the specificity of elastase may depend on more subsites than P1. In this case, only small structural perturbations in CI-2MA59 could cause an overall decrease in the binding strength. This conclusion supports the finding of Jallat *et al.* (1986), in the context of the $\alpha$-1-antitrypsin

**Table 13.2**   Kinetically determined inhibition constants

| Inhibitor | $K_i$ (nM) | | | |
| --- | --- | --- | --- | --- |
| | Subtilisin BPN'[a] | Chymotrypsin[b] | Trypsin[c] | Elastase[d] |
| CI2wt(Met59) | 0.0029 | 1.6[g] | hydrolysis[e] | ~30 |
| CI2MK59 | hydrolysis[e] | ≫10 | 5.6 | ND |
| CI2MY59 | ND[f] | 0.49[h] | ND | ND |
| CI2MA59 | ND | ND | ND | ~600 |

[a] 25°C, 0.1 M Tris (pH 8.6), 0.05% Tween, succinyl-Ala-Ala-Pro-Phe-*p*-nitroanilide as substrate; [b] 25°C, 144 mM Tris (pH 7.78), sAAPFpNA as substrate; [c] as [b] but L-BAPNA as substrate, [d] 200 mM Tris (pH 8.0), succinyl-Ala-Ala-Ala-*p*-nitroanilide as substrate; [e] non-specific hydrolysis; [f] not determined; [g] $k_{on} = 3.8 \times 10^6\,\mathrm{M}^{-1}\,\mathrm{s}^{-1}$, $k_{off} = 6.1 \times 10^{-3}\,\mathrm{s}^{-1}$; [h] $k_{on} = 2.0 \times 10^6\,\mathrm{M}^{-1}\,\mathrm{s}^{-1}$, $k_{off} = 9.3 \times 10^{-4}\,\mathrm{s}^{-1}$.

inhibitor, where an elastase inhibitor was created by placing a valine at the P1 position. The P3 residue in this case was an isoleucine, but when it was replaced with alanine the degree of inhibition was greatly reduced, indicating that other members of the reactive site loop are involved in inhibition.

## Inhibitory mechanism

Many slow binding inhibitors have been found to result from a very low value of $k_{on}$ and it has been suggested that these inhibitors induce a conformational change upon binding (Morrison and Walsh 1988). Analysis of slow binding data can, therefore, be used to provide information on the mechanism of binding (Longstaff *et al.* 1990). The association rate constant for inhibition by CI-2 is not very low, in fact it is within two orders of magnitude of the diffusion controlled limit for association of large molecules. Results at the inhibitor concentrations studied suggest that a simple bimolecular association of enzyme and inhibitor occurs:

$$E + I \underset{k_{off}}{\overset{k_{on}}{\rightleftharpoons}} EI$$

However, it is possible that the complex undergoes conformational change upon association of enzyme and inhibitor, in which case the following reaction scheme would apply:

$$E + I \underset{k_{-1}}{\overset{k_1}{\rightleftharpoons}} EI \underset{k_{-2}}{\overset{k_2}{\rightleftharpoons}} EI'$$

This would give the same kinetics as the former mechanism if the inhibitor concentrations were not saturating. To verify this, further experiments would need to be designed because it was not possible to increase the inhibitor concentrations any further in the methods described.

It has long been accepted that inhibitors bind to enzymes in the manner of a good substrate (Read *et al.* 1983; McPhalen and James 1988; Bode *et al.* 1989). They can bind tightly, forming a network of interactions within the active site. The process subsequent to binding, the very slow turnover of the inhibitor, remains to be addressed.

During enzymatic catalysis, hydrolysis of the substrate is facilitated by a lowering of the energy barrier for this reaction (Fersht 1985). This is done by distorting the conformation of the substrate, through interactions formed between substrate and enzyme residues in the active site. The substrate adopts the labile tetrahedral transition state, which is more easily attacked by the nucleophilic hydroxyl residue (Fig. 13.8a). However, when the inhibitor binds in the active site of the serine protease a complex hydrogen bonding network is established and this may mean that favourable non-covalent bonds must be broken to form the enzyme–transition state complex. This is energetically unfavourable and so may account for the slow hydrolysis (Fig. 13.8b). This theory is supported by most of the crystallographic structures solved for the serine protease–inhibitor complexes, where the inhibitors are shown to be structurally similar in conformation to their unbound state (Huber and Bode 1978; Fujinaya *et al.* 1982; Chen and Bode 1983; McPhalen and James 1987).

The mechanism of hydrolysis involves a series of

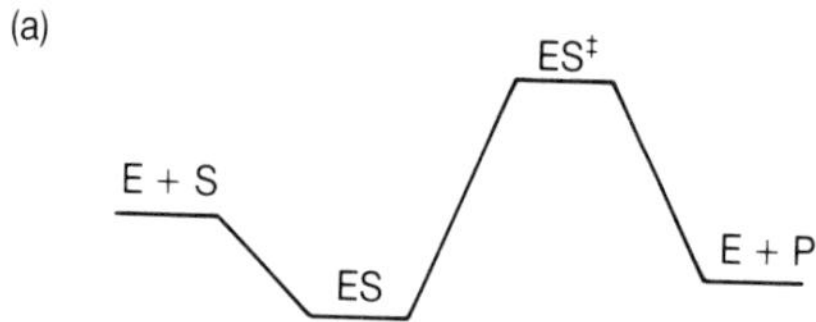



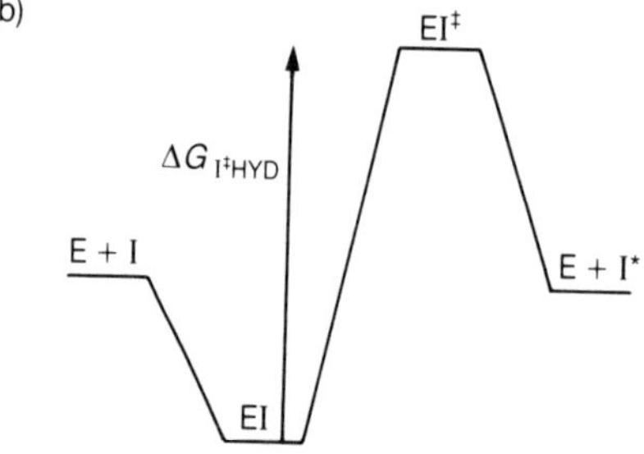



**Fig. 13.8**   Reaction profile for the interaction of a serine protease with (a) substrate and (b) protein inhibitor. The inhibitor binds tightly, leading to a stable EI complex, equivalent to the Michaelis complex. The energy of the transition state (EI⁺) is raised due to inherent inhibitor stability. Thus the energy barrier to hydrolysis (ΔG⁺hyd) is large, and production of cleaved inhibitor (I*) is not favoured. Thus inhibition can be achieved through two mechanisms: (i) lowering the energy of the Michaelis complex through enzyme–inhibitor complementarity; (ii) raising the transition state through inflexibility of the peptide in the substrate binding cleft. The EI complex is thus in a 'thermodynamic pit' with a large barrier preventing hydrolysis.

intermediates. These lead to the formation of the transition state and are also formed between the transition state and the release of product (Fersht 1985). The rigidity of the inhibitor within the complex could aid the slow rate of hydrolysis. Upon peptide cleavage, an amino group, which moves to another site in the complex, is released. When the inhibitor is cleaved, the released amino group could remain attached to the enzyme inhibitor complex in such a position that it then could attack the next intermediate in the pathway of hydrolysis, the acyl enzyme. In this way, an equilibrium between cleaved and virgin inhibitor would be established, which favours the intact inhibitor and hence slows the overall rate of hydrolysis. These proposals were investigated by the study of active site mutants of subtilisin inhibited by CI-2. Possessing broad specificity (Kraut 1977), a residue within the active site of subtilisin has been identified as important in substrate binding (Wells *et al.* 1987b). This residue is found at position 156, and is glutamate in subtilisin BPN', but serine in subtilisin Carlsberg. A mutant of subtilisin BPN' (ES156) also possesses a serine at position 156 (Russell and Fersht 1987). Interactions of CI-2wt (Met59) and CI-2MK59 with these three forms of the enzyme were studied, using the kinetic methods detailed earlier.

Slow binding type inhibition was found in most cases and equilibrium constants were determined (Table 13.3). CI-2MK59 was a poorer inhibitor of all three enzymes, although the inhibitor was no longer hydrolysed when complexed with subtilisin Carlsberg or subtilisin BPN'ES156. Deviations from the steady-state rate were observed for the interaction of CI-2MK59 with subtilisin BPN', suggesting that hydrolysis of the inhibitor was occurring. This was confirmed upon PAGE of the complex, and a discrete fragment, of molecular weight 6000, was visualized. Hydrolysis cannot be due to a structural alteration in the inhibitor because it only occurs in the presence of subtilisin BPN'. It has been shown structurally (Poulos *et al.* 1976) and kinetically (Kraut 1977; Wells *et al.* 1987a) that substrates with hydrophobic residues at P1 bind differently to subtilisin BPN' than those possessing a lysine at the P1 position. The hydrophobic side chains point towards a glycine at position 166 in the enzyme active site, whilst the lysine side chain can form a salt bridge with Glu156. So, on binding to subtilisin BPN', the P1 residue of CI-2MK59 can interact with Glu156, introducing stress or strain into the binding loop of the inhibitor. The subsequent small structural alterations in the enzyme-bound complex could perturb the difference in energy levels between the enzyme–inhibitor complex and the transition state. This would reduce the energy barrier to hydrolysis and/or allow the released amino group to move further from the acyl enzyme, thus shifting the equilibrium towards the cleaved inhibitor.

The importance of loop rigidity is apparent from inhibition studies using CI-1, a protein closely related to CI-2 and which lacks one of the two stabilizing arginines (Arg67), responsible for holding the reactive site loop in place (Boisen *et al.* 1981). CI-1 is a poorer inhibitor of subtilisin than CI-2 and is more readily cleaved, presumably due to loss of rigidity and stability (Svendsen *et al.* 1982).

# Conclusions

This protein engineering study on CI-2 is the first to redesign the specificity of a plant protease inhibitor. The results indicate that one amino acid change can significantly alter the specificity towards serine proteases, thus confirming the importance of the residue at the P1 position in the reactive site. A clear example of this effect is the creation of a stronger inhibitor of chymotrypsin upon altering the P1 residue from methionine to tyrosine, and so reducing the rate of dissociation of the enzyme–inhibitor complex.

It was possible to target a new serine protease, trypsin, with the mutant CI-2MK59, designed through consideration of known substrate preferences of this enzyme. In addition to creating a strong trypsin inhibitor, non-specific hydrolysis was eliminated, although the inhibitor was still found to be specifically cleaved, presumably at the reactive site peptide bond.

Tailoring the inhibitory specificity may not always be so straightforward as mutagenesis of the P1

**Table 13.3**  Inhibition of subtilisins with CI-2wt and CI-2MK59

| Subtilisin | CI-2wt | | CI-2MK59 | |
| --- | --- | --- | --- | --- |
|  | $K_i$ (nM) | $k_{on}$ (s$^{-1}$ M$^{-1}$ × 10$^{-6}$) | $K_i$ (nM) | $k_{on}$ (s$^{-1}$ M$^{-1}$ × 10$^{-6}$) |
| BPN' | 2.9 | 1.9 | hydrolysis | |
| Carlsberg | 12.6 | a | 25 | 1.1 |
| BPN'ES156 | 3.2 | 3.2 | 66 | 0.5 |

[a] True slow binding kinetics were not observed.

residue. This residue obviously plays a significant part in determining specificity but additional reactive site residues may be involved in subsite binding to the enzyme. For example, effective inhibitors of elastase require a valine at the P3 position (Morrison and Walsh 1988). The CI-2 work concentrated only on modifications at the P1 site, it would be anticipated that, with further changes within the reactive site loop, more subtle alteration in inhibitory properties could be achieved.

It must also be appreciated that, when placed at the P1 site, certain residues may cause a slight perturbation in the structure of the reactive site loop, which in turn will lead to a decrease in the strength of binding and inhibition of the target enzyme. This phenomenon occurred with the mutant CI-2MA59, which was designed as an effective inhibitor of elastase, but which proved to be a weaker inhibitor of this enzyme than CI-2wt. Thus, the structural integrity of the reactive site loop must be maintained if a more potent inhibitor is desired.

In addition to tailoring specificity the work on CI-2 has enabled a more detailed understanding of the mechanism of serine protease inhibition. The evidence would suggest that strong inhibition is a reflection of the stability of the inhibitor within the reactive site of the enzyme, such that peptide bond hydrolysis becomes energetically unfavourable. Cleavage of the inhibitor within the active site may occur, but it is not a prerequisite for strong inhibition.

This work may have clinical implications. Of therapeutic interest are the plasma serine protease inhibitors (serpins) such as $\alpha$-1-antiprotease, antithrombin III, $\alpha$-1-antichymotrypsin and C1 inhibitor. It is known that some of these inhibitors are cleaved upon binding and that the structure of the virgin and modified (cleaved) inhibitor can be very different with the termini far apart (Bruch *et al.* 1988). However, it is possible that the serpin–protease complexes would have similar structures to other serine protease inhibitor–protease complexes, as cleavage is a slow event and it is the initial binding and inhibition prior to cleavage that is biologically important. The initial binding can be accounted for by the processes of energy barriers described previously. Crystal structures will be needed to prove this theory, but if it is true then activity modifications could be designed based on studies such as outlined for CI-2.

# References


Bode, W., Meyer, E.J. Jr. and Powers, J.C. (1989). Human leukocyte and porcine pancreatic elastase: X-ray crystal structures, mechanism, substrate specificity and mechanism-based inhibitors. *Biochemistry.* **28**: 1951–63.

Boisen, S., Anderson, C.Y. and Heigaard, J. (1981). Inhibitors of chymotrypsin and microbial serine proteases in barley grains. *Physiologia Plantarum.* **52**: 167–76.

Bruch, M., Weiss, V. and Engel, J. (1988). Plasma serine proteinase inhibitors (serpins) exhibit major conformational changes and a large increase in conformational stability upon cleavage at their reactive sites. *Journal of Biological Chemistry.* **263**: 16626–30.

Carrell, R.W., Jeppsson, J.-O., Laurell, C.B., Brennan, S.D., Owen, M.C., Vaughan, L. and Boswell, D.R. (1982). Structure and variation of human $\alpha$1-antitrypsin. *Nature.* **298**: 329–34.

Cha, S. (1975). Tight-binding inhibitors. I. Kinetic behaviour. *Biochemical Pharmacology.* **24**: 2177–85.

Chen, Z. and Bode, W. (1983). Refined 2.5A X-ray crystal structure of the complex formed by porcine kallikrein A and the bovine pancreatic trypsin inhibitor. Crystallisation, Patterson search, structure determination, refinement, structure and comparison with its components and the bovine trypsin–pancreatic trypsin inhibitor complex. *Journal of Molecular Biology.* **164**: 283–311.

Clore, G.M., Gronenborn, A.M., James, M.N.G., Kjaer, M., McPhalen, C.A. and Poulsen, F.M. (1987a). Comparison of the solution and X-ray structures of barley serine proteinase inhibitor 2. *Protein Engineering.* **1**: 313–18.

Clore, G.M., Gronenborn, A.M., Kjaer, M. and Poulsen, F.M. (1987b). The determination of the three dimensional structure of barley serine proteinase inhibitor 2 by nuclear magnetic resonance, distance geometry and restrained molecular dynamics. *Protein Engineering.* **1**: 305–11

Courtney, M., Jallat, S., Tessier, L.-H., Benavente, A., Crystal, R.G. and Lecocq, J.-P. (1985). Synthesis in *E. coli* of $\alpha$1-anti-trypsin variants of therapeutic potential for emphysema and thrombosis. *Nature.* **313**: 149–51.

Fersht, A.R. (1985). In *Enzyme Structure and Mechanism*, Vol. 2, (2nd edn). W.H. Freeman and Co., New York.

Fujinaya, M., Read, R.J., Sielecki, A., Ardelt, W., Laskowski, M. Jr. and James, M.N.G. (1982). Refined crystal structure of the molecular complex of *Streptomyces griseus* protease B, a serine protease, with the third domain of the ovomucoid inhibitor from turkey. *Proceedings of the National Academy of Sciences, USA.* **79**: 4868–72.

Ghrayeb, J., Kimura, H., Takahara, M., Hsiung, H., Masui, Y. and Inouye, M. (1984). Secretion cloning vectors in *E. coli*. *European Molecular Biology Organisation (EMBO) Journal*. **3**: 2437–42.

Graham, J.S., Pearce, G., Merryweather, J., Titani, K., Ericsson, L. and Ryan, C.A. (1985). Wound-induced proteinase inhibitors from tomato leaves. *Journal of Biological Chemistry*. **260**: 6555–60.

Haynes, R. and Feeney, R.E. (1968). Transformation of active site lysine in naturally occurring trypsin inhibitors. A basis for a general mechanism for inhibition of proteolytic enzymes. *Biochemistry*. **7**: 2879–85.

Huber, R. and Bode, W. (1978). Structural basis of activation and action of trypsin. *Accounts of Chemical Research*. **11**: 114–22.

Jallat, S., Carvallo, D., Tessier, L.-H., Roecklen, D., Reitsch, C., Oguski, F., Crystal, R.G. and Courtney, M. (1986). Altered specificities of genetically engineered α1-anti-trypsin variants. *Protein Engineering*. **1**: 29–35.

Jonassen, I. and Svendsen, I. (1982). Identification of the reactive sites in two homologous serine proteinase inhibitors isolated from barley. *Carlsberg Research Communications*. **47**: 199–203.

Kramer, B., Kramer, W. and Fritz, H.-J. (1984). Differential base/base mismatches are correlated with different efficiencies by the methyl-directed mismatch-repair system of *E. coli*. *Cell*. **38**: 879–87.

Kraut, J. (1977). Serine proteases: structure and mechanism of catalysis. *Annual Reviews of Biochemistry*. **46**: 331–58.

Laskowski, M. Jr. and Kato, I. (1980). Protein inhibitors of proteinases. *Annual Reviews of Biochemistry*. **49**: 593–626.

Lehnhardt, S., Pollitt, S. and Inouye, M. (1987). The differential effect on two hybrid proteins of deletion mutations within the hydrophobic region of the *E. coli* OmpA signal peptide. *Journal of Biological Chemistry*. **262**: 1716–19.

Longstaff, C., Campbell, A.F. and Fersht, A.R. (1990). Recombinant chymotrypsin inhibitor 2 (CI2): expression, kinetic analysis of inhibition with α-chymotrypsin and wild-type and mutant subtilisin BPN', and protein engineering to investigate inhibitory specificity and mechanism. *Biochemistry*. **29**: 7339–47.

Maniatis, T., Fritsch, E.F. and Sambrook, J. (1982). In *Molecular Cloning: a laboratory manual*. Cold Spring Harbor Laboratory, New York.

McPhalen, C.A. and James, M.N.G. (1987). Crystal and molecular structure of the serine proteinase inhibitor CI2 from barley grains. *Biochemistry*. **26**: 261 9.

McPhalen, C.A. and James, M.N.G. (1988). Structural comparison of two serine proteinase–protein inhibitor complexes: eglin c-subtilisin Carlsberg and CI2-subtilisin Novo. *Biochemistry*. **27**: 6582–98.

McPhalen, C.A., Svendsen, I., Jonassen, I. and James, M.N.G. (1985). Crystal and molecular structure of chymotrypsin inhibitor 2 from barley seeds in complex with subtilisin Novo. *Proceedings of the National Academy of Sciences, USA*. **82**: 7242–6.

Morrison, J.F. and Walsh, C.T. (1988). The behaviour and significance of slow-binding enzyme inhibitors. *Advances in Enzymology*. **61**: 201–301.

Neu, H.C. and Heppel, L.A. (1965). The release of enzymes from *E. coli*. by osmotic shock and during the formation of spheroplasts. *Journal of Biological Chemistry*. **240**: 3655–92.

Ouchterlony, O. (1958). In *Progress in Allergy*. **3**: 1–78. (ed. P. Kallos). Karger, Basel.

Owen, M.C., Brennan, S.O., Lewis, J.H. and Carrell, J.W. (1983). Structure and variation of human α1-anti-trypsin. *New England Journal of Medicine*. **309**: 694–8.

Poulos, T.L., Alden, R.A., Freer, S.T., Birktoft, J.J. and Kraut, J. (1976). Polypeptide halomethyl ketones bind to serine proteases as analogs of the tetrahedral intermediate. X-ray crystallographic comparison of lysine- and phenylalanine-polypeptide chloromethyl inhibited subtilisin. *Journal of Biological Chemistry*. **251**: 1097–103.

Read, R.J., Fujinaga, M., Sielecki, A.R. and James, M.N.G. (1983). Structure of the complex of *Streptomyces griseus* protease B and the third domain of the turkey ovomucoid inhibitor at 1.8 Å resolution. *Biochemistry*. **22**: 4420–33.

Russell, A.J. and Fersht, A.R. (1987). Rational modification of enzyme catalysis by engineering surface charge. *Nature*. **328**: 496–500.

Ryan, C.A. (1989). Proteinase inhibitor gene families: strategies for transformation to improve plant defenses against herbivores. *Bioessays*. **10**: 20–4.

Schecter, T. and Berger, A. (1967). On the size of the active site in proteases, I. Papain. *Biochemical and Biophysical Research Communications*. **27**: 157–62.

Seemuller, U., Meier, M., Ohlsson, K., Muller, H.-P. and Fritz, H. (1977). Isolation and characterisation of a low molecular weight inhibitor (of chymotrypsin and human granulocytic elastase and cathepsin G) from leeches. *Hoppe-Seyler's Zeitschrift für Physiologische Chemie*. **358**: 1105–17.

Shotton, D.M. and Hartley, B.S. (1970). Three-dimensional structure of tosyl-elastase. *Nature*. **225**: 802–6.

Stroud, R.M., Kay, L.M. and Dickerson, R.E. (1974). The structure of bovine trypsin: electron density maps of the inhibited enzyme at 5 Å and at 2.7 Å resolution. *Journal of Molecular Biochemistry*. **83**: 185–208.

Svendsen, I. and Jonassen, I., Hejgaard, J. and Boisen, S (1980). Amino acid sequence homology between a serine protease inhibitor from barley and potato inhibitor I. *Carlsberg Research Communications*. **45**: 389–95.

Svendsen, I., Boisen, S. and Hejgaard, J. (1982). Amino acid sequence of serine protease inhibitor CI1 from barley, homology with barley inhibitor CI2, potato inhibitor I and leech eglin. *Carlsberg Research Communications*. **47**: 45–53

Wells, J.A., Cunningham, B.C., Graycar, T.P. and

Estell, D.A. (1987a). Recruitment of substrate specificity properties from one enzyme into a related one by protein engineering. *Proceedings of the National Academy of Sciences, USA.* **84:** 5167–71.

Wells, J.A., Powers, D.B., Bott, R.R., Graycar, T.P. and Estell, D.A. (1987b). Designing substrate specificity by protein engineering of electrostatic interactions. *Proceedings of the National Academy of Sciences, USA.* **84:** 1219–23.

Williams, J.W. and Morrison, J.F. (1979). The kinetics of reversible tight-binding inhibition. *Methods in Enzymology.* **63:** 437–67.

Williamson, M., Forde, J., Buxton, B. and Kreis, M. (1987). Nucleotide sequence of barley chymotrypsin inhibitor 2 (CI2) and its expression in normal and high-lysine barley. *European Journal of Biochemistry.* **165:** 99–106.

Zoller, M. and Smith, M. (1984). Oligonucleotide-directed mutagenesis: a simple method using two oligonucleotide primers and a single-stranded DNA template. *DNA.* **3:** 478–88.

# 14
# Redesigning Ricin for Therapeutic Purposes

J. Michael Lord, Jane H. Gould, Peter T. Richardson, Robert A. Spooner, Richard Wales and Lynne M. Roberts

## Introduction

Many plants contain ribosome inactivating proteins (RIPs) (Stirpe and Barbieri 1986). Type 1 RIPs are single chain polypeptides, which are often glycosylated, of molecular weight about 30 000 Da. These proteins catalytically inactivate protein synthesis on eukaryotic ribosomes by hydrolysis of the N-glycosidic bond of a specific adenine residue present in 26–28S ribosomal RNA (Endo and Tsurugi 1987; Endo et al. 1988). The target adenine residue (adenine 4324 in the case of rat liver 28S rRNA) is present in a highly conserved stem-loop structure that is found in all 23, 26 or 28S rRNAs. By analogy with prokaryotic ribosomes (Moazed et al. 1988) this loop interacts with cofactors and removing the specific adenine residue renders it unable to bind elongation factors, thereby preventing protein synthesis.

In certain plants an RIP (in this case called the A-chain) is part of a heterodimeric protein in which it is covalently linked to a second polypeptide (the B-chain) via a single disulphide bond. These heterodimers are called type 2 RIPs and in all examples characterized to date the B-chain is a galactose-specific lectin (Olsnes and Pihl 1982). Type 2 RIPs, of which ricin is the best known example, are amongst the most potently cytotoxic compounds known. Examples of type 1 and type 2 RIPs are given in Table 14.1.

## Structure and cytotoxicity of ricin

Ricin is abundantly present in castor bean (*Ricinus communis*) seeds. Since its initial characterization over a century ago (Stillmark 1888), ricin has been

**Table 14.1** Examples of plant RIPs (Stirpe and Barbieri 1986)

| RIP | Type | Host plant |
| --- | --- | --- |
| Pokeweed antiviral protein | 1 | *Phytolacca americana* (pokeweed) |
| Saporin | 1 | *Saponaria officinalis* (soapwort) |
| Trichosanthin | 1 | *Trichosanthes kirilowii* (Chinese cucumber) |
| Dianthin | 1 | *Dianthus caryophyllius* (carnation) |
| Tritin | 1 | *Triticum aestivum* (wheat) |
| Ricin | 2 | *Ricinus communis* (castor bean) |
| Abrin | 2 | *Abrus precatorius* (Indian licorice) |
| Viscumin | 2 | *Viscum album* (mistletoe) |
| Modeccin | 2 | *Adenia digitata* (kilyambiti) |

sequenced (Funatsu et al. 1978; Funatsu et al. 1979), cloned (Halling et al. 1985; Lamb et al. 1985) and its quaternary structure and cytotoxicity (Nicolson et al. 1974) and X-ray structure (Montfort et al. 1987) have been determined. The potent toxicity of ricin towards mammalian cells is mediated by the B-chain. The two galactose-specific sugar binding sites on the ricin B-chain (Villafranca and Robertus 1981) allow the holotoxin to bind to galactose-terminating glycoproteins or glycolipids on the surface of target cells (Sandvig et al. 1976). Galactose-containing components that can act as receptors for plant toxins such as ricin are normally abundant on the surface of mammalian cells; HeLa cells, for example, contain $3 \times 10^7$ receptors (Sandvig et al. 1976). After binding to the cell surface, ricin is internalized via endocytosis primarily, although probably not exclusively (Moya et al.

1985), in coated pits and coated vesicles (van Deurs *et al.* 1986). All the internalized ricin appears to be transported into endosomes but translocation of the toxin into the cytosol does not occur from this low pH compartment. From the endosome, vesicular transport returns some of the endocytosed ricin to the cell surface or delivers the bulk of the toxin to lysosomes, where it is presumably destroyed. A small proportion (5%) of the endocytosed ricin is transported from the endosomes to the trans Golgi network. Transport to the trans Golgi network appears to be the crucial intracellular route for ricin intoxication (van Deurs *et al.* 1986; van Deurs *et al.* 1988). Mammalian cells, which are extremely sensitive to ricin at 37°C, are markedly less sensitive to the toxin at 19°C when the transport of material from endosomes to the trans Golgi network is blocked (Sandvig *et al.* 1986). Ricin A-chain is translocated across the membrane of trans Golgi elements into the cytosol where it has access to its ribosomal substrate (Fig. 14.1). Ricin B-chain is believed to play a key role in facilitating A-chain translocation (Youle and Neville 1982) although it

is not known whether it has a direct role in A-chain translocation (through a translocation domain present in the B-chain) or whether it simply concentrates the toxin in the trans Golgi by interacting with newly galactosylated glycoproteins present in this intracellular compartment.

Purified ricin A-chain is not cytotoxic because, in the absence of the B-chain, it is unable to bind to or to enter the cytosol of target cells. By the same criterion, many plant tissues containing active single chain RIPs, such as wheat germ or barley (Roberts and Stewart 1979; Coleman and Roberts 1982), can be safely ingested in large amounts.

## Ricin biosynthesis

Ricin, and the closely related lectin *Ricinus communis* agglutinin (RCA) (Nicolson *et al.* 1974), are encoded by a small multigene family (Halling *et al.* 1985; Roberts *et al.* 1990). Gene expression in *Ricinus* is tissue- and development-specific and takes place in the endosperm tissue of the seeds during the later stages of seed maturation (Roberts and Lord 1983). The A- and B-chains of ricin (Butterworth and Lord 1984), in common with the corresponding subunits of RCA (Roberts *et al.* 1985), are encoded together within a single transcript that generates a preproprotein precursor. Prepro-ricin, whose structure is illustrated schematically in Figure 14.2, contains 576 amino acid residues. An N-terminal sequence (35 residues), which includes but is not entirely accounted for by a signal peptide (Roberts *et al.* 1987), precedes the A-chain (267 residues), which is separated by a 12 residue linker peptide from the B-chain (262 residues). During ricin biosynthesis on membrane-bound ribosomes, the N-terminal signal peptide directs the cotranslational discharge of nascent pro-ricin into the lumen of the rough endoplasmic reticulum (Roberts and Lord 1981). This segregation step is accompanied by N-glycosylation of pro-ricin, which is folded into a conformation stabilized by 5 intrachain disulphide bonds. Pro-ricin is transferred from the endoplasmic reticulum by vesicular transport via the Golgi complex, to the protein bodies (Lord 1985), where an acid endo-

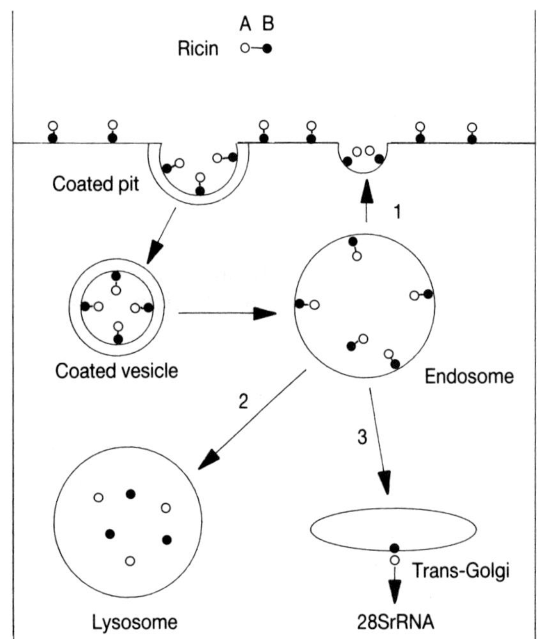



**Fig. 14.1**  Ricin uptake by endocytosis. The first intracellular station for endocytosed ricin is the acidic endosome. From there ricin may be (1) recycled to the cell surface, (2) be delivered to lysosomes and presumably degraded, or (3) be delivered to the trans Golgi network from where ricin A-chain crosses the membrane and enters the cytosol.

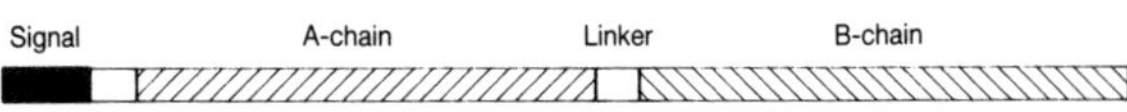



**Fig. 14.2**  Schematic representation of the structure of prepro-ricin.

protease removes the linker peptide to produce native ricin (Harley and Lord 1985).

## Therapeutic potential of ricin

In 1970, Lin *et al.* demonstrated that ricin, and the structurally related cytotoxic plant lectin abrin, had strong antitumour effects on Ehrlich ascites tumour in mice. Several subsequent studies confirmed that tumour cells were more sensitive to ricin than normal cells, and this led to a phase 1 study in which ricin was administered to 54 cancer patients with advanced disease (Fodstad *et al.* 1984). However, despite its initial promise, the *in vivo* usefulness of ricin was constrained by the major limitation of all cytotoxic antitumour reagents: lack of specificity. Clearly, while ricin B-chain might preferentially target the holotoxin into rapidly dividing tumour cells, unwanted delivery to normal cells is unavoidable. Such is the cytotoxic potency of ricin that minimal interaction with normal cells is life-threatening.

Since their introduction in 1975 (Kohler and Milstein 1975), the potential of monoclonal antibodies as anticancer agents has been increasingly explored. As a better understanding of the surface properties of different cell types has emerged, certain tumour cells have been found to express specific antigens that are either not present on the surface of normal cells or are expressed at dramatically reduced levels (Greaves 1982; Levy and Miller 1983). The antigens, therefore, represent tumour-specific targets for appropriate antibodies that selectively bind to the tumour cells. It was hoped that this interaction would lead to the elimination of antigen-bearing target cells. Unfortunately this has not proved to be the case in practice and, possibly with some rare exceptions, antibody molecules on their own are not effective in killing tumour cells (Freeman and Mayhew 1986). The heterogeneity of tumour cells is a problem here; those cells not recognized by the antibody are able to proliferate undisturbed, and the cells to which antibody does bind can internalize surface-bound components by endocytosis and digest them intracellularly. In spite of these and other limitations, antibodies against tumour-associated antigens do specifically recognize and interact with tumour cells (Sikora *et al.* 1983), giving them a distinct advantage over most available drugs used in cancer chemotherapy.

In recent years, conjugates between monoclonal antibodies and toxins such as ricin (called immunotoxins) have been prepared yielding hybrid molecules that combine the specificity of the antibody with the cytotoxic potency of the toxin. The idea of using antibodies raised against defined cell surface antigens to direct toxins to specific cells goes back to the beginning of this century, when Paul Ehrlich proposed his 'magic bullet' concept. In this way, the B-chain of ricin, which confers non-specificity, can be replaced with a cell type-specific monoclonal antibody to produce an immunotoxin capable of binding and entering antigen-bearing target cells. As will be discussed later, it may be desirable to retain ricin B-chain in the conjugate (i.e. an immunotoxin containing intact ricin) but in this case steps must be taken to eliminate non-specific cell binding.

While immunotoxins were initially designed as anti-cancer agents, which relied on an antibody moiety for cell targeting, several effective variations on this theme have also emerged. For example, any ligand, lymphokine, growth factor, hormone, etc. whose receptor is specifically present on potential target cells can be linked to a toxin such as ricin A-chain to produce a specific cytotoxic conjugate.

## Immunotoxin construction and cytotoxicity

Ricin-based immunotoxins usually consist of ricin A-chain chemically linked to an antibody or antibody fragment (Vitetta *et al.* 1987). A variety of other plant cytotoxin A-chains or type 1 RIPs have also been used. Certain bacterial toxins, which kill target cells in a manner similar to ricin, have also been used. One such bacterial toxin, *Pseudomonas* exotoxin A, is being actively utilized as a component of specifically cytotoxic conjugates (Pastan and FitzGerald 1989). From the large number of plant toxins, ricin has the advantage of being well characterized, easily purified and, in contrast to, for example, diphtheria toxin, humans rarely show prior immunity to this plant lectin.

Ricin is purified from crude extracts of *Ricinus communis* seeds by affinity chromatography on a matrix containing immobilized β-galactoside. Elution from the matrix with N-acetylgalactosamine removes ricin and effectively separates it from RCA, which remains bound. After reducing the interchain disulphide bond, the A- and B-chains of ricin can be easily separated by ion exchange chromatography (Saltvedt 1976).

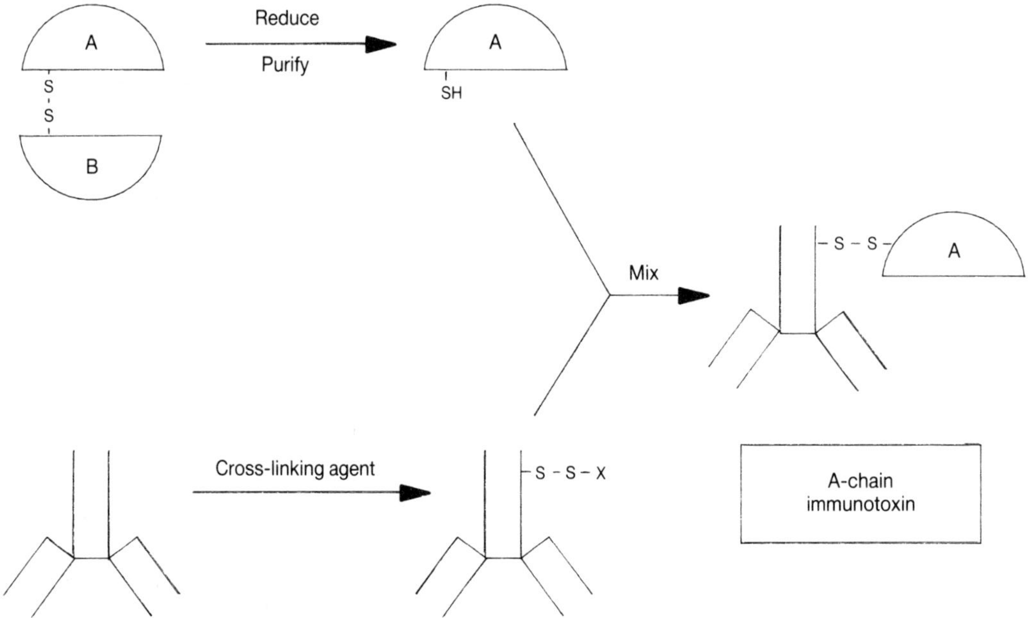


**Fig. 14.3**   Preparation of a ricin A-chain-containing immunotoxin.

The coupling of purified ricin A-chain to antibody can be achieved by a variety of chemical methods, but is normally achieved by means of a reducible disulphide bond (Thorpe and Ross 1982). The most common strategy is to use a heterobifunctional cross-linking agent carrying a disulphide bond, such as N-succinimidyl-3-(2-pyridyldithio) propionate (SPDP), which reacts with free amino groups on the antibody (Fig. 14.3). The derivatized antibody is mixed with ricin A-chain at neutral pH to allow conjugation by disulphide exchange. The resulting immunotoxins are separated from free A-chains and uncoupled antibody by gel filtration and affinity chromatography using immobilized anti-A-chain antibodies and immobilized antigen.

For preparing immunotoxins containing intact ricin, the toxin is normally derivatized with a bifunctional cross-linking agent before being mixed with antibody prepared for cross-linking by partial reduction of its disulphide bonds (Youle and Neville 1980).

Immunotoxins containing intact ricin are invariably toxic to cells bearing the appropriate antigens, often surpassing the native toxin in potency (Uhr 1984). The major limitation of these conjugates is lack of specificity, because non-specific cell binding mediated by ricin B-chain overrides the target cell specificity conferred by the antibody. For *in vitro*

applications, such as bone marrow purging, this is not a serious limitation because opportunistic cell binding by the B-chain can be readily prevented by competition with high concentrations of free galactose or lactose (Youle and Neville 1980). However, this kind of protective approach cannot be used for *in vivo* use and a permanent blockade of the cell-binding domain of the toxin is essential. Because such a blockade has not yet been effectively achieved, most researchers have used immunotoxins containing purified or, more recently, recombinant ricin A-chain. Such conjugates are extremely selective in their toxic effect. The major disadvantage of ricin A-chain immunotoxins is that cytotoxic potency is variable and unpredictable (Vitetta and Uhr 1985). Some A-chain immunotoxins are as toxic to target cells as their intact toxin counterparts, while the toxicity of others can be weak and occasionally non-existent (Thorpe 1985).

It is generally believed that the principal reason for the lack of cytotoxicity of many A-chain immunotoxins is the absence of ricin B-chain. Thus, while the antibody can replace the cell binding function of the B-chain, and can do so in a highly selective manner, it cannot substitute for the second putative function of the B-chain, facilitating the translocation of the A-chain across an intracellular membrane. Because many ricin A-chain immunotoxins are cytotoxic, the A-chain on its own must possess a limited ability to cross the membrane, although the rate of translocation may be orders of magnitude greater in the presence of B-chain (Weil-Hillman *et al.* 1985).

## Therapeutic targets for ricin-containing conjugates

It is beyond the scope of this chapter to provide a comprehensive review of the ricin-containing conjugates produced and evaluated to date. This area has been extensively reviewed recently (e.g. Vitetta *et al.* 1987; Blakey and Thorpe 1988; Byers and Baldwin 1988; Lord *et al.* 1988; FitzGerald and Pastan 1989). Table 14. 2 lists some ricin A-chain conjugates and indicates their therapeutic targets. In general, immunotoxins and related conjugates readily and selectively kill cancer cells growing in culture, while sparing normal cells. Whole animal cancer models provide a much greater challenge and have served to highlight some of the problems likely to be encountered when these conjugates are used in the clinic.

Many toxins, including ricin A-chain, are N-glycosylated and this can cause problems during *in vivo* use. For example, ricin A-chain immunotoxins are cleared from the bloodstream of experimental animals much more rapidly than the antibody alone. Clearance is due primarily to the interaction of mannose and fucose residues on the A-chain with receptors on host cells, in particular the liver (Skilleter and Foxwell 1986).

The stability of the immunotoxin in the bloodstream is crucial. Clearly, if the disulphide bond linking the antibody and the toxin is unstable, the amount of toxin specifically delivered to the target

**Table 14.2**  Examples of some immunotoxins and related conjugates containing ricin A-chain (Frankel 1988)

| Conjugate | Therapeutic target |
| --- | --- |
| Ricin A-anti-CD5 | T lymphocytes, leukaemic cells (T and B) |
| Ricin A-anti melanoma antigen | Malignant melanoma |
| Ricin A-anti-CD22 | B cell lymphomas and leukaemias |
| Ricin A-anti-Tac | HTLV-I-induced adult T cell leukaemia |
| Ricin A-anti-Id antibody to acetylcholine receptor | Myasthenia gravis |
| Ricin A-anti-transferrin receptor | Ovarian cancer |
| Ricin A-CD4 | HIV-infected cells |
| Ricin A-IgE | Mast cells/allergies |

cells will be reduced. In addition, free antibody would compete with the immunotoxin for target cell antigen. Antigen distribution is obviously critical for antibody targeting. If the tumour cell-specificity of the antigen is not absolute, cross-reaction between the immunotoxin and normal cells may be a problem. Normally, the antigen will be much more abundant on the tumour cells, which, accordingly, will have a much greater sensitivity to the immunotoxin. Antigen shedding by target cells can also result in released antigen intercepting the immunotoxin before it reaches the target cells, thereby reducing the amount of potentially effective immunotoxin.

In spite of these and other limitations, several phase 1 clinical trials on immunotoxins are currently under way. The conjugates are being evaluated in the treatment of several cancers, including chronic lymphocytic leukaemia, malignant melanoma, colon cancer, breast cancer, ovarian cancer and B-cell lymphoma. The murine monoclonal antibodies being used in these clinical trials provoke a human antimurine monoclonal antibody response (Byers and Baldwin 1988). These include human antibody responses against the framework, the isotype and the idiotype of the murine antibodies. In spite of intense effort, therapeutically useful human monoclonal antibodies have not emerged (James and Bell 1987). Recombinant DNA technology and cloned antibody genes offer one possible solution to this problem. Site-directed mutagenesis has been used to impose the hyper-variable regions that determine antigen binding for a mouse antibody onto a human variable-region framework (Jones *et al.* 1986). The resulting engineered antibody molecule had the antigen-binding specificity of the mouse antibody within an essentially human antibody protein. Ricin A-chain and other toxins are also immunogenic. As a result, immunotoxin treatment is often combined with immunosuppression, using agents such as cyclosporin.

Human bone marrow transplantation is an area in which ricin-based immunotoxins are making a significant impact by destroying unwanted cells. In autologous bone marrow transplantation, the patient undergoing treatment is also the marrow donor. For treating bone marrow cancer, a sample of the marrow is removed during remission and stored. During a subsequent relapse, cancer cells are killed by completely destroying the patient's bone marrow by chemotherapy and radiotherapy. The marrow removed earlier is reinfused, bringing back sufficient viable stem cells to reconstitute the haematopoietic system. To ensure that tumour cells are not reintroduced during transplantation, the marrow is purged before reinfusion. Ricin-based immunotoxins have been used to purge marrow of malignant B cells (Bregni *et al.* 1986) and T cells (Strong *et al.* 1985). Intact ricin immunotoxins can be used for this purpose by eliminating non-specific cytotoxicity using excess free galactose.

In allogeneic transplantation, patients receive bone marrow cells from a healthy donor after destruction of their own bone marrow. Unless the immunological compatibility between the donor and the patient is perfect, graft-versus-host disease (GvHD) results as the graft perceives its new host as foreign and reacts against it. The use of immunotoxins that kill the T cells in the donor marrow greatly reduces the incidence of GvHD (Filipovich *et al.* 1985).

GvHD is a serious complication for recipients of allogeneic bone marrow transplants. A ricin A-chain immunotoxin has been used successfully to treat a patient with severe grade III–IV, steroid-resistant, acute GvHD—after an allogeneic, human-leucocyte antigen-identical bone marrow transplant for acute myelogenous leukaemia. The monoclonal antibody used recognizes the CD5 antigen that is present on 95% of peripheral blood T lymphocytes. Immunotoxin therapy produced a complete clinical response and, after therapy, acute GvHD did not recur (Kernan *et al.* 1988).

In an exciting recent development, soluble recombinant CD4, the T cell receptor to which the gp120 glycoprotein of the human immunodeficiency virus (HIV) binds, specifically interacts with HIV-infected cells that are expressing gp120 on their surface and on budding HIV particles. CD4 does not kill the infected cells to which it binds and, in an attempt to confer potency, the rCD4 has been linked via a gene fusion approach to *Pseudomonas* exotoxin A (Chaudhary *et al.* 1988) and, via a disulphide bond, to ricin A-chain (Till *et al.* 1988). The CD4 toxin conjugates selectively kill HIV-infected cells in the laboratory and will shortly enter clinical trials.

## Modifying ricin for therapeutic purposes

The most obvious way ricin has been modified for therapeutic purposes has already been discussed, namely, replacing the B-chain of whole ricin with an alternative cell binding moiety, usually a monoclonal antibody. In addition, ricin A- and B-chains are themselves being modified. Traditionally, ricin A-chain used in immunotoxin construction has been purified from *Ricinus* seed extracts. Native ricin A-chain has two major disadvantages: (i) it is difficult to completely remove all traces of B-chain and (ii) ricin A-chain is N-glycosylated and the structure of its oligosaccharide side chains promotes rapid hepatic clearance of A-chain-containing conjugates *in vivo*. Because of this, purified ricin A-chain is often chemically or enzymically deglycosylated before use (Blakey and Thorpe 1988). Recombinant ricin A-chain will increasingly be used for conjugate preparation. DNA encoding ricin A-chain has been excised from prepro-ricin DNA and, using appropriate vectors, expressed in *Escherichia coli* (O'Hare *et al.* 1987; Piatak *et al.* 1988). *E. coli* is a particularly appropriate host for this purpose because its ribosomes are insensitive to ricin A-chain and it does not N-glycosylate proteins. Recombinant ricin A-chain purified from *E. coli* extracts is soluble and has full biological activity.

A major potential problem during the *in vivo* use of conjugates containing ricin A-chain may be lack of specificity. Such is the potency of ricin A-chain that penetration of just one, or at most a few, molecules into the cytosol of a cell may be sufficient to kill that cell. This means that specificity must be virtually absolute to prevent unwanted cytotoxic effects. It may be useful to genetically modify ricin A-chain to deliberately reduce, without eliminating, its catalytic activity. The selection and characterization of ricin A-chain mutants in yeast has confirmed the feasibility of this approach (Frankel *et al.* 1989). Ricin A-chain DNA was placed under the control of an inducible yeast promoter (GAL1) and transformed into yeast. Induction of ricin A-chain expression was lethal. This was used as the basis for selection of ricin A-chain mutants that inactivated the toxin. Several mutant A-chains had single amino acid changes involving residues located in the proposed active site (Montfort *et al.* 1987). Glu177 was identified in this way as one of the key active site residues. Site-directed mutagenesis has been used to convert Glu177 to either Asp or Ala. The mutant proteins were expressed in *E. coli*, purified and quantitatively assayed for enzymic activity. The mutant A-chains were catalytically active but their activity was two to three orders of magnitude lower than wild type recombinant A-chain. The use of such mutant ricin A-chains in immunoconjugates may be beneficial if lowering the toxicity of the A-chain essentially eliminates non-specific toxicity, provided that the toxin can still be delivered to the target cells in sufficient quantity to cause their destruction.

Protein engineering is also being employed to eliminate the lectin activity of ricin B-chain. It is hoped that mutant B-chain will retain its ability to translocate ricin A-chain across an intracellular membrane but will no longer be capable of cell binding. This may give rise to whole ricin conjugates suitable for *in vivo* use. Ricin B-chain is a bilobal molecule possessing two sugar binding sites. The X-ray structure has identified the key sugar binding residues as Lys40 and Asn46 in the first site and Asn255 in the second site (Montfort *et al.* 1987). Site-directed mutagenesis of ricin B-chain DNA has been used to convert these residues to Met40, Gly46 and Gly255 (Richardson, Roberts, Woodland and Lord, unpublished results). Mutant ricin B-chain has been expressed in *Xenopus* oöcytes, using an approach that was successful for wild type B-chain (Richardson *et al.* 1988). The recombinant product was completely soluble but its ability to bind galactose had been virtually eliminated. This recombinant product is currently being evaluated to see if it retains the ability to translocate A-chain.

## Recombinant chimeric molecules

Now that cloned genes encoding several toxins, lymphokines, growth factors, peptide hormones, ligands and antibody heavy and light chains are available, recombinant chimeric molecules are increasingly being produced by heterologous expression of gene fusions. In the case of the bacterial toxins *Pseudomonas* exotoxin A and diphtheria toxin, the portion of the toxin gene encoding the cell binding domain has been detected and replaced with DNA encoding an alternative cell binding protein. In this way recombinant chimeric molecules have been produced that retain the ability of the native toxin to enter and kill cells but whose targets are now limited to those cells with the surface component that interacts with the new binding function. These chimeric molecules have been the subject of a recent review (Pastan and FitzGerald 1989).

To date, no fusion proteins containing ricin A-chain have been reported. We have used gene fusions to produce several ricin A-chain-containing fusion proteins, including staphylococcal protein A-ricin A-chain, interleukin-2-ricin A-chain and antibody $V_L$-ricin A-chain. Although both components of such fusion proteins are biologically active, the ricin A-chain moiety is catalytically active and the other component is able to bind to its natural protein target, the fusion proteins are not cytotoxic. We believe this results from the failure of the ricin A-chain moiety, in contrast to its bacterial toxin counterparts, to cross an intracellular membrane when part of a fusion protein. It seems that ricin A-chain must be coupled to the carrier protein by a disulphide bond equivalent to that which naturally links the A- and B-chains in intact ricin. The solution to this problem may be to engineer a cleavable linker peptide contained within a disulphide-linked loop into ricin A-chain-fusion proteins. This will allow the intracellular reductive release of the ricin A-chain moiety.

## Future developments

In spite of improvements to current immunoconjugates that might result from genetic engineering, several limitations may reduce their usefulness for *in vivo* therapy. These limitations include host immune response, target cell specificity, tumour cell heterogeneity, non-specific toxicity, and doubtless several others. A major limitation could be the inability of toxin conjugates to reach and effectively permeate solid tumours. The molecular size of the conjugate will be a key factor here, and much future work will be directed towards reducing this size. This will be accomplished by deleting antibody and toxin genes to the smallest size that will allow the expressed protein products to retain the required biological function. The antibody position may take the form of a single recombinant polypeptide containing the heavy and light chain variable regions joined via a linker peptide (Huston *et al.* 1988) fused to a toxin (Chaudhary *et al.* 1989), or a heavy chain variable domain fragment (Ward *et al.* 1989). Likewise, as toxin catalytic sites and membrane translocation regions are better defined, minimum-sized functional domains will be generated.

The search for conjugates that are smaller, more specific and less immunogenic will continue for some time yet.

# References

Blakey, D.C. and Thorpe, P.E. (1988). An overview of therapy with immunotoxins containing ricin or its A chain. *Antibody Immunoconjugates and Radiopharmaceuticals*. **1**: 1–16.

Bregni, M., Fabritis, D., Raso, V., Greenberger, J., Lipton, J., Nadler, L., Rothstein, L., Ritz, J. and Barst, R.C. (1986). Elimination of clonogenic tumour cells from human bone marrow using a combination of monoclonal antibody-ricin A chain conjugates. *Cancer Research*. **46**: 1208–13.

Butterworth, A.G. and Lord, J.M. (1983). Ricin and *Ricinus communis* agglutinin subunits are all derived from single sized precursors. *European Journal of Biochemistry*. **137**: 57–65.

Byers, V.S. and Baldwin, R.W. (1988). Therapeutic strategies with monoclonal antibodies and immunoconjugates. *Immunology*. **65**: 329–35.

Chaudhary, V.K., Mizukami, T., Fuerst, T.R., FitzGerald, D.J., Moss, B., Pastan, I. and Berger, E.A. (1988). Selective killing of HIV-infected cells by recombinant human CD4-*Pseudomonas* exotoxin fusion protein. *Nature*. **335**: 369–72.

Chaudhary, V.K., Queen, C., Junghams, R.P., Waldmann, T.A., FitzGerald, D.J. and Pastan, I. (1989). A recombinant immunotoxin consisting of two antibody variable domains fused to *Pseudomonas* exotoxin. *Nature*. **339**: 394–7.

Coleman, W.H. and Roberts, W.K. (1982). Inhibitors of animal cell-free protein synthesis from grains. *Biochimica et Biophysica Acta*. **696**: 239–44.

Endo, Y. and Tsurugi, K. (1987). RNA N-glycosidase activity of ricin A chain. Mechanism of action of the toxic lectin ricin on eukaryotic ribosomes. *Journal of Biological Chemistry*. **262**: 8128–30.

Endo, Y. and Tsurugi, K. and Lambert, J.M. (1988). The site of action of six ribosome-inactivating proteins from plants on eukaryotic ribosomes: the RNA N-glycosidase activity of the proteins. *Biochemical and Biophysical Research Communications*. **150**: 1032–6.

Filipovitch, A.H., Vallera, D.A., Youle, R.J., Neville, D.M. and Kersey, J.H. (1985). Ex vivo T-cell depletion with immunotoxins in allogeneic bone marrow transplantation: the pilot clinical study for prevention of graft-versus-host disease. *Transplantation Proceedings*. **17**: 442–4.

FitzGerald, D. and Pastan, I. (1989). Targeted toxin therapy for the treatment of cancer. *Journal of the National Cancer Institute*. **81**: 1455–63.

Fodstad, O., Kvalheim, G., Godal, A., Lotsberg, J., Aamdal, S., Host, H. and Pihl, A. (1984). Phase 1 study of the plant protein ricin. *Cancer Research*. **44**: 862–5.

Frankel, A. (1988). *Immunotoxins*. Kluwer Academic Publishers, Dordrecht, The Netherlands.

Frankel, A., Schlossman, D., Welsh, P., Hertler, A.,

Withers, D. and Johnston, S. (1989). Selection and characterization of ricin toxin A-chain mutations in *Saccharomyces cerevisiae*. *Molecular and Cellular Biology*. **9**: 415–20.

Freeman, A.I. and Mayhew, E. (1986). Targeted drug delivery. *Cancer*. **58**: 573–83.

Funatsu, G., Yoshitake, S. and Funatsu, M. (1978). Primary structure of the Ile chaine of ricin D. *Agricultural and Biological Chemistry*. **42**: 501–3.

Funatsu, G., Kimura, M. and Funatsu, M. (1979). Primary structure of the Ala chain of ricin D. *Agricultural and Biological Chemistry*. **43**: 2221–4.

Greaves, M.F. (1982). Target structures and target cells for cancer therapy with monoclonal antibodies: finding the candidates. *Cancer Surveys*. **1**: 451–66.

Halling, K.C., Halling, A.C., Murray, E.E., Ladin, B.F., Houston, L.L. and Weaver, R.F. (1985). Genomic cloning and characterization of a ricin gene from *Ricinus communis*. *Nucleic Acids Research*. **13**: 8019–33.

Harley, S.M. and Lord, J.M. (1985). *In vitro* endoproteolytic cleavage of castor bean lectin precursors. *Plant Science*. **41**: 111–16.

Huston, J.S., Levingson, D., Mudgett-Hunter, M. *et al.* (1988). Protein engineering of antibody binding sites: recovery of specific activity in an anti-digoxin single-chain $F_v$ anologue produced in *Escherichia coli*. *Proceedings of the National Academy of Sciences, USA*. **85**: 5879–83.

James, K. and Bell, G.T. (1987). Human monoclonal antibody production—current status and future prospects. *Journal of Immunological Methods*. **100**: 5–20.

Jones, P.T., Dear, P.H., Foote, J., Neuberger, M.S. and Winter, G. (1986). Replacing the complementary-determining region in a human antibody with that from a mouse. *Nature*. **231**: 522–4.

Kernan, N.A., Byers, V., Scannon, P.J., Mischat, R.P., Brochstein, J., Flomenberg, N., Dupont, B. and O'Reilly, R.J. (1988). Treatment of steroid-resistant acute graft-vs-host disease by *in vivo* administration of an anti-T-cell ricin A chain immunotoxin. *Journal of the American Medical Association*. **259**: 3154–7.

Kohler, G. and Milstein, C. (1975). Continuous cultures of fused cells secreting antibody of predefined specificity. *Nature*. **256**: 495–7.

Lamb, F.I., Roberts, L.M. and Lord, J.M. (1985). Nucleotide sequence of cloned cDNA coding for preproricin. *European Journal of Biochemistry*. **145**: 265–9.

Levy, R. and Miller, R.A. (1983). Tumour therapy with monoclonal antibodies. *Federation Proceedings*. **42**: 2650–6.

Lin, J.-Y., Tserng, K.-Y., Chen, C.-C., Lin, L.-T. and Tung, T.-C. (1970). Abrin and ricin: new antitumour substances. *Nature*. **227**: 292–3.

Lord, J.M. (1985). Precursors of ricin and *Ricinus communis* agglutinin. Glycosylation and processing during synthesis and intracellular transport. *European Journal of Biochemistry*. **146**: 411–16.

Lord, J.M., Spooner, R.A., Hussain, K. and Roberts,

L.M. (1988). Immunotoxins: properties, applications and current limitations. *Advanced Drug Delivery Reviews*. **2**: 297–318.

Moazed, D., Robertson, J.M. and Noller, H.F. (1988). Interaction of elongation factors EF-G and EF-Tu with a conserved loop in 23 S RNA. *Nature*. **334**: 362–4.

Montfort, W., Villafranca, J.E., Monzungo, A.F., Ernst, S.R., Katzin, B., Rutenber, E., Xuong, N.H., Hamil, R. and Robertus, J.D. (1987). The three-dimensional structure of ricin at 2.8 Å. *Journal of Biological Chemistry*. **262**: 5398–403.

Moya, M., Dantry-Varsat, A., Gond, B., Louvard, D. and Boquet, P. (1985). Inhibition of coated pit formation in Hep2 cells blocks the cytotoxicity of diphtheria toxin but not that of ricin. *Journal of Cell Biology*. **101**: 548–59.

Nicolson, G.L., Blaustein, J. and Etzler, M. (1974). Characterization of two plant lectins from *Ricinus communis* and their quantitative interaction with a murine lymphoma. *Biochemistry*. **13**: 196–204.

O'Hare, M., Roberts, L.M., Thorpe, P.E., Watson, G.J., Prior, B. and Lord, J.M. (1987). Expression of ricin A chain in *Escherichia coli*. *FEBS Letters*. **216**: 73–8.

Olsnes, S. and Pihl, A. (1982). Toxic lectins and related proteins. In *Molecular Action of Toxins and Viruses* (ed. P. Cohen and S. van Heyningen), pp. 51–105. Elsevier Press, Amsterdam.

Pastan, I. and FitzGerald, D.J. (1989). *Pseudomonas* exotoxin: chimeric toxins. *Journal of Biological Chemistry*. **264**: 15157–60.

Piatak, M., Lane, J., Laird, W., Bjorn, M., Wang, A. and Williams, M. (1988). Expression of soluble and fully functional ricin A chain in *E. coli* is temperature sensitive. *Journal of Biological Chemistry*. **263**: 4837–43.

Richardson, P.T., Gilmartin, P., Roberts, L.M., Colman, A. and Lord, J.M. (1988). Expression of functional ricin B chain in *Xenopus* oocytes. *Bio/Technology*. **6**: 565–70.

Roberts, W.K. and Stewart, T.S. (1979). Purification and properties of a translation inhibitor from wheatgerm. *Biochemistry*. **18**: 2615–21.

Roberts, L.M. and Lord, J.M. (1981). Protein biosynthetic capacity in the endosperm tissue of ripening castor bean seeds. *Planta*. **152**: 420–7.

Roberts, L.M., Lamb, F.I., Pappin, D. and Lord, J.M. (1985). Primary sequence of *Ricinus communis* agglutinin. Comparison with ricin. *Journal of Biological Chemistry*. **260**: 15682–6.

Roberts, L.M., Lamb, F.I. and Lord, J.M. (1987). Biosynthesis and molecular cloning of ricin and *Ricinus communis* agglutinin. In *Membrane Mediated Cytotoxicity* (ed. B. Bonavida and J.R. Collier), pp. 73–82. Alan R. Liss Inc., New York.

Roberts, L.M. Tregear, J. and Lord, J.M. (1990). Molecular cloning of ricin. In *Genetically Engineered Toxins*, (ed. A. Frankel), Marcel Dekker Inc., New York. In press.

Saltvedt, E. (1976). Structure and toxicity of purified *Ricinus* agglutinin. *Biochimica et Biophysica Acta*. **451**: 536–46.

Sandvig, K., Olsnes, S. and Pihl, A. (1976). Kinetics of binding of the toxic lectins abrin and ricin to surface receptors of human cells. *Journal of Biological Chemistry*. **251**: 3977–84.

Sandvig, K., Tonnessen, T.I. and Olsnes, S. (1986). Ability of inhibitors of glycosylation and protein synthesis to sensitize cells to abrin, ricin, shigella toxin and *Pseudomonas* toxin. *Cancer Research*. **46**: 6418–22.

Sikora, K., Smedley, H. and Thorpe, P.E. (1983). Tumour imaging and drug targeting. *British Medical Bulletin*. **40**: 233–9.

Skilleter, D.N. and Foxwell, B.M.J. (1986). Selective uptake of ricin A chain by hepatic non-parenchymal cells *in vitro*. Importance of mannose oligosaccharide in the toxin. *FEBS Letters*. **196**: 344–8.

Stillmark, H. (1888). Uber Ricin, ein gifiges Fermentans den Samen vor *Ricinus communis* L. und anderen Euphorbiaceen. Inaugural Dissertation, University of Dorpat, Estonia.

Stirpe, F. and Barbieri, L. (1986). Ribosome-inactivating proteins up to date. FEBS Letters. **195**: 1–8.

Strong, R.C., Uckun, F., Youle, R.J., Kersey, J.-H. and Vallera, D.A. (1985). Use of multiple T-cell directed intact ricin immunotoxins for autologous bone marrow transplantation. *Blood*. **66**: 627–35.

Thorpe, P.E. (1985). Antibody carriers of cytotoxic agents in cancer therapy: a review. In *Monoclonal Antibodies '84: biological and clinical applications*, (ed. A. Pinchera, G. Doria, F. Dammacco and A. Bargellesi), pp. 475–506. Editrice Kurtis, Rome.

Thorpe, P.E. and Ross, W.C.J. (1982). The preparation and cytotoxic properties of antibody–toxin conjugates. *Immunological Reviews*. **62**: 119–58.

Till, M.A., Ghetie, V., Gregory, T., Patzer, E.J., Porter, J.P., Uhr, J.W., Capon, D.J. and Vitetta, E.S. (1988). HIV-infected cells are killed by rCD4-ricin A chain. *Science*. **242**: 1166–8.

Uhr, J.W. (1984). Harnessing nature's poisons. *Journal of Immunology*. **133**: 1–10.

van Deurs, B., Petersen, O.W., Olsnes, S. and Sandvig, K. (1986). Delivery of internalized ricin from endosomes to cisternal Golgi elements is a discontinuous, temperature-sensitive process. *Experimental Cell Research*. **171**: 137–52.

van Deurs, B., Sandvig, K., Petersen, O.W., Olsnes, S., Simons, K. and Griffiths, G. (1988). Estimation of the amount of internalized ricin that reaches the trans Golgi network. *Journal of Cell Biology*. **106**: 253–67.

Villafranca, J.E. and Robertus, J.D. (1981). Ricin B chain is a product of gene duplication. *Journal of Biological Chemistry*. **256**: 554–6.

Vitetta, E.S. and Uhr, J.W. (1985). Immunotoxins: redirecting Nature's poisons. *Cell*. **41**: 653–4.

Vitetta, E.S., Fulton, R.J., May, R.D., Till, M. and Uhr, J.W. (1987). Redesigning Nature's poisons to create anti-tumour reagents. *Science*. **238**: 1098–104.

Ward, E.S., Güssow, D., Griffiths, A.D. *et al.* (1989). Binding activities of a repertoire of single immunoglobulin variable domains secreted from *Escherichia coli*. *Nature*. **341**: 544–6.

Weil-Hillman, G., Runge, W., Jansen, F.K. and Vallera, D.A. (1985). Cytotoxic effect of anti-Mr 67000 protein immunotoxins on human tumours in a nude mouse model. *Cancer Research*. **45**: 1328–36.

Youle, R.J. and Neville, D.M. (1980). Anti-Thyl 1-2 monoclonal antibody linked to ricin is a potent cell-type-specific toxin. *Proceedings of the National Academy of Sciences, USA*. **77**: 5483–7.

Youle, R.J. and Neville, D.M. (1982). Kinetics of protein synthesis inactivation by ricin-anti-Thy 1 antibody hybrids. *Journal of Biological Chemistry*. **257**: 1598–1600.

# Part V  CROP PROTECTION

# 15

# An Enzymatic Basis for Herbicide Resistance: Cytochrome P450 Mono-oxygenases

Daniel P. O'Keefe, Reijer Lenstra and Charles A. Omer

## Introduction

The agricultural use of herbicides is based upon their ability to kill undesirable plant species either in a selective or non-selective manner. Biochemically, herbicides function by inhibiting an enzyme that is crucial to a metabolic pathway essential for the survival of the plant. Some plant species are insensitive to certain herbicides on their first exposure, and in some plants, particularly weeds, resistance occurs within a population as a result of continued selection pressure from a particular herbicide, or class of herbicides. Both situations illustrate the phenomenon of resistance, yet they are generally the result of two distinct mechanisms. In the first, enzymes already in the plant metabolize the herbicidal compound to a metabolite or metabolites that are much less phytotoxic. This is often the basis for the species selectivity of the herbicide. The crop plant metabolizes the compound to a non-phytotoxic metabolite, and the weeds, incapable of this metabolism, are killed. In the second resistance mechanism, the enzyme that is inhibited by the herbicide is less sensitive to it, either because of a pre-existing natural variation in the population or as a result of mutation/selection. A third general mechanism of resistance arises from poor delivery of the herbicide to its primary site of action as a result of other factors, such as sequestration within the plant or decreased uptake. The major emphasis of this chapter is to focus on the resistance conferred by a class of xenobiotic metabolizing enzyme systems known as cytochrome P450 mono-oxygenases. While widely studied in animals, the characteristics of these enzymes in plant herbicide metabolism are only beginning to be understood biochemically. The interested reader is directed to previous reviews for an overview of herbicide resistance and metabolism (Cole 1983; O'Keefe et al. 1987; Mazur and Falco 1989; Wallnofer and Engelhardt 1989).

## Detoxification or target site resistance: the chlorsulphuron example

The herbicide chlorsulphuron (2-chloro-N-[(4-methoxy-6-methyl-1,3,5 triazin-2-yl)aminocarbonyl]benzenesulphonamide), can be used to provide a more concrete example of the two mechanisms of resistance outlined above. Chlorsulphuron is a member of the class of sulphonylurea herbicides, characterized by their extremely low use rates. The sulphonylureas, and another class of herbicides—the imidizolinones—function by inhibiting acetolactate synthase, an enzyme in the branched chain amino acid biosynthetic pathway essential for the synthesis of valine, leucine and isoleucine (Chaleff and Mauvais 1984; Shaner et al. 1984; Mazur and Falco 1989).

Originally produced as a herbicide to control weeds in wheat, chlorsulphuron exerts its selectivity because wheat has an extremely effective enzymatic system for its detoxification. This detoxifying system is absent from the predominant weed species that chlorsulphuron is used to control. It is thought that a cytochrome P450-based mono-oxygenase initially hydroxylates the benzene ring of chlorsulphuron in an overall rate-limiting reaction. This metabolite, which is still phytotoxic, is subsequently conjugated to glucose by a glucosyl transferase enzyme, generating a non-phytotoxic metabolite (Sweetser *et al.* 1982). Although none of the enzymes involved in chlorsulphuron metabolism in wheat have been further characterized or purified, an enzymatically analogous mono-oxygenase system from the bacterium *Streptomyces griseolus* has been extensively studied (Romesser and O'Keefe 1986; O'Keefe *et al.* 1987, 1988, 1991; Omer *et al.* 1990). This will be the topic for extensive discussion later.

A number of amino acid positions in the plant acetolactate synthase can be altered in such a fashion that the enzyme still functions but its susceptibility to sulphonylureas is decreased. This was originally observed by selection of resistant tobacco cells from tissue culture (Chaleff and Ray 1984) and subsequently by random and site-directed mutagenesis (Sebastian and Chaleff 1987; Mazur and Falco 1989). This approach, unlike the inherent resistance of wheat to chlorsulphuron, is not based on actively performing any function to eliminate the herbicide from the plant. The short term concentrations of herbicide in active site resistant plants are probably no different than in susceptible plants. The long term fate of the applied herbicide in the plant is of some concern from both an environmental and a health standpoint. The ultimate agronomic success and acceptability of this approach will probably be dependent on the removal of herbicide residues by both endogenous and engineered herbicide degrading enzymes.

## Transformation as a basis of engineering

The use of herbicide resistance or sensitivity phenotype has some very clear, strong advantages for the study of engineered plant species. Foremost among these is that selection for the phenotype, clearly defined by the gene under study, is simply a matter of screening for sensitivity to the herbicide. The availability of a resistant form of a target enzyme can in some ways be limiting if one needs to wait for nature to select one. However, once available (and unless sexual crosses into desirable species is possible) it may be necessary to integrate the resistant form of the gene into the species desired by some transformation technique. Since this integration into the genome is random, it does not affect the expression of the unaltered gene, and the plant will be expressing both the sensitive and resistant target enzyme. The physiological assump-

tion made here is that the higher level of total enzyme will have no undesirable effects.

Alternatively, the strategy of adding an entirely new enzyme to a plant has some inherent advantages. The ideal candidate for this would be a highly specific herbicide detoxifying enzyme that completely degraded the herbicide and altered the endogenous metabolism of the plant in no other way. The currently available candidates for this function are probably not ideal. Unknown, and essentially untested, is the possibility that the added enzyme will be able to function with other substrates normally found in the cell, and metabolize them with quite unpredictable consequences. None the less, herbicide metabolizing genes have been successfully introduced into plants in several instances. Tobacco resistant to bromoxynil (Stalker *et al.* 1988), tobacco, potato and tomato resistant to phosphinothricin or bialophos (DeBlock *et al.* 1987; Thompson *et al.* 1987) and tobacco resistant to 2,4-D (Streber *et al.* 1987, Streber and Willmitzer 1989) have been produced. In each case, the plants were transformed to produce a single enzyme that metabolizes the herbicide to an inactive form (bromoxynil nitrilase, phosphinothricin acetyltransferase, and 2,4-D mono-oxygenase). Notably, the original source of all these detoxifying genes was bacterial.

## Plant cytochrome P450 mono-oxygenases

There have been many reports on the involvement of cytochromes P450 in different aspects of plant metabolism. These reports range in substantiation from an inference that hydroxylated metabolites were generated by a cytochrome P450, to descriptions of purified enzymes of this type (Cole 1983; O'Keefe *et al.* 1987, and references therein). We will confine our discussion here to the few cytochromes P450 that have actually been purified and the one for which both the protein and cDNA have been isolated. But first, some clarification of the biochemical characteristics of this type of enzyme will serve to introduce the crucial features known about all cytochromes P450.

### Cytochrome P450 mono-oxygenases: general features

Cytochrome P450-based mono-oxygenase systems

are enzymes that use reducing equivalents from reduced pyridine nucleotides and molecular dioxygen to incorporate a single atom of molecular oxygen into the substrate. The cytochromes P450 from mammalian liver have been widely studied for many years because of their broad substrate specificity and importance in xenobiotic, particularly pharmaceutical, metabolism. The cytochrome P450 is a monomeric haem protein of approximately 50 kDa, and takes its name from the characteristic absorption maximum of the CO coordinated ferrocytochrome at around 450 nm. In eukaryotes, cytochromes P450 are intrinsic membrane proteins found predominantly in the microsomal fraction (endoplasmic reticulum) of the cell; the prokaryotic P450s are soluble. The reducing equivalents are delivered via a flavoprotein reductase in the microsomal cytochromes and generally via an electron transfer system composed of a soluble ferredoxin and NAD(P)H:ferredoxin reductase in the prokaryotic cytochromes. A notable exception to this generalization about prokaryotic cytochromes P450 is the 110 kDa catalytically self-sufficient cytochrome $P450_{BM-3}$ from *Bacillus megaterium*, which contains both reductase and mono-oxygenase (P450) function in a single polypeptide (Narhi and Fulco 1987).

Functionally, the insertion of a hydroxyl group derived from one atom of molecular oxygen onto an aryl or alkyl carbon of the substrate is the primary product of the mono-oxygenase reaction. Mechanistically, this is carried out by reduction of a haem-coordinated molecular oxygen to an $[FeO]^{3+}$ intermediate (activated oxygen), followed by hydrogen abstraction from the substrate, and finally radical recombination of this $[FeOH]^{3+}$ intermediate with the substrate carbon radical, restoring the $[Fe]^{3+}$ resting state of the enzyme. When the hydroxyl is inserted onto a carbon adjacent to a secondary or tertiary amine (or adjacent to an ether or ester oxygen) subsequent non-enzymatic rearrangement leads to an N-dealkylated, O-dealkylated, or free acid form, respectively, of the substrate (for reviews, see Ortiz de Montellano 1986; Guengerich and Macdonald 1990).

There is one three-dimensional structure available for a cytochrome P450, the camphor hydroxylase cytochrome $P450_{CAM}$ from *Pseudomonas putida* (Poulos *et al*. 1985; Poulos *et al*. 1987). The protein is divided into a helix-rich and helix-poor domain, with the substrate binding site roughly at the interface of these domains, buried within the protein adjacent to one face of the haem. There are

no obvious access points in the static X-ray diffraction structure either for substrate into the active site or for the transfer of reducing equivalents from the physiological reductant, putidaredoxin. The structure clearly illustrates how the protein aligns the substrate in the appropriate orientation such that the $[FeO]^{3+}$ intermediate is in closest proximity to the hydroxylation site on the substrate. Because no other three-dimensional cytochrome P450 structures have been determined, the $P450_{CAM}$ structure has been widely used and modelled in studies of other, often distantly related, proteins.

## Plant cytochrome P450 mono-oxygenases: properties

Higher plant cytochromes P450 have been purified from tulip (*Tulipa gesneriana*) bulbs (Higashi *et al*. 1985), Jerusalem artichoke (*Helianthus tuberosus*) tubers (Gabriac *et al*. 1985), avocado (*Persea americana*) fruit mesocarp (O'Keefe and Leto 1989) and soybean (*Glycine max*) cell culture (Kochs and Grisebach 1989). While herbicide detoxification is generally studied in leaves, we are not aware of any cytochrome P450 purified to homogeneity from this tissue. None of the cytochromes that have been purified are clearly involved in herbicide metabolism. The tulip bulb protein has had no enzymatic activity attributed to it, the Jerusalem artichoke cytochrome is quite clearly a trans cinnamic acid hydroxylase terminal oxidase, and the avocado cytochrome can carry out xenobiotic metabolism in the general sense (*p*-chloro-N-methyl aniline demethylase). Nonetheless, the properties of these proteins serve as the archetypes for the entire plant kingdom until further data become available.

All of the cytochromes P450 purified from plants are integral membrane proteins found in the microsomal membrane fraction. Immunocytochemical results with antibody raised against the avocado cytochrome P450 have further confirmed that this protein is localized in the endoplasmic reticulum (Bourett *et al*. 1989). Because of the requirement for reducing equivalents, and the inability of these proteins to utilize reduced pyridine nucleotides directly, mono-oxygenase activity requires a reductase capable of transferring electrons from reduced pyridine nucleotides to the cytochrome P450. This protein has been successfully purified from several plant sources (Benveniste *et al*. 1986; Kochs and Grisebach 1989). In the analogous systems from animals, a single reductase protein services several cytochromes P450 and the specificity of the mono-oxygenase reaction resides solely on the substrate binding site found on the cytochrome (Peterson and Prough 1986). This implies that a rather broad specificity reductase in plant endoplasmic reticulum could interact with many different P450s, a point not to be overlooked for considerations of transferring the gene for a desirable activity from one species to another.

How do plant cytochromes P450 differ from other eukaryotic P450s? Virtually all the sequence information for eukaryotic cytochromes P450 comes from mammalian hepatic proteins. Recently, the sequence of a ripening-associated mRNA in avocado mesocarp was shown to encode a protein identical to the cytochrome P450 purified from the same tissue (Bozak *et al.* 1990). Avocado P450 has a very obvious membrane anchor at the amino terminal end (residues 1–19 of ARP2 are rich in hydrophobic amino acids and charged residues are totally absent, Fig. 15.1). This is followed by a region of charged amino acids, and a subsequent region very rich in prolines. This 'proline cluster' is a common, but functionally unexplained, feature near the amino terminus of many cytochromes

P450 (Black and Coon 1986). No other obvious membrane spanning segment can be found in the remainder of the avocado P450 primary sequence when it is analysed by the hydropathy algorithm of Kyte and Doolittle (1982). In this procedure, a running hydrophilic/hydrophobic score is calculated for each amino acid over a user-determined window of neighbouring amino acids and one searches the primary sequence for hydrophobic stretches of approximately 20 amino acids that could define transmembrane α-helices.

The existence of a membrane anchoring sequence in the avocado P450 is consistent with current interpretations of the sequences of animal (mostly mammalian) cytochromes P450. Most investigators now believe that these proteins have one or two membrane spanning segments near the amino terminus (Kalb and Loper 1988; Nelson and Strobel 1988; Szczesna-Skorupa *et al.* 1988; Brown and Black 1989; Edwards *et al.* 1989; Vergeres *et al.* 1989) and that the remainder of the protein resembles the structure determined for the soluble bacterial cytochrome P450$_{\mathrm{CAM}}$. The most strongly conserved region in all cytochromes P450 is that which forms a pocket around the cysteine that provides the proximal sulphydryl ligand for the

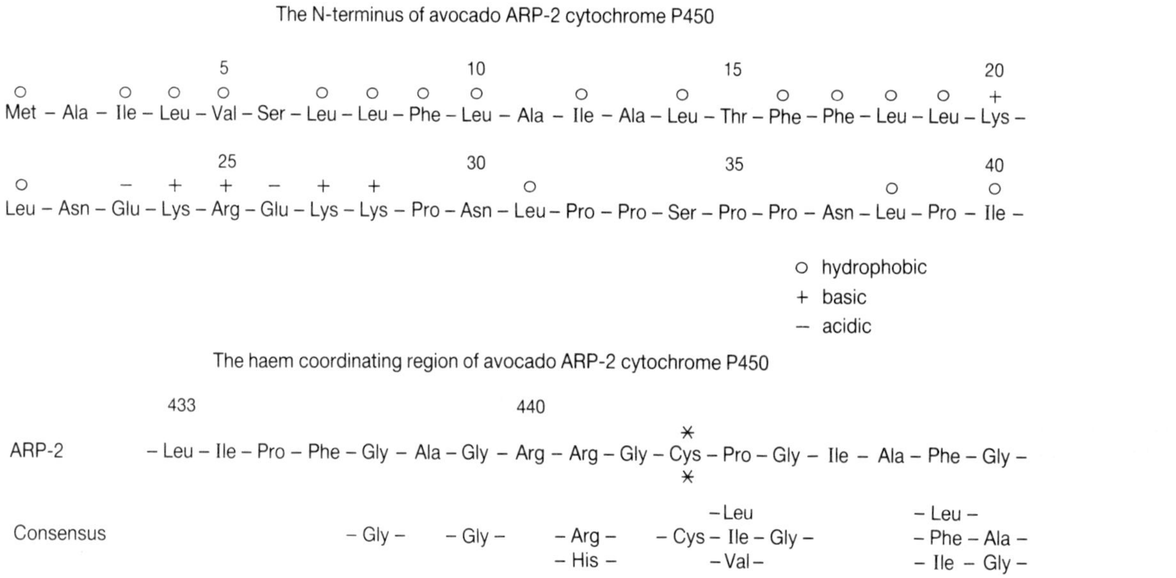

The N-terminus of avocado ARP-2 cytochrome P450

```
           5                    10                    15                     20
  O       O   O   O      O   O   O   O       O         O        O   O   O   O   +
Met – Ala – Ile – Leu – Val – Ser – Leu – Leu – Phe – Leu – Ala – Ile – Ala – Leu – Thr – Phe – Phe – Leu – Leu – Lys –

           25                   30                    35                     40
  O       –   +   +   –   +   +           O                         O        O
Leu – Asn – Glu – Lys – Arg – Glu – Lys – Lys – Pro – Asn – Leu – Pro – Pro – Ser – Pro – Pro – Asn – Leu – Pro – Ile –
```

O hydrophobic
+ basic
− acidic

The haem coordinating region of avocado ARP-2 cytochrome P450

```
         433                                    440
                                                          *
ARP-2         – Leu – Ile – Pro – Phe – Gly – Ala – Gly – Arg – Arg – Gly – Cys – Pro – Gly – Ile – Ala – Phe – Gly –
                                                          *
```

| | | | | |
|---|---|---|---|---|
| | | | −Leu | − Leu − |
| Consensus | − Gly − | − Gly − | − Arg − | − Cys − Ile − Gly − | − Phe − Ala − |
| | | | − His − | −Val− | − Ile − Gly − |

**Fig. 15.1**   Sequence segments of avocado cytochrome P450. At the top is the amino terminal portion of the protein designated ARP-2, which is identical to the ripening associated cDNA sequenced by Bozak *et al.* Note the abundance of hydrophobic residues and the absence of any charged residues in the first 19 amino acids. At the bottom is the sequence in the region of the putative cysteine haem ligand (Bozak *et al.* 1990). The consensus sequence for this region from other P450 sequences is also shown.

haem (Poulos *et al.* 1985; Poulos *et al.* 1987). The sequence of avocado cytochrome P450 (Avo LXXIA1) is no exception to this (Fig. 15.1). This region is also strongly conserved in the plant protein, and the substitution of a proline in the position after the haem coordinating cysteine (a position where Leu, Ile or Val has been found in all other described sequences) is notable. The consequences of having such a strongly sterically-constrained amino acid like Pro in this position have yet to be explored.

## Plant cytochromes P450: how many?

A major unanswered question concerning plant cytochromes P450 is just how many of them there are. This topic has been considered at length previously (O'Keefe *et al.* 1987), yet new data accumulated in the interim merit some further discussion. At the core of this discussion is the fact that mammalian liver is a rich source of a wide variety of xenobiotic metabolizing cytochromes P450 (Nebert and Gonzalez 1987), yet the need for such a diverse array and for broad substrate specificity in autotrophic plants is arguable. All of the cytochrome P450 purification work in plants has indicated that more than one form of P450 is present in the tissue studied (Gabriac *et al.* 1985; Higashi *et al.* 1985; O'Keefe and Leto 1989; Kochs and Grisebach 1990), although the avocado study indicated that the two forms were very similar (> 92% identical primary sequence).

Some recent reports indicate that under conditions where cytochrome P450-dependent enzyme activity is induced by many-fold, the net increase in total spectrally detectable cytochrome P450 can be only a few per cent (Stewart and Schuler 1989; Fonne-Pfister *et al.* 1990; Frear and Mardaus, unpublished). The suggestion from these results is that the P450 dependent activity is coming from large changes in the amount of a very small fraction of the total cytochrome P450, or the alternative possibility that the 'induction' is modulating the activity of the P450 complement already there. Spectroscopic studies of the substrate binding properties of the P450 (O'Keefe and Leto 1989) could resolve these two possibilities. Nonetheless, this first interpretation supports the hypothesis of the presence of at least one other cytochrome P450 in substantial excess over the active one in the tissues studied.

Considering the difficulties in purifying cytochromes P450 from low-yielding plant tissues, a number of investigators have used an alternative strategy of selective inhibition of putative P450s by mechanism-based inhibitors (Reichhart *et al.* 1982; Cabanne *et al.* 1987; Hallahan *et al.* 1988; McFadden *et al.* 1989). Differential inhibition by mechanism-based inhibitors, particularly 1-aminobenzotriazole, have proved extremely useful in discerning between different P450 activities. Even with very good kinetic data, and very large differences in inhibitory efficacy, this approach can only resolve small numbers of discrete cytochrome P450 activities. So it remains that, in spite of evidence that there is often more than a single cytochrome P450 in plant tissues, there is still no compelling evidence to suggest that the number of distinct P450 proteins is very large. The final resolution of this issue will undoubtedly require the application of molecular genetic techniques to determine the number of P450 genes in the plant genome.

## The herbicide mono-oxygenase systems of *Streptomyces griseolus*

The ability of whole cells of the actinomycete, *Streptomyces griseolus*, to metabolize a wide variety of herbicidal sulphonylureas led to the discovery that this organism had both constitutive and herbicide-inducible cytochromes P450 (Romesser and O'Keefe 1986; O'Keefe *et al.* 1988). Two herbicide-inducible cytochromes P450, known as $P450_{SU1}$ and $P450_{SU2}$ (P450CVA1 and P450CVB1, respectively, by the nomenclature of Nebert *et al.* 1987), have been purified to homogeneity and the genes cloned and sequence (Omer *et al.* 1990). These proteins function as terminal oxidases of herbicide mono-oxygenase systems, in concert with low molecular weight ferredoxins and NAD(P)H:ferredoxin oxidoreductase (O'Keefe *et al.* 1987; O'Keefe *et al.* 1991). While both $P450_{SU1}$ and $P450_{SU2}$ have a similar affinity for sulphonylurea molecules, they exhibit remarkably different primary structures. Metabolically, the difference between them is most clearly demonstrated by their metabolism of chlorimuron ethyl (N-[(4-chloro-6-methoxypyrimidine-2yl) aminocarbonyl]-2-ethoxycarbonylbenzenesulphonamide).

$P450_{SU1}$ metabolism leads to the formation of two metabolites in mutually exclusive reactions of nearly equal rates, a demethylation of the methoxy substituent on the pyrimidine ring and oxidative cleavage of the ethyl ester (see Fig. 15.3). P450$_{SU2}$ primarily (approximately 90%) carries out only the demethylation reaction (Guengerich 1987; O'Keefe *et al.* 1987; Harder *et al.* in preparation). A possible explanation for the different metabolite partitioning by the two *S. griseolus* cytochromes P450 is that a single conformation of the substrate in the active site (see Fig. 15.3) can lead to the formation of either of two distinct metabolites, both of which are excluded from any subsequent metabolism by the same enzyme. This type of 'folded' conformation may be more readily achievable in P450$_{SU1}$ than in P450$_{SU2}$, leading to the distinct patterns of metabolite formation.

## Comparison of P450$_{SU1}$ and P450$_{SU2}$ to P450$_{CAM}$

We have compared the amino acid sequences of P450$_{SU1}$ and P450$_{SU2}$ with other cytochromes P450 (Omer *et al.* 1990). An excellent P450 to compare with P450$_{SU1}$ and P450$_{SU2}$ is cytochrome P450$_{CAM}$, from *Pseudomonas putida*, for which a high resolution (1.63 Å) crystal structure has been determined (Poulos *et al.* 1985; Poulos *et al.* 1987). Such comparisons may enable possible identification of regions of the P450$_{SU1}$ and P450$_{SU2}$ proteins involved in sulphonylurea recognition and suggest amino acids that could be targeted for site-directed mutagenesis in order to change the specificity of these, or other, cytochromes P450. The aforementioned comparison of the amino acid sequences of these three proteins (Omer *et al.* 1990) is shown here, along with additional interpretations (Fig. 15.2).

Sequence alignments show that the amino acid sequence of cytochrome P450$_{CAM}$ is 29.7% positionally identical to that of cytochrome P450$_{SU1}$ and 29.4% identical to cytochrome P450$_{SU2}$, with overall similarities (including conservative amino acid changes) of 50.3% and 50.1%, respectively. Cytochromes P450$_{SU1}$ and P450$_{SU2}$ are 44.4% identical to each other, with an overall similarity of 65.2%. However, the amino termini of the three proteins are quite dissimilar for about the first 60 amino acids. Overall, several clusters of amino acids appear to be similar among the three proteins, particularly the regions 165–173, 244–268, 282–290 and 347–385, using the P450$_{CAM}$ numbering.

The regions that are the most conserved among the three proteins align with those that interact with the haem in the P450$_{CAM}$ protein. The sulphydryl group of Cys357 of P450$_{CAM}$ provides an axial ligand to the Fe atom of the haem in the protein and amino acids 350–360 of P450$_{CAM}$ form a pocket consisting of a β-pair, β-bulge and hairpin turn around this cysteine residue. Eight of eleven amino acids in this region are identical among the three proteins, including the haem ligand cysteine. Two α-helices (helix I and L) of P450$_{CAM}$ sandwich the haem in the protein.

Helix L of P450$_{CAM}$ (amino acids 359–378) begins after the pocket that surrounds the Fe ligating Cys357. Seven of the twenty amino acids in this helix are identical at the corresponding residues in the three proteins, with an additional three amino acids where conservative substitutions occur in the corresponding positions of cytochromes P450$_{SU1}$ and P450$_{SU2}$. In P450$_{CAM}$ a Pro residue terminates helix L and this amino acid is conserved in all three proteins.

Helix I of P450$_{CAM}$ (amino acids 234–267) has 10 out of 34 residues identical with P450$_{SU1}$ and an additional 12 that differ by conservative changes, while 9 out of 34 residues of P450$_{SU2}$ are identical with those of P450$_{CAM}$, an additional 10 differing by conservative substitutions. Internal to helix I is a region of α-helical distortion, which is necessary to make room for $O_2$ coordination to the haem of P450$_{CAM}$. A consensus '$O_2$ binding' amino acid sequence of -A/G-G-X-D/E-T- has been proposed for this feature by comparing a number of cytochromes P450 (Poulos *et al.* 1987). The amino acid sequence -A-G-H-E-T-, which fits the consensus, is found in cytochromes P450$_{SU1}$ and P450$_{SU2}$ at the same position as the '$O_2$ binding' site in P450$_{CAM}$.

The two haem propionates in P450$_{CAM}$ are hydrogen (H)-bonded to five amino acids of the protein, with a sixth H-bond provided by a water molecule. Three of these five amino acids, Arg112, Arg299 and His355 are in identical locations in the aligned sequences of the three proteins. A fourth amino acid, Gln108 of P450$_{CAM}$ has a His residue at the corresponding aligned positions of P450$_{SU1}$ and P450$_{SU2}$ that could substitute for Gln in H-bonding. The fifth H-bond donor from P450$_{CAM}$, Asp297, has a corresponding Asn in P450$_{SU2}$ and an Asp residue one position away in the aligned sequence of P450$_{SU1}$.

The striking homology of these proteins in the regions involved with interaction between the apoprotein and the haem indicate that the three proteins are quite likely to have considerable over-

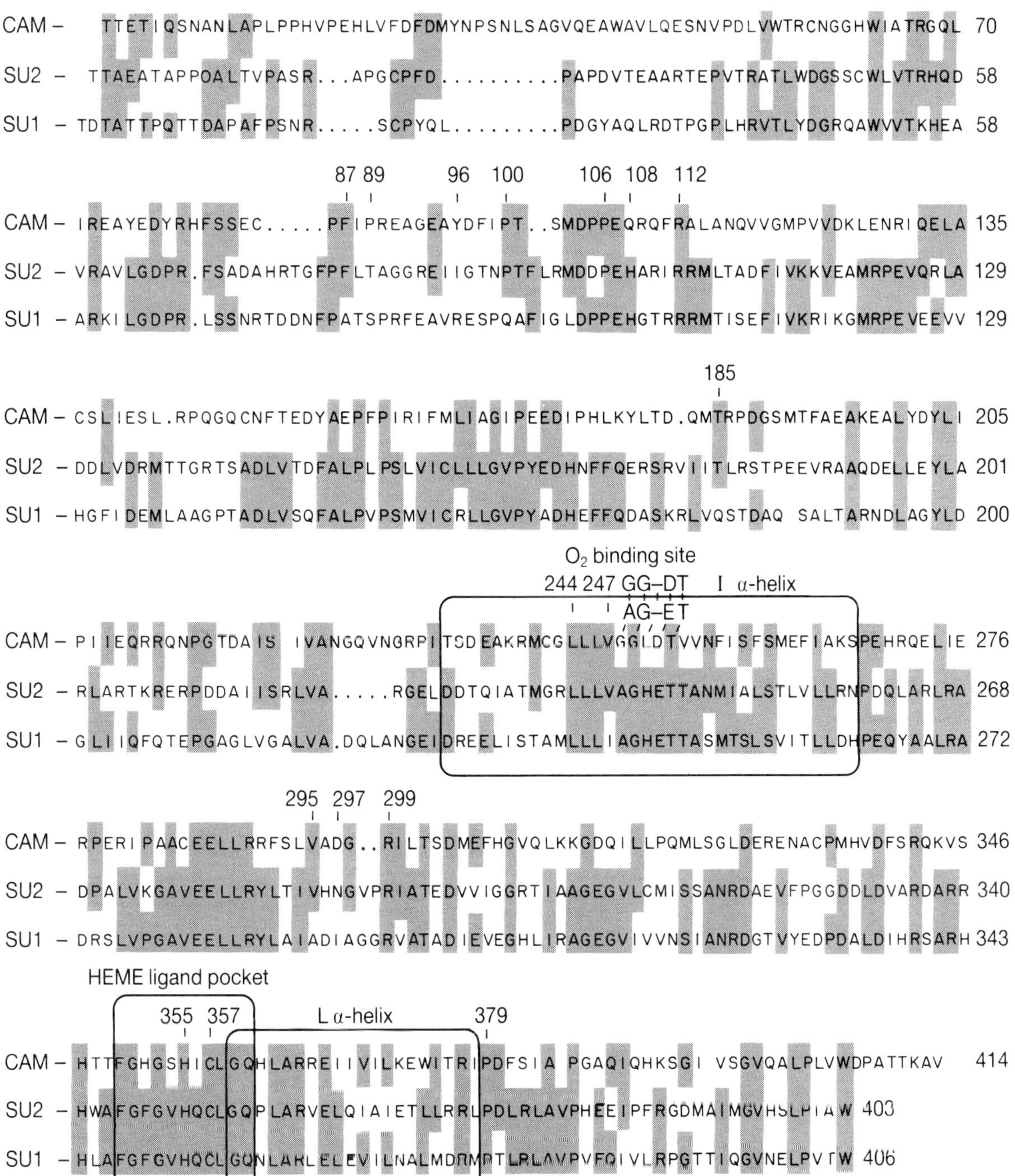


**Fig. 15.2** A comparison of the primary amino acid sequences of cytochrome P450$_{CAM}$ with the sequences of *S. griseolus* cytochromes P450$_{SU1}$ and P450$_{SU2}$. Features from the three-dimensional structure of cytochrome P450$_{CAM}$ are shown. (Taken from Omer *et al.* 1990.)

all structural similarity, perhaps including portions of the substrate binding sites. All three proteins, like most P450s, utilize hydrophobic substrates. Yet there are quite dramatic differences between the substrate specificity of cytochrome $P450_{CAM}$ compared to both cytochromes $P450_{SU1}$ and $P450_{SU2}$ (Omer *et al.* 1990). Aside from predominantly hydrophobic interactions, the appropriate orientation of camphor in the active site is strongly dependent on a hydrogen bond between the camphor carbonyl and Tyr96 (Fisher and Sligar 1985). This interaction is spectrally detectable by a shift in the UV absorption spectrum of the protein. There are no similar spectral shifts in $P450_{SU1}$ and $P450_{SU2}$, suggesting that hydrogen bonding to an analogous tyrosine is not important for sulphonylurea binding to these proteins. Alternatively, we expect that the negative charge on the sulphon-

amide nitrogen of a sulphonylurea at neutral pH may be important for binding to the protein. Obvious factors that also differentiate between the binding sites of $P450_{CAM}$ and $P450_{SU1}$ are simply the size of the molecules, as shown in Figure 15.3, and their flexibility. Camphor is a fused ring system and is relatively rigid, whereas sulphonylureas can rotate around several bonds, leading to orientations where the two sites of hydroxylation are adjacent (Fig. 15.3). While it is possible that a single conformation of substrate in the active site leads to the production of two metabolites, as is observed for chlorimuron ethyl with $P450_{SU1}$ (Harder *et al.* in preparation), we have not experimentally resolved whether this is more likely than two distinct binding orientations leading to the two products.

Because the haem itself forms one entire side of

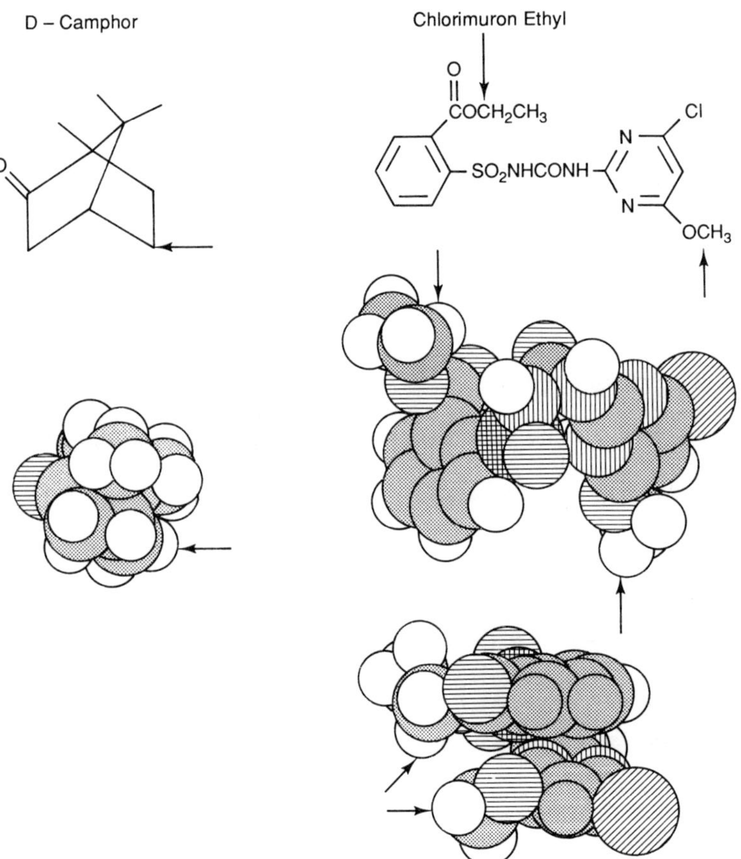


**Fig. 15.3**   A comparison of the substrate for $P450_{CAM}$ (camphor) with chlorimuron ethyl, a substrate for the *S. griseolus* cytochromes. Arrows indicate the locations of hydroxylation reactions by $P450_{CAM}$ and $P450_{SU1}$. Note the two conformations of chlorimuron ethyl, the lower one indicating a stacked conformation of the two rings, one of several in which the two different sites of hydroxylation by $P450_{SU1}$ are adjacent.

the active site pocket, many of the hydrophobic contacts of camphor within the active site of P450$_{CAM}$ occur with the haem. From the above comparative analysis, the haem is similarly placed in all three proteins and probably plays an important role in providing a hydrophobic environment for the sulphonylurea substrates of P450$_{SU1}$ and P450$_{SU2}$. Similarly, because the highly conserved helix I also borders one side of the substrate binding site, there are already substantial constraints on the structure of the P450$_{SU1}$ and P450$_{SU2}$ substrate binding sites. Additional contacts between cytochrome P450$_{CAM}$ and camphor occur through a number of residues. The two amino acids of P450$_{CAM}$ in closest contact with camphor, Leu244 and Val295, are either identical or have conservative substitutions in the aligned P450$_{SU1}$ and P450$_{SU2}$ sequences. These residues are in the helix-rich, more invariant end of the binding site (Poulos *et al*. 1985). Other amino acids of P450$_{CAM}$ that have close contacts with camphor are Phe87, Tyr96, Thr185, Val247 and Asp297. As already discussed, Tyr96 is not conserved in either of the *S. griseolus* cytochromes. The analogous region around Tyr96 has also been implicated as important in progesterone binding to P450IIC4 and P450IIC5 (Kronbach *et al*. 1989). This region in P450$_{CAM}$ contains three *cis* prolines (89, 100 and 106) that are important for the alignment of residues 87, 96 and 101 (Poulos *et al*. 1987). Prolines in this vicinity of both P450$_{SU1}$ and P450$_{SU2}$, while not exactly aligned, could be serving a similar function. However, the entire region, from residue 79 to 93 (in SU1 numbering), is very poorly conserved between SU1 and SU2, making it difficult to recognize it as being important for herbicide binding.

Of the remaining residues involved in hydrophobic contact with camphor, Thr185 is conserved in P450$_{SU2}$ and has a Ser one residue away in P450$_{SU1}$. Val247 is conserved in P450$_{SU2}$ with an Ile at the corresponding P450$_{SU1}$ residue. Asp297 has an Asn at that location in P450$_{SU2}$ with an Asp one residue away in P450$_{SU1}$. It is not surprising that the active site residues of P450$_{CAM}$ are poorly conserved in P450$_{SU1}$, but it is somewhat surprising that they are so strongly conserved in P450$_{SU2}$. Moreover, given the similarity of substrates for the two *S. griseolus* cytochromes, our expectations were that the substrate binding regions could be found by looking for sequence similarities. This has not been the case. As Figure 15.2 shows, some of the strongest similarities between P450$_{SU1}$ and P450$_{SU2}$

are in regions that are remote from the aligned regions of substrate binding in P450$_{CAM}$ (see 143–173 and 276–290 of the SU1 numbering, for example). However, recent findings demonstrate that site-directed mutations of the eukaryotic P450$_{coh}$ in residues that are clearly remote from the active site (based on the P450$_{CAM}$ crystal structure) can strongly influence the substrate specificity of the enzyme (Lindberg and Negishi 1989). Findings such as these serve to emphasize that the analysis of structure based on sequence comparisons must be viewed with a very critical eye.

## Conclusions

Earlier, we described several cases where herbicide resistance was conferred by the introduction into plants of genes encoding for enzymes capable of herbicide detoxification. The sources for all of these resistance genes were bacteria. The existence of a variety of herbicide metabolizing cytochrome P450 mono-oxygenase activities within the entire Plant Kingdom suggests that plants will provide an alternative and more closely related source of detoxification genes. Nonetheless, this avenue has not yet been explored, and for comparatively simple reasons. Several successful attempts at purification of the P450 protein, and an unknown number of unsuccessful attempts at purification, have only recently led to the availability of antibodies and sequence information that would be useful to identify the cytochrome P450 genes. This has led to the identification of an authentic cytochrome P450 cDNA sequence from a plant (Bozak *et al*. 1990). It is probable that this will lead to the identification of P450 genes in other species and, given the variety of P450 substrate specificities we know to be present in the Plant Kingdom, this could lead to an explosion of transformations of agronomically and horticulturally useful plants using P450-mediated traits.

As evidenced by the majority of successes reported so far, further exploration for bacterial genetic material encoding for herbicide metabolizing genes is also a valuable approach. The genus *Streptomyces* has long been known as a source of 'unusual' chemical metabolites, and this ability is also evidenced by the fact that one species, *S. griseolus*, can express two distinct cytochromes P450 capable of metabolizing a wide range of sulphonylurea (and some other) herbicides. A possibility worth exploring is that certain groups of bacteria may be able to produce a wide variety of herbicide metabolizing enzymes on demand; strategies may then be designed to retrieve the detoxifying gene(s) from these organisms.

Several laboratories have reported recent progress in modifying the substrate specificity of several cytochromes P450 by site-directed mutagenesis. These studies have been stimulated by knowledge of the three-dimensional structure of cytochrome P450$_{CAM}$. In one report the modified residues in P450$_{CAM}$ were selected from their known involvement in the orientation of the substrate molecule in the active site of the protein (Atkins and Sligar 1989). Other studies have been based on the inferred homology in the active site region of the eukaryotic P450s to cytochrome P450$_{CAM}$, and also on the well resolved substrate specificity differences between some closely related (high sequence similarity) mammalian cytochromes (Furuya *et al.* 1989; Imai and Nakamura 1989; Kronbach *et al.* 1989; Lindberg and Negishi 1989). The finding that residues predicted by inference to be involved in substrate binding are indeed involved, adds further support to the conclusion that most eukaryotic cytochromes P450 may be modelled by adding a hydrophobic anchor to the amino terminus of P450$_{CAM}$. More importantly, though, the alteration of substrate specificity suggests that it may be possible to reconfigure cytochrome P450 molecules to optimize reactions on a xenobiotic of choice ('designer detox'). Coupled with the already promiscuous substrate specificity of these proteins, this seems an obvious route for future design of herbicide resistant plants.

## Acknowledgements

The authors are grateful to Kristin Bozak and Rolf Christoffersen for access to their data prior to publication, and to their many colleagues at DuPont and DNA Plant Technology who have been involved in this work, particularly: Ken Leto, Caroline Dean, Jim Tepperman, Jim Romesser, Pat Harder and Nina Patel.

## References

Atkins, W.M. and Sligar, S.G. (1989). Molecular recognition in cytochrome P-450: alteration of regioselective alkane hydroxylation via protein engineering. *Journal of the American Chemical Society*. **111**: 2715–17.

Benveniste, I., Gabriac, B. and Durst, F. (1986). Purification and characterization of the NADPH-cytochrome P450 (cytochrome c) reductase from higher plant microsomal fraction. *Biochemical Journal*. **235**: 365–73.

Black, S.D. and Coon, M.J. (1986). Comparative structures of P-450 cytochromes. In *Cytochrome P-450 Structure, Mechanism, and Biochemistry* (ed. P.R. Ortiz de Montellano), pp. 161–216. Plenum Press, New York.

Bourett, T.M., Howard, R.J. and O'Keefe, D.P. (1989). Immunocytochemical localization of plant cytochromes P-450. *Plant Physiology*. **89s**: 913.

Bozak, K.R., Yu, H., Sirevag, R. and Christofferson, R.E. (1990). Cloning and sequence analysis of ripening related cytochrome P-450 cDNAs from avocado fruit. *Proceedings of the National Academy of Sciences, USA*. **87**: 3904–8.

Brown, C.A. and Black, S.D. (1989). Membrane topology of mammalian cytochromes P-450 from liver endoplasmic reticulum. *Journal of Biological Chemistry*. **264**: 4442–9.

Cabanne, F., Hoby, D., Gaillardon, P. and Durst, F. (1987). Effect of the cytochrome P-450 inactivator 1-aminobenzotriazole on the metabolism of chlortoluron and isoproturon in wheat. *Pesticide Biochemistry and Physiology*. **28**: 371–80.

Chaleff, R.S. and Mauvais, C.J. (1984). Acetolactate synthase is the site of action of two sulfonylurea herbicides in higher plants. *Science*. **224**: 1143–5.

Chaleff, R.S. and Ray, T.B. (1984). Herbicide resistant mutants from tobacco cell culture. *Science*. **223**: 1148–51.

Cole, D. (1983). Oxidation of xenobiotics in plants. In *Progress in Pesticide Biochemistry and Toxicology*, Vol. 3 (ed. D.H. Hutson and T.R. Roberts), pp. 199–254. John Wiley and Sons Ltd., New York.

DeBlock, M., Botterman, J., Vandewiele, M. *et al.* (1987). Engineering herbicide resistance in plants by expression of a detoxifying enzyme. *EMBO Journal*. **6**: 2513–18.

Edwards, R.J., Murray, B.P., Boobis, A.R. and Davies, D.S. (1989). Identification and location of α-helices in mammalian cytochromes P-450. *Biochemistry*. **28**: 3762–70.

Fisher, M.T. and Sligar, S.G. (1985). Tyrosine motions in relation to the ferric spin equilibrium of cytochrome P-450$_{CAM}$. *Biochemistry*. **24**: 6696–701.

Fonné-Pfister, R., Gaudin, J., Kreuz, K. *et al.* (1990). Hydroxylation of primisulfuron by an inducible cytochrome P-450-dependent mono-oxygenase system from maize. *Pesticide Biochemistry and Physiology*. **37**: 165–73.

Furuya, H., Shimizu, T., Hirano, K. *et al.* (1989). Site directed mutagenesis of rat liver cytochrome P-450$_d$: catalytic activities toward benzphetamine and 7-ethoxycoumarin. *Biochemistry*. **28**: 6848–57.

Gabriac, B., Benveniste, I. and Durst, F. (1985). Isolation and characterization of cytochrome P-450 from higher plants (*Helianthus tuberosis*). *Comptes Rendus*. **301 Serie III**: 753–8.

Guengerich, F.P. (1987). Oxidative cleavage of carboxylic acids by cytochrome P-450. *Journal of Biological Chemistry*. **262**: 8459–62.

Guengerich, F.P. and Macdonald, T.L. (1990). Mechanisms of cytochrome P-450 catalysis. *FASEB Journal*. **4**: 2453–9.

Hallahan, D.L., Heasman, A.P., Grossel, M.C. *et al.* (1988). Synthesis and biological activity of an azido derivative of paclobutrazole, an inhibitor of gibberellin biosynthesis. *Plant Physiology*. **88**: 1425–9.

Higashi, K., Ikeuchi, K., Obara, M. *et al.* (1985). Purification of a single form of microsomal cytochrome P-450 from tulip bulbs (*Tulipa gesneriana* L.). *Agricultural and Biological Chemistry*. **49**: 2399–405.

Imai, Y. and Nakamura, M. (1989). Point mutations at threonine-301 modify substrate specificity of rabbit liver microsomal cytochromes P-450 (laurate ($\omega$-1)-hydroxylase and testosterone 16$\alpha$-hydroxylase). *Biochemical and Biophysical Research Communications*. **158**: 717–22.

Kalb, V.F. and Loper, J.C. (1988). Proteins from eight eucaryotic cytochrome P-450 families share a segmented region of sequence similarity. *Proceedings of the National Academy of Sciences, USA*. **85**: 7221–5.

Kochs, G. and Grisebach, H. (1989). Phytoalexin synthesis in soybean: purification and reconstitution of cytochrome P-450 3,9-dihydroxypterocarpan 6a-hydroxylase and separation from cytochrome P450 cinnamate 4-hydroxylase. *Archives of Biochemistry and Biophysics*. **273**: 543–53.

Kronbach, T., Larabee, T.M. and Johnson, E.F. (1989). Hybrid cytochromes P-450 identify a substrate binding domain in P-450IIC5 and P-450IIC4. *Proceedings of the National Academy of Sciences, USA*. **86**: 8262–5.

Kyte, J. and Doolittle, R.F. (1982). A simple method for displaying the hydrophobic character of a protein. *Journal of Molecular Biology*. **157**: 105–32.

Lindberg, R.L. and Negishi, M. (1989). Alteration of mouse chytochrome P-450$_{coh}$ substrate specificity by mutation of a single amino-acid residue. *Nature*. **339**: 632–4.

Mazur, B.J. and Falco, S.C. (1989). The development of herbicide resistant crops. *Annual Review of Plant Physiology and Plant Molecular Biology*. **40**: 441–70.

McFadden, J.J., Frear, D.S. and Monsager, E.R. (1989). Aryl hydroxylation of diclofop by a cytochrome P-450 dependent monooxygenase from wheat. *Pesticide Biochemistry and Physiology*. **34**: 92–100.

Narhi, L.O. and Fulco, A.J. (1987). Identification and characterization of two functional domains of cytochrome P-450$_{BM-3}$, a catalytically self sufficient monooxygenase induced by barbiturates in *Bacillus megaterium*. *Journal of Biological Chemistry*. **262**: 6683–90.

Nebert, D.W., Adesnik, M., Coon, M.J. *et al.* (1987). The P450 gene superfamily: recommended nomenclature. *DNA*. **6**: 1–11.

Nebert, D.W. and Gonzalez, F.J. (1987). P450 genes: structure, evolution and regulation. *Annual Review of Biochemistry*. **56**: 945–93.

Nelson, D.R. and Strobel, H.W. (1988). On the membrane topology of vertebrate cytochrome P-450 proteins. *Journal of Biological Chemistry*. **263**: 6038–50.

O'Keefe, D.P., Gibson, K.J., Emptage, M.H. *et al.* (1991). Ferredoxins from two sulfonylurea herbicide mono-oxygenase systems in *Streptomyces griseolus*. *Biochemistry*. **30**: 447–55.

O'Keefe, D.P. and Leto, K.J. (1989). Cytochrome P-450 from the mesocarp of avocado (*Persea americana*). *Plant Physiology*. **83**: 1141–9.

O'Keefe, D.P., Romesser, J.A. and Leto, K.J. (1987). Plant and bacterial cytochromes P-450: involvement in herbicide metabolism. In *Phytochemical Effects of Environmental Compounds* (ed. J.A. Saunders, L. Kosack-Channing and E.E. Conn), pp. 151–73. Plenum Press, New York.

O'Keefe, D.P., Romesser, J.A. and Leto, K.J. (1988). Identification of constitutive and herbicide inducible cytochromes P-450 in *Streptomyces griseolus*. *Archives of Microbiology*. **149**: 406–12.

Omer, C.A., Lenstra, R., Litle, P.J. *et al.* (1990). Genes for two herbicide inducible cytochromes P-450 from *Streptomyces griseolus*. *Journal of Bacteriology*. **172**: 3335–45.

Ortiz de Montellano, P.R. (1986). Oxygen activation and transfer. In *Cytochrome P-450 Structure, Mechanism and Biochemistry* (ed. P.R. Ortiz de Montellano), pp. 217–72. Plenum Press, New York.

Peterson, J.A. and Prough, R.A. (1986). Cytochrome P-450 reductase and cytochrome b$_5$ in cytochrome P-450 catalysis. In *Cytochrome P-450 Structure, Mechanism and Biochemistry* (ed. P.R. Ortiz de Montellano), pp. 89–118. Plenum Press, New York.

Poulos, T.L., Finzel, B.C., Gunsalus, I.C. *et al.* (1985). The 2.6 Å crystal structure of *Pseudomonas putida* cytochrome P-450$_{CAM}$. *Journal of Biological Chemistry*. **260**: 16122–30.

Poulos, T.L., Finzel, C.C. and Howard, A.J. (1987). High resolution crystal structure of cytochrome P-450$_{CAM}$. *Journal of Molecular Biology*. **195**: 687–700.

Reichhart, D., Simon, A., Durst, F. *et al.* (1982). Autocatalytic inactivation of plant cytochrome P-450 enzymes: selective inactivation of cinnamic acid-4-hydroxylase from *Helianthus tuberosus* by 1-aminobenzotriazole. *Archives of Biochemistry and Biophysics*. **216**: 522–9.

Romesser, J.A. and O'Keefe (1986). Induction of cytochrome P-450 dependent sulfonylurea metabolism in *Streptomyces griseolus*. *Biochemical and Biophysical Research Communications*. **140**: 650–9.

Sebastian, S.A. and Chaleff, R.S. (1987). Soybean mutants with increased tolerance for sulfonylurea herbicides. *Crop Science*. **27**: 948–52.

Shaner, D.L., Anderson, P.C., Stidham, M.A. (1984). Imidazolinones (potent inhibitors of acetohydroxyacid synthase). *Plant Physiology*. **76**: 545–6.

Stalker, T.M., McBride, K.E. and Malyj, L.D. (1988). Herbicide resistance in transgenic plants expressing a bacterial detoxification gene. *Science*. **242**: 419–23.

Stewart, C.B. and Schuler, M.A. (1989). Antigenic crossreactivity between bacterial and plant cytochrome P-450 monooxygenases. *Plant Physiology*. **90**: 534–41.

Streber, W.R. and Willmitzer, L. (1989). Transgenic tobacco plants expressing a bacterial detoxifying enzyme are resistant to 2,4-D. *Bio/technology*. **7**: 811–16.

Streber, W.R., Timmis, K.N. and Zenk, M. (1987). Analysis, cloning, and high level expression of 2,4-dichlorophenoxyacetate monooxygenase gene *tfdA* of *Alcaligenes eutrophus* JMP134. *Journal of Bacteriology*. **169**: 2950–5.

Sweetser, P.B., Schow, G.S. and Hutchison, J.M. (1982). Metabolism of chlorsulfuron by plants. *Pesti-cide Biochemistry and Physiology*. **17**: 18–23.

Szczesna-Skorupa, E., Browne, N., Mead, D. and Kemper, B. (1988). Positive charges at the $NH_2$ terminus convert the membrane anchor signal-peptide of cytochrome P-450 to a secretory signal peptide. *Proceedings of the National Academy of Sciences, USA*. **85**: 738–42.

Thompson, C.J., Raomouva, N., Tizard, R. *et al.* (1987). Characterization of the herbicide resistance gene *bar* from *Streptomyces hygroscopicus*. *EMBO Journal*. **6**: 2519–23.

Vergeres, G., Winterhalter, K.H. and Richter, C. (1989). Identification of the membrane anchor of microsomal rat liver cytochrome P-450. *Biochemistry*. **28**: 3650–5.

Wallnofer, P.R. and Engelhardt, G. (1989). Microbial degradation of pesticides. In *Chemistry of Plant Protection 2* (ed. W.S. Bowers, W. Ebing, D. Martin, I. Yamamoto and R. Wegler), pp. 1–115. Springer-Verlag, Berlin.

# 16

# Protein Engineering of *Bacillus thuringiensis* δ-Endotoxins and Genetic Manipulation for Plant Protection

Donald H. Dean and Michael J. Adang

## Introduction

Biorational pesticides have the potential of becoming the major means of pest control, replacing chemicals as a pesticide against susceptible insects (Wilcox *et al.* 1986). Among the many microbial pest control agents, *Bacillus thuringiensis* is the most widely used. The advantages of *B. thuringiensis* include its low environmental impact, low persistence and high specificity against target insects with low or no activity against non-target insects and animals. There is a low development of resistance in the target insects (especially when compared to chemical pesticides). The ease of making derivatives of the bioactive proteins through genetic engineering, and the ease with which the genes may be cloned into plants, makes them ideal for future development. Even the economic picture is becoming positive, with the rising costs of developing chemical pesticides.

2.3 million kg of *B. thuringiensis* are used worldwide (Rowe and Magaritis 1987). In forestry, *B. thuringiensis* made up 74% of the total pesticide used in Canada against spruce bud worm in 1986 (Morris *et al.* 1986) and is in even greater use today. In the United States *B. thuringiensis* surpassed other control measures for gypsy moth in 1989 (Twardus 1989).

Over 450 uses and formulations of *B. thuringiensis* are registered by the EPA (P. Hutton, personal communication) for pest insect control. The bulk of the market is controlled by three producers: Abbott Laboratories (North Chicago, Illinois, USA), Solvay (Brussels, Belgium), and Sandoz (Basel, Switzerland). Opportunities for discovering novel strains and for genetically improving existing strains have encouraged more companies to develop *B. thuringiensis* biopesticides. Novo BioKontrol Division of Novo Laboratories, Mycogen and Ecogen are new entrants to the agricultural *B. thuringiensis* market. The dominant commercial product since the 1970s is based on the subspecies *kurstaki* HD-1 (serotype 3a3b). It is effective on over 200 crops and against more than 55 lepidopteran species. It is one of the products of choice against forest insects, spruce budworm and gypsy moth and is increasingly used for early season control of *Heliothis* species on cotton, where worms are resistant to pyrethroid insecticides, and on vegetables against diamondback moths, which are resistant to all known chemical controls.

More recently, the discovery of subsp. *tenebrionis* by Krieg *et al.* (1983) and subsp. *san diego* by Herrnstadt *et al.* (1986) broadened the host range for biopesticides to include *Coleoptera*. The first use of this strain is for the control of Colorado potato beetles. In some areas of the United States, Colorado potato beetles are resistant to all chemical pesticides and alternative methods of control are needed. The Mycogen Corporation is marketing a product, M1, and other companies (Abbott Laboratories, Sandoz and Ecogen) are near commercial production with similar formulations. However, it is clear that the use of microbial pest control measures on agricultural crops lags behind their use on forests, but this too may change because of recent developments; namely, overexpression of the *B. thuringiensis* insecticidal crystal proteins (ICPs) and introduction of ICP genes into agricultural crop plants. The present review is intended to address the potential for protein engineering of the *B. thuringiensis* ICPs and other insecticidal proteins for plant protection, including their introduction into the plant genome.

This review will cover the rather diverse research that has been done in the area of biochemistry and molecular genetics (i.e. protein engineering) of the *B. thuringiensis* crystal protein genes and plant molecular genetics pertaining to the introduction of these genes into plants. Since the major interest of this book is plants and crop protection, we shall consider the mode of action of those ICPs that affect *Lepidoptera*. ICPs that affect mosquitoes and blackflies are of no less interest academically and economically, but are less relevant in the context of plant protection. Those ICPs that affect coleopterans are less well understood *vis-à-vis* their mode of action and biochemistry. Consequently, we will add only a few words at the appropriate places.

The two authors have had specific experience in the area each is addressing in this review. Protein engineering will be reviewed by DHD and genetic manipulation in plants and plant epiphytes by MJA. One particular caveat should be made about this liaison: the structural model proposed in the protein engineering section is wholly the responsibility of DHD.

## *Bacillus thuringiensis* and its insecticidal crystal proteins

### Biological properties and mode of action

*B. thuringiensis* is a spore-forming bacterium that forms insecticidal proteinaceous crystals *in situ* (hence the protoxin has been called the δ-endotoxin (Heimpel and Angus 1959)) during the latter stages of its life cycle. There are many varieties (serotypes) of *B. thuringiensis*, some of which are active against certain dipterans (specifically mosquitoes and blackflies) or *Coleoptera* (e.g. Colorado potato beetle), but most of which are active against lepidopterans (for general reviews, see Dean 1984; Aronson *et al.* 1986; Whiteley and Schnepf 1986; Andrews *et al.* 1987). During the sporulation stage, *B. thuringiensis* forms a parasporal crystal, the shape of the crystal depending upon the type of gene encoding the crystal protein. Lepidopteran crystal proteins of the cryIA class are bipyramidal; dipteran crystal proteins of the cryIV class are amorphous; coleopteran-active crystals of the cryIIIA class are flat and rectangular and crystals of the cryIIA class, which are active against both *Lepidoptera* and *Diptera* are cuboidal.

The mode of action of the lepidopteran ICPs has been included in other recent reviews (Milne *et al.* 1990) and will only be briefly listed here:

(1) The gene product occurs as an inactive protoxin in the crystal.

(2) The crystal and subunits are dissolved at the alkaline pH of the insect midgut, cleaved by insect gut proteases and reduced to activated toxins.

(3) The midgut columnar epithelial cells and goblet cells are attacked by the toxin and damaged, causing an immediate loss of the active potassium transport system and ultimate lysis. In susceptible insects, paralysis and death may occur with the toxin alone. In less sensitive insects, damage to the midgut allows an avenue for invasion of the haemolymph by *B. thuringiensis* from the germinated spore (Heimpel and Angus 1959).

(4) The mechanism of cell lysis appears to involve binding to specific receptors on the cell membrane (Hofmann *et al.* 1988a, b; Van Rie *et al.* 1989, 1990a, b) and a toxic ionophoric lesion of the membrane (Wolfersberger *et al.* 1986; Knowles and Ellar 1987).

(5) The intracellular action of the toxin has become a point of debate. Fast *et al.* (1978) reported that enzyme-digested toxin, covalently bound to Sephadex beads, has the same effect on Cf-124 cells as free toxin; that is, the cells swell and burst. This indicates that action at the cell membrane is sufficient for all activity, at least on tissue culture cells, but differences in intracellular action on different insects (Travers *et al.* 1976; Ebersold *et al.* 1977; Nishiitsutsuji-Uwo and Endo 1979; Endo and Nishiitsutsuji-Uwo 1980; Tojo 1986) suggest an intercellular role for the activated toxin, or a part of it.

### Genetics of δ-endotoxin genes

#### Genetic nomenclature

The first ICP gene to be isolated was cloned by Schnepf and Whiteley (1981). Since then, a number of groups have cloned and sequenced ICP genes, which now total fourteen. Several genetic nomenclatures were adopted, including *icp* (insecticidal crystal protein, Shibano *et al.* 1985; Ge *et al.* 1989), *kurhd1*, from a combination of the name of the *B. thuringiensis* serotype, *kurstaki*, and the strain name, HD-1 (Geiser *et al.* 1986), *D-entox* (Dean 1982), and *cry* (Held *et al.* 1982). Recently, a unifying nomenclature for the *B. thuringiensis* ICP genes has been suggested (Hofte and Whiteley 1989), utilizing *cry* as the mnemonic (Table 16.1). The fundamental basis for the taxonomy is the specificity as to insect order: *cry*I is active against *Lepidoptera*; *cry*II against *Lepidoptera* and *Diptera*; *cry*III against *Diptera*; and *cry*IV against *Coleoptera*. The secondary criterion is amino acid sequence homology: *cry*IA(*a*), *cry*IA(*b*), and *cry*IA(*c*) show greater than 80% sequence homology to *cry*IA(*a*), while *cry*IB shows less than 80% homology to *cry*IA(*a*).

Several exceptions to this nomenclature immediately presented themselves: *cry*IA(*b*) exhibits high homology to the other *cry*IA genes but is known to be active against both *Lepidoptera* and *Diptera*, due to 1–3 amino acid alterations (Haider *et al.* 1986; Haider *et al.* 1987; Haider and Ellar 1989; Haider *et al.* 1989), hence should be classified as *cry*II by the fundamental criterion. The classification of *cry*IIB is a double anomaly. It is greater than 80% homologous to *cry*IIA and therefore these two genes should be designated *cry*IIA(*a*) and *cry*IIA(*b*), comparable to the *cry*IA series. However, *cry*IIB is active only against *Lepidoptera* (also due to a few amino acid changes) and therefore

should be classified as a *cry*I gene by the fundamental criterion. Clearly, insecticidal activity may be dramatically altered by a few mutations. *cry*IA(*a*), *cry*IA(*b*), and *cry*IA(*c*), show more than an order of magnitude difference in activity against certain lepidopterans (see below), despite their sequence similarity; indeed, there is no significant reason why these genes should not be classified by separate locus letters rather than the clumsy parentheses. Despite the many shortcomings of this nomenclature system, it has provided a rough organization for the naming of *B. thuringiensis* ICP genes.

## Location of genes

The genes that encode the ICPs are sometimes located on the chromosome (Klier *et al.* 1983; Kronstad *et al.* 1983) but are usually located on plasmids (Kronstad *et al.* 1983). It has been shown that plasmids may be transmitted by a conjugation-like mechanism in *B. thuringiensis* and other bacilli (Gonzalez *et al.* 1982). There may also be more than one gene per cell (Kronstad *et al.* 1983; Wilcox *et al.* 1986). Given these facts, it is not surprising that new forms of the ICP gene have arisen and are found ubiquitously among the serotypes, perhaps shaped by selective pressure to respond to a new insect host. In this manner, *B. thuringiensis* has been able to adapt as an insect pathogen.

**Table 16.1** Insecticidal crystal protein genes and their specificity

| Gene | Specificity | | References |
| | Order | Species | |
|------|-------|---------|------------|
| *crylA(a)* | *Lepidoptera* | *Bombyx mori* | Ge *et al.* 1989 |
| *crylA(b)* | *Lepidoptera* | *Lymantria dispar* | Wolfersberger 1990 |
| | *Diptera* | *Aedes aegyptii* | Haider *et al.* 1987 |
| *crylA(c)* | *Lepidoptera* | *Trichoplusia ni* | Ge *et al.* 1990 |
| | | *Heliothis virescens* | Ge *et al.* 1990 |
| | | *Pieris brassicae* | Hofte and Whiteley 1989 |
| *crylB* | *Lepidoptera* | *Pieris brassicae* | Hofte and Whiteley 1989 |
| *crylc* | *Lepidoptera* | *Mamestra brassicae* | Hofte and Whiteley 1989 |
| | | *Spodoptera littoralis* | |
| *crylD* | *Lepidoptera* | *Manduca sexta* | Hofte and Whiteley 1989 |
| *crylIA* | *Lepidoptera* | *Lymantria dispar* | Widner and Whiteley 1990 |
| | *Diptera* | *Aedes aegyptii* | Widner and Whiteley 1990 |
| *crylIB* | *Lepidoptera* | *Lymantria dispar* | Widner and Whiteley 1990 |
| *crylIIA* | *Coleoptera* | *Leptinotarsa decemlinoata* | Kreig *et al.* 1983 |
| *crylIIB* | *Coleoptera* | not given | Sick *et al.* 1990 |
| *crylVA* | *Diptera* | *Culex* | Delecluse *et al.* 1988 |
| *crylVB* | *Diptera* | *Aedes* | Delecluse *et al.* 1988 |
| *crylVC* | *Diptera* | *Aedes* | Hofte and Whiteley 1989 |
| *crylVD* | *Diptera* | *Aedes* | Hofte and Whiteley 1989 |

## Two size classes of ICPs

Comparison of the derived amino acid sequences of the ICP genes reveals several remarkable features that may help locate functional domains of the protein. Two major structural groups exist: Type I, the large ICPs of approximately 130 kDa (*cry*I series, and *cry*IV*A* and *B*); and Type II, the small ICPs of approximately 65 kDa (*cry*II series; *cry*III*A*; *cry*IV*C* and *D*). The COOH-terminal halves of Type I are moderately conserved. The high cysteine content in this region suggests that the COOH-terminus is necessary for protein crystal formation *in vivo*, although this has not been rigorously demonstrated. This COOH-terminal region is digested away, along with a few residues from the NH₂-terminus, during the proteolytic activation of the protoxin. Hence, Type I ICPs consist of a NH₂-terminal toxic region and a COOH terminal putative crystal-forming region. Type II ICPs lack this COOH-terminus and appear to be naturally truncated at about the same position as the activated Type I toxins. How the Type II ICPs form crystals is not known, but we know that CryIIIA lacks disulphide bridges (Bernhard 1986).

## Composition of the toxin regions

The first half of the toxic region of the CryIA protein is highly conserved (Fig. 16.1). The second half is highly variable (i.e. less than 50% amino acid homology) with small interspersed tracts of conserved amino acids (Geiser *et al.* 1986; Hofte and Whiteley 1989; Table 16.2). The hypervariable region may be further divided into two sections based on differences in amino acid sequences among the three *cry*IA genes. In the first section, the *cry*IA(*b*) and *cry*IA(*c*) genes are similar, and the *cry*IA(*a*) gene differs. In the second section, the *cry*IA(*a*) and *cry*IA(*c*) genes are similar and the *cry*IA(*b*) gene differs. The actual boundaries of these portions of the toxicity region are: conserved region, amino acids 28–282; hypervariable region, section I, amino acids 283–461; and hypervariable region, section II, amino acids 462–610. We shall see that these sections may represent functional domains.

The tracts of highly conserved amino acids shown in Table 16.2 are also observed in all of the 14 identified ICP genes. These are identified as five (Hofte and Whiteley 1989) or more (personal

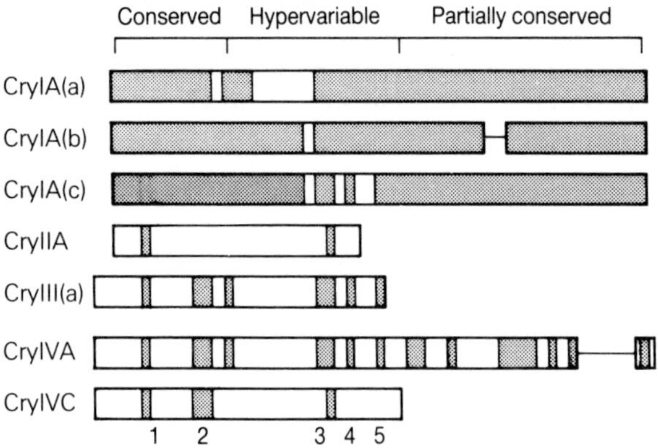

Fig. 16.1 Tracts of similar amino acids = ▨ in representative Cry proteins. The terms 'conserved', 'hypervariable' and 'partially conserved' refer only to the CryIA proteins. The sequences of the conserved tracts (1–5) are given in Table 16.2.

**Table 16.2** Conserved tracts of amino acids of the Cry proteins shown in Fig. 16.1

| Postition[a] | Sequence[b] |
| --- | --- |
| 1. 153–182 | Y – V – L L s v Y v Q A A N I H L – v L R D v – v F – – – W |
| 2. 225–272 | W v – y N – y R R e I T L t V L D i v A L F P n Y D – R – Y P – – – – I – – – s – L T R E i Y T |
| 3. 452–500 | f s W – H R S a e f – – N i l – – – – I T Q I P a V K – – – L – – – – – V V – G P G f T G G D l v |
| 4. 517–530 | S t S t r Y R v R v r Y A S |
| 5. 596–607 | V y l D r i E f i P v t |

[a] Position (amino acid residue) is relative to the sequence in the CryIA(a) proteins; [b] upper case letters are highly conserved amino acids; lower case are conservative changes; – indicates non-conservative changes.

observation) tracts, depending on how one sorts them. The importance of some of these tracts becomes even more obvious when the *cry*IIIA coleopteran gene is compared to the other genes. The *cry*IIIA gene is generally non-homologous with the other *cry* genes, yet it shares some homology with the 'conserved' region of the *cry*IA lepidopteran genes, and it shares three highly conserved tracts with the 'hypervariable' region of the *cry*I genes. Interestingly, the last highly conserved tract is at the boundary of the toxicity region, and any deletions into it from the COOH-terminal end completely remove toxicity (Hofte *et al.* 1986). This tract is also the terminus of the *cry*IIIA protein and the activated Type I toxins. Highly conserved tracts are considered to be important to protein function or structure (Bowie *et al.* 1990).

Genetic deletion studies (Adang *et al.* 1985; Wabiko *et al.* 1985; Schnepf and Whiteley 1985; Vaeck *et al.* 1986; Hofte *et al.* 1987) show that the gene sequence near the NH$_2$-terminal half of the toxin is required for toxicity. This predicts that amino acid residues 29–610 form the active toxin. Small deletions in the structural gene at the NH$_2$- and COOH-terminal ends with the exonuclease Bal31 (Hofte *et al.* 1986; Vaeck *et al.* 1986) demonstrate that toxicity requires the amino acids around residue 27 at the NH$_2$-terminal end and amino acid 607 at the C-terminal end. This analysis verifies the previous deletion studies, which demonstrated that only part of the δ-endotoxin gene is necessary for full toxicity (Adang *et al.* 1985; McLinden *et al.* 1985; Schnepf and Whiteley 1985; Wabiko *et al.* 1985). These genetic deletion studies confirmed biochemical studies showing by proteolytic cleavage of the toxin (Nagamatsu *et al.* 1984) that the NH$_2$-terminal half of the protein forms the active toxin.

The COOH-terminal part of the toxicity region comprises a highly conserved tract (no. 5, Hofte and Whiteley 1989). Hofte *et al.* (1986) have proposed that this tract forms a specific three-dimensional structure that is required for toxicity. Chou and Fasman's (1978) predictions would indicate a slight possibility of an α-helix for this tract.

Among the five highly conserved tracts of amino acids, four (nos 1, 2, 4 and 5) have significant structures of hydrophobic residues. The application of the principles of Bowie *et al.* (1990) to this combination of properties, i.e. conserved and hydrophobic, would suggest that these tracts play the structural role of forming the hydrophobic core of the toxin.

One of the most interesting conserved tracts, observed in virtually all insecticidal crystal proteins, is the 'alternating arginine' tract (Tract 4, residues 520–25 on *cry*IA). This has the sequence SQRYRVRIRYASTT which, according to Chou and Fasman, is a strong β-sheet that would form the structure:

This presents a strongly charged face and a strongly hydrophobic face, suggesting a very active region well anchored to a hydrophobic core of the toxin. Interestingly, this arginine-rich region (and all the arginines of the toxin) is resistant to trypsin (except in certain modest structural changes) (Ge *et al.* in preparation). Such a sequence is unique in the NRL protein database, but amphiphilic α-helices consisting of arginines on one face and hydrophobic residues on the opposite face are found in all K$^+$, Na$^+$ and Ca$^{2+}$ channels (Catterall 1988). Since ICPs inhibit K$^+$ transport, this homology should not be ignored. A very interesting possibility is that the arginine face may interact with the opposite charges (aspartic and glutamic acids) of conserved tract no. 5 in the folded toxin. Haider and Ellar (1989) have proposed that the sequences of tract no. 4 interact with other sequences in the NH$_2$-terminal 250 amino acids of the toxin. These models provide targets for experimental exploration.

## Specificity

A recent review (Milne *et al.* 1990) covers the subject of specificity in great detail. Table 16.1 summarizes the known specificities of the *cry* genes that have been studied to date. As observed, some of the classes of *cry* genes have subclasses that show considerable insect species specificity, for example *cry*IA(*a*), *cry*IA(*b*) and *cry*IA(*c*) have non-overlapping insecticidal activity on certain insects, but most classes of *cry* genes have only one representative gene and therefore the specificity range is limited to the insect order.

Wilcox *et al.* (1986) first showed that a particular gene plays a role in specific action against a particular insect. In analysing a number of formulations of *B. thuringiensis* used as biological pesticides, this group discovered that the *B. thuringiensis* HD-1 strain in a particular formulation, the 1980 Standard Preparation, was missing one of the three genes it normally carries, the '5.3 kb' gene, or *cryIA(b)*. After testing this and other formulations on a number of insects, they noted that *Spodoptera exigua* was significantly less affected by this formulation. This implies that the *cryIA(b)* gene has greater specificity for *S. exigua* than *cryIA(a)* or *cryIA(c)*.

Kondo *et al.* (1987) recently cloned and sequenced *cryIA(a)* and *cryIA(b)* from *B. thuringiensis* HD-1. In performing bioassays using the *Escherichia coli* clones against *Bombyx mori*, they observed a significant difference between the two genes. The *cryIA(a)* gene product killed half of the larvae at 4 µg/dish, whereas 51 µg/dish of the *cryIA(c)* gene product was needed to kill the same number. Ge *et al.* (1989) confirmed these results, showing a 300× lower LC$_{50}$ value for *cryIA(a)*.

Hofte and Whiteley (1989) and Ge *et al.* (1989) give specificity data for several *cry* genes (Table 16.1). These data indicate that the activity spectrum of *B. thuringiensis* is largely due to the specificity of ICPs, which is often controlled by a few amino acid differences. Milne *et al.* (1990) discuss several other factors that affect *B. thuringiensis* specificity, such as the spore, insect gut proteases and *B. thuringiensis* proteases, but as a first approximation, the insecticidal spectrum of a *B. thuringiensis* strain is determined by its ICPs.

midgut (Hofmann *et al.* 1988a, b; Van Rie *et al.* 1989, 1990a) but nonetheless ignore toxin effects on goblet cells, which are the major site of K$^+$ transport (Dow 1986).

*Specificity domains*

Ge *et al.* (1989) located regions on CryIA proteins that determine insect specificity by homologue scanning. With this protein engineering technique (Cunningham *et al.* 1989) regions of two proteins with high structural homology but significant functional differences are exchanged. Thus CryIA(a) and CryIA(c) are 82% homologous but differ more than 100-fold in insecticidal specificity. By judiciously adding or removing restriction enzyme sites from the respective genes, Ge *et al.* (1989) were able to exchange different fragments of the toxic regions of the two protoxin genes. Using *B. mori*, which showed particular sensitivity to CryIA(a) and virtually none to CryIA(c), they located the *B. mori* specificity domain at amino acid residues 330–450 of CryIA(a) (Fig. 16.2a). In subsequent work with the same set of derived mutants, Ge *et al.* (1991) have located the specificity domains for *Tricopulsia ni* and *Heliothis virescens* on the CryIA(c) protein (Fig. 16.2b). To date, there is no correlation between specificity domains and other functional or structural features.

Van Rie *et al.* (1990a, b) have concluded that there is a family of different toxin receptors on insect midgut epithelial cells. Furthermore, a par-

## Protein engineering of insecticidal proteins: structure–function studies

### Functional domains

The known functions of the ICPs are: (i) binding to insect midgut tissue (Hofmann *et al.* 1988a, b); (ii) rapid inhibition of potassium transport (Harvey and Wolfersberger 1979); (iii) creation of pores in insect cells (Knowles and Ellar 1986); and (iv) specificity to various insects (Milne *et al.* 1990).

Of course, we might expect that some of these functions result from the same action, i.e. the binding to a K$^+$ channel may result in its opening. However, some of these functions have been established using insect tissue culture cells (Fast and Morrison 1972; Knowles and Ellar 1986) that are not from midgut epithelium. Brush border membrane vesicles from insect midguts provide a more realistic picture of toxin binding to the insect

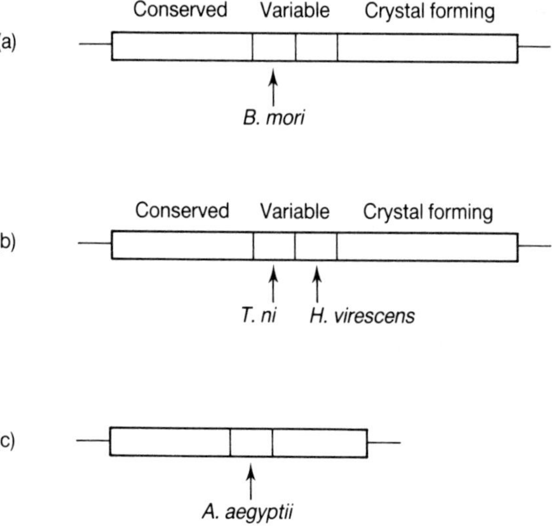


**Fig. 16.2** Specificity domains of (a), CryIA(a); (b), CryIA; (c), CryIIA insecticidal crystal proteins.

ticular toxin may bind to more than one receptor. Specificity has been proposed to be an attribute of receptor binding (Ge *et al.* 1989; Van Rie *et al.* 1989, 1990a). Experimental data support this view in the case of distantly related ICPs, such as CryIA compared to CryIC (Van Rie *et al.* 1989), but are contradicted by recent data on closely related toxins (CryIA(b) and CryIA(c)). Wolfersberger (1990) has shown that these two toxins have over 400-fold difference in activity against gypsy moth, but the most active toxin, CryIA(b), actually binds significantly less than the least active toxin, CryIA(c). With these new data, the role of receptor binding in ICP specificity is still an open question.

Widner and Whiteley (1990) also used homology scanning to locate the *Aedes aegyptii* specificity domain on CryIIA (lepidopteran/dipteran active) by exchanging regions of the gene with the highly homologous (87%) CryIIB (lepidopteran). The mosquito-specific domain is located in a 76-amino acid segment (residues 307–82) in the COOH portion of cryIIA. This region somewhat overlaps the *B. mori* specificity domain of CryIA(a) and the *T. ni* specificity domain of CryIA(c), but not the *H. virescens* specificity domain of CryIA(c) (Fig. 16.2c).

*Cytolytic domains*
The general concept of the insecticidal toxins is that they have a cytolytic domain or set of amino acids (not necessarily contiguous) responsible for cell lysis, either by formation of a pore or by opening a channel (or both). Several authors have identified regions of the toxin that have membrane spanning potential (Hefford *et al.* 1987; Haider and Ellar 1989; Hofte and Whiteley 1989). The region from residues 45–65 on CryIA toxins is particularly hydrophobic and has high membrane-spanning potential. This region has a predicted secondary structure of two or three β-sheets interrupted by a turn. Most of the well studied membrane-spanning structures of toxins or membrane proteins have an α-helical structure. Bacterial porin is an exception in having a β-sheet transmembrane region. There is no sequence homology between ICP toxins and bacterial porin.

Ellar and co-workers have proposed several models for insecticidal cytolytic proteins. *cytA* is a small, 25 kDa, general cytolytic protein found in *B. thuringiensis* var. *israelensis*. This protein is a haemolytic, non-specific cytolytic toxin and shares no homology with the *cry* toxins. Ellar *et al.* (1985) have proposed a model for the membrane insertion and pore formation of this protein. Haider and Ellar (1989) have also proposed a general model for the CryIA(b) toxin, which attempts to describe the dual lepidopteran/dipteran activity of this toxin by proteolytic processing. The cytolytic domain is described as containing two amphiphilic α-helices (residues 75–96 and 126–135) and requiring six toxin molecules to form a pore. We have examined these sequences and find that they are neither hydrophobic nor amphiphilic (one face hydrophobic, one face hydrophilic).

*Two functional domains*
ICP toxins are currently considered to have two functional regions: an $NH_2$-terminal cytolytic domain and a COOH-terminal specificity (or receptor binding) domain. Several authors have proposed two functional domains, one for binding and one for cytolysis (Andrews *et al.* 1987; Brousseau and Masson 1988; Haider and Ellar 1989; Convents *et al.* 1990) based on examination of the DNA sequence of several ICPs. The exact location of these domains has not yet been determined.

**Structural domains**

*Protoxin*

*Overall organization*   A major structural feature of the Type I, 135 kDa, protoxin molecule is that all of the 14–18 cysteines are involved in intermolecular disulphide bridges (Dastedar and Nickerson 1979; Bietlot *et al.* 1990). These connect the monomers into dimers, which are structural units forming a bipyramidal crystal. The crystal is composed of dimeric units of about 230 kDa (Luthy and Ebersold 1981). These units remain associated, even in the presence of 8 M urea. 'Swelling' of the crystal may be observed under these conditions and probably represents disorganization of some regions of the monomers.

The lack of intramolecular disulphide bridges and the particular folding properties give the protoxin another unusual feature. After solubilization (high pH in the presence of reducing agents), it may be almost completely denatured in 8 M urea or 6 M GuHCl and can refold upon dilution to restore complete biological activity and physical characteristics (Choma and Kaplan 1990b). Type I protoxins may be among the largest proteins able to completely refold from the denatured state.

*Crystal-forming domains*   As discussed above, it is widely assumed by reviewers that the C-terminal halves of the protoxin function in formation of the crystal. There is now strong circumstantial evidence to support this hypothesis:

(1) The fact that the crystal is formed by interstrand disulphide bridges (Dastidar and Nickerson 1979; Bietlot *et al.* 1990) and that the majority of these lie in the C-terminal half of the protoxin is strong evidence for the role of the COOH end in crystal formation. Two cysteines ($C_{10}$ and $C_{15}$) are present in the N-terminal 28 fragment, and this region may also be involved in crystal formation.

(2) While it is true that removal of the $NH_2$-terminal 29 amino acids or the COOH-terminal half of the protoxin does not reduce the toxicity of the toxin, it has not been demonstrated that removal of the C-terminal half prevents crystal formation. This has been difficult to demonstrate because of the failure to recrystallize bipyramidal crystals from native solubilized protoxin. The ability to produce native crystals by overexpressing cloned ICP in *E. coli* provides a way to study the crystal-forming properties of the C-terminal half of the protoxin. Ge *et al.* (1990) have explored the parameters of *E. coli* regulatory sequences for transcription and translation and have obtained overexpression of CryIA(c) to the extent of 48% total cell protein. Bona fide crystals similar to *B. thuringiensis* crystals in crystal lattice, solubility and biological activity are produced in *E. coli*. These are unlike the insoluble amorphous aggregates produced by *E. coli* from non-crystal overexpressed genes (Lim *et al.* 1989). However, the specific role of individual disulphide bridges must await further biochemical and genetic research.

*Processing the protoxin*  After (or during) solubilization of the crystal in the insect gut, monomer protoxins are cleaved by trypsin-like insect gut proteases to form the active toxin. This process is itself interesting and reveals something of the overall structure of the protoxin. Choma *et al.* (1990b) have observed that the protoxin is processed in six steps by successive cleavage of approximately 10 kDa fragments from the C-terminus towards the N-terminus until approximately residue 620, the C-terminal boundary of the toxin. These 10 kDa fragments are rapidly degraded to smaller peptides. This implies that the C-terminal half of the protoxin is wrapped around the toxic core in an *escargot*

manner, such that the most accessible tryptic site, 10 kDa from the C-terminus, is cleaved first, exposing the second trypsin site. This view of the protoxin structure is also supported by general biochemical characteristics of the protoxin. The C-terminal half of the protoxin is considerably more hydrophilic than the toxic core, suggesting a more superficial position.

*Toxin: two structural domains*

*Biochemical properties*  Predictions of protein secondary structure from the primary amino acid sequence are based upon the association of amino acid sequences with secondary structures in proteins of known structure. The accuracy of such predictions is only about 60% (Chou and Fasman 1978; Garnier *et al.* 1978). Predictions of hydrophobicity (Hopp and Woods 1981) and turns have higher predictive success.

The CryIA protein toxic core is highly hydrophobic, insoluble at neutral pH in the absence of chaotropic agents, and lacks cysteines (Choma and Kaplan 1990a; Convents *et al.* 1990). The $NH_2$-terminal half of the toxin is predicted to be the more hydrophobic, consisting of two–three long stretches of amino acids. This would suggest a hydrophobic core that folds early upon synthesis and primes further folding of the toxin and protoxin. This agrees with denaturation/renaturation studies, which are even more dramatic for the toxin than for the protoxin. The toxin rapidly refolds from 8 M urea (pH 8.0; 5 min upon dilution) but refolds more slowly from 6 M GuHCl—3 h (Choma and Kaplan 1990b). Some structural features that allow for the rapid renaturation remain in 8 M urea. Detailed unfolding studies with varying GuHCl at different pH conditions yield a biphasic (or multiphasic) denaturation curve. Convents *et al.* (1990) propose two or more folding domains for the *cryIA* toxins based on differential denaturation characteristics. The $NH_2$-terminal half of the toxin unfolds at pH 8.0–11.0 in 4 M GuHCl, while the COOH-terminal half unfolds at pH 7.0 down to pH 4.0 in 4 M GuHCl.

The topology of the two domain structure fits nicely with the natural division of the *cryIA* toxins into conserved $NH_2$-terminal and hypervariable COOH-terminal regions. It also fits well with recent protease studies on the toxins. Several authors have found that the tryptic limit toxin may be further cleaved into two peptides of approximately equal size with selective proteases; namely papain, subtilisin, chymotrypsin or thermolysin (Convents *et al.* 1990; Choma *et al.* 1990a; G.G. Chestukhina personal communication; D. Witt personal communication; Lee and Dean personal observations). These studies indicate that the two structural domains are linked by a protease-sensitive bridge.

The secondary structures of the two domains are quite different. The $NH_2$-terminal domain is quite rich in α-helices, as determined by circular dichromatic spectroscopy (Convents *et al.* 1990; Choma *et al.* 1990a), Raman spectroscopy, and infrared spectroscopy (Choma *et al.* 1990b). This empirical determination agrees well with the predictions of Chou and Fasman (1978). The COOH-terminal structural domain has a very low α-helical content and is predominantly β-sheets (Convents *et al.* 1990; Choma *et al.* 1990a, b), as determined by the above empirical method and as predicted by the Chou and Fasman (1978) method.

*A model for the structure for the cryIA toxin*
The structural and functional data of the ICP toxins are beginning to fit together into a rather nice picture. Although most of the empirical data relate to the *cryIA* toxins, the similarity of all of the toxins in their hydrophobicity patterns (Geiser *et al.* 1986; Andrews *et al.* 1987), and alignment of predicted secondary structure (Convents *et al.* 1990) suggests that they all follow the same structural motif. The picture is one of a toxin composed of two structural domains connected by a bridge, with each structural domain supporting a functional domain. The model is that the $NH_2$-terminal, α-helix-rich domain is the cytolytic domain and the COOH-terminal domain is the specificity or receptor binding domain. The important caveat in this model is that we are far from certain about the location of the functional domain.

### Modification of ICPs

The truly 'engineering' aspects of ICPs relate to the cognitive design of better pesticides based on what we know about toxic activity. The problem at present is, of course, that we don't know enough about the mode of action of ICPs to design better toxins *a priori*. Often, at this stage, something interesting results from exploratory research. For example, in our examination of homology switch mutants, we discovered that one had significantly greater activity than expected for *H. virescens* (Ge *et al.* 1991). Another example is the finding of 'up' mutants in a random mutagenesis producing 15 000

candidates (Rusche *et al.* 1988).

The best example, to date, of cognitive design of better ICPs is the elegant work of Fischoff *et al.* (1990). In modifying CryIA proteins for expression in plants, they made three significant improvements as a result of: (i) removing AT-rich regions of the genes; (ii) altering the codon bias to one more favourable to plant expression; and (iii) exchanging a 25-amino acid protease inhibitor for a portion of the $NH_2$-terminal 29 non-toxin region of the *cryIA* gene. The third improvement increased activity six–eight-fold. The second improvement increased expression in plants 500-fold.

## Other insecticidal protein toxins

Several other insecticidal proteins or peptides are of great interest as biological pesticides. Scorpion toxins are small peptides of 36 and 65 residues and are insect-specific, $Na^+$ channel-blocking neurotoxins (Grishin *et al.* 1982). These have the advantage of being more amenable to manipulation by protein engineering because they are small and at least one of them has had its tertiary structure determined (Arseniev *et al.* 1984). Spider toxins are also potential insecticides of interest (Bindokas and Adams 1989). Full exploitation of these insecticidal proteins will require innovative genetic engineering and plant molecular genetic approaches to reach their full potential. One of the problems that must be overcome with these neuropeptides is that they must be introduced into the insect haemolymph to have full activity. Finally, a whole series of environmentally friendly natural insecticides from plants have been identified (Arnason *et al.* 1989), which may be employed for plant protection.

In summary, it is hoped that the foregoing description of the biochemistry and genetics of ICPs provides the structural and functional basis for understanding the most appropriate implementation of these toxins for plant protection covered in the remaining section of this review.

## Plant protection

### Current and imminent application methods

Considering the range of insects killed, why is the use of *B. thuringiensis* and other microbial insecticides not more widespread? Carlton (1988) estimates that microbial pesticides are only 0.5% of the pesticide market. Part of the answer to this question lies in the commercial perception of disadvantages. They are more expensive than conventional insecticides and are short lived. This short life is environmentally attractive, but results in increased applications and higher cost to the farmer. They are specific for some insects, making broad insect control more difficult than with chemical pesticides. If a farmer has several classes of insects attacking his field at one time, several microbial pesticides are needed. They are generally toxic to young larvae, so careful monitoring is necessary and *B. thuringiensis* treatments must adequately cover plant parts eaten by the pest. Some of these limitations can be overcome through improvements in formulation and application of genetic technologies.

It is generally accepted that the use of *B. thuringiensis* and other biopesticides will increase greatly in the next five to ten years. However, these projections depend on predicted improvements in insect control with *B. thuringiensis* biopesticides. The *B. thuringiensis* subsp. *kurstaki* HD-1 products of today are more potent than those of five years ago, mostly due to improved manufacturing and formulation. The challenge for a new series of *B. thuringiensis* biopesticides will be to accomplish improved pest control while maintaining the desirable qualities of *B. thuringiensis*, particularly its safety to non-target organisms. Commonly cited insect pests as targets for new biopesticides are *Heliothis* species on cotton and European corn borer and corn rootworms. The development of new ways of delivering *cry* proteins to pest insects using transgenic plants and micro-organisms spurs the search for new strains and insecticidal genes.

The process of *B. thuringiensis*'s discovery and improvement depends heavily on the tools of molecular biology and the knowledge of *cry* genes and how they work. This applies to strain isolation as well as to the genetic modification and transfer of insecticidal genes. An advantage of this bacterium is the decades of experience in fermentation, processing and formulation that allow the delivery of a stable biopesticide to the plant surface. Also important is the experience that entomologists have in the use of *B. thuringiensis* in pest management practices. Wilcox *et al.* (1986) and Shieh (1988) emphasize the need for improved spore/crystal formulations that provide better coverage and are more stable on plant surfaces. These practical considerations must be integrated into strategies for improvement because the commercial success of this research will finally be measured by pest control in the field.

Common approaches are emerging for the discovery and development of improved strains. They can be considered a continuum that should provide novel biopesticides starting in 1990 and continue over the next five to ten years. Each of these steps will be discussed below.

## Strain and gene discovery

Biopesticide products developed using methods other than recombinant DNA encounter fewer regulatory hurdles than products employing those technologies. Faced with the common commercial pressures of today, this tends to bias short-term research and development towards natural isolates and non-recombinant technologies.

The discovery of novel *B. thuringiensis* strains and insecticidal genes provides a basis for new biopesticides. Systematic searches for novel *B. thuringiensis* strains are routine in many industrial laboratories. These screening programmes are proprietary, but the *B. thuringiensis* literature suggests several approaches to isolate and identify strains with unique biological specificities. Success is likely to come from a combination of approaches and technologies.

The source of samples is of obvious importance in the isolation of *B. thuringiensis* strains. Historically, new isolates are most often found in diseased and dead insects. *B. thuringiensis* was first isolated by Ishiwata (1901) in Japan, where it was a pathogen in a silkworm colony. Next, it was reported by Berliner (1911) in a flour moth infestation in Germany. This trend continued with the isolation of the commercial strain *B. thuringiensis* subsp. *kurstaki* HD-1 from a mass-reared colony of pink bollworms (Dulmage 1970). More recently, *B. thuringiensis* subsp. *tenebrionis* was found in a dead common meal beetle (*Tenebrio molitor*). Studies investigating the ecological distribution of *B. thuringiensis* now direct our attention to the isolation of *B. thuringiensis* from natural environments. Researchers utilize the bacterium's ability to form spores and the spores' heat resistance (Ohba and Aizawa 1978). Ohba and Aizawa (1986) collected 136 soil samples from throughout Japan, heat-treated suspensions at 65° for 30 min, plated the mixtures on nutrient agar and chose colonies for examination based on morphology. They obtained an average of 50 *B. thuringiensis–Bacillus cereus* colony types per gram of soil and, of these, 2.7% produced crystal inclusions. Of the 189 isolates from this study, 48 were *Lepidoptera*-toxic, 20 were mosquito-toxic and 121 were toxic to neither (Ohba and Aizawa 1986). A study of the distribution of *B. thuringiensis* in soils in the United States utilized an enrichment medium containing polymyxin B and penicillin (Saleh *et al.* 1969) to reduce the number of non-*Bacillus* bacteria in samples (Delucca *et al.* 1981). In a 2-year evaluation of 115 fields, they found *B. thuringiensis* in 7% of the soils tested where it comprised 0.5% of *Bacillus* species. Travers *et al.* (1987) describe a novel method of enriching for *B. thuringiensis*. They selectively inhibited the germination of *B. thuringiensis* spores by sodium acetate, while allowing the germination of other *Bacillus* species. This was followed by pasteurization at 80° to kill non-*B. thuringiensis* organisms and plating on rich medium. They reported that 20–96% of the colonies picked from agar plates were crystal-forming *Bacillus* species.

Recently, Martin and Travers (1989) applied their sodium acetate selection method to over 1100 soil samples collected worldwide. They found *B. thuringiensis* to be ubiquitous occurring in 70% of the samples tested. *B. thuringiensis* also abounds in grain elevator dusts. Delucca *et al.* (1982) analysed 20 samples of settled dust and 53 of respirable dust from four grain elevators and found *B. thuringiensis* in 55% of the settled samples and 17% of the respirable samples. Using this information on *B. thuringiensis* occurrence, Donovan *et al.* (1988) isolated a beetle-toxic strain from soybean grain dust in Kansas and Sick *et al.* (1990) reported a new coleopteran-active strain from the variety *tolworthi*. The above studies demonstrate that *B. thuringiensis* is found in many ecological niches, most often in relatively low numbers, and can be isolated using standard microbiological enrichment techniques.

After a panel of strains is isolated, the task is to identify those strains harbouring insecticidal proteins that may have potential for commercialization. Some biotypes of *B. thuringiensis* have unique crystal shapes that are discernible under phase contrast microscope observation. For instance, coleopteran-active cuboidal crystals are easily distinguished from the more common lepidopteran-active bipyramidal crystals. Insect bioassays are the foundation for strain evaluation. Unfortunately, the maintenance of insect colonies is expensive and bioassays are tedious. The simplest method is to grow *B. thuringiensis* to sporulation and use a crude spore/crystal mixture (Dulmage *et al.* 1981) in

feeding tests. Others have dissolved insecticidal crystals in alkali containing reducing agents and tested insects against solubilized protoxins (McLinden *et al.* 1985; Jaquet *et al.* 1987). Most insecticidal crystals consist of several toxic protein subunits and crystal potency results from their combined action. This complicates basing strain selection on bioassay potency. For example, *kurstaki* HD-1 has three *cryIA* genes and two *cryII* genes. These multiple proteins have widely different insecticidal specificities that contribute to the insecticidal profile of strain HD-1 (Hofte *et al.* 1988; MacIntosh *et al.* 1990). Strains *B. thuringiensis* subsp. *israelensis* and *morrosoni* PG-14 need the combined action of several insecticidal proteins to kill mosquitoes optimally (Wu and Chang 1985). These examples illustrate how a strain may have marginal activity against an important pest, yet merit further study using individual insecticidal proteins.

Novel strains and insecticidal genes may be revealed through a combination of molecular techniques. One method is to use DNA hybridization of gene-specific probes to search for homologous crystal protein genes in unknown strains. Kronstad *et al.* (1983) first used a *cry* gene probe to demonstrate the diversity and distribution of insecticidal genes among *B. thuringiensis* strains. As more *cry* genes were cloned and sequenced, the probes became more specific and useful as screening tools. Prefontaine *et al.* (1987) used oligonucleotide probes specific for *cryIA(a)*, *cryIA(b)* and *cryIA(c)* lepidopteran-active genes to determine the distribution of these gene types among 13 *B. thuringiensis* serotypes. In a similar study, Visser (1989) cloned restriction fragments homologous to four *cryI*-type genes and screened DNA from 25 strains (18 serotypes) by Southern hybridization analysis. By comparing homologous fragments in test strain DNA with the expected size fragments in the cloned genes, he was able to first, identify strains harbouring the four gene types and, second, where the size of the homologous fragment differed, identify potential *cry* gene variants. These hybridization surveys are quick and identify variants that might be useful for optimizing a crystal protein for activity against a pest insect.

A second molecular tool for identifying insecticidal proteins uses antibodies for identifying individual crystal components. Krywienczyk *et al.* (1981), Lynch and Baumann (1985) and Smith (1987) used antisera to compare crystal serotypes among the lepidopteran active *B. thuringiensis* strains. Each study contributed to the understanding that *B. thuringiensis* crystals contain serologically distinct components and that a strain's crystal serotype often does not correspond to its flagellar antigen serotype. Polyclonal antisera discriminate between the major classes of Cry proteins. Samasanti *et al.* (1986) showed that CryI (bipyramidal toxin) and CryII (cuboidal toxin) proteins were antigenically distinct and their presence was detected by ELISA. Polyclonal antisera are very useful in the quantitative detection of crystal proteins, but it is difficult to discriminate between closely related Cry proteins.

In contrast to polyclonal antisera, monoclonal antibodies have the specificity to distinguish closely related *cry* proteins. Huber-Lukac *et al.* (1986) first used monoclonal antibodies to study k-1 (*kurstaki*-1; see Krywienczyk *et al.* 1981) type crystal proteins and found similar crystal antigens in other subspecies. Monoclonal antibodies are best used when the gene product they are binding to is characterized. Adang *et al.* (1987) used an mAb, which binds to CryIA(c) toxin, and not CryIA(a) or CryIA(b), to isolate the *cryIA(c)* gene from a *B. thuringiensis* subsp. *kurstaki* HD-1 plasmid library. Hofte *et al.* (1988) used monoclonal antibodies that react with CryIA, CryIB, and CryIC lepidopteran-active proteins to measure their distribution among *B. thuringiensis* strains. They found that CryIA proteins are the most widespread of the *B. thuringiensis* proteins among the serovars. They also observed that CryIB, the 'type' protein of subsp. *thuringiensis* strain 4412, and CryIC(a) subspp. *aizawai* and *entomocidus Spodoptera*-toxic proteins have a relatively restricted distribution. Monoclonal antibodies have the specificity and, when used in ELISA tests, the speed necessary for use in large screening programmes.

Molecular screening, with both DNA and monoclonal antibody probes, is a powerful tool for recognizing variants of crystal protein genes and for selecting isolated strains for further development or gene manipulation.

### B. thuringiensis *plasmid curing and exchange*

There are other methods for improving *B. thuringiensis* biopesticides that employ biotechnology and recombinant DNA, but the final product is not considered 'genetically engineered' by regulatory agencies. The process of plasmid exchange allows the transfer of plasmids between related *Bacillus* species (Gonzalez *et al.* 1982). This process is useful in the improvement of *B. thuringiensis*

biopesticides. *B. thuringiensis* crystals constitute approximately 30% of the total weight of sporulating cells and there is probably little opportunity to change crystal yield per cell. Due to the differences in *cry* protein action, crystals frequently contain proteins that are ineffective against a particular species. Plasmid exchange allows a greater amount of the desired *cry* protein to accumulate in the strain's crystals. This occasionally results in increased effective range or potency (Carlton 1988). Although not well documented, patent applications claim that the combination of *cry* gene types in a strain can have surprising synergistic activity or toxicity to a pest not killed by parental gene types. More expected is the result seen when a lepidopteran gene and coleopteran-active gene are harboured in the same bacterium, producing a strain with combined insect toxicity. This permits the fermentation and marketing of a single strain. Biopesticides produced by conjugation of *B. thuringiensis* are now being field-tested and will be registered for use in 1990.

### Non-living delivery systems for ICP

In most cases, the *B. thuringiensis* spore itself is not critical to the ability of the bacteria to kill insects. Exceptions are with marginally susceptible insects, such as meal moth larvae, *Plodia interpunctella* and *Ephestia kuhniella*. This permits the development of *B. thuringiensis* protein insecticides which are independent of the normal host bacterium. An advantage is that insecticidal gene systems that will not replicate in the environment can be used.

Mutants of *B. thuringiensis* that do not produce spores may be used for the production of modified insecticidal proteins. This would permit the manufacture of a product containing insecticidal crystals but not viable bacteria or spores, obviating the concern regarding the deliberate release of genetically engineered micro-organisms.

Scientists at Mycogen have developed a method of encapsulating *B. thuringiensis* proteins within a protective bacterial cell wall coat. They express insecticidal protein in *Pseudomonas* bacteria, which are killed after fermentation. Dead bacteria/insecticidal protein sprayed on plants is more stable to environmental factors than wild type insecticidal crystals, yet the gene encoding the protein does not replicate. Furthermore, wild type insecticidal proteins have been field-tested for the past two summers with encouraging results. Future tests, with

USDA/EPA approval, will utilize genetically modified insecticidal proteins.

## Introduction of insecticidal protein genes into plant epiphytes/endophytes

The spectrum of activities might also be increased by transferring insecticidal genes into micro-organisms that multiply on, or in, plants. This would (i) deliver *B. thuringiensis* proteins to plant parts that are not covered by conventional spraying techniques and (ii) increase persistence of insecticidal protein for longer control. At the Monsanto Company, a *B. thuringiensis* insecticidal gene was cloned into a corn root colonizer, *Pseudomonas fluorescens*, using transposase deletion derivatives of the transposon Tn5 containing a *cry* gene (Obukowicz *et al.* 1986). This biopesticide was insecticidal to lepidopteran insects in growth chamber tests. When threatened with litigation from environmental advocacy groups, field trials with this organism were cancelled. Subsequently, a sensitive marker system was developed for following fluorescent pseudomonads in the environment (Drahos *et al.* 1986) and field trials were conducted with such a strain in 1988. Tests with the marked pseudomonad established a data base on survival and movement of a genetically engineered micro-organism in the environment.

Crop Genetics International have developed a novel biopesticide by inserting a *B. thuringiensis cry* gene into the chromosome of *Clavibacter xylii* subsp. *cynodontis*. *C. xylii* is a cornyeform bacterium that inhabits the xylem of bermuda grass. Scientists at CGI found that it can be transferred to corn via seed treatment. After germination, bacteria colonize xylem and express *B. thuringiensis* protein. If sufficient insecticidal protein is produced it will protect the stem from damage by European corn borers (*Ostrinia nubialis*). The second small-scale field trial was performed in 1989 and a registered biopesticide is scheduled for early 1990s.

There is certainly potential for developing numerous biopesticides based on producing insecticidal proteins in plant-associated micro-organisms. The major questions to be answered relate to environmental issues of gene stability, microbial persistence and impact on non-target invertebrates and animals. These issues must be addressed and the probability of public acceptance must be sufficiently high before there is expanded research in using *B. thuringiensis* proteins in plant-associated micro-organisms.

## Introduction of insecticidal protein genes directly into plants

The list of genes encoding proteins conferring insect resistance upon transfer to plants is short. The qualifications of high activity against target insects upon feeding and safety for non-target invertebrates and animals are met by few known proteins. The candidate insecticidal protein needs to be stable in the plant cell and the insect midgut. Because the exoskeleton, foregut and hindgut of insects is sclerotized cuticle, the most likely route of toxicity is through the midgut. *B. thuringiensis* insecticidal proteins satisfy these criteria and experiments began in the early 1980s to develop insect-resistant transgenic plants based on the introduction of *B. thuringiensis* toxin genes into plants. The first *B. thuringiensis* genes to be cloned and sequenced were from *B. thuringiensis* subsp. *kurstaki* (for review see Hofte and Whiteley 1989) and were expressed in tobacco (Vaeck *et al.* 1987) and tomato (Fischoff *et al.* 1987; Adang *et al.* 1988). Tobacco hornworms were killed after feeding on transgenic plants expressing low levels of *B. thuringiensis* protein.

Critical to these successes were discoveries in the field of plant transformation. *Agrobacterium tumefaciens* is a bacterium that transfers a segment of its DNA into the genome of plants. Scientists have used the nucleotide sequences necessary for DNA transfer to design vectors, called Ti vectors, that allow foreign genes to be engineered into crop plants. With the aid of Ti vectors, *B. thuringiensis* genes were introduced into the nuclear genomes of plant cells. Tissues harbouring the transferred genes were regenerated into whole plants, unchanged except for the introduced genes. The transfer of *B. thuringiensis* genes using Ti vectors has now been accomplished for other dicotyledonous species, including potato, soybean and cotton. Further research is directed towards refining the expression for improved insect control.

## Potential problems and solutions

Pesticide resistance in insects is an issue of growing importance to agriculture. Over 500 species of insects have acquired resistance to chemical pesticides (Croft 1987). Fortunately, there have been few reports of resistance to microbial insecticides. For many years, attempts to select resistant strains to *B. thuringiensis* spore–crystal mixtures were unsuccessful (Boman 1981). In spite of this good record, researchers predicted that, over a period of time, resistance to *B. thuringiensis* would be a serious challenge to the industry (Aronson *et al.* 1986). McGaughey (1985) reported resistance to Dipel, a commercial formulation of *B. thuringiensis* in Indian meal moths (*Plodia interpunctella*). Lower levels of resistance have been found in the almond moth (*Cadra cautella*) (McGaughey and Beeman 1988), the tobacco budworm (*Heliothis virescens*) (Stone *et al.* 1989) and the diamondback moth (*Plutella maculinennis*) (B. Tabashnik, personal communication). These reports have led to an awareness of the potential for insects becoming resistant to *B. thuringiensis* biopesticides.

Since the primary report in 1985, biological, genetic and molecular aspects of Indian meal moth resistance have been investigated. McGaughey, in his original report (1985), fed *P. interpunctella* (Indian meal moth) larvae doses of *B. thuringiensis* subsp. *kurstaki* HD-1 spores and crystals sufficient to produce 70–90% mortality. This resulted in populations of larvae with nearly 30-fold resistance in two generations. He also tested native populations from treated and untreated grain bins for susceptibility to *B. thuringiensis* subsp. *kurstaki* HD-1 in diet bioassays. Colonies from treated bins were significantly more tolerant to *B. thuringiensis*. Resistance appeared to be inherited as a recessive trait and was stable once resistance levels reached a plateau (McGaughey and Beeman 1988). Resistant *P. interpunctella* larvae were still susceptible to other *B. thuringiensis* strains, particularly subsp. *aizawai* (McGaughey and Johnson 1987). This is important because crystals from subsp. *aizawai* strains contain CryI proteins distinct from the Dipel crystals (Hofte and Whiteley 1989). Dipel contains subsp. *kurstaki* HD-1 as the active ingredient, whose crystals are a mixture of CryIA and CryII ICPs. Possible mechanisms for resistance included changes in gut proteases that cleave protoxins to toxins and changes in brush border receptors. Johnson *et al.* (1990) tested the first hypothesis by comparing resistant and susceptible strains of the Indian meal moth. There were no significant quantitative or qualitative differences between gut proteases of the two strains. Both strains were similar in their ability to process crystals to toxin. Van Rie *et al.* (1990b) tested the

second model by using [$^{125}$I]-labelled toxin and brush border membrane vesicles from resistant and susceptible strains to analyse toxin receptors. They found that resistance in *P. interpunctella* is correlated with a 50-fold reduction in affinity of the membrane receptor. The resistant strain was more susceptible to CryIC protoxin and toxin than the sensitive strain and labelled CryIC toxin bound with higher affinity to the resistant than to the susceptible strain. These results represent an adaptation by Indian meal moths to toxin by altering protein levels in midgut membranes. As Van Rie *et al.* (1990b) note, the insect compensates for a reduction in the amount of one receptor by the increased level of another.

Insect resistant plants, recombinant microorganisms with crystal protein genes and improved potency of crystal protein will contribute to the increased presence of *B. thuringiensis* crystal proteins in the environment. Steps to delay the onset of resistance are being discussed in academia and industry. Commercial concerns have established a 'B.t. Management Working Group' to address these issues.

One practical solution to the problem of resistance is to employ a combination of insecticidal proteins that associate with different receptors (Mani 1985).

Genetic manipulation of toxin proteins and the multitude of application methods will, in the near future, allow us to develop the most cytolytic toxins active on a particular pest with a variety of receptor binding domains. The major limiting factor at the present time is our caution in applying this powerful new technology.

# References

Adang, M.J., Staver, M.J., Rocheleau, T.A., Leighton, J., Barker, R.F. and Thompson, D.V. (1985). Characterized full-length and truncated plasmid clones of the crystal protein of *Bacillus thuringiensis* ssp. *kurstaki* HD-73 and their toxicity to *Manduca sexta*. *Gene*. **36**: 289–300.

Adang, M.J., Idler, K.F. and Rocheleau, T.A. (1987). Structural and antigenic relationships among three insecticidal crystal proteins of *Bacillus thuringiensis* subsp. *kurstaki*. In *Biotechnology in Invertebrate Pathology and Cell Culture* (ed. K. Maramorosch), pp. 85–99. Academic Press, San Diego.

Adang, M., DeBoer, D., Endres, J., Firoozatady, E., Klein, J., Merlo, A., Merlo, D., Murray, E., Rashka, K. and Stock, C. (1988). Manipulation of *Bacillus thuringiensis* genes for pest insect control. In *Biotech-*

nology, *Biological Pesticides and Novel Plant–Pest Resistance for Insect Pest Management* (ed. D.W. Roberts and R.R. Granados), pp. 31–7. Boyce Thompson Institute, Ithaca, NY.

Andrews, R.E., Jr., Faust, R.M., Wabiko, H., Raymond, K.C. and Bulla, L.A. (1987). The biotechnology of *Bacillus thuringiensis*. *CRC Critical Reviews in Biotechnology*. 6: 163–232.

Arnason, J.T., Philogene, B.J.R. and Morand, P. (1989). *Insecticides of Plant Origin*. American Chemical Society Books.

Aronson, A.I., Beckman, W. and Dunn, P. (1986). *Bacillus thuringiensis* and related insect pathogens. *Microbiology Reviews*. 50: 1–24.

Arseniev, A.S., Kondakov, V.I., Malorov, V.N., and Bystrov, V.F. (1984). NMR solution spatial structure of 'short' scorpion insectotoxin, I$_5$A. *FEBS Letters*. 165: 57–62.

Berliner, E. (1911). Uber die schlaffsucht der mehlmottenraupe. *Zeitschrift Gesamte Getreidewes*. 252: 3160–2.

Bernhard, K. (1986). Studies on the delta-endotoxin of *Bacillus thuringiensis* var. *tenebrionis*. *FEMS Microbiology Letters*. 33: 261–5.

Bietlot, H.P.L., Vishnubhatla, I., Carey, P.R., Pozcgay, M. and Kaplan, H. (1990). Characterization of the cysteine residues and disulfide linkages in the protein crystal of *Bacillus thuringiensis*. *Biochemical Journal*. In press.

Bindokas, V.P. and Adams, M.E. (1989). Aga-1: a presynaptic calcium channel antagonist from venom of the funnel web spider, *Agelenopsis aperta*. *Journal of Neurobiology*. 20: 171–88.

Boman, H.G. (1981). Insect responses to microbial infections. In *Microbial Control of Pests and Plant Diseases 1970–1980* (ed. H.D. Burges), pp. 769–95. Academic Press, London.

Bowie, J.U., Reidhaar-Olson, J.F., Lim, W.A. and Sauer, R.T. (1990). Deciphering the message in protein sequences: tolerance to amino acid substitutions. *Science*. 247: 1306–10.

Brousseau, R. and Masson, L. (1988). *Bacillus thuringiensis* insecticidal crystal protein: gene structure and mode of action. *Biotechnology Advances*. 6: 697–724.

Carlton, B.C. (1988). Development of genetically improved strains of *Bacillus thuringiensis*: a biological insecticide. In *Biotechnology for Crop Protection* (ed. P.A. Hedin, J.J. Menn and R.M. Hollingworth), pp. 260–79. American Chemical Society, Washington DC.

Catterall, W.A. (1988). Structure and function of voltage-sensitive ion channels. *Science*. 242: 50–61.

Choma, C.T. and Kaplan, H. (1990a). *Bacillus thuringiensis* crystal protein: effect of chemical modification on the cysteine and lysine residues. *European Journal of Biochemistry*. 189: 523–7.

Choma, C.T. and Kaplan, H. (1990b). Folding and enfolding of the protoxin from *Bacillus thuringiensis*: evidence that the N- and C-terminal halves fold independently. Submitted.

Choma, C.T., Surewicz, W.K., Carey, P.K., Pozcgay, M., Raynor, T. and Kaplan, H. (1990a). Unusual proteolysis of the protoxin and toxin from *Bacillus thuringiensis*: structural implication. *European Journal of Biochemistry*. In press.

Choma, C.T., Surewicz, W.K., Mantsch, H.H., Carey, P.R., Pozcgay, M. and Kaplan, H. (1990b). Secondary structure of the entomocidal toxin from *Bacillus thuringiensis* subsp. *kurstaki* HD-73. *Journal of Protein Chemistry*. 9: 87–94.

Chou, P.Y. and Fasman, G.D. (1978). Prediction of the secondary structure of proteins from their amino acid sequence. *Advances in Enzymology*. 47: 45–148.

Convents, D., Houssier, C., Lasters, I. and Lauwereys, M. (1990). The *Bacillus thuringiensis* δ-endotoxin evidence for a two domain structure of the minimal toxic fragment. *Journal of Biological Chemistry*. 265: 1369–75.

Croft, B.A. (1987). *Chemical and Engineering News*. March.

Cunningham, B.C., Jhuraic, P., Ng, P. and Wells, J.A. (1989). Receptor and antibody epitopes in human growth hormone identified by homolog-scanning mutagenesis. *Science*. 243: 1330–6.

Dastidar, P.G. and Nickerson, K.W. (1979). Interchain crosslinks in the entomocidal *Bacillus thuringiensis* protein crystals. *FEBS Letters*. 108: 411–14.

Dean, D.H. (1982). Molecular biology and molecular genetics of toxin biogenesis in *Bacillus thuringiensis*. In *Basic Biology of Microbial Larvicides of Vectors of Human Diseases* (ed. F. Michal). UNDP/WORLD BANK/WHO Geneva, Switzerland.

Dean, D.H. (1984). Biochemical genetics of the bacterial insect control agent *Bacillus thuringiensis*: basic principles and prospects for genetic engineering. *Biotechnology and Genetic Engineering Reviews*. 2: 341–63.

Delécluse, A., Bourgouin, C., Klier, A. and Rapoport, G. (1988). Specificity of action on mosquito larvae of *Bacillus thuringiensis israeliensis* toxins encoded by two different genes. *Molecular and General Genetics*. 214: 42–7.

Delucca, A.J., Simonson, J.G. and Larson, A.D. (1981). *Bacillus thuringiensis* distribution in soils of the United States. *Canadian Journal of Microbiology*. 27: 865–70.

Delucca, A.J., Palmgren, M.S. and Ceigler, A. (1982). *Bacillus thuringiensis* in grain elevator dusts. *Canadian Journal of Microbiology*. 28: 452–6.

Donovan, W.P., Gonzalez, J.M., Pearce Gilbert, M. and Dankocsik, C. (1988). Isolation and characterization of EG2158, a new strain of *Bacillus thuringiensis* toxic to coleopteran larvae, and nucleotide sequence of the toxin gene. *Molecular General Genetics*. 214: 365–72.

Dow, J.A.T. (1986). Insect midgut function. *Advances Insect Physiology*. 19: 187–328.

Drahos, D.J., Hemming, B.C. and McPherson, S. (1986) Tracking recombinant organisms in the environment: β-galactosidase as a selectable non-antibiotic marker for fluorescent pseudomonads. *Bio/Technology*. 4: 439–44.

Dulmage, H.T. (1970). Insecticidal activity of HD-1, a new isolate of *Bacillus thuringiensis* var. *alesti*. *Journal of Invertebrate Pathology*. **15**: 232–9.

Dulmage, H.T. *et al.* (1981). Insecticidal activity of isolates of *Bacillus thuringiensis* and their potential for pest control. In *Microbial Control of Pests and Plant Diseases, 1970–1980* (ed. H.D. Burges), p. 193. Academic Press, London.

Ebersold, H.R., Luthy, P. and Muller, M. (1977). Changes in the fine structure of the gut epithelium of *Pieris brassicae* by the delta-endotoxin of *Bacillus thuringiensis*. *Bulletin of the Society of Entomology, Suisse*. **50**: 269–76.

Ellar, D.J., Thomas, W.E., Knowles, B.H., Ward, S., Todd, J., Drobniewski, F., Lewis, J., Sawyer, T., Last, D. and Nichols, C. (1985). Biochemistry, genetics, and mode of action of *Bacillus thuringiensis* δ-endotoxins. In *Molecular Biology of Microbial Differentiation* (ed. J.A. Hoch and P. Setlow), pp. 230–40. American Society for Microbiology, Washington DC.

Endo, Y. and Nishiitsutsuji-Uwo, J. (1980). Mode of action of *Bacillus thuringiensis* δ-endotoxin: histopathological changes in the silkworm midgut. *Journal of Invertebrate Pathology*. **36**: 90–103.

Fast, P.F. and Morrison, I. (1972). The δ-endotoxin of *Bacillus thuringiensis* IV. The effect of δ-endotoxin on ion regulation by midgut tissue of *Bombyx mori* larvae. *Journal of Invertebrate Pathology*. **20**: 280–4.

Fast, P.F., Murphy, D.W. and Sohi, S.S. (1978). *Bacillus thuringiensis* δ-endotoxin: evidence that toxin acts at the surface of susceptible cells. *Experientia*. **34**: 762–3.

Fischoff, D.A., Bowdis, K.S., Perlak, F.J., Marrone, P.G., McCormick, S.M., Niedermeyer, J.G., Dean, D.A., Kusano-Kretzmer, K., Mayer, E.J., Rochester, D.E., Rogers, S.G. and Fraley, R.T. (1987). Insect tolerant transgenic tomato plants. *Bio/Technology*. **5**: 807–13.

Fischoff, D.A., Perlak, F.J., Fuchs, S.C., MacIntosh, W.R., Deaton, X. and Sims, S.R. (1990). Progress in development of insect resistant crops. *Journal of Cellular Biochemistry, Supplement*. **14E**: 267, (Abstract).

Garnier, J., Osguthorpe, D.J. and Robson, B. (1978). Analysis of the accuracy and implications of simple method for predicting the secondary structure of globular proteins. *Journal of Molecular Biology*. **120**: 97–120.

Ge, A.Z., Shivarova, N.I. and Dean, D.H. (1989). Location of the *Bombyx mori* specificity domain on a *Bacillus thuringiensis* delta-endotoxin. *Proceedings of National Academy of Sciences, USA*. **86**: 4037–41.

Ge, A.Z., Rivers, D., Milne, R. and Dean, D.H. (1991). The specificity domains of *Bacillus thuringiensis* δ-endotoxin *cryIA(c)* for lepidopteran insects *Heleothis virescens* and *Tricoplusia ni*. In preparation.

Geiser, M., Schweitzer, S. and Grimm, C. (1986). The hypervariable region in the genes coding for entomopathogenic crystal proteins of *Bacillus thuringiensis*: nucleotide sequence of the *kurhd1* gene of subsp. *kurstaki* HD1. *Gene*. **48**: 109–18.

Gonzalez, J.M., Jr., Brown, B.J. and Carlton, B.C. (1982). Transfer of *Bacillus thuringiensis* plasmids coding for δ-endotoxin among strains of *B. thuringiensis* and *B. cereus*. *Proceedings of the National Academy of Sciences, USA*. **79**: 6951–5.

Grishin, E.V., Volkova, T.M. and Soldatova, L.N. (1982). A study of the toxic component of the venom of the caucasian subspecies of the scorpion *Buthus eupeus* (transl.). *Bioorganicheskaya Khimiya*. **8**: 155–65.

Haider, M.Z. and Ellar, D.J. (1989). Mechanism of action of *Bacillus thuringiensis* insecticidal delta-endotoxin: interaction with phospholipid vesicles. *Biochemie Biophysics Acta*. **978**: 216–22.

Haider, M.Z., Knowles, B.H. and Ellar, D.J. (1986). Specificity of *Bacillus thuringiensis* var. *colmeri* insecticidal delta-endotoxin is determined by differential proteolytic processing of the protoxin by larval gut proteases. *European Journal of Biochemistry*. **156**: 531–40.

Haider, M.Z., Ward, E.S. and Ellar, D.J. (1987). Cloning and heterologous expression of an insecticidal delta-endotoxin gene from *Bacillus thuringiensis* var. *aizawai* IC1 toxic to both lepidoptera and diptera. *Gene*. **52**: 285–90.

Haider, M.Z., Smith, G.P. and Ellar, D.J. (1989). Delineation of the toxin coding fragments and an insect-specificity region of a dual toxicity *Bacillus thuringiensis* crystal protein gene. *FEMS Letters*. **58**: 157–64.

Harvey, W.R. and Wolfersberger, M.G. (1979). Mechanism of inhibition of active potassium transport in isolated midgut of *Manduca sexta* by *Bacillus thuringiensis* endotoxin. *Journal of Experimental Biology*. **83**: 293–304.

Hefford, M.A., Brousseau, R., Prefontaine, G., Hanna, Z., Condie, J.A. and Lau, P.C.K. (1987). Sequence of a lepidopteran toxin gene of *Bacillus thuringiensis* subsp. *kurstaki* NRD-12. *Journal Biotechnology*. **6**: 307–22.

Heimpel, A. and Angus, T. (1959). The site of action of crystalliferous bacteria in lepidoptera larvae. *Journal of Insect Pathology*. **1**: 152–70.

Held, G.E., Bulla, L.A., Jr., Ferrari, E., Hoch, J., Aronson, A.I. and Minnich, S.C. (1982). Cloning and localization of the lepidopteran protoxin gene of *Bacillus thuringiensis* subsp. *kurstaki*. *Proceedings of the National Academy of Sciences, USA*. **79**: 6065–9.

Herrnstadt, C., Soares, G., Wilcox, E.R. and Edwards, D.L. (1986). A new strain of *Bacillus thuringiensis* with activity against coleopteran insects. *Bio/Technology*. **4**: 305–8.

Hofmann, C., Luthy, P., Hutter, R. and Pliska, V. (1988a). Binding of the delta-endotoxin from *Bacillus thuringiensis* to brush-border membrane vesicles of the cabbage butterfly (*Pieris brassicae*). *European Journal of Biochemistry*. **173**: 85–91.

Hofmann, C., Vanderbruggen, H., Hofte, H., Van Rie, J., Jansens, S. and van Mellaert, H. (1988b). Specificity of *Bacillus thuringiensis* delta-endotoxins is correlated with the presence of high-affinity binding sites in the brush border membrane of target insect midguts. *Proceedings of National Academy of Sciences, USA.* **85:** 7844–8.

Hofte, H. and Whiteley, H.R. (1989). Insecticidal crystal proteins of *Bacillus thuringiensis. Microbiology Reviews.* **53**(2): 242–55.

Hofte, H., deGreve, H., Seurinck, J., Jansens, S., Mahillon, J., Ampe, C., Vandekerckhove, J., Vanderbruggen, H., vanMontagu, M., Zabeau, M. and Vaeck, M. (1986). Structural and functional analysis of a cloned delta endotoxin of *Bacillus thuringiensis berliner* 1715. *European Journal of Biochemistry.* **161:** 273–80.

Hofte, H., Seurinck, J., Van Houtven, A. and Vaeck, M. (1987). Nucleotide sequence of a gene encoding an insecticidal protein of *Bacillus thuringiensis* var. *tenebrionis* toxic against *Coleoptera. Nucleic Acids Research.* **15:** 7183.

Hofte, H., Van Rie, J., Jansens, S., Van Houtven, A., Vanderbruggen, H. and Vaeck, M. (1988). Monoclonal antibody analysis and insecticidal spectrum of three types of lepidopteran-specific insecticidal crystal proteins of *Bacillus thuringiensis. Applied and Environmental Microbiology.* **54:** 2010–17.

Hopp, T.P. and Woods, K.R. (1981). Prediction of protein antigenic determinants from amino acid sequences. *Proceedings of National Academy of Sciences, USA.* **78:** 3824–8.

Huber-Lukac, M., Jaquet, F., Luthy, P., Hutter, R. and Braun, D.G. (1986). Characterization of monoclonal antibodies to a crystal protein of *Bacillus thuringiensis* subsp. *kurstaki. Infection and Immunity.* **54:** 228–32.

Ishiwata, S. (1901). On a kind of severe flacherie (sotto disease) (I.) *Dainihon Sanshi Keiho.* **9:** 1–5.

Jaquet, F., Hütter, R. and Lüthy, P. (1987). Specificity of *Bacillus thuringiensis* delta-endotoxin. *Environmental Microbiology.* **53:** 500–4.

Johnson, D.E., Brookhart, G.L., Kramer, K.J., Barnett, B.D. and McGaughey, W.H. (1990). Resistance to *Bacillus thuringiensis* by the Indian meal moth *Plodia interpunctella*: comparison of midgut proteinases from susceptible and resistant larvae. *Journal of Invertbrate Pathology.* **55:** 235–44.

Klier, A., Parsot, C. and Rapoport, G. (1983). *In vitro* transcription of the cloned chromosomal crystal gene from *Bacillus thuringiensis. Nucleic Acids Research.* **11**(12): 3973–87.

Knowles, B.H. and Ellar, D.J. (1986). Characterization and partial purification of a plasmid membrane receptor for *Bacillus thuringiensis* VAR. *kurstaki* lepidopteran specific δ-endotoxin. *Journal Cell Science.* **83:** 89–101.

Knowles, B.H. and Ellar, D.J. (1987). Colloid-osmotic lysis is a general feature of the mechanism of action of *Bacillus thuringiensis* delta-endotoxins with different insect specificities. *Biochemica Biophysica Acta.* **924:** 509–18.

Kondo, S., Tamura, N., Kunitate, A., Hattori, M., Akashi, A. and Ohmori, I. (1987). Cloning and nucleotide sequencing of two insecticidal δ-endotoxin genes from *Bacillus thuringiensis* var. *kurstaki* HD-1 DNA. *Agricultural and Biological Chemistry.* **51**(2): 455–63.

Krieg, V.A., Huger, A.M., Langenbruch, G.A. and Schnetter, W. (1983). *Bacillus thuringiensis* var. *tenebrionis*: a new pathotype effective against larvae of *Coleoptera. Zeitschrift Angewandte Entomology.* **96:** 500–8.

Kronstad, J.W., Schnepf, H.E. and Whiteley, H.R. (1983). Diveristy of locations for *Bacillus thuringiensis* crystal protein genes. *Journal of Bacteriology.* **154**(1): 419–28.

Krywienczyk, J., Dulmage, H.T., Hall, I.M., Beegle, C.C., Arakawa, K.Y. and Fast, P.G. (1981). Occurrence of *kurstaki* k-1 crystal activity in *Bacillus thuringiensis* subsp. *thuringiensis* serovar (H1). *Journal of Invertebrate Pathology.* **37:** 62–5.

Lim, W.K., Smith-Somerville, H.E. and Hartmen, J.K. (1989). Solubilization and renaturation of overexpressed aggregates of mutant tryptophan synthase α-subunits. *Applied and Environmental Microbiology.* **55:** 1106–11.

Luthy, P. and Ebersold, H.R. (1981). *Bacillus thuringiensis* delta-endotoxin: histopathology and molecular mode of action. In *Pathogenesis of Invertebrate Microbial Diseases* (ed. E. Davidson), pp. 235–68. Allanheld Osmun, Totowa, New Jersey.

Lynch, M.J. and Baumann, P. (1985). Immunological comparisons of the crystal protein from strains of *Bacillus thuringiensis. Journal of Invertebrate Pathology.* **46:** 47–57.

Mani, G.S. (1985). Evolution of resistance in the presence of two insecticides. *Genetics.* **109:** 761–83.

MacIntosh, S.C., Stone, T.B., Sims, S.R., Hunst, P.L., Greenplate, J.T., Marrone, P.G., Perlak, F.J., Fischoff, D.A. and Fuchs, R.L. (1990). Specificity and efficacy of purified *Bacillus thuringiensis* proteins against agronomically important insects. *Journal of Invertebrate Pathology.* **56:** 258–66.

Martin, P.A. and Travers, R.S. (1989). Worldwide abundance and distribution of *Bacillus thuringiensis* isolates. *Applied and Environmental Microbiology.* **55:** 2437–42.

McGaughey, W.H. (1985). Insect resistance to the biological insecticide *Bacillus thuringiensis. Science.* **229:** 193–5.

McGaughey, W.H. and Johnson, D.E. (1987). Toxicity of different serotypes and toxins of *Bacillus thuringiensis* to resistant and susceptible Indian meal moths (*Lepidoptera:Pyralidae*). *Journal of Economic Entomology.* **80:** 1122–6.

McGaughey, W.H. and Beeman, R.W. (1988). Resistance to *Bacillus thuringiensis* in colonies of Indian meal moth and almond moth (*Lepidoptera:Pyralidae*). *Journal of Economic Entomology*. **81**: 28–33.

McLinden, J.H., Sabourin, J.R., Clark, B.D., Gensler, D.R., Workman, W.E. and Dean, D.H. (1985). Cloning and expression of an insecticidal k-73 type crystal protein gene from *Bacillus thuringiensis* var. *kurstaki* into *Escherichia coli*. *Applied and Environmental Microbiology*. **50(3)**: 623–8.

Milne, R., Ge, A.Z., Rivers, D. and Dean, D.H. (1990). Specificity of insecticidal crystal proteins: implications for industrial standardization. In *Analytical Methods for Native and Genetically Engineered Microbial Pesticides* (ed. L. Hickle and B. Fitch), pp. 22–35. American Chemical Society Books.

Morris, O.N., Cunningham, J.C., Finney-Crawley, J.R., Jaques, R.P. and Kenoshita, G. (1986). Microbial insecticides in Canada: their registration and use in agriculture forestry and public and animal health. *Bulletin of Entomology Society of Canada*. **18(2)**: Suppl. 1–43.

Nagamatsu, Y., Itai, Y., Hatanaka, C., Funatsu, G. and Hayashi, K. (1984). A toxic fragment from the entomocidal crystal protein of *Bacillus thuringiensis*. *Agricultural and Biological Chemistry*. **486**: 6110–90.

Nishiitsutsuji-Uwo, J. and Endo, Y. (1979). Mode of action of *Bacillus thuringiensis* δ-endotoxin: effect on TN-368 cells. *Journal of Invertebrate Pathology*. **34**: 267–75.

Obukowicz, M.G., Perlak, F.J., Kusano-Kretzmer, K., Mayer, E.J., Bolten, S.L. and Watrud, L.S. (1986). Tn5-mediated integration of the delta-endotoxin gene from *Bacillus thuringiensis* into the chromosome of root-colonizing pseudomonas. *Journal of Bacteriology*. **168**: 982–9.

Ohba, M. and Aizawa, K. (1978). Serological identification of *Bacillus thuringiensis* and related bacteria isolated in Japan. *Journal of Invertebrate Pathology*. **33**: 387–8.

Ohba, M. and Aizawa, K. (1986). Distribution of *Bacillus thuringiensis* in soils of Japan. *Journal of Invertebrate Pathology*. **47**: 277–82.

Prefontaine, G., Fast, P., Lau, P.C.K., Hefford, M., Hanna, Z. and Brousseau, R. (1987). Use of oligonucleotide probes to study the relatedness of delta-endotoxin genes among *Bacillus thuringiensis* subspecies and strains. *Applied and Environmental Microbiology*. **53**: 2808–14.

Rowe, G.E. and Margaritis, A. (1987). Bioprocess developments in the production of bioinsecticides by *Bacillus thuringiensis*. *Critical Reviews in Biotechnology*. **6**: 87–127.

Rusche, J., Jellis, C.L., Farrell, K., Farley, J., Hodgon, J., Carson, H. and Witt, D. (1988). Expression and alteration of a *Bacillus thuringiensis* endotoxin gene in *Escherichia coli*. In *Biotechnology for Crop Protection* (ed. J.J. Menn and R.M. Hollingworth), p. 454. American Chemical Society, Washington DC.

Saleh, S.M., Harris, R.F. and Allen, O.N. (1969). Method for determining *Bacillus thuringiensis* var. *thuringiensis* Berliner in soil. *Canadian Journal of Microbiology*. **15**: 1101–4.

Samasanti, W., Tojo, A. and Aizawa, K. (1986). Insecticidal activity of bipyramidal and cuboidal inclusions of delta-endotoxin and distribution of their antigens among various strains of *Bacillus thuringiensis*. *Agricultural and Biological Chemistry*. **50**: 1731–5.

Schnepf, H.E. and Whiteley, H.R. (1981). Cloning and expression of the *Bacillus thuringiensis* crystal protein gene in *Escherichia coli*. *Proceedings of the National Academy of Sciences, USA*. **78(5)**: 2893.

Schnepf, H.E. and Whiteley, H.R. (1985). Delineation of a toxin-encoding segment of a *Bacillus thuringiensis* crystal protein gene. *Journal of Biological Chemistry*. **260(10)**: 6273–80.

Shibano, Y., Yamagata, A., Nakamura, N., Iizuka, T., Sugisaki, H. and Takanami, M. (1985). Nucleotide sequence coding for the insecticidal fragment of the *Bacillus thuringiensis* crystal protein. *Gene*. **34**: 243–51.

Shieh, T.R. (1988). *Bacillus thuringiensis* biological insecticide and biotechnology. In *The Impact of Chemistry on Biotechnology: multidisciplinary discussions*, (ed. M.P. Phillips, S.P. Shoemaker, R.D. Middlekauf and R.M. Ottenbrite), pp. 207–16. American Chemical Society, Washington DC.

Sick, A., Gaertner, F. and Wong, A. (1990). Nucleotide sequence of a coleopteran active toxin gene from a new isolate of *Bacillus thuringiensis* subsp. *tolworthi*. *Nucleic Acids Research*. **18**: 1305.

Smith, R.A. (1987). Use of crystal serology to differentiate among varieties of *Bacillus thuringiensis*. *Journal of Invertebrate Pathology*. **50**: 1–8.

Stone, T.B., Sims, S.R. and Marrone, P.G. (1989). Selection of tobacco budworm for resistance to a genetically engineered *Pseudomonas fluorescens* containing the δ-endotoxin of *Bacillus thuringiensis* subsp. *kurstaki*. *Journal Invertebrate Pathology*. **53**: 228–34.

Tojo, A. (1986). Mode of action of bipyramidal delta-endotoxin of *Bacillus thuringiensis* subsp. *kurstaki* HD-1. *Applied and Environmental Microbiology*. **51**: 630–3.

Travers, R.S., Faust, R.M. and Reichelderfer, C.F. (1976). Effects of *Bacillus thuringiensis* var. *kurstaki* on isolated lepidopteran mitochondria. *Journal of Invertebrate Pathology*. **28**: 351–6.

Travers, R.S., Martin, P.A. and Reichelderfer, C.F. (1987). Selective process for efficiency isolation of soil *Bacillus* species. *Applied and Environmental Microbiology*. **53**: 1263–6.

Twardus, D. (ed.) (1989). USDA forest service gypsy moth aerial suppression/eradication projects—1989. *Gypsy Moth News*. **20**: 2.

Vaeck, M., Hofte, H., Reynaerts, A., Leemans, J., Van Montagu, M. and Zabeau, M. (1986). Engineering of insect resistant plants using a *B. thuringiensis* gene. In

*Molecular Strategies for Crop Protection*, (ed. C.J. Arntzen and C.A. Ryan). DuPont–UCLA Symposium.

Vaeck, M., Reynaerts, A., Hofte, H., Jansens, S., DeBeuckeleer, M., Dean, C., Zabeau, M., Van Montagu, M. and Leemans, J. (1987). Transgenic plants protected from insect attack. *Nature*. **328**: 33–7.

Van Rie, J., Jansens, S., Hofte, H., Degheele, D. and Van Mallaert, H. (1989). Specificity of *Bacillus thuringiensis* delta-endotoxins. *European Journal of Biochemistry*. **186**: 239–47.

Van Rie, J., Jansens, S., Hofte, H., Degheele, D. and Van Mellaert, H. (1990a). Receptors on the brush border membrane of the insect midgut as determinants of the specificity of *B. thuringiensis* delta-endotoxins. *Applied Environmental Microbiology*. **56**: 1378–85.

Van Rie, J., McGaughey, W.H., Johnson, D.E., Barnett, B.D. and Van Mallaert, H. (1990b). Mechanism of insect resistance to the microbial insecticide *Bacillus thuringiensis*. *Science*. **247**: 72–4.

Visser, B. (1989). A screening for the presence of four different crystal protein gene types in 25 *Bacillus thuringiensis* strains. *FEMS Letters*. **58**: 121–4.

Wabiko, H., Held, G.A. and Bulla, L.A., Jr. (1985). Only part of the protoxin gene of *Bacillus thuringiensis* subsp. *berliner* 1715 is necessary for insecticidal activity. *Applied and Environmental Microbiology*. **49**: 706–8.

Whiteley, H.R. and Schnepf, H.E. (1986). The molecular biology of parasporal crystal body formation in *Bacillus thuringiensis*. *Annual Reviews of Microbiology*. **40**: 549–76.

Widner, W.R. and Whiteley, H.R. (1990). Location of the dipteran specificity region in a lepidopteran-dipteran crystal protein from *Bacillus thuringiensis*. *Journal of Bacteriology*. **172**: 2826–32.

Wilcox, D.R., Shivakumar, A.G., Melin, B.E., Miller, M.F., Benson, T.A., Schopp, C.W., Casuto, D., Gundling, G.J., Bolling, T.J., Spear, B.B. and Fox, J.L. (1986). Genetic engineering of bio-insecticides. In *Protein Engineering: applications in science, industry, and medicine* (ed. M. Inouye and R. Sarma), pp. 395–413. Academic Press, New York.

Wolfersberger, M.G. (1990). The toxicity of two *Bacillus thuringiensis* δ-endotoxins to gypsy moth larvae is inversely related to the affinity of binding to brush border membranes for the toxins. *Experientia*. **46**: 475–7.

Wolfersberger, M., Hofmann, C. and Luthy, P. (1986). Interaction of *Bacillus thuringiensis* delta-endotoxin with membrane vesicles isolated from lepidopteran larval midgut. *Zentralblatt Bakteriologie Mikrobiologie Hygiene*. **15**: 237–8.

Wu, D. and Chang, F.N. (1985). Synergism in mosquitocidal activity of 26 and 65 Kdal proteins from *Bacillus thuringiensis* subsp. *israelensis* crystal. *FEBS Letters*. **190**: 232–6.

# Part VI  NOVEL PLANT PROTEINS

# 17

# Prospects and Progress in the Production of Foreign Proteins and Peptides in Plants

Enno Krebbers, Dirk Bosch and Joel Vandekerckhove

## Introduction

The other chapters in this volume have focused on the modification of plant proteins, with the aim of altering some functional aspect of the plant or to change the quality of a plant product. This chapter will consider the use of plants as a production system for foreign proteins or peptides. The potential of using biological systems for the production of proteins was recognized from the earliest days of recombinant DNA and transformation technology, and today a wide variety of systems are available, including *Escherichia coli* and other prokaryotes (Josephson and Bishop 1988; Schein 1989), yeast (Ratner 1989; Sabin *et al.* 1989), insect cell culture (Cameron *et al.* 1989), filamentous fungi (Saunders *et al.* 1989), mammalian cell culture (Rhodes and Birch 1988), transgenic animals (Van Brunt 1988) and, for peptides, chemical synthesis. Each of these systems has advantages and disadvantages, both technical and economic in nature. Prokaryotic systems do not always process or modify eukaryotic proteins in the correct fashion, or may make insoluble products (Mitraki and King 1989; Schein 1989). Systems dependent on cell culture, either prokaryotic or eukaryotic, are inherently limited in capacity by the need for fermentation equipment and the cost thereof. There is a variety of proteins for which the demand is, or will become, sufficient to require production systems that are economical at large scale. These include proteins or peptides used in the pharmaceutical, agricultural or veterinary industries and proteins, particularly enzymes, used in industrial processes. Advances in transformation techniques have now extended the potential production systems to higher eukaryotes, including transgenic animals and plants.

Plants offer particular advantages. Perhaps foremost is the low cost of growing plants on a large scale and the flexibility in how many plants are grown. If one desires to increase production by a given amount, extra hectares of the appropriate plant can be sown; one is not limited by fermenter capacity or the increments by which the latter can be increased. Plants have natural storage organs (seeds, tubers, etc.) where high levels of specific classes of proteins are already stockpiled. The technology for harvesting and processing specific plant parts at medium to very large scales already exists, as anyone who has visited a rapeseed or sugar beet processing facility, where tons of material are processed per day, can confirm. Transformation procedures for a variety of dicotyledonous crop plants are also well established, or will be soon.

Plants clearly have disadvantages too, as illustrated by the relative paucity of publications on the subject compared to those related to other production systems. Chief among these is the slowness of most transformation systems. Once a gene is cloned behind an appropriate promoter its expression in *E. coli* can be tested within a day, in yeast within a few days. Depending on the species, the plant molecular biologist may need to wait from 4 months to a year to obtain adult transformed plants and, if initial experiments are positive, several more months are needed to multiply the transformed plant sufficiently to be able to do field trials. Little is known about the modifications that foreign proteins may undergo, and even less about the difficulties involved in the downstream processing of proteins produced in plants on a large scale. Finally, a variety of legal and regulatory issues remain to be resolved.

This chapter is entitled 'Progress and Prospects'. At the time of writing there is distinctly more of the latter than the former, but the number of publications appears to be set for an exponential increase. This chapter will describe the technical basis for the use of plants as a production system, consider some of the criteria that transgenic plants will have to meet and discuss some of the experimental possibilities. It will then consider some of the progress made so far, including work from the authors' laboratories, and will finally summarize the prospects. It will be interesting to see, in one or two years time, how accurate any predictions are; the rate of growth in plant cell and molecular biology and biotechnology is such that the only accurate prediction may be that expectations will be exceeded.

# Technical prerequisites

The expression of a foreign protein in plants requires, first of all, the ability to transform the species of choice. The list of crop plants amenable to transformation, either mediated by *Agrobacterium* or otherwise, is constantly expanding. Prominent on the current list are tobacco, potato, tomato, rapeseed, cotton and various vegetable crops (Gasser and Fraley 1989). Initial reports suggest that the routine transformation of other species, including soybean (Hinchee *et al.* 1988; McCabe *et al.* 1988) will soon be possible. Not all of these will, of course, be useful as protein production systems; the choice of species will be considered further below. The cereal crops have until recently been recalcitrant to transformation but recent work suggests that this is finally changing (Gordon-Kamm *et al.* 1990). Transformation of monocotyledons will greatly expand the potential list of useful plants because processing techniques for the cereals are very well established.

A variety of proteins have been expressed in transgenic plants for purposes other than production; most of these are plant proteins or prokaryotic proteins conferring resistance to antibiotics (Reiss *et al.* 1984) or herbicides (De Block *et al.* 1987). Many of these are complex enzymes and their expression in most plant tissues and, equally important, the demonstration (in routine activity tests, Reiss *et al.* 1984; De Block *et al.* 1987) that they can be isolated from these tissues in an active form, demonstrates that there is no *a priori* reason why this should not be possible with other proteins. However, work with the crystal proteins of *Bacillus thuringiensis* suggests that not all proteins are so easily expressed in plant systems (Vaeck *et al.* 1987).

Although the expression of these prokaryotic proteins in plants is promising, it is not clear if such results can be generalized to most eukaryotic, particularly mammalian, proteins. Given the time involved in most transformation experiments, a rapid tool to determine if a mammalian or other message is translatable in plant cells, and if the protein product is stable, would be useful. Transient expression systems (Fromm *et al.* 1987; Denecke *et al.* 1989) offer such a tool. The introduction of heterologous DNA into protoplasts via electroporation offers the possibility of gathering information about mRNA stability, protein synthesis and protein stability. Many proteins are made as prepro-polypeptides and the exact role of pro-sequences is often unknown. Again, transient expression systems can be used to see if a given protein can be made without its natural pro-sequence and, if this is not possible, may provide a means to check if plant cells are capable of any necessary processing. Unfortunately, most transient expression systems described use leaf protoplasts and the results may not always be applicable to other plant tissues. Their use may also require that two separate constructs be made, one using a tissue specific promoter for expression in the desired tissue of the plant and one with a promoter suitable for use in the transient expression system. However, work in the authors' laboratories has demonstrated the utility of such experiments in determining the relative stability of mature and precursor proteins in plant cells.

# Practical approaches

## Criteria for a useful plant production system

The principle of expression of a foreign protein in plants is straightforward. However, the details of how to do this in a fashion resulting in an economically viable production system require further consideration. It is worth setting out some of the criteria that a plant production system should satisfy.

(1) The transgenic plant should be phenotypically normal and healthy and its growth rate should not differ from that of normal plants.
(2) The part or parts of the plant in which the protein accumulates should be easy to harvest and process.
(3) It would be advantageous if the harvested plant material could be stored for some time before processing takes place.
(4) The protein should be relatively easy to purify from the harvested plant material and should be free of contaminating products.
(5) The protein should not be modified in an undesired way.
(6) Expression levels must be high enough to be economically viable.

Points (4) and (5), in particular, may dictate that different proteins should be isolated from different parts of the plant or even from specific tissues. If a given tissue of a particular species contains specific modifying enzymes or proteases, or products that copurify with the protein of interest, this combination will not be suitable. Points (2) and (3) dictate that expression levels should be high in parts of the plant for which harvesting and processing technology exist or are easily adaptable. Thus, the seeds of bean or rape species, the tubers of potatoes, specialized roots such as in carrot plants, and the leafy tissue of some vegetable crops are examples of tissues for which harvesting, storage, and/or processing technologies are well developed.

## Tissue-specific or constitutive expression?

One approach could be to use promoters that have been shown to give medium to high levels of expression in all or most tissues of the plant, such as the 35S promoter of the cauliflower mosaic virus (CaMV) (Benfey *et al.* 1989). There are three potential disadvantages to such an approach. First, a plant expressing a foreign protein in all its tissues may be less fit under field conditions, both because of the energy required for such production and because some proteins may interfere with metabolic processes. If the presence of the protein of interest results in less efficient photosynthesis, for example, production levels may well be lower than

if the protein is expressed only in harvestable non-photosynthetic tissues, such as the seed or tuber. Expression in such storage tissues is unlikely to affect the overall growth of the plant. Secondly, tissue-specific promoters may give higher levels of expression in the tissue in which they are active than constitutive (or almost constitutive) promoters. Examples of tissue-specific promoters include the ribulose 1,5-bisphosphate carboxylase/oxygenase (Rubisco) small subunit promoter in photosynthetic tissues and many storage protein gene promoters for developing seeds (Dean *et al.* 1989; Okamuro and Goldberg 1989). Finally, as most farming technology is set up to harvest given parts of the plant, depending on the crop, the problem of disposal of the remaining parts arises. Regulatory guidelines associated with transgenic plants expressing chimeric genes may prevent disposal in the usual way of those portions of the plant not used for production purposes. These may be eased if it can be demonstrated that the foreign protein or peptide is present only in the plant parts that have been harvested.

Thus, for several reasons the expression of foreign proteins in specific plant organs or tissues is preferable. Much of the work on plant gene expression in the past 10 years has been directed at understanding the developmentally-regulated and tissue-specific expression of plant genes (Okamuro and Goldberg 1989) and many genes are known to be driven by specifically regulated promoters. The following are of particular interest in the present context.

## Seed specific promoters

Seed storage proteins have been discussed at length elsewhere in this volume (see Chapters 7–10). The promoters of these genes are of particular interest for the production of foreign proteins. At specific stages of seed development, seed protein mRNAs can reach very high proportions of total cellular RNA (Walling *et al.* 1986) and a given class of seed protein can represent a correspondingly high portion of total protein. This offers the potential of producing similar levels of an introduced protein driven by the seed protein promoter. Seeds also offer other advantages as protein production organs. They may be stored for long periods of time without degradation of the storage proteins. This is presumably due in part to the nature of the proteins, some of which are remarkably stable (Ampe *et al.* 1986) but also in part to the structure and metabolic state of mature seeds, raising the hope that introduced proteins will be similarly stable. For many crop plants, the technology for harvesting and processing seeds on a large scale is routine. Seeds and seed storage protein gene promoters thus appear to satisfy several of the criteria discussed above.

### Tuber-specific promoters

Like seeds, the tubers of potato plants are natural storage organs that are grown on a large scale and for which the technology for harvesting and processing exists. Harvested potatoes can be stored for some months under appropriate conditions. Again, as in seeds, tubers contain large quantities of storage proteins whose synthesis is driven by promoters that are expressed largely or exclusively in tubers (Rocha-Sosa *et al.* 1989). Thus, potato plants meet many of the criteria for commercial protein production.

### Leaf-specific promoters

Several proteins are expressed at extremely high levels in green tissues, the most prominent of which is ribulose 1,5-bisphosphate carboxylase (Rubisco). The large subunit is encoded on the chloroplast genome, the small subunit by the nuclear genome. The promoters of Rubisco genes produce what is often called the most abundant protein on earth, and the use of the small subunit promoters for expression of genes in transgenic plants has been well studied (De Almeida *et al.* 1989, for example). However, most of the promoters in this class are, in fact, not actually leaf-specific, but are expressed in photosynthetic tissues. This means that a protein expressed using such a promoter will be present in many parts of the plant which, as outlined above is potentially disadvantageous. Leaves are not usually considered to be storage organs (but see Staswick 1988) and expression of foreign proteins in large quantities might interfere with basic metabolic processes. However, there is no direct evidence for this; laboratory tobacco plants expressing phosphinothricin acetyl transferase (PAT) at 0.1% of total protein appear normal, and field trials with such plants show no reduction in yield (De Greef *et al.* 1989). Green tissue generally cannot be stored after harvesting in the same way that seeds or tubers can. However, for proteins or peptides that can resist drying, or where production can be adjusted to match harvesting times, leaves may provide an alternative to seeds or tubers.

### Other tissues

In the future, the enlarged roots present in some plants may prove to be useful as sites for protein production. Sugar beets and carrots are two obvious examples that meet several of the criteria discussed above. Appropriate promoters remain to be isolated. The use of fruit tissue is more problematic, but cannot be excluded.

## Intracellular localization

In addition to the tissue in which the protein is expressed, the intracellular (or extracellular) location of the protein may affect the fitness of the plant, the stability of the protein or the ease of purification of the protein. If the foreign protein can be targeted away from the location of essential cellular processes, the likelihood of the protein interfering with these processes is reduced. The protein is best targeted away from cellular compartments where specific proteases may be present and preferably to one where the electrochemical (redox potential, pH) and biochemical (foldases or chaperones) environments are such that the protein can fold into its proper conformation without undesired modifications. If such targeting also makes the protein easier to purify, the ideal situation is met. At the moment it is difficult to predict which compartment would be most suitable for a given protein, although one might guess, for example, that proteins normally located in the cytoplasm will be more stable there, while those normally secreted would correctly pass through the endoplasmic reticulum if the construct included an appropriate signal peptide. Assuming that the basic rules of intracellular targeting are similar in plant and animal cells, the protein would be expected to reach an environment more or less similar to its natural one.

At least four intracellular locations can be envisaged. The first of these is the cytoplasm itself. This does not require targeting signals, but may result in an unwanted methionine residue (from an introduced initiation codon) being left at the amino terminus of the protein. It is not yet clear if the ubiquitin fusion approach (Sabin *et al.* 1989) can also be applied to plant systems to circumvent the problem posed by this additional methionine.

Secondly, the construct could include a signal peptide, so that the protein would pass into the endoplasmic reticulum and, if it did not contain any specific signals causing it to be retained in the intracellular transport system or vacuole (Dorel *et al.* 1989), be secreted. This would result in compartmentalization between the cell membrane and wall. Whether the protein would be stored stably here, degraded for some reason, or somehow pass into the plant transport system is not clear, and would probably depend on the protein. Since cleavage of the signal peptide is, in most cases, relatively independent of the following sequence (but not always; see Andrews *et al.* 1988) a protein with the desired amino terminus would be produced.

It is now clear that there are specific targeting signals that cause seed storage proteins to be targeted to the vacuole rather than be secreted (Dorel *et al.* 1989). The signals responsible for this in yeast have been identified (Valls *et al.* 1987), and similar work is under way in plants. If such signals could be included in the chimeric constructs, the protein could be directed to vacuoles. This may or may not be an advantage in leaf tissue, but is not unlikely to be advantageous in seeds; most seed storage proteins are stored in specialized vacuoles called protein bodies, which should in principle be ideal storage organelles. The problem then becomes how to remove the targeting signal from the introduced protein. Some storage proteins, such as the 2S albumins, undergo extensive post translational processing, which appears to take place after arrival of the protein in the protein bodies (Chrispeels *et al.* 1982). If the targeting signal is present in one of the fragments removed during such processing, and if the processing is similar to the signal peptide cleavage in that it is relatively independent of the sequence that follows, the problem might be solved.

Finally, one could envisage targeting proteins to plastids. De Boer *et al.* (1988) have demonstrated that chloroplast transit peptides can direct passenger molecules not only to chloroplasts, but also to chromoplasts. Conversely, a transit peptide derived from an amyloplast-located protein can direct a protein to the chloroplast (Klösgen *et al.* 1989). Some organs contain plastids that are specialized storage organs, such as the amyloplasts present in potato tubers, and these might be a suitable location for foreign proteins. Here again, the processing of transit peptides appears to be relatively independent of the following sequence, so that correct amino termini could be obtained.

To determine whether a protein, if equipped with the appropriate targeting signals, can indeed be transported to the chosen cellular compartment, the transient expression system can again be used. Signal peptide-mediated secretion of a passenger protein out of a plant cell and its correct processing have been studied (Denecke *et al.* 1990). The function of retention signals can also be analysed in such a system. Both transient expression or *in organello* systems can be used to determine if a transit peptide is able to route a given passenger molecule into plastids, and whether correct processing takes place (Teeri *et al.* 1989; Smeekens *et al.* 1986).

Unfortunately, few data are yet available and intitially the choice of both tissue and intracellular location will have to be determined empirically. To some extent this will probably always be necessary when a new class of proteins is to be produced. The available data are considered in the next section.

## Recent progress

Hiatt *et al.* (1989) have presented the most complete picture so far of the expression of foreign proteins in plants. Their results bear out several of the above suppositions. Using a constitutive promoter, they expressed the mouse immunoglobulin heavy and light chains in separate lines of tobacco plants. Constructs that lacked or included signal peptides were made. The constructs without signal peptides resulted in the accumulation of very little of the desired products, while those including signal peptides resulted in an average yield of 0.14% of total protein, with a maximum of 0.24% (for the heavy chain). The yield of the light chain reached only 0.06% of total protein. When plants carrying the two genes were crossed, the antibodies assembled and the yields were much higher, reaching a maximum of 1.3% of total protein. These results demonstrate that intracellular location is a critical issue, due either to the presence of chaperonins or to some other characteristic of the environment in which the protein is expressed. Antibodies are normally assembled in the e.r./Golgi system, not in the cytoplasm. The increased yield upon crossing the two lines, so that both chains are present, suggests that stable accumulation in this case depends on normal assembly and folding (transcriptional effects were excluded), a lesson that will doubtless recur among other multimeric complexes. The results cited above were obtained using leaf tissue; no data were reported for other parts of the plants. No phenotypic abnormalities were reported. Characterization of the proteins produced with respect to modifications, exact processing, or subcellular location was not reported.

While this paper was in the editing process, Düring *et al.* (1990) reported the results of similar experiments. They used a construction in which both chains were introduced simultaneously under the control of a different constitutive promoter than that used by Hiatt *et al.* They included signal peptides for both chains, and were able to demonstrate the presence of assembled immunoglobulins in the endoplasmic reticulum. Curiously, immunogold labelling experiments also showed some of the introduced protein associated with the chloroplasts, but the significance of this observation is not clear. Expression levels appear to be significantly lower than those reported by Hiatt *et al.*

Sijmons *et al.* (1990) have produced Human Serum Albumin (HSA) in potato plants. Again, using a constitutive promoter, they demonstrated that HSA made without its natural six amino acid pro-sequence, but with a signal peptide, could be accumulated in the extracellular fluid of leaf tissue and that when protoplasts were made from the transgenic plants, HSA was secreted into the culture medium. If the sequences encoding the pro-sequence were included in the construct, the pro-peptide was made and secreted, but the pro-sequence was not removed. These results are encouraging in that they demonstrate that, at least in one case, portions of a protein usually removed post-translationally in a mammalian system need not be included for expression in a plant system. No results were reported concerning expression of HSA in other tissues. In leaves, the total amount of protein reached only 0.02%, significantly less than that reported by Hiatt *et al.* (1989) for immunoglobulin chains. As similar promoters were used, the reason for the difference must lie either in the choice of plant species or be inherent to the chimeric gene constructs or to the proteins expressed. As a basis for comparison, the small subunit (SSU) of Rubisco may reach 50% of total protein in tobacco (Kung 1976) (resulting from the expression of a five or more member multigene family) and De Almeida *et al.* (1989) reported that an *Arabidopsis* Rubisco SSU promoter fused to the gene encoding PAT resulted in the gene product representing 1% of total protein in tobacco leaves. As in the case of the immunoglobulin chains (Hiatt *et al.* 1989), HSA could not be detected when

expressed without a signal peptide.

De Zoeten *et al.* (1989) and Grill *et al.* (1989) have reported the use of a system not discussed so far. They used infection with modified viral genomes to produce a 'transient' expression of foreign proteins in leaves. As the plant is not transformed, the protein is produced for a matter of only a few weeks. De Zoeten *et al.* (1989) produced human interferon in turnip (*Brassica rapa*) in this fashion, and demonstrated that the product was localized in the viral inclusion bodies normally formed during infection. Crude extracts of the infected plants were shown to have activity against Vesicular Stomatitis Virus (VSV) at a level corresponding to $2 \times 10^{-4}$ g of active interferon per 100 g fresh weight. This would probably correspond to 0.02–0.04% of total protein. The result suggests that at least some of the interferon was properly folded. Prolonged presence of interferon in the plant resulted in loss of activity, presumably caused by proteolytic breakdown. It was shown that ubiquitin was also present in the inclusion bodies. This may be taken as an indication that at least some of the proteins in the viral inclusion bodies are destined for degradation.

Grill *et al.* (1989, reviewed in Knight 1989) produced melanin in plants using a system not yet completely documented. Expression levels of up to 20% of cell mass were reported. Such transient systems will clearly be less broadly applicable than stably transformed plants, and the methods needed to infect large numbers of plants in the field may lead to regulatory problems. The advantage of the system is its speed; no plant transformation or tissue culture steps are required. In cases where the amounts of protein needed are not great but product flexibility is important, this system may prove useful.

## Production of peptides in 2S albumins

The discussion thus far has focused on full sized proteins, but production of peptides is also of increasing interest as the list of peptides with useful biological activities increases (see references in Krebbers and Vandekerckhove 1990). Peptides have also been produced using a variety of microbiological systems, usually as parts of fusion proteins (which is how many are produced *in vivo*) (Josefson and Bishop 1988). Work in the authors' laboratory (Vandekerckhove *et al.* 1989) has demonstrated that such an approach can also be adopted in plants. The advantages that seeds offer as a possible location for production of proteins and peptides have been enumerated above. In addition, seed storage proteins have long been considered potential targets for modification because most have no apparent enzymatic function. Thus it was reasoned that sequences encoding the peptide could be embedded within the open reading frame, flanked by sequences encoding protease cleavage sites (Fig. 17.1). As the peptide is then 'piggybacking' on a complete storage protein, the latter will carry the peptide through all the processing and targeting steps discussed above. The storage protein can then be purified and the peptide cleaved

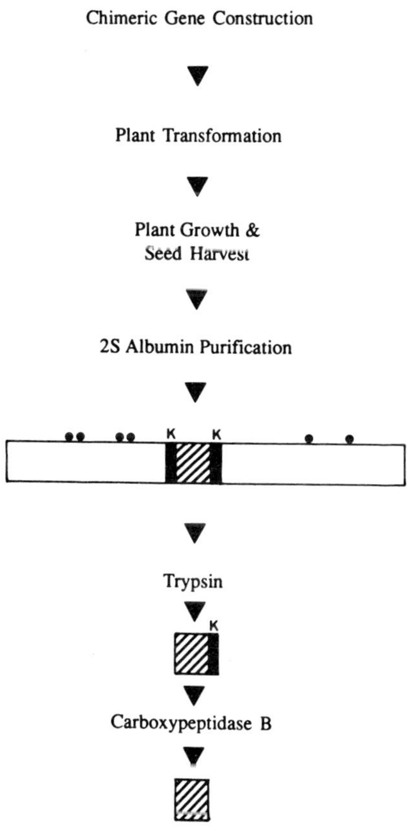



**Fig. 17.1**  An overview of the procedure used in the authors' laboratories to produce Leu-enkephalin in plant seeds. Only the large subunit of the modified 2S albumin is shown. Dots represent cysteine residues; the striped region represents the peptide and the black bars represent the tryptic cleavage sites. In principle, any proteolytic cleavage site, enzymatic or chemical, which does not occur in the peptide of interest can be used at these positions. If the peptide happens to end in such a residue, the final step is unnecessary. Taken from Krebbers and Vandekerckhove, 1990.

out. The problem is to define a region of a storage protein that can tolerate such a change. This becomes a protein engineering problem, bringing this chapter back to the main theme of this volume. Despite their having no enzymatic function, it is becoming clear that not just any change can be made in seed storage protein structures. Hoffman *et al.* (1988) have demonstrated that a modification in the β-phaseolin gene results in the modified protein being produced, but then degraded during transport. Sonnewald (1989) reported similar results when modifications were made in the tuber storage protein patatin which, while in a region thought to be on the outside of the protein, changes its pI. Presumably, if a change that interferes with the correct folding, post-translational processing or transport of the protein to the storage vacuoles is made, the modified seed storage protein may be liable to degradation.

The 2S albumin family of seed storage proteins have characteristics that suggested that, even in the absence of a crystal structure, intelligent guesses could be made as to what portions of the protein might tolerate significant alterations. The 2S albumins are among the smallest of storage proteins, consisting of two subunits of approximately 9 and 3 kDa, linked by disulphide bridges. Comparisons of gene and protein sequences reveal that the proteins undergo significant post-translational processing (Krebbers *et al.* 1988). The sequences of the protein chains remaining after processing are somewhat conserved, but the most striking feature is the conservation of the number and position of the cysteine residues, of which there are eight. One intercysteine region, between cysteine residues 6 and 7 (counting from the first cysteine in the small subunit), shows some variation in length. It was reasoned that modifications could be made either in this region or at the end of the large subunit, beyond the final cysteine. Since the pro-peptide undergoes the removal of a carboxyl terminal fragment, the latter was considered risky because the introduced peptide might be removed, and so the internal site was chosen.

The first peptide produced in this way was Leu-enkephalin, a 5-amino acid neuropeptide with opiate activity (Hughes 1975). A portion of the sequence encoding intercysteine region 6–7 of the *Arabidopsis thaliana* 2S albumin gene was replaced with sequences encoding the peptide flanked by codons encoding lysine residues as tryptic cleavage sites. The modified gene was transformed into both *Arabidopsis* and *Brassica napus* (rapeseed) and 2S albumins from the seeds of transformed seeds were isolated. 2S albumins are relatively simple to separate from most other proteins because of their small size and solubility in low salt solutions. After tryptic cleavage, the peptide, with an extra lysine residue still attached to the carboxyl terminus, was isolated. The extra lysine was subsequently removed using carboxypeptidase B.

The yields from this first example were encouraging. In both *Arabidopsis* and *Brassica*, plants producing up to 200 nmol of peptide per gram of seed have been isolated. For enkephalin, this corresponds to approximately 110 µg peptide per gram of seed. As *Brassica* is capable of producing 3000 kg of seed per hectare, the isolation of hundreds of grams of peptide per hectare are, in theory, possible. It should be noted that these are the absolute yields obtained after all purification steps, not adjusted for losses during the purification. These yields can probably be optimized by increasing the efficiency of extraction, by using plant varieties that produce more protein or by enhancing gene expression (Krebbers and Vandekerckhove 1990). Recent work (Guerche *et al.* 1990) have demonstrated that the promoter used in these experiments is not the most active of the four 2S albumin promoters in the *Arabidopsis* gene family, already suggesting a possible means to increase yields. It was estimated that, in *Arabidopsis*, the chimeric 2S albumin represented almost 3% of total seed protein. Compared to the levels obtained by expressing other proteins using constitutive promoters (see above), these results are consistent with the idea that using promoters expressed at high levels in certain tissues or organs at given stages of development can result in higher levels of expression.

If similar yields can be achieved with larger peptides, the weight per hectare will increase. Five other modified 2S albumins have been stably expressed so far, containing either peptides or modifications designed to alter the amino acid composition of the 2S albumin (De Clercq *et al.* 1990 and unpublished data). Quantification of the yields is currently underway. Unfortunately, as no crystal structure of the 2S albumin is yet available, little is possible in the way of prediction, and each modification must be tested individually. The range of modifications possible within the confines of the 2S albumin is not yet known. Given the apparently important structural role of the cysteine residues, producing peptides containing cysteines may prove difficult in 2S albumins. The general principle is,

of course, extendable to other seed storage proteins; in each case, regions suitable for modification must be identified.

The questions remaining before this encouraging laboratory result can be converted to a commercial production system have been discussed (Krebbers and Vandekerckhove 1990). Some are unique to the peptide production system while others are applicable to both peptide and protein production in plants. Specific to the case of peptides are the unknown downstream processing costs related to cleavage of the peptide from the storage protein. The longer the peptide, the greater the chance the cleavage sites for the more inexpensive proteases or chemical reagents will occur within the sequence, necessitating the use of more expensive ones. The internal location of the peptide means that after the initial cleavage, it will often be necessary to remove the remaining carboxy terminal residue (Fig. 17.1). The extra costs of this step (when necessary) will have to be weighed against the cost advantages of other steps in the procedure when compared to other peptide production systems. Special modifications, such as amidation of carboxyl termini, will have to be carried out after purification (see Krebbers and Vandekerckhove 1990 for alternatives).

## Prospects

A series of technical, economic and regulatory questions remain for both proteins and peptides. First, how difficult will it be to purify the product (or, restated, what will the process cost at large scales of production)? This will be related to the expression levels; the variable results obtained in the experiments carried out so far suggest that further work is needed, both in the laboratory and ultimately under field conditions. Will plant tissues provide unusual contaminants, making downstream processing unexpectedly difficult? Will unexpected modifications occur? For these reasons it is important that purification be attempted from different tissues.

Little attention has been paid to the problem raised by the differing patterns of glycosylation known to exist between mammalian and plant cells. These can be summarized as follows (reviewed in Faye *et al.* 1989). The initial steps of glycosylation are similar or identical in plants and animals, and in both cases take place mainly at the sequence Asn-

X-Ser/Thr. However, there is a crucial difference in the conversion of high mannose glycans to complex glycans. This takes place in the Golgi complex and is carried out by different enzymes in plants and mammals. The fucose linkage is $\alpha$-1,3 in plants and $\alpha$-1,6 in mammals, plant proteins have no mannose-6-phosphate groups, and plant complex glycans may include a xylose moiety not present in mammalian cells. The latter may prove to be the most important difference, as such xylose groups are known to be antigenic (Lauriere *et al.* 1989). It will be of interest to determine if mammalian glycoproteins made in plant systems are indeed so modified, and if so how this affects their biological activity. The transfer of xylose from the xylose diphosphate donor to the glycan acceptor is catalysed by xylosyl transferase and is a final step in the glycan modification pathway (Johnson and Chrispeels 1987). Therefore, it may be possible to block expression of the xylosyl transferase gene (using antisense or other techniques) in the tissues where the heterologous protein is expressed without affecting earlier steps in the glycosylation pathway. This would depend on the development of the tissue involved not being adversely affected by such a modification.

There are also non-technical issues related to registration and regulatory issues to be overcome. It is unclear if a pharmaceutical product produced in plants but already produced by other means will be required to go through a completely new set of clinical trials, or if it will suffice to demonstrate that the protein is chemically identical to the product already approved. If it is necessary to go through new trials, the costs of production will have to be orders of magnitude lower than current technology before pharmaceutical companies will be willing to make such an investment. Other regulatory problems are also apparent. One of the major advantages of plants as a production system would appear to be the ability to produce large quantities of material. In order to be able to do this, plants will have to be grown in the field as well as the greenhouse. A great deal of public education will be necessary before it will be politically acceptable for regulatory agencies to allow the growth of plants producing peptides (even embedded within a storage protein, and thus inactive), not to mention plants producing enzymes or protein hormones, in fields. Appropriate precautions to prevent the unwanted spread of introduced genes, such as the use of male sterile plants (Mariani *et al.* 1990), will be necessary.

In summary, the first efforts at producing foreign proteins or peptides in plants have demonstrated that the process is technically feasible. A great deal of further investigation is necessary to determine how such expression can be optimized and what the problems involved in purification will be. However, the results obtained so far are encouraging and progress on the technical side should be rapid in the next few years. If the results obtained indicate that the economic criteria can be satisfied, then within the not too distant future, assuming the regulatory issues can be resolved, plants grown for commercial production of proteins and peptides will be a reality.

## Acknowledgements

The authors thank L. Willmitzer and A. Hoekema for communication of unpublished data and M. Chrispeels for helpful discussions. The authors warmly acknowledge the contributions of their collaborators in the work performed at PGS and the University of Gent, which are supported in part by a grant from the Belgian NFWO to JV. DB was supported by a grant from the European Communities.

## References

Ampe, C., Van Damme, J., de Castro, L., Sampaio, M.J., van Montagu, M. and Vandekerckhove, J. (1986). The amino-acid sequence of the 2S sulfur-rich proteins from seeds of Brazil nut (*Bertholletia excelsa* H.B.K.). *European Journal of Biochemistry.* **159**: 597–604.

Andrews, D.W., Perera, E., Lesser, C. and Lingappa, V.R. (1988). Sequences beyond the cleavage site influences signal peptide function. *The Journal of Biological Chemistry.* **263**: 15791–8.

Benfey, P.N., Ren, L. and Chua, N. (1989). The CaMV 35S enhancer contains at least two domains which can confer different developmental and tissue specific expression patterns. *EMBO Journal.* **8**: 2195–202.

Cameron, I.R., Possee, R.D. and Bishop, D.H.L. (1989). Insect cell culture technology in baculovirus expression systems. *Trends in Biotechnology.* **7**: 66–70.

Chrispeels, M.J., Higgins, T.J. and Spencer, D. (1982). Assembly of storage protein oligomers in the endoplasmic reticulum and processing of the polypeptides in the protein bodies of developing pea cotyledons. *The Journal of Cell Biology.* **93**: 306–13.

De Almeida, E.R.P., Gosselé, V., Muller, C.G., Dockx, J., Reynaerts, A., Botterman, J., Krebbers, E. and

Timko, M.P. (1989). Transgenic expression of two marker genes under the control of an Arabidopsis rbcS promoter: sequences encoding the Rubisco transit peptide increase expression levels. *Molecular and General Genetics.* **218:** 78–86.

Dean, C., Pichersky, E. and Dunsmuir, P. (1989). Structure, evolution and regulation of RbcS genes in higher plants. *Annual Review of Plant Physiology and Plant Molecular Biology.* **40:** 415–39.

De Block, M., Botterman, J., Vandewiele, M., Dockx, J., Thoen, C., Gosselé, V., Movva, N., Thompson, C., Van Montagu, M. and Leemans, J. (1987). Engineering herbicide resistance in plants by expression of a detoxifying enzyme. *EMBO Journal.* **6:** 2513–18.

De Boer, D., Cremers, F., Teertstra, R., Smits, L., Hille, J., Smeekens, S. and Weisbeek, P. (1988). *In vivo* import of plastocyanin and a fusion protein into developmentally different plastids of transgenic plants. *EMBO Journal.* **9:** 2631–5.

De Clercq, A., Vandewiele, M., Van Damme, J., Guerche, P., Van Montagu, M., Vandekerckhove, J. and Krebbers, E. (1990). Stable accumulation of modified 2S albumin seed storage proteins with higher methionine contents in transgenic plants. *Plant Physiology.* **94:** 970–9.

De Greef, W., Delon, R., De Block, M., Leemans, J. and Botterman, J. (1989). Evaluation of herbicide resistance in transgenic crops under field conditions. *Bio/technology.* **7:** 61–4.

Denecke, J., Gosselé, V., Botterman, J. and Cornelissen, M. (1989). Quantitative analysis of transiently expressed genes in plant cells. *Methods in Molecular and Cellular Biology.* **1:** 19–27.

Denecke, J., Botterman, J. and Deblaere, R. (1990). Protein secretion in plant cells can occur via a default pathway. *The Plant Cell.* **2:** 51–9.

De Zoeten, G.A., Penswick, J.R., Horisberger, M.A., Ahl, P., Schultze, M. and Hohn, T. (1989). The expression, localization, and effect of a human interferon in plants. *Virology.* **172:** 213–22.

Dorel, C., Voelker, T.A., Herman, E.M. and Chrispeels, M.J. (1989). Transport of proteins to the plant vacuole is not by bulk flow through the secretory system, and requires positive sorting information. *The Journal of Cell Biology.* **108:** 327–37.

Düring, K., Hippe, S., Kreuzaler, F. and Schell, J. (1990). Synthesis and self-assembly of a functional monoclonal antibody in transgenic *Nicotiana tabacum. Plant Molecular Biology.* **15:** 281–93.

Faye, L., Johnson, K.D., Sturm, A. and Chrispeels, M.J. (1989). Structure, biosynthesis, and function of asparagine-linked glycans on plant glycoproteins. *Physiologia Plantarum.* **75:** 309–14.

Fromm, M., Callis, J., Tayler, L.P. and Walbot, V. (1987). Electroporation of DNA and RNA in plant protoplasts. *Methods in Enzymology.* **153:** 351–66.

Gasser, C.S. and Fraley, R.T. (1989). Genetically engineering plants for crop improvement. *Science.* **244:** 1293–9.

Gordon-Kamm, W.J., Spencer, T.M., Mangano, M.L., Adams, T.R., Daines, R.J., Start, W.G., O'Brien, J.V., Chambers, S.A., Adams, W.R., Willetts, N.G., Rice, T.B., Mackey, C.J., Krueger, R.W., Kausch, A.P. and Lemaux, P.G. (1990). Transformation of maize cells and regeneration of fertile transgenic plants. *Plant Cell.* **2:** 603–18.

Grill, L., Erwin, R.L., Berliner, D.L. and Hubbard, E.T. (1989). PCT patent application WO 8908145.

Guerche, P., Tire, C., Grossi De Sa, F., De Clercq, A., Van Montagu, M. and Krebbers, E. (1990). Quantitation and localization of expression of the four 2S albumin genes of *Arabidopsis thaliana* and analysis of the effect of increasing gene family size. *Plant Cell.* **2:** 469–78.

Hiatt, A. Cafferkey, R. and Bowdish, K. (1989). Production of antibodies in transgenic plants. *Nature.* **342:** 76–8.

Hinchee, M.A.W., Conner-Ward, D.V., Newell, C.A., McDonnell, R.E., Sato, S.J., Gasser, C.S., Fischhoff, D.A., Re, D.B., Fraley, R.T. and Horsch, R.B. (1988). Production of transgenic soybean plants using Agrobacterium-mediated DNA transfer. *Bio/technology.* **6:** 915–22.

Hoffman, L.M., Donaldson, D.D. and Herman, E.M. (1988). A modified storage protein is synthesized, processed, and degraded in the seeds of transgenic plants. *Plant Molecular Biology.* **11:** 717–29.

Hughes, J. (1975). Isolation of an endogenous compound from the brain with pharmacological properties similar to morphine. *Brain Research.* **88:** 295–308.

Johnson, K.D. and Chrispeels, M.J. (1987). Substrate specificities of *N*-acetylglucosaminyl-, fucosyl-, and xylosyltransferases that modify glycoproteins in the Golgi apparatus of bean cotyledons. *Plant Physiology.* **84:** 1301–8.

Josephson, S. and Bishop, R. (1988). Secretion of peptides from *E. coli:* a production system for the pharmaceutical industry. *Trends in Biotechnology.* **6:** 218–24.

Klösgen, R.B., Saedler, H. and Weil, J.H. (1989). The amyloplast-targeting transit peptide of the waxy protein of maize also mediates protein transport in vitro into chloroplasts. *Molecular and General Genetics.* **217:** 155–61.

Knight, P. (1989). Recombinant melanin expressed in plants. *Bio/technology.* **7:** 20.

Krebbers, E. and Vandekerckhove, J. (1990). Production of peptides in plant seeds. *Trends in Biotechnology.* **8:** 1–3.

Krebbers, E., Herdies, L., De Clercq, A., Seurinck, J., Leemans, J., Van Damme, J., Segura, M., Gheysen, G., Van Montagu, M. and Vandekerckhove, J. (1988). Determination of the processing sites of an Arabidopsis 2S albumin and characterization of the complete gene family. *Plant Physiology.* **87:** 859–66.

Kung, S.D. (1976). Tobacco fraction I protein: a unique genetic marker. *Science.* **191:** 429–34.

Lauriere, M., Lauriere, C., Chrispeels, M.J., Johnson,

K.D. and Sturm, A. (1989). Characterization of a xylose-specific antiserum that reacts with the complex asparagine-linked glycans of extracellular and vacuolar glycoproteins. *Plant Physiology*. **90**: 1182–8.

Mariani, C., De Beuckeleer, M., Truettner, J., Leemans, J. and Goldberg, R.B. (1990). Inductions of male sterility in plants by a *chimaeric ribonuclease* gene. *Nature*. **347**: 737–41.

McCabe, D.E., Swain, W.F., Martinell, B.J. and Christou, P. (1988). Stable transformation of soybean (*Glycine max*) by particle acceleration. *Bio/technology*. **6**: 923–6.

Mitraki, A. and King, J. (1989). Protein folding intermediates and inclusion body formation. *Bio/technology*. **7**: 690–7.

Okamuro, J.K. and Goldberg, R.B. (1989). Regulation of plant gene expression: general principles. *The Biochemistry of Plants*. **15**: 1–82.

Ratner, M. (1989). Protein expression in yeast. *Bio/technology*. **7**: 1129–33.

Reiss, B., Sprengel, R.R., Will, H. and Schaller, H. (1984). A new sensitive method for quantitative and qualitative assay of neomycin phosphotransferase in crude cell extracts. *Gene*. **30**: 211–18.

Rhodes, M. and Birch, J. (1988). Large-scale production of proteins from mammalian cells. *Bio/technology*. **6**: 518–23.

Rocha-Sosa, M., Sonnewald, U., Frommer, W., Schell, J. and Willmitzer, L. (1989). Both developmental and metabolic signals activate the promoter of class I patatin gene. *EMBO Journal*. **8**: 23–9.

Sabin, E.A., Lee-Ng, C.T., Shuster, J.R. and Barr, P.J. (1989). High level expression and *in vivo* processing of chimeric ubiquitin fusion proteins in *Saccharomyces cerevisiae*. *Bio/technology*. **7**: 750–9.

Saunders, G., Picknett, T.M., Tuite, M.F. and Ward, M. (1989). Heterologous gene expression in filamentous fungi. *Trends in Biotechnology*. **7**: 283–7.

Schein, C.H. (1989). Production of soluble recombinant proteins in bacteria. *Bio/technology*. **7**: 1141–9.

Sijmons, P.C., Dekker, B.M.M., Schrammeijer, B., Verwoerd, T.C., Van den Elzen, P.J.M. and Hoekema, A. (1990). Production of correctly processed human serum albumin in transgenic plants. *Bio/technology*. **8**: 217–21.

Smeekens, S., Bauerle, C., Hageman, J., Keegstra, K. and Weisbeek, P. (1986). The role of the transit peptide in the routing of precursors toward different chloroplast compartments. *Cell*. **46**: 365–75.

Sonnewald, U. (1989). Konstruktion und Analyse von Chimären Patatin-Genen unter Besonderer Berücksichtigung der Stabilitat und Subzellulären Kompartmentierung der Patatin Proteine. Thesis, University of Cologne.

Staswick, P.E. (1988). Soybean vegetative storage protein structure and gene expression. *Plant Physiology*. **87**: 250–4.

Teeri, T.H., Patel, G.K., Aspegren, K. and Kauppinen, V. (1989). Chloroplast targeting of neomycin phosphotransferase II with a pea transit peptide in electroporated barley mesophyll protoplasts. *Plant Cell Reports*. **8**: 187–90.

Vaeck, M., Reynaerts, A., Hofte, H., Jansens, S., De Beuckeleer, M., Dean, C., Zabeau, M., Van Montagu, M. and Leemans, J. (1987). Transgenic plants protected from insect attack. *Nature*. **328**: 33–7.

Valls, L.A., Hunter, C.P., Rothman, J.H. and Stevens, T.H. (1987). Protein sorting in yeast: the localization determinant of yeast vacuolar carboxypeptidase Y residues in the propeptide. *Cell*. **48**: 887–97.

Van Brunt, J. (1988). Molecular farming: transgenic animals as bioreactors. *Bio/technology*. **6**: 1149–54.

Vandekerckhove, J., Van Damme, J., Van Lijsebettens, M., Botterman, J., De Block, M., Vandewiele, M., De Clercq, A., Leemans, J., Van Montagu, M. and Krebbers, E. (1989). Enkephalins produced in transgenic plants using modified 2S seed storage proteins. *Bio/technology*. **7**: 929–32.

Walling, L., Drews, G.N. and Goldberg, R.B. (1986). Transcriptional and post-transcriptional regulation of soybean seed protein mRNA levels. *Proceedings of the National Academy of Sciences, USA*. **83**: 2123–7.

# Index

Page numbers in *italics* refer to figures and tables.